The Pineapple

Botany, Production and Uses

SECOND EDITION

MIX
Paper from
responsible sources
FSC® C007785
FSC
www.fsc.org

The Pineapple

Botany, Production and Uses

SECOND EDITION

Edited by

Garth M. Sanewski

Queensland Department of Agriculture and Fisheries, Australia

Duane P. Bartholomew

University of Hawaii at Manoa, Honolulu, USA

and

Robert E. Paull

University of Hawaii at Manoa, Honolulu, USA

CABI

CABI is a trading name of CAB International

CABI
Nosworthy Way
Wallingford
Oxfordshire OX10 8DE
UK

Tel: +44 (0)1491 832111
Fax: +44 (0)1491 833508
E-mail: info@cabi.org
Website: www.cabi.org

CABI
745 Atlantic Avenue
8th Floor
Boston, MA 02111
USA

Tel: +1 (617)682-9015
E-mail: cabi-nao@cabi.org

A catalogue record for this book is available from the British Library, London, UK.

Library of Congress Cataloging-in-Publication Data

Names: Sanewski, G., editor. | Bartholomew, D. P. (Duane Phillip), editor. | Paull, Robert E., editor.
Title: The pineapple : botany, production and uses / edited by Garth M. Sanewski, Duane P. Bartholomew, Robert E. Paull.
Description: Second edition. | Boston, MA : CABI, [2018] | Includes bibliographical references and index.
Identifiers: LCCN 2018007061| ISBN 9781786393302 (hbk : alk. paper) | ISBN 9781786393319 (epdf)
Subjects: LCSH: Pineapple.
Classification: LCC SB375 .P47 2018 | DDC 634/.774--dc23 LC record available at https://lccn.loc.gov/2018007061

ISBN-13: 9781786393302 (hbk)
9781786393319 (ePDF)
9781786393326 (ePub)

Commissioning editor: Rachael Russell/Rebecca Stubbs
Editorial assistants: Alexandra Lainsbury and Tabitha Jay
Production editor: Marta Patiño

Typeset by AMA DataSet Ltd, Preston, UK
Printed and bound in the UK by Bell & Bain Ltd, Glasgow

Contents

Contributors

Duane P. Bartholomew, Department of Tropical Plant and Soil Science, University of Hawaii at Manoa, Honolulu, HI 96822, USA. Email: duaneb@hawaii.edu

José Ramon Botella, Plant Genetic Engineering Laboratory, School of Agriculture and Food Sciences, University of Queensland, St Lucia, Queensland 4072, Australia. Email: J.Botella@uq.edu.au

Ching-Cheng Chen, Department of Horticulture, National Chung Hsing University, 250 Kuo Kuang Road, Taichung, Taiwan, Republic of China. Email: ccc@dragon.nchu.edu.tw

Geo Coppens d'Eeckenbrugge, CIRAD, UMR AGAP, Avenue Agropolis, 34398 Montpellier, France. Email: geo.coppens@cirad.fr

Sandra Aparecida de Assis, Universidade Estadual de Feira de Santana, 44036-900, Feira de Santana, Bahia, Brazil. Email: sandrinhaassis@yahoo.com.br

Aristoteles Pires de Matos, Embrapa Mandioca e Fruticultura (Embrapa Cassava and Fruits), Rua Embrapa s/n, 44380-000 Cruz das Almas, Bahia, Brazil. Email: aristotles.matos@embrapa.br

Everton Hilo de Souza, Universidade Federal do Recôncavo da Bahia, 44380-000, Cruz das Almas, Bahia, Brazil. Email: hilosouza@gmail.com

Fernanda Vidigal Duarte Souza, Embrapa Mandioca e Fruticultura (Embrapa Cassava and Fruits), Rua Embrapa s/n, 44380-000, Cruz das Almas, Bahia, Brazil. Email: fernanda.souza@embrapa.br

Marie-France Duval, CIRAD, UMR AGAP, Campus agro-environnemental Caraïbe BP 214-97285 Le Lamentin Cedex, Martinique. Email: marie-france.duval@cirad.fr

Sharon Hamill, Queensland Department of Agriculture and Fisheries, Maroochy Research Facility, Nambour, Queensland 4560, Australia. Email: Sharon.hamill@daf.qld.gov.au

Jhonny Vásquez-Jiménez, Consultant, Florencia, San Carlos, Alajuela, Costa Rica. Email: jvasquez@proagrocr.com

Davi T. Junghans, Embrapa Mandioca e Fruticultura (Embrapa Cassava and Fruits), PO Box 007, 44380-000, Cruz das Almas, Bahia, Brazil. Email: davi.junghans@embrapa.br

Hian-Lien Ko, Queensland Department of Agriculture and Fisheries, Maroochy Research Facility, Nambour, Queensland 4560, Australia. Email: Lien.Ko@daf.qld.gov.au

Freddy Leal, Universidad Central de Venezuela, Facultad de Agronomia, Maracay, Aragua, Venezuela. Email: flealpinto@hotmail.com

José Manoel Marconcini, Embrapa Instrumentação, 13560-970, São Carlos, São Paulo, Brazil. Email: Jose.marconcini@embrapa.br

Tracie Matsumoto, United States Department of Agriculture. ARS Daniel K. Inouye US Pacific Agricultural Research Center, Hilo, HI, USA. Email: Tracie.Matsumoto@ars.usda.gov

Tullio Raphael Pereira de Pádua, Embrapa Mandioca e Fruticultura (Embrapa Cassava and Fruits), Rua Embrapa s/n, 44380-000 Cruz das Almas, Bahia, Brazil. Email: tullio.padua@embrapa.br

Robert E. Paull, Department of Tropical Plant and Soil Sciences, College of Tropical Agriculture and Human Resources, University of Hawaii at Manoa, 3190 Maile Way, Honolulu, HI 96822-2279, USA. Email: paull@hawaii.edu

Domingo Haroldo Reinhardt, Embrapa Mandioca e Fruticultura (Embrapa Cassava and Fruits), Rua Embrapa s/n, 44380-000 Cruz das Almas, Bahia, Brazil. Email : Domingo.Reinhardt@embrapa.br

Garth M. Sanewski, Queensland Department of Agriculture and Fisheries, Maroochy Research Facility, Nambour, Queensland 4560, Australia. Email: garth.sanewski@daf.qld.gov.au

Alfredo Rodrigues Sena Neto, Universidade Federal de Lavras, 37200-000, Lavras, Minas Gerais, Brazil. Email: arsenaneto@gmail.com

Brent Sipes, Department of Tropical Plant and Soil Sciences, College of Tropical Agriculture and Human Resources, University of Hawaii at Manoa, 3190 Maile Way, Honolulu, HI 96822-2279, USA. Email: Sipes@hawaii.edu

Alain Soler, CIRAD, UPR GECO, F-97455 Saint-Pierre, Réunion, France. GECO, Univ Montpellier, CIRAD, Montpellier, France. Email: Alain.Soler@cirad.fr

Preface

Of all the domesticated fruit crops, pineapple holds a unique place because of its specialized physiology, its somewhat bizarre appearance, its place of origin and its fantastic journey from there to the wider world. In recent years the world pineapple industry has been on a remarkable second journey, with shifts in the balance of production from processing to fresh market and new centres of production emerging. New cultivars, principally one, MD-2, have reshaped markets and fresh pineapple consumption and driven the pineapple industry in some countries to change or collapse. The pineapple genome has now been sequenced three times. Many of these changes or developments have occurred since the publication of the first edition of *The Pineapple*. The second edition builds on the accumulated experience of the first edition authors by including new authors from a wider number of producing countries, including new chapters and new developments and by taking advantage of new publishing technology. Each chapter has been fully revised, updated and expanded. New images have been included where available. Like the first edition, the second edition is intended to be an all-inclusive publication for those interested in pineapple production or just wanting to better understand the plant. The editors have recognized that those who might use this book include producers as well as researchers. Consequently, the writing style straddles the academic and practical in some chapters.

The first edition relied on a solid background of published material on 'Smooth Cayenne', the then predominant cultivar. Research in pineapple continues but is no longer focused on the archetypical 'Smooth Cayenne'. Unfortunately there is still a large gap in our understanding of newer cultivars and their performance in different environments. Our basic understanding of pineapple is still very much based on 'Smooth Cayenne'. New scientists with an interest in pineapple will find no shortage of issues needing urgent attention.

While the second edition is a combined contribution from many authors, the reader will find that some, because of their lifetime of experience in pineapple and their high standing within the scientific community, have contributed to multiple chapters. Geo Coppens d'Eeckenbrugge and Duane P. Bartholomew deserve special mention for their global contributions to our understanding of pineapple and to this book.

The editors would like to thank all authors and contributors for their efforts, especially given that language has been a barrier for some. We would also like to thank the publisher CAB International for their friendly assistance and advice.

Garth M. Sanewski
Duane P. Bartholomew
Robert E. Paull
Brisbane and Honolulu, 2018

1 History, Distribution and World Production*

Freddy Leal[1] and Geo Coppens d'Eeckenbrugge[2]
[1]*Universidad Central de Venezuela, Aragua, Venezuela;* [2]*CIRAD, UMR AGAP, Montpellier, France*

Pre-Columbian Importance and Distribution, and Early Post-contact Diffusion

On 4 November 1493, Cristobal Colón (Christopher Columbus) reached Guadeloupe in the Lesser Antilles and, according to the chronicles of Pedro Mártyr de Anglería (1530), found the plant and fruit of the pineapple in a small Indian village in the southern part of the island. This first contact of the Europeans with the plant and fruit was confirmed by Michele da Cuneo in 1495: 'There were some (plants) like artichoke plants, but four times as tall, which gave a fruit in the shape of a pine cone, twice as big, which is excellent, and it can be cut with a knife like a turnip, and it seems to be wholesome' (Morrison, 1963: 216).

The Portuguese also discovered the pineapple with their share of the New World, obtained through the Treaty of Tordesillas of 1494, involving Castilla and Portugal. On 22 April 1500, Pedro Alvares Cabral landed in Porto Seguro, in the south of the Brazilian state of Bahia, officially for the first time. Cabral's fleet continued to India, opening a major commercial route. This route was further improved in 1502 by the discovery of Saint Helen Island, midway between Porto Seguro and the Cape of Good Hope, and the pineapple was introduced there in 1505 (Collins, 1960). However, the earliest published description of the pineapple in South America comes from the observations of Pigafetta in the Rio de Janeiro area in 1519: 'This fruit resembles a pine cone and is extremely sweet and of exquisite taste' (Pigafetta, 1801: 14, authors' translation). This report was followed by many others in areas corresponding to Mexico (Ciudad Real, 1584); Nicaragua and Costa Rica (López de Velazco, 1574); Colombia (Cieza de León, 1553); Venezuela, Brazil and Paraguay (Muratori, 1743); and Ecuador (Collins, 1951; Leal, 1989). Pineapple was also observed far inland by Gaspar de Carvajal, who accompanied Francisco de Orellana, the Amazon river discoverer, from 1540 to 1542, and declared: 'The land is very large and beautiful, and very abundant of meals and fruits, like pineapples and [avocado] pears' (Carvajal, 1542: 41, authors' translation).

The pineapple (*Ananas comosus* var. *comosus*) was an essential plant for the Amerindians of the Antilles and the northern coast of South America (*terra firma*) before the arrival of Colón; they knew that pineapples were brought to the Antilles islands from the Orinoco and Amazon basins many years before by the dreaded Carib Indians, who dominated the islands. In their travels, the Carib usually took with them seeds, roots and plants to places already invaded with their large canoes. To them pineapple was a delicious fruit which could produce liquor for the

* A revision of the chapter by K.G. Rohrbach, F. Leal and G. Coppens d'Eeckenbrugge.

occasional festivities, reduce swellings, cure scratches and wounds; and the most important thing, rotten pineapples could be used to obtain venomous substances to smear arrows and spears for war (Haughton, 1978; Leal and Coppens d'Eeckenbrugge, 1996).

Gonzalo Fernández de Oviedo (1535) observed that the pineapple was very common in the Caribbean basin as well as on the South America mainland, where it was known under different names; he also made the first drawing and a good description:

> Each pine grows on a very sharp thorny thistle with long prickly leaves, very wild; from the middle of this thistle emerges a round stalk which bears only one pine, which takes about 10 months or a year to ripen. Once a fruit is cut, this thistle produces no more fruit and serves for nothing except to litter the ground [. . .] In order to plant other thistle-pines these tufts are the seed or succession of this fruit because by taking this tuft which the pine has on top of it or any of the others which are joined to the stem, and thrusting them in to the earth, two or three fingers deep, leaving one half of the tuft uncovered, they soon take very well and in the course of time I have mentioned each tuft will produce a new thistle and bear another pine as I have said [. . .] The pineapples from the mainland of South and Central America are better than the ones found in the islands.
>
> (Fernández de Oviedo, 1535: fol. 77, authors' translation)

As shown by this wide distribution and cultivation, Native Americans had domesticated and dispersed the plant well before the arrival of Colón, and had a thorough knowledge of the plant, differentiating cultivars, wild types, related taxa and their cultivation. The words 'nana' or 'anana' have designated the pineapple in languages of the three major Amazonian families (Arawak, Carib and Tupi), all through the Orinoco and Amazon basins, as well as in *terra firma*. Wild pineapples are often called 'nanai' or 'ananai' (Leal and Coppens d'Eeckenbrugge, 1996). In Guarani, 'nana' is the plant and 'anana' is the fruit (Alvarado, 1939). The Brazilian name 'abacaxi', originally designating particular cultivars, is derived from the Guarani word for the maize ear (Bertoni, 1919). The Spanish 'piña' and the English 'pineapple' came from the comparison with the exotic pinecone. The Portuguese, instead, diffused the word 'ananas' together with the plant all along the tropical shores from western Africa to the Indian Ocean, so the root word 'nana' has acquired a pantropical distribution (Dalgado, 1913, 1919).

In addition to the fresh fruit use, the native Americans used pineapple for the preparation of alcoholic beverages (pineapple wine) (Raleigh, 1596), for fibre production, emmenagogue, abortifacient, anti-amoebic, vermifuge, correction of stomach disorders and poisoning of arrow heads and spears. Most of these medicinal uses are related to the proteolytic enzyme of the pineapple. (Leal and Coppens d'Eeckenbrugge, 1996; Maurer, 2001). The Amerindians also domesticated two fibre plants from *A. comosus*. Along the Amazon River and to the north, they cultivated the *curagua* (*A. comosus* var. *erectifolius* (L.B. Smith) Coppens & Leal, a smooth-leaved type with a high yield of long and strong fibres, used for making ropes, fishing lines, fishing nets, hammocks and loin cloths (Leal and Amaya, 1991). The curagua has found a new economic use in the production of cut flowers, and more recently in biocomposites for the automotive industry (Zah *et al.*, 2007; Neves Monteiro *et al.*, 2013). South and south-east of the Amazon, they obtained fibres from the wild *A. macrodontes* Morren, as well as from the cultivated *crauatá/caroatá de rede* or *ananás bravo* (*A. comosus* var. *bracteatus* (Lindley) Coppens & Leal) (Arruda da Câmara, 1810). Today, variegated forms of this plant are used as tropical garden ornamentals or for the cut-flower market, thanks to the long and bright red bracts of their inflorescences.

The Europeans were particularly fascinated by the pineapple. Since their first travels, Spaniards have imported pineapple fruits, which were consumed in Europe when the trip was fast enough to make them edible; one of them was presented to the Emperor Charles V, who found it very pretty but refused to taste it (Humboldt, 1816). The pineapple not only travelled to Europe but was also carried in the great voyages of the 16th and 17th centuries. Unlike the fruit, the plant and its vegetative propagules are tough, durable and very resistant to drought, which greatly facilitated its diffusion around the world. The Portuguese introduced the pineapple to St Helen Island in 1505, and to Madagascar and southern India before 1548. It was also reported in the Philippines in 1558, coming from China, and naturalized in Java in 1599. Its cultivation was reported in Nepal in 1601, in Guinea in 1602, in Singapore in 1637 and in Formosa (Taiwan) in 1650 (Laufer, 1929, cited by Collins, 1960; Chadha and Pareek, 1988).

Pineapple was readily accepted as an outstanding new fruit and other uses were recognized throughout the world. In the gardens of Mozambique, where he arrived in 1586, Frei João dos Santos (1891: 49) observed 'many pineapples', both on the coast (province of Sofala) and more than 400 km inland (province of Tete). Van Linschoten (1610: 136), who worked in India between 1583 and 1592, noted:

> The pineapple is not native there, instead it was brought from Brazil by the Portuguese. They were much prized at the beginning . . . but now they are of little value as a result of their abundance . . . they are the size of a melon, the shape of a distaff or a pine-apple, easy to cut, of a red color admixed with green, and grow to the height of a cubit.
>
> (Van Linschoten, 1610: fol. 136, authors' translation)

In 1571, the people of the Philippines were already making their traditional piña cloth from pineapple leaves fibres, and the natives of Malaysia used the fruit to regulate human reproduction (Gimlette, 1915, cited by Laszlo and Henshaw, 1954). The diffusion of the pineapple by the Portuguese explains the geographic distribution of the cultivars Singapore Canning and Selangor Green, which diffused from Porto Seguro into West Africa (at least between Côte d'Ivoire and Angola) and the Indo-Pacific basin (from East Africa and South Asia to the Philippines) in the early 16th century. Slightly later, as they had explored Brazil further south, the Portuguese also diffused the cultivar Pérola to the Gulf of Guinea, which explains its current presence in West Africa, and the recent development of a small specialized export to Europe, where its delicious fruits are esteemed by connoisseurs.

The pineapple was grown in greenhouses in Europe, becoming a fashionable plant for kings, aristocrats, well-to-do and educated people, botanists and horticulturists. According to Loudon (1822), the plant was introduced to England in 1690 by Bentick, later Count of Portsmouth; but the first attempts at cultivation in Europe date to the end of the 16th century, when Le Court (or La Court), a very rich Flemish trader, grew it in Drieoeck, close to Leyden. Later, he published a Horticulture Treatise, where he included the flower induction of pineapple. From Holland, in 1719, Matthew Decker sent pineapple plants to England, even though the plant had been introduced in 1690 as a botanical sample. The pineapple was introduced to France in 1730 by King Louis XV, who ordered the construction of greenhouses at Versailles to grow them (Coppens d'Eeckenbrugge *et al.*, 1997).

The Colonial and Post-colonial Germplasm Diffusion and Improvement

The introduction of pineapple cultivars was very active until the late 19th century. Griffin (1806) pointed out that it would be an endless and unnecessary work to enumerate all the cultivars in England, because many of them are worthless and their cultivation cumbersome; he described the ten most interesting ones, mentioning as the best the 'Oval pine-apple' or 'Queen-pine'. This cultivar, brought from Barbados, was famous in England before 1661 (Evelyn, 1661, cited by Collins, 1960). Later, Loudon (1822), Munro (1835) and Beer (1857), described many cultivars and stated that their list was still increasing regularly.

In 1819, France dispatched an expedition to its colonies in America and the Pacific to collect seeds and plants for its botanical gardens in Paris and Versailles. At Cayenne, the capital of French Guiana, Samuel Perrottet (1824) collected a new pineapple, which he named *Bromelia mai-pouri*. This epithet, which means 'tapir', is still used for many large-fruited cultivars in the Amazon and Orinoco basins. The five plants collected by Perrottet were multiplied and sent from Paris to England, Belgium, Holland and the Azores, and then, from England, to Florida, Jamaica, Sri Lanka and Australia. In the late 19th century, they arrived in Hawaii, from where they were distributed to the Greater Antilles, Mexico, the Philippines, Taiwan and Kenya in the first half of the 20th century (Collins, 1960). It was thus distributed and known all around the world as the cultivar Cayenne Lisse (Smooth Cayenne) that dominated the industry for many years (Rohrbach *et al.*, 2003; Okihiro, 2009).

With the notable exceptions of 'Smooth Cayenne' and 'Queen', most early cultivars disappeared as commercial cultivation in Europe declined and pineapples were imported from the West Indies. Thanks to its high yield and good canning characteristics, 'Smooth Cayenne' became the pillar of the 20th-century pineapple industry worldwide, with particular developments in Hawaii, South-East Asia and West Africa, whereas 'Queen' and 'Singapore Canning' were relegated to regional roles in the Old World tropics. In the Americas, a few other cultivars had retained regional importance too, such as: 'Española Roja' (*syn.* 'Red Spanish'), which was popular in the Caribbean basin, 'Pérola', the favourite Brazilian cultivar, which still compensates its less appealing green conical white-fleshed fruits by its rusticity and a very sweet and aromatic flavour; the spineless 'Perolera', and its sports, 'Manzana' in Colombia and 'Capachera' in Venezuela, particularly adapted to highland cultivation in the Andean valleys.

But the largest diversity of cultivars from the Amazonian cradle of the pineapple remained ignored until the end of the 20th century. Efforts towards varietal diversification, initiated at the Pineapple Research Institute (PRI) of Hawaii, focused on hybrid breeding to develop a cultivar that would surpass 'Smooth Cayenne' in several characteristics. However, even the best hybrids failed in the final evaluations, due to the combination of some fatal flaw and the 'Cayenne conservatism' that hampered effective varietal innovation in an industry that, for decades, had been tailored by and for 'Smooth Cayenne'. When the PRI closed in 1975, its hybrids were turned over to the founding companies. One of them, called 'MD-2' was selected in 1973, but only diffused by Del Monte in 1996, under the commercial name 'Del Monte Gold' (Bartholomew *et al.*, 2010). This cultivar was so well accepted by the North American and European consumers that it boosted the world fresh pineapple market. Sadly, this final success story did not inspire more diversification efforts. 'MD-2' substituted for 'Smooth Cayenne' as the leading fresh fruit cultivar; however, none of the interesting hybrids created in later breeding programmes (Brazil, Malaysia, Martinique, Ivory Coast) was seriously tested to challenge this new hegemony. On the contrary, 'MD-2' even took a share from regional cultivars in the markets of tropical countries, reducing global pineapple diversity.

Development of the World Production and Trade

Because of the short life of the fruit, the economic importance of pineapple developed along with efficient transportation and preservation. Commercial transportation of pineapples started in the 19th century. Loudon (1822: 5) stated that

> of late years the Pine Apple had been sent to England in abundance, attached to the entire plant, and a cargo has arrived from Providence Island, in the Bermudas, in 6 weeks. This facility of cultivation, and their more general culture, has greatly lessened their price and rendered them common.

These lower prices led to a decline in European greenhouse cultivation. In any event, the massive commercialization of fresh pineapple had to wait for better means of transportation: modern refrigerated ships and cargo planes. Preservation techniques were similarly improved, from the first jams exported by the European colonists of Brazil and New Spain (México) (Thevet, 1558; Acosta, 1590) to the canned pineapple at the end of the 19th century. Commercialization during the mid-19th century developed based on the shortest trade routes rather than an optimum pineapple production environment. Production in Florida, the Bahamas, Cuba and Puerto Rico supplied the North American market, while the Azores supplied the European market. The latter maintained their monopoly on the European fresh fruit market until the Second World War, when production and imports shifted to African countries (Py *et al.*, 1987).

Commercial production of pineapple started in Hawaii at the end of the 19th century using primarily cultivar 'Smooth Cayenne' for both fresh fruit and canning. There, pineapple development was boosted by the invention of a Dole company engineer, Henry Gabriel Ginaca: a machine that removed the pineapple's rough skin, cored and sized its body, and sliced the flesh. The *ginaca* was capable of delivering 35 pineapples per minute to an assembly line of trimmers, usually women, who sat at tables to inspect and remove any imperfections on the fruit before sending it for canning. With improvements, the ginaca was capable of handling more than 100 pineapples per minute (Okihiro, 2009). All these improvements determined that, for many years, until the early 21st century, the international exchanges were largely superior for the canned fruit than for the fresh fruit. Furthermore, other important canning operations started around the same period in South-East Asia (Malaysia in 1888, Taiwan in 1902, Philippines in 1920), Australia, South Africa, France (Martinique), Cuba, Puerto Rico and Kenya. The Second World War ruined the South-East Asian industry and destroyed international trade. Hawaii gained a strong leadership position that lasted until the development of new competitors (Côte d'Ivoire, Philippines and Kenya) between 1950 and the early 1960s.

In these same years, refrigerated sea transportation developed and diminished the importance of proximity to the market; Hawaii, Côte d'Ivoire and Taiwan shifted part of their production to the fresh fruit market, exporting, respectively, to the North American, European and Japanese markets. The Philippines expanded greatly their production in the 1970s, exporting canned products and fresh fruits to Japan (Py *et al.*, 1987). Today the canned-product market remains important but with the introduction of cultivar MD-2, the fresh fruit has taken a leading position on the international market, and it has grown steadily since then. However, this impressive trend must not make us forget that two-thirds of the pineapple produced in the world is consumed as a fresh fruit in the country of production. For example, Brazil, China and India, three top producers (Fig. 1.1) do not participate significantly in exports. Globally, there are around 30 countries producing more than 100,000 tonnes (t), plus 50 or so with a production of less than 100,000 t (Loeillet, 2015).

In the long term, the evolution of world pineapple production during the last five decades (Fig. 1.2) shows three phases in its progression. Until the late 1970s, most of the increase came from fresh fruit markets within tropical countries; the comparatively limited international trade increased slowly, and the participation of fresh fruit remained marginal. In the 1980s, the production increase was mostly boosted by the strong growth of pineapple juice, while the fresh fruit production maintained its slower but more regular growth for both national and international markets. In the late 1990s, the global growth rate was boosted again, this time by the release of 'MD-2' on the international fresh fruit market in 1996.

The introduction of 'MD-2' has had an impact, not only in traded volumes, but on geography of pineapple production and commercialization as well. In 1996, fresh pineapple imports,

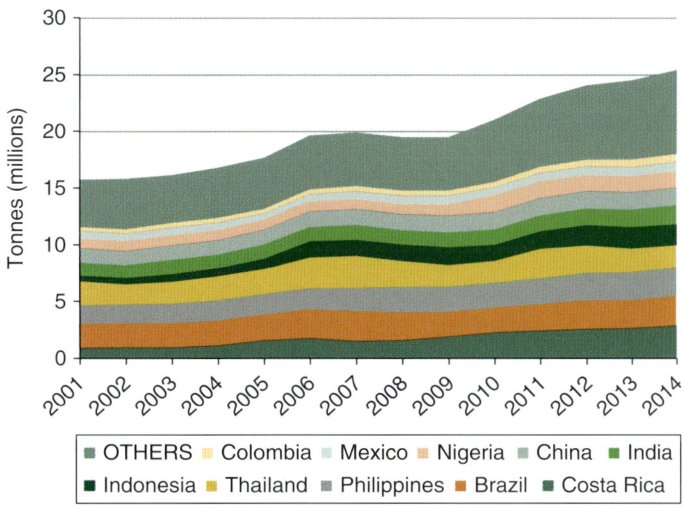

Fig. 1.1. World pineapple production, with shares of the top nine producing countries (million tonnes). Data from Cirad-FruiTrop.

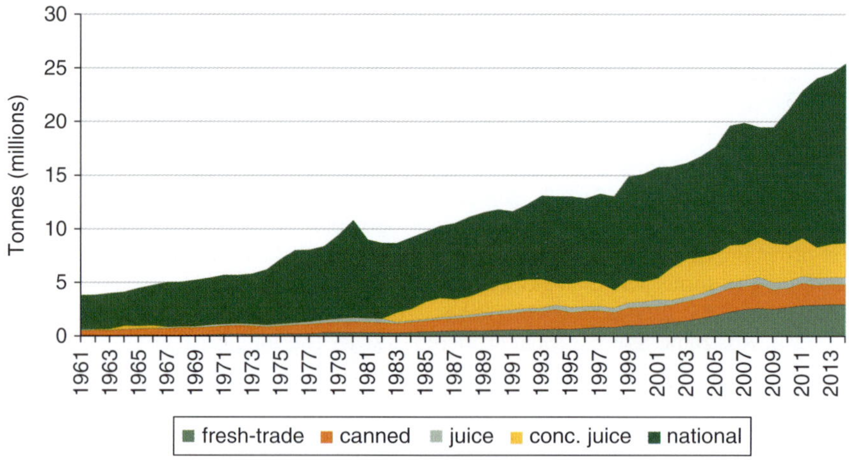

Fig. 1.2. Evolution of world pineapple production (million tonnes) from 1961 to 2013, for the fresh fruit (national and international markets) or for processed products (fresh fruit equivalent). Data from Cirad-FruiTrop.

essentially from 'Smooth Cayenne', were led by the EU and the USA with 283,258 and 135,255 t respectively. The former market was dominated by Côte d'Ivoire, with a share around 55%. Costa Rica came second, with 22%, whereas its share of the US market was 63%. Well supported by a powerful logistical and commercial organization, the introduction of 'MD-2' by Del Monte in Costa Rica was an immediate success. More regular and better suited to consumer expectations, thanks to its lower acidity, it sold at distinctly higher prices than 'Smooth Cayenne'. This success soon attracted many new growers, first in Costa Rica. The US fresh fruit market reacted immediately, and its volume quadrupled in a decade, exceeding 500,000 t before 2005 (Fig. 1.3).

Volumes stabilized in 2014, at their highest level (more than 1 million tonnes). The relative increase was slower in the EU, as the market was then more developed and 'MD-2' had to compete with 'Smooth Cayenne' from Côte d'Ivoire (Fig. 1.4). In this transition phase, it benefited from price bonuses of up to more than 350 euros/t (Loeillet, 2014), supplanting its rival while the market volume reached a growth of 47% between 1996 and 2003 (417,583 t). From then on, the growth curve became much steeper, and imports more than doubled in five years, exceeding 930,000 t in 2008, before stabilizing, with the consequent loss of the price bonus in favour of 'MD-2'. Thus the market is saturated at above 1 million tonnes for the USA, 850,000 t for the EU, and further volumes can be absorbed only by a fall in prices. In the end, the fresh pineapple has become a mere agricultural commodity, despite its favourable exotic image, and its market has become a mass market where price is more important than quality (Loeillet, 2016; Dawson and Loeillet, 2017). There is no product differentiation, except for marginal air-freight pineapple market niches, fed with very distinct cultivars such as 'Queen', from the islands of La Réunion and Mauritius, or 'Sugar Loaf' ('Pérola'), from West Africa, in Europe. Exceptionally, some very small operators have managed to introduce small quantities of new hybrids, as was the case for the Embrapa's 'Imperial' in France (Nicolas Reuse, personal communication).

The Current Picture and Beyond

With a world production of above 25 million tonnes, having duplicated in less than 25 years, pineapple is the third most important tropical fruit, behind bananas and mango (considering that citrus cultivation is mostly subtropical). Although the general growth trend looks positive, particularly on domestic markets (Figs 1.1–1.4), the world market is not free of problems

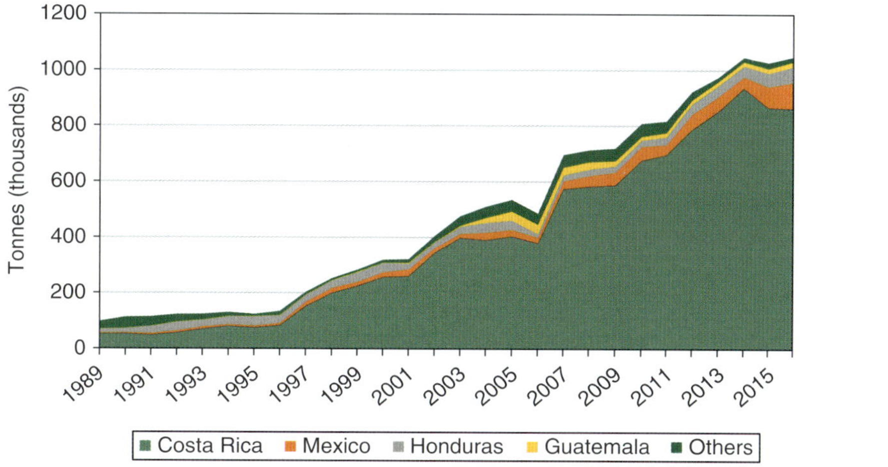

Fig. 1.3. Evolution of fresh pineapple imports (thousand tonnes) of the USA (including the small fraction re-exported to Canada). Data from Cirad-FruiTrop.

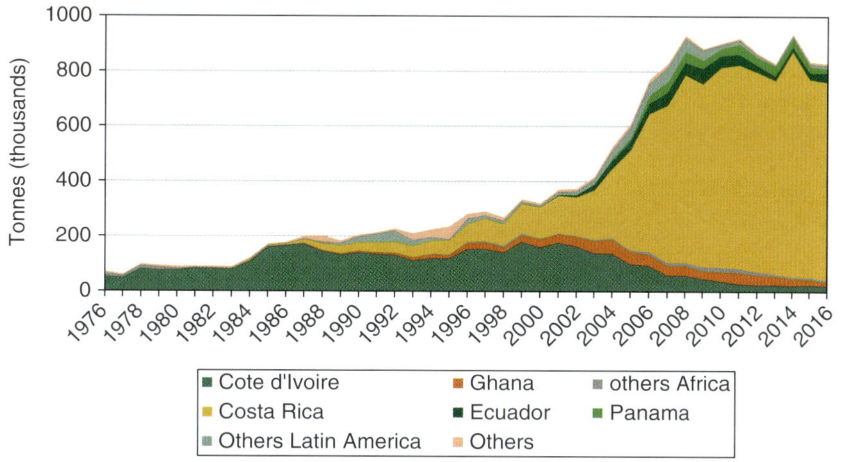

Fig. 1.4. Evolution of European fresh pineapple imports (thousand tonnes). Data from Cirad-FruiTrop.

and threats. In fact, there are two international pineapple markets, because of the double dichotomy underlined by Loeillet (2017): one related to the product, between fresh and processed fruits; and a geographic one, between America, more particularly Central America, and Asia. We shall follow Loeillet's analysis in our presentation of these two markets.

The fresh fruit market, totalling 2.6–3 million tonnes, is extremely simple, as it is focused on the USA and the EU (Japan accounting for less than 200,000 t), and fed by an overwhelmingly dominant provider, Costa Rica, with respective market shares of 82–90% and 86–87% on the 2014–2016 period. In terms of genetic diversity, the monopoly of 'MD-2' is similar. In Costa Rica, 43,000 ha are dedicated to it. Considering also product from neighbouring countries between southern Mexico and Panama, this region of production comprises 85% of the world fresh pineapple export, selling to North America, Europe, Russia, temperate countries of southern South America, and very recently China (den Herder, 2017). The second region of importance is in the Philippines, with a market share around 10%, feeding the markets of Asian neighbours, in particular Japan (Loeillet, 2017). The success story is likely behind us, as the impressive growth of the two major markets has come to an end. In a mass market where volumes are tied to prices, there are no incentives for differentiation on fruit quality.

The processed fruit production is based on 'Smooth Cayenne' and concentrated in South-East Asia, led by Thailand, Indonesia and the Philippines, which account for 89% of the world offer of canned pineapple and 69% of all processed products. Less important players are Vietnam, China, Kenya, South Africa and Swaziland. In America, Costa Rica is an important actor of the pineapple juice market, with a production based on 'MD-2'. In volume, juice is more important (58%) than the canned fruit (42%); however, it is considered a subproduct of the canning industry, so the latter is more important in terms of value. After growing to the equivalent of 5 million tonnes of fruit in 2011, the processed pineapple market decreased by 10% (4.6 million tonnes in 2015). Among other factors, climatic conditions influence production levels. In the last year, the market has been quite tense, with a high price volatility for all types of products (Loeillet, 2017).

After the introduction of 'MD-2', the fresh pineapple world market has gone through two decades of strong growth, at the expense of the canned pineapple sector. Globally, the result looks positive; however, most indicators point to a loss of momentum. Furthermore, the overall picture is one of fragility, related to increased geographic concentration and specialization, lack of innovation and product differentiation, particularly in terms of quality and varietal diversity, and exposure to environmental hazards. Geographic concentration and genetic uniformity make the pineapple industry highly vulnerable to environmental and biological hazards, particularly when accelerating global change increases the frequency and importance of climate variations. The pineapple industry has seriously considered neither the phytosanitary risk (e.g. the potential impact of a severe disease such as the Brazilian fusariosis under current cultivation standards) nor the consumer demands for organic production, lesser environmental impact and social responsibility.

The lack of innovation seems to be due in good part to conservatism. This is particularly obvious in the varietal domain. For example, the much acclaimed introduction of 'MD-2' was a commercial innovation, not really a genetic one, as this hybrid had been selected 23 years earlier. This long delay shows how reluctant the profession was to change; it also shows that pineapple experts had not realized that the acidity of 'Smooth Cayenne' had long deterred consumers from fresh pineapples. So strong was the industry's addiction to this cultivar that the initial PRI objective of varietal diversification against the phytosanitary risk (Williams and Fleisch, 1993) was perverted to the creation of a 'Super Cayenne'. And the fact is that 'MD-2' is genetically close to 'Smooth Cayenne' (see Chapter 4 in this volume), so it offers no better perspective for the pineapple industry resilience. For example, its susceptibility (*Dickeya* sp.) has caused outbreaks of bacterial heart rot in Hawaii (Marrero *et al.*, 2013), while its susceptibility to fusariosis undermined a Del Monte attempt to cultivate it in Brazil, because of production losses reaching 45% (Jonathas, 2010).

Another example of conservatism comes from the latter country, where Embrapa conducted an efficient breeding programme to control this disease, producing excellent hybrids with fusariosis resistance, high sugar content and full bright coloration at harvest, such as the above-mentioned 'Imperial' released in 2003 (Bartholomew *et al.*, 2010). More than a decade later, its adoption by Brazilian growers has been limited to a very high-priced niche market, while 'Pérola' and 'Smooth Cayenne' have remained overwhelmingly dominant, despite the extra costs and yield losses due to fusariosis. More generally, in the last decades, many new hybrids have been produced in breeding programmes of several countries, mostly for the fresh fruit market (Bartholomew *et al.*, 2010, 2016; see Chapter 4 in this volume), but they have rarely been given a serious chance. If varietal differentiation is neglected, qualitative differentiation (e.g. organic pineapple, social and green labels) becomes extremely difficult, limiting the possibilities that the king of fruits can soon recover its crown.

References

Acosta, J. (1590) *Historia natural y moral de las Indias.* Mexico: Fondo de Cultura Económica.

Alvarado, L. (1939) *Glosario de voces indígenas. Obras Completas,* 2 vols. Caracas, Venezuela: Fundación la Casa de Bello, 1984.

Anglería, P.M. de (1530) *Décadas del nuevo mundo.* Colección de Fuentes para la Historia de América. Buenos Aires, Argentina: Ed. Bajel, 1944.

Arruda da Câmara, M. (1810) *Dissertação sobre as plantas do Brazil que podem dar linhos proprios para muitos usos da sociedade, e suprir a falta de canhamo.* Rio de Janeiro, Brazil: Impressão Regia, pp. 3–7, 13–22.

Bartholomew, D.P., Coppens d'Eeckenbrugge, G. and Chen, C.C. (2010) Pineapple. In: Clark, J.R. and Finn, C.E. (ed.) Register of new fruit and nut cultivars list 45. *HortScience* 45, 740–742.

Bartholomew, D.P., Sanewski, G.M. and Chen, C.C. (2016) Pineapple. In: Gasic, K. Preece, J.E. and Karp, D. (ed.) Register of new fruit and nut cultivars list 48. *HortScience* 51, 640–641.

Beer, J.G. (1857) Die Familie der Bromeliaceen. In: Morren, E. (ed.) *Monographie des Ananas. Belgique Horticole (Liège)* 28, 144–172.

Bertoni, M.S. (1919) Contributions à l'étude botanique des plantes cultivées. I. Essai d'une monographie du genre *Ananas. Anales Científicos Paraguayos (Ser. II)* 4, 250–322.

Carvajal, G. de (1542) Relación del nuevo descubrimiento del famoso Rio Grande de las Amazonas. In: Pinillos Iglesias, M.N. (ed.) (2011) *Descubrimiento del río de las Amazonas. Relación de Gaspar de Carvajal.* Babelia.doc. Available at: http://blogs.elpais.com/papeles-perdidos/pdf/gaspardecarvajal.pdf (accessed 22 March 2018).

Chadha, K.L. and Pareek, O. (1988) Genetic resources of fruit crops: achievements and gaps. *Indian Journal of Plant Genetic Resources* 1, 43–48.

Cieza de León, P. (1553) *La crónica del Perú.* Biblioteca de Autores Españoles. Madrid: Ediciones Atlas, 1958.

Ciudad Real, A.D. de (1584) *Tratado curioso y docto de la grandeza de la Nueva España.* Universidad Autónoma de México, 1976.

Collins, J.L. (1951) Antiquity of the pineapple in America. *Southwest Journal of Antrhopology* 7, 145–155.

Collins, J.L. (1960) *The Pineapple.* London: Leonard Hill.

Coppens d'Eeckenbrugge, G. Leal, F. and Duval, M.F. (1997) Germplasm resources of pineapple. *Horticultural Reviews* 21, 133–175.

Dalgado, S.R. (1913) *Portuguese Vocables in Asiatic Languages.* English trans. by A.X. Soares (1936), Baroda, India: Baptist Mission Press.

Dalgado, S.R. (1919) *Glossario Luso–Asiático.* Coimbra, Portugal: Imprensa da Universidade.

Dawson, C. and Loeillet, D. (2017) European pineapple market. Pineapple vs. pineapple: the sole issue. *Fruitrop* 249: 22–25.

den Herder, K. (2017) Costa Rica: increased production necessary to supply Chinese market. *Fresh Plaza,* 28 September. Available at: http://www.freshplaza.com/article/182245/Costa-Rica-Increased-production-necessary-to-supply-Chinese-market (accessed 28 February 2018).

dos Santos, J. (1891) *Ethiopia Oriental (Varia historia de cousas notaveis do Oriente),* vol. 1. Lisbon, Portugal: Escriptorio da Empreza.

Fernández de Oviedo, G. (1535) *La historia general de las Indias.*

Griffin, W. (1806) *A Treatise on the Culture of the Pine-apple.* Newark, Notts: S. & J. Ridge.

Haughton, C.S. (1978) *Green Immigrants – The Plants that Transformed America.* New York: Harcourt Brace Jovanovich Inc.

Humboldt, A. (1816) *Viaje a las regiones equinocciales del nuevo continente,* 5 vols. Ministerio de Educación. Caracas, Venezuela: Dirección de Cultura y Bellas Artes, 1972.

Jonathas, Andreh (2010) Del Monte tem bons resultados na exportação da banana produzida em Limoeiro do Norte. Available at: http://limoeirodonorte.blogspot.fr/2010/08/del-monte-tem-bons-resultados-na.html (accessed 24 September 2017).

Laszlo, H. de and Henshaw, P.S. (1954) Plant materials used by primitive peoples to affect fertility. *Science* 119, 626–631.

Leal, F. (1989) On the history, origin and taxonomy of the pineapple. *Interciencia* 14, 235–241.

Leal, F. and Amaya, L. (1991) The curagua (*Ananas lucidus*, Bromeliaceae) crop in Venezuela. *Economic Botany* 45, 216–224.

Leal, F. and Coppens d'Eeckenbrugge, G. (1996) Pineapple. In: Janick, J. and Moore, J.N. (ed.) *Fruit Breeding*. John Wiley & Sons, New York, pp. 565–606.

Loeillet, D. (2014) World pineapple market. On a volcano. *Fruitrop* 228, 18–28.

Loeillet, D. (2015) World fresh pineapple market. Crash test! *Fruitrop* 236, 43–41.

Loeillet, D. (2016) World fresh pineapple market. A market of uncommon simplicity. *Fruitrop* 228, 40–48.

Loeillet, D. (2017) Ananas. In: Chalmin, P. and Jégourel, Y. (ed.) *CyclOpe. Les marchés mondiaux 2017.* Paris: Vent d'Est, Vent d'Ouest, Economica.

López de Velazco, J. (1574) *Geografía y descripción universal de las Indias.* Biblioteca de Autores Españoles. Madrid: Ediciones Atlas, 1971.

Loudon, J.C. (1822) *The Different Modes of Cultivating the Pine-Apple, from its First Introduction to Europe to the Late Improvements of T.A. Knight Esq.* London: Houlgman Hurst Resorme Brown.

Marrero, G., Schneider, K.L., Jenkins, D.M. and Alvarez A.M. (2013) Phylogeny and classification of *Dickeya* based on multilocus sequence analysis. *International Journal of Systematic and Evolutionary Microbiology* 63, 3524–3539.

Maurer, H. (2001) Bromelain: biochemistry, pharmacological and medical use. *Cell and Molecular Life Sciences* 58, 1234–1245.

Morrison, S.E. (1963) *Journals and other documents of the life and voyages of Christopher Columbus.* New York: Heritage Press.

Munro, D. (1835) Classification of pineapple varieties. *Transactions of the Horticultural Society of London* (Ser.2) 1, 1–34.

Muratori, L. (1743) *Relation des missions du Paraguay.* Paris: La Découverte-Maspero, 1983.

Neves Monteiro S., Duarte Lopes F.P., Oliveira Nascimento D.C., da Silva Ferreira, A. and Gundappa Satyanarayanac, K. (2013) Processing and properties of continuous and aligned curaua fibers incorporated polyester composites. *Journal of Materials Research and Technology* 2, 2–9.

Okihiro, G.Y. (2009) Pineapple culture. A history of the tropical and temperate zones. Berkeley: University of California Press.

Perrottet, S. (1824) *Catalogue raisonné des plantes introduites dans les colonies francaises de Bourbon et de Cayenne, et de celles rapportées vivantes des mers d'Asie et de la Guyane, au Jardin du Roi à Paris.* Paris: Imprimerie Lebel, Imprimeur du Roi.

Pigafetta, A. (1801) *Premier voyage autour du monde par le Chevallier Pigafetta sur l'escadre de Magellan, pendant les années 1519, 20, 21 et 22.* French translation by C. Amoretti, publisher of the Italian manuscript. Paris: Jansen.

Py, C., Lacoeuilhe, J.J. and Teisson, C. (1987) *L'ananas, sa culture, ses produits.* Paris: G.P. Maisonneuve & Larose.

Raleigh, W. (1596) *Discovery of the Large, Beautiful Empire of Guiana*, ed. V.T. Harlow. London: The Argonaut Press.

Rohrbach, K.G., Leal, F., Coppens d'Eeckenbrugge, G. (2003) History, distribution and world production. In: Bartholomew, D.P., Paull, R.E. and Rohrbach, K.G. (eds) *The Pineapple: Botany, Production and Uses.* CAB International, Wallingford (Great Britain), pp. 1–12.

Thevet, A. (1558) *Les singularitez de la France antarctique, autrement nommée Amerique & de plusieurs Terres et Isles decouvertes de nostre temps.* Paris: Maurice de La Porte.

Van Linschoten, J.H. (1610) *Histoire de la navigation de Jean Hughes de Linscot Hollandois et de son voyage es Indes Orientales. Avec annotations de Bernard Paludanus.* Amsterdam: Imprimerie de Théodore Pierre.

Williams, D.D.F. and Fleish, H. (1993) Historical review of pineapple breeding in Hawaii. *Acta Horticulturae* 334, 67–76.

Zah R., Hischier R., Leão A.L. and Braun, I. (2007) Curauá fibers in the automobile industry – a sustainability assessment. *Journal of Cleaner Production* 15, 1032–1040.

2 Morphology, Anatomy and Taxonomy

Geo Coppens d'Eeckenbrugge[1] and Freddy Leal[2]
[1]CIRAD, UMR AGAP, Montpellier, France; [2]Universidad Central de Venezuela, Aragua, Venezuela

Morphology

Ananas comosus is a herbaceous perennial of the *Liliopsidae* (Monocotyledonous) whose terminal inflorescence gives origin to a multiple fruit (sorose). After maturation of the first fruit, the plant develops new shoots from axillary buds, so producing new growth axes capable of producing another fruit. In most commercial plantings, the plants are not allowed to produce more than two to three crops, due to a reduction in fruit size and uniformity. A new plantation is then established using vegetative propagules. This vegetative reproduction is also dominant in wild pineapples where, in addition to lateral shoots, the crown and slips contribute to propagation as they resume rapid growth at fruit maturity. The long peduncle then bends because of this mass, the crowns and slips reach the ground and may root. Thus most natural populations appear to consist of a single clone, expanding as if propagating by stolons.

The adult plant is 1–2 m high and 1–2 m wide, and has the general shape of a spinning top. The main morphological structures to be distinguished are the stem, the leaves, the peduncle, the multiple fruit or syncarp, the crown, the shoots and the roots (Fig 2.1). The following description is mainly focused on the cultivated pineapple and is partly based on the anatomical studies of Krauss (1948, 1949a, 1949b) and Okimoto (1948).

Fig. 2.1. Main morphological structures of the pineapple plant. Photo G. Sanewski.

Stem

The pineapple stem is club-shaped, with a length of 25–50 cm and a width of 2–5 cm at the base and 5–8 cm at the top. Its aerial part is straight and erect, while the shape of the earthed part depends on the material used for planting. It is markedly curved when coming from a slip, as the stem of these propagules is comma-shaped, less so when coming from a stem shoot, or erect when coming from a crown. Nodes can be seen by the leaf scars left after stripping the leaves from the stem. Internodes are short (1–10 mm according to their position), so the whole rosette appears dense and compact. Flattened shoot buds, 3–5 mm high and about 5 mm wide, occur in the leaf axils. In the central portion of the stem, they are larger because of an increase in the size of their prophyll (the first leaf of the shoot, which encloses it). A striking feature of the pineapple stem is the presence of adventitious roots breaking through the epidermis, and growing flattened and distorted, tightly wound around the stem, between the leaves. Their older portion is suberized. These aerial roots rarely produce laterals. They are elongated from a few millimetres in the subapical region to 10 cm or more near the stem base. Thus, the underground portion of the stem is covered with many adventitious fibrous roots.

The stem (Fig 2.2) is constituted by a central cylinder, or stele, and a cortex, separated by a thin layer of vascular bundles produced by the dome-shaped apical meristem. The dense network of vascular tissue separating cortex and stele consists chiefly of xylem with very little phloem. In this tissue, areas of nonvascular tissue, or leaf gaps, are disposed at intervals, allowing leaf-trace bundles to pass from the cortex into the stele. This vascular cylinder is thicker

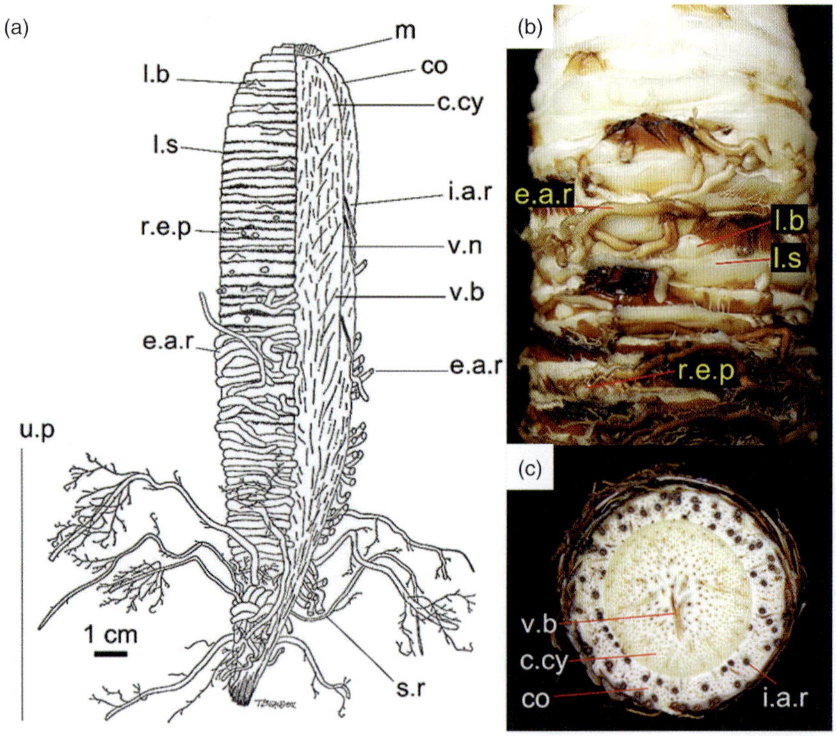

Fig. 2.2. Plant Stem Anatomy. (a) Illustration of the stem of a mature plant with the leaves removed and a longitudinal quarter cut from one side. (b) Image of the external mid-portion of the stem with leaves removed. (c) Cross-section image of the stem. co: cortex; c.cy: central cylinder; e.a.r: external portion of adventitious root; i.a.r: internal portion of adventitious root; l.b: leaf bud; l.s: leaf scar; m: meristem; r.e.p: root emergence point; s.r: soil root; u.p: underground portion; v.b: vascular bundle of leaf; v.n: vascular network. Photos b and c by G. Sanewski; Illustration a by T. Beaudou from Py *et al.*, 1987 with permission courtesy of C. Teisson.

and suberized at the stem base. On the cortical side, a narrow layer of long, thin-walled cells bound it. Vascular bundles are very numerous throughout the stem but less so in the cortex than in the stele. The latter is mainly constituted by a compact parenchyma with abundant starch. It contains large cells with raphides of calcium oxalate crystals. The cortex is composed of a parenchymatous tissue, crossed by the isolated vascular bundles going to the leaves, by the adventitious roots originating at the boundary with the central cylinder, and circumferential small bands of vascular tissue lying just above the leaf attachment to the stem. The inner parenchyma of this cortex is also rich in starch and contains raphide cells. Limiting the stem externally is the epidermis, with peltate trichomes in the nodal regions.

Leaves

The sessile leaves enclose the stem on two-thirds of its circumference. The phyllotaxy varies, being 5/13 in large-fruited cultivated pineapples and 3/8 in small-fruited wild pineapples (Kerns *et al.*, 1936). Leaf number is variable between cultivars but is generally around 40–80. The lower ones, originating from the planting material or produced soon after planting, are smaller (5–20 cm) as compared with the younger ones, which can reach more than 1.6 m in length and 7 cm in width, depending on the cultivar and environmental conditions. The apical ones are short and erect. The leaves are ensiform and, except for the young apical ones, broader at their base that forms a non-chlorophyllous sheath around the stem. The blades then taper progressively to a sharply pointed indurated tip. The constriction between the sheath and the blade is more marked in certain wild pineapples. Temporary stress during leaf growth may cause variations in width or spininess, or both, along the blade. Leaves are semi-rigid, thanks to their crescent-shaped section. As in other bromeliads, this allows the plant to collect water in the rosette, where it can be absorbed by the aerial roots present along the stem or through the epidermis of the sheath. The concave adaxial face is green to dark green, with some anthocyanins, to dark red or purple according to the variety and conditions. The abaxial side is convex, with a surface corrugated by longitudinal grooves. Both sides are covered by peltate trichomes, particularly the abaxial one, which is densely furfuraceous and silvery (see Fig 6.10 in this volume). The leaf margins are usually spiny, however certain varieties are partially or totally inermis. In some smooth varieties, the lower epidermis is folded over the leaf edge and extended over the upper surface, so producing a narrow silvery stripe, a trait called 'piping' by Collins (1960) (Fig 2.3).

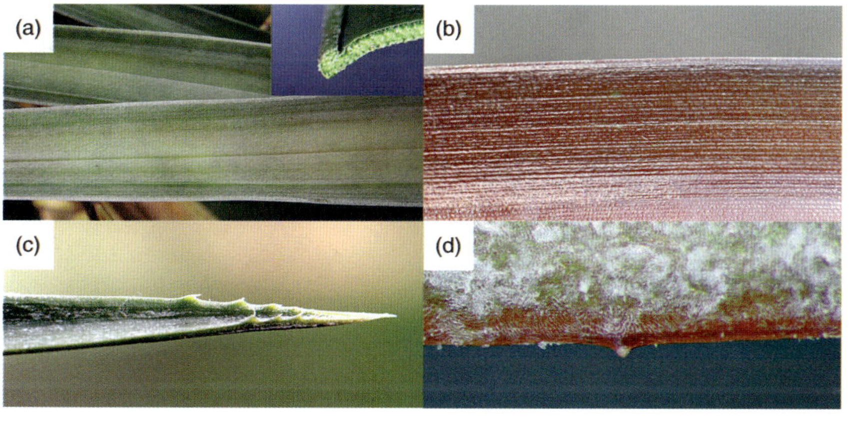

Fig. 2.3. Smooth leaf margin phenotypes in pineapple. (a) Piping (cv. 'Manzana'). Insert shows fold in piping leaf margin which is present in some varieties. (b) Smooth (var. *erectifolious*). (c) Spiny tip (cv. Smooth Cayenne). (d) Sand paper, an incomplete expression of the piping gene, with atrophied, quite invisible, spines shown at high magnification. Photos G. Sanewski.

A section across the leaf (Fig 2.4) shows successively a thick and smooth cuticle; a particular upper epidermis, which consists of a single layer of cells, each containing a silica body, oriented perpendicularly to the leaf axis, and rigidified by thick and undulated lateral and inner cell walls; the hypodermis; the water-storage tissue, consisting of various layers of thin-walled cells, which accounts for one-quarter to one-half of the leaf thickness, depending on the water status of the plant; the chloroplast-rich mesophyll with the vascular bundles, fibre strands and aerating canals; the lower hypodermis and epidermis, with the stomata arranged longitudinally along the characteristic grooves of the abaxial leaf side, covered by numerous trichomes giving it a silvery appearance and increasing reflectance. The fibre strands confer a high tear resistance to the pineapple leaves. Stomatal density is low at about 80 stomata mm^{-2}. The trichomes, present in most all the known bromeliads, are flat and shield-shaped, parallel to the leaf surface. They consist of a central disc of live cells, an outer ring of dead cells, and a pluricellular stalk arising from the epidermis and subepidermis. Like the stomata, they are nested in small cavities along the bottom of the underside furrows, with their broad head spreading out to virtually cover the entire leaf surface. In many bromeliad species, the dead cells can absorb water and nutrients, and the stalk can carry them into the inner leaf tissues. The shield also acts as a plug, closing the concavity where the stalk is inserted and protecting it from drying out. These trichomes function as one-way valves, playing an important role in the capacity of bromeliads to improve and maintain their water status (Benzing, 1980). However, the pineapple trichomes appear to be hydrophobic and do not absorb water

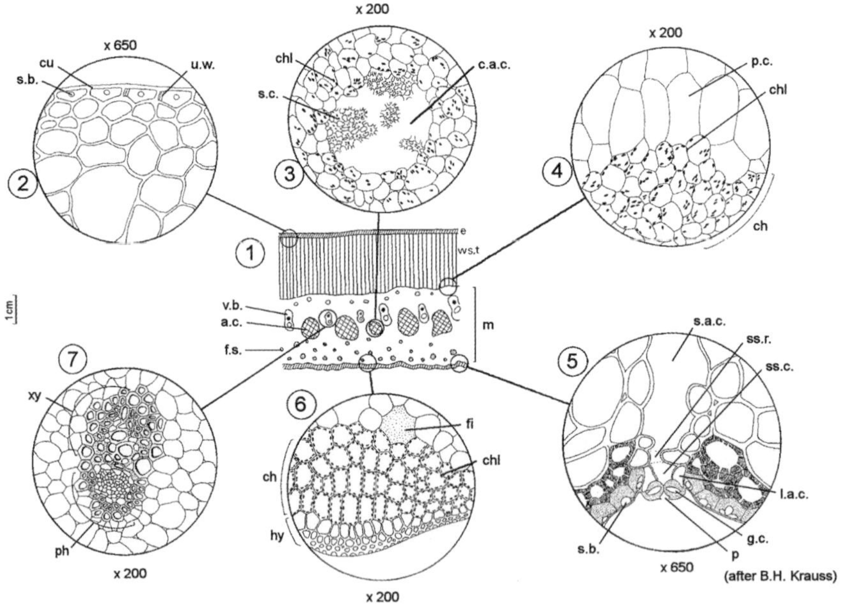

Fig. 2.4. (1) Diagram of a transverse section of a 'Smooth Cayenne' leaf showing: a.c., aerating canal; f.s., fibre strand; e, epidermis; v.b., vascular bundle; m, mesophyll; ws.t., water-storage tissue. (2) Outer epidermis showing: cu, cuticle; s.b., silica body; u.w., undulating wall. (3) Aerating canal showing: c.a.c., central aerating canal; s.c., stellate cell; chl, chloroplast. (4) Lower boundary of water-storage tissue showing: ch, chlorenchyma; chl, chloroplast; p.c., palisade cells of water-storage tissue. (5) Lower epidermis showing: ss.r., substomatal ring; s.a.c., secondary aerating canal; l.a.c., lateral accessory cell; g.c., guard cell; ss.c, subtomatal chamber; s.b, silica body; p, pore. (6) Hypodermis and mesophyll (trichomes not shown) showing: ch, chlorenchyma; chl, chloroplast; fi, fibres; hy, hypodermis. (7) Vascular bundle (mesophyll without chloroplasts) showing: ph, phloem; xy, xylem. (After Py *et al.*, 1987, with permission courtesy of C. Teisson.)

(Krauss, 1948, 1949a). The presence of large numbers of mitochondria in the stalk cells gave indirect evidence for an important role in uptake of dissolved nutrients (Sakai and Sanford, 1980). Another important role is to protect the plant from excessive transpiration and intense sunlight. More generally, the thick cuticle, the water-storage tissue, the disposition of the stomata, the trichomes, as well as the CAM metabolism, all contribute to the remarkable water economy of pineapple.

Roots

Primary roots are only found in very young seedlings. They die soon after germination and are replaced by the adventitious roots. These form a short and compact system at the stem base, with numerous strong roots and limited branching. Under ideal conditions, the soil root system may spread up to 1–2 m laterally and 0.85 m in depth. The number of roots produced after planting is positively correlated with shoot weight and crowns produce more roots than do shoots. The root internal anatomy is typical of monocots, with towards the centre: the epidermis, with root hairs cells; the cortex comprising the exodermis, the outer cortex with sclerenchyma and aerating canals, and the inner cortex with a lacunar parenchyma; the endodermis, pericycle, vessels and pith. Their most characteristic trait is the medullar structure given by the aerating canals formed by the tip-to-tip junction of raphide cells in the outer cortex, and by the air lacunae formed by the disparition of thin-walled cell groups. Branch roots originate in the pericyclic region of the main roots.

Inflorescence and fruit

The peduncle and inflorescence develop from the apical meristem, whose diameter is suddenly increased until the initiation of the peduncle (Kerns *et al.*, 1936). The stage of inflorescence emergence is called 'red heart' because of the 5–7 reddish peduncle bracts at its base. These bracts are shorter and narrower than the ordinary leaves. The peduncle elongates after flower formation. Its length varies widely with the botanical varieties or even cultivars. In addition to its bracts, it bears, in many cultivars, a variable number of slips (up to a dozen or more), which can be positioned more or less regularly between the stem and the fruit, at the axis of the peduncle bracts, or grouped just beneath the fruit. These slips can be considered as dwarfish 'aborted' fruits with a relatively large crown (Collins, 1960). They may constitute an appreciated source of planting material in extensive cultivation systems.

The inflorescence consists of less than 50 (in some wild clones) to more than 200 (in some cultivars) individual flowers; it is capped by a crown, composed of numerous short leaves (up to 150) on a short stem. The flowers or individual fruits are disposed around the central axis according to a 8/21 phyllotaxy in most pineapples, particularly large cultivated varieties (Linford, cited by Kerns *et al.*, 1936) and a 5/13 phyllotaxy for some small-fruited wild pineapples, or for young cultivated pineapples flowering prematurely (Kerns *et al.*, 1936). The fibrous axis containing the many vascular bundles that supply the flowers is continuous with the peduncle and with the short stem of the crown (Fig. 2.5). Between the uppermost flower and the crown is a transition region with bracts but no flowers. The edible part of the fruit consists chiefly of the ovaries and of the bases of sepals and bracts and of the cortex of the axis. The fruit shell is composed chiefly of sepal and bract tissues and the apices of the ovaries (Okimoto, 1948). Anthesis normally takes place within a day. Flowering lasts 10–15 days and occurs in a more or less acropetal succession along the inflorescence axis, but some cultivars flower very disorderly.

Flowers are hermaphroditic and trimerous, with three sepals, three petals, six stamens in two whorls of three, and one tricarpellate pistil (Fig 2.6). The anthers are bilobed, introrse and dorsifixed. The hollow, trilobed and trifid style is almost as long as the petals and equal or longer than the stamens. At anthesis, each stylar canal is an unobstructed open channel from the stigma to the locule directly above the placenta. Petals are ligulate, free, each bearing at its base two slender funnelform scales or, more rarely, lateral folds that overlap the filaments. Petals are white at their base to violet-blue at their tip. They are so close together at their outer end that only small

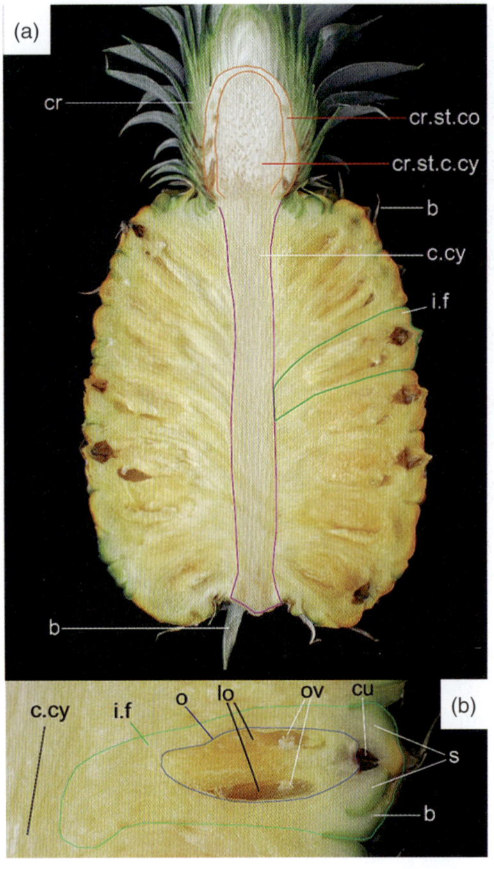

Fig. 2.5. Longitudinal vertical sections of infrutescence (a), and individual fruitlet (b). b: bract; c.cy: central cylinder; cr: crown; cr.st.c.cy: crown stem central cylinder; cr.st.co: crown stem cortex; cu: cupule; i.f: individual fruitlet; lo: locule; o: ovary; ov: ovule; s: sepal. Photos G. Sanewski.

insects can enter the flower. This narrow tubular flower and the abundant nectar production are particularly adapted to hummingbird pollination. Indeed, the three large nectary glands are so productive that nectar often fills the corolla and seeps out. The sepals are deltoid and appear similar to the bracts in colour and texture. Each flower is surrounded and subtended at its base by a pulpous and thick bract, covered by trichomes, which becomes pointed and papiraceous at its tip. Parts of three other bracts complete the enclosure of the flower. Bract spininess is correlated with leaf spininess.

In the syncarpic inflorescence resulting from the fusion of the basal part of the flowers and their axis, ovaries of adjacent flowers are separated by the parenchymatous tissue of the calyx and bract bases. The ovary is inferior, tricarpellate, and trilocular, with the three septa forming an inverted Y when seen in tangential section of the inflorescence. The placentae and ovules are located in the upper part of three deep cavities called locules, which are separated by the nectary glands. The ovules are caudate and arranged in two single or double rows. The number of ovules per flower varies with the cultivars, from 16 to 71 (Coppens d'Eeckenbrugge *et al.*, 1993). The occurrence of two types of ovules (unitegmic orthotropous and bitegmic anatropous) within the same ovary is common and orthotropous ovules are fertilized (Okimoto, 1948; Rao and Wee, 1979; F. Van Miegroet, 1993, personal communication). Orthotropous ovules are much less frequent than anatropous ovules and their presence and numbers are a varietal characteristic (Duval and Coppens d'Eeckenbrugge, unpublished results). Pollen grains are prolate and spheroidal, biconvex, isopolar and bilaterally symmetrical, diaperturate with circular to slightly elongated apertures situated at the poles. The equatorial (36–51 μm) and polar (46–59 μm) dimensions are variable. The exine is reticulate, and the polar areas show finer reticulation than the rest of the surface (Wee and Rao, 1979). Male and female gametogenesis, as well as embryogenesis, are reviewed by Sanewski *et al.*, in Chapter 4 in this volume.

There is no floral abscission, and, except for the withering of the style, stamens and petals, the entire blossom develops parthenocarpically into a berry-like fruitlet. In the cultivated pineapple, growth from blossoming inflorescence to mature fruit can result in a 20-fold increase in weight. The enlargement of the calyx results from continued growth by cellular division, in stages up to flowering, and cell enlargement, in the later stages. During this size increase, cell walls get thinner. The bract, sepal and ovary tissues are prominent structures in the mature fruit. The large, conspicuous bract is fleshy and widened at its base and bends over the flattened calyx surface, covering one-half of the fruitlet. Its papery tip dries during maturation. Internally, the locules get longer but relatively narrower and less conspicuous in the developed fruit because of

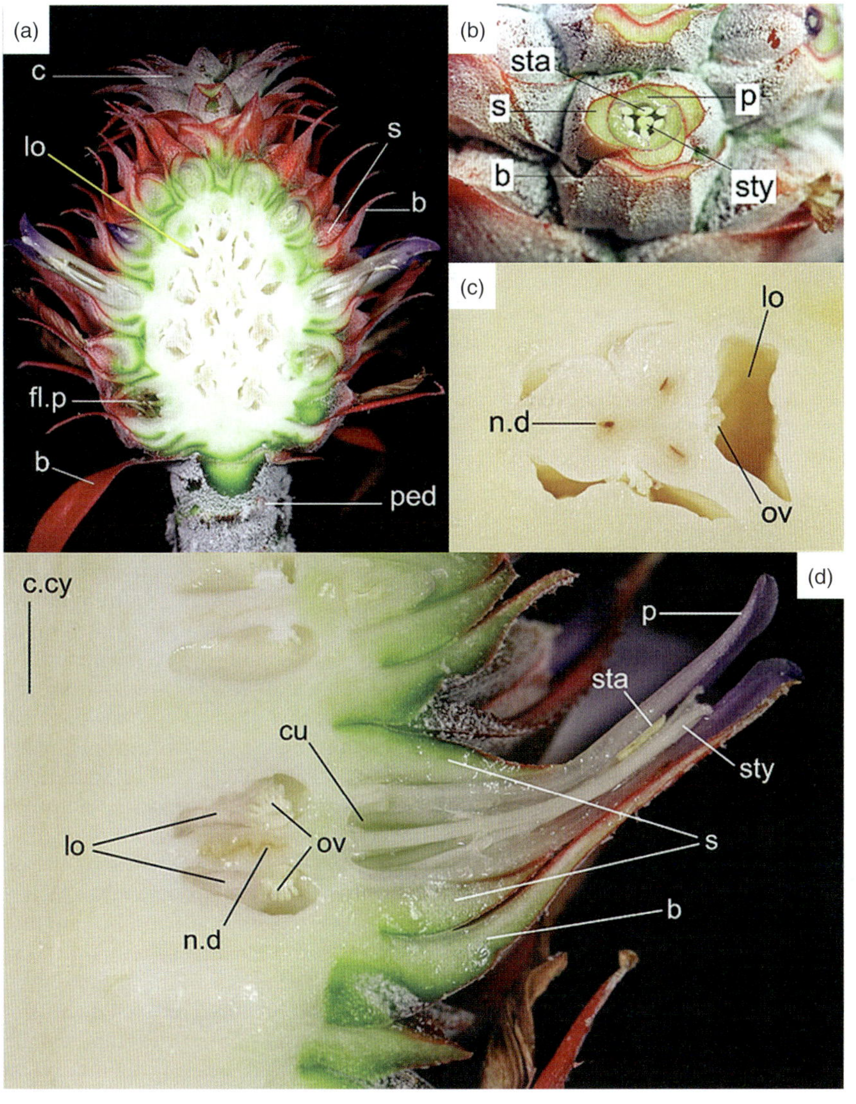

Fig. 2.6. Longitudinal vertical section of an inflorescence (a), shallow cross-section of a flower (b), deep cross-section of a flower (c) and longitudinal vertical section mid-way through a flower at anthesis (d) showing detail of the inflorescence and flower. b: bract; c: crown; c.cy: central cylinder; cu: cupule; fl.p: floral parts; lo: locule; n.d: nectary duct; ov: ovules; p: petal; ped: peduncle; s: sepal; sta: stamen; sty: style. Photos G. Sanewski.

the expansion of adjacent tissues, especially of the septa. Placentas show some enlargement but far less than the septal tissues, unless they bear mature seeds. The seeds are approximately 3–5 mm long and 1–2 mm wide, flat on one side and curved on the other, with a pointed end. They contain a hard flinty endosperm and a minute embryo enclosed in a brown-to-black coat, extremely tough and leathery, and roughened by numerous longitudinal ridges (Miles Thomas and Holmes, 1930). In the mature fruit, the stylar canals get completely closed first by a mucilaginous plug, soon after anthesis, and, a week or two later, by cellular occlusion.

Vegetative propagules

Vegetative propagules are classified according to their position on the plant. Ground suckers appear on the underground part of the stem. Plant suckers occur on the aerial part. Slips appear on the peduncle and are often grouped near the base of the fruit. Sometimes, they are produced from the basal eyes of the fruit (collar-of-slips). The crown can be used for planting when the pineapple fruit is processed. Some plants may lack a crown or, on the contrary, produce multiple crowns. Also, crownlets may grow at the base of the main crown or from some of the upper fruitlets.

Taxonomy

The Bromeliaceae

Pineapple belongs to the order *Bromeliales*, family *Bromeliaceae*, subfamily *Bromelioideae*. With more than 50 genera and 3140 species (Givnish *et al.*, 2011), this is the largest and most ecologically diverse family whose natural distribution is restricted to the New World, with the exception of *Pitcairnia feliciana* (Aug. Chev.) Harms & Mildbr., which is native to Guinea. Their unified geographic distribution and their strong adaptation toward an epiphytic mode of life indicate that it is a quite young family. On the other hand, morphologically they seem older than *Rapateaceae*, which led Smith (1934a) to accept that their restriction to the Americas does not indicate extreme youth. The *Bromeliaceae* have adapted to a very wide range of habitats, ranging from terrestrial to epiphytic, deep shade to full sun, mesic to extremely xeric, sea level to alpine, and from the hot and humid tropics to the cold and dry subtropics. They thus cover a wide area, from the centre of the USA to the northern regions of Argentina and Chile (Smith, 1934a). The *Bromeliaceae* are set apart from other monocots by the unique, stellate or scale-like multicellular hairs and the unusual conduplicate, spiral stigmas (Gilmartin and Brown, 1987). They are also characterized by a short stem, a rosette of narrow stiff leaves, terminal inflorescences in the form of racemes or panicles, hermaphroditic and actinomorphic trimerous flowers with well differentiated calyx and corolla, six stamens, and superior to inferior trilocular ovary, with axile placentation and numerous ovules. Fruits are capsules or berries, and contain small naked, winged or plumose seeds, with a reduced endosperm and a small embryo. Most species are epiphytic or saxicolous, but some are terrestrial. They are particularly adapted to water economy based on: (i) rosette structure; (ii) ability to absorb water and nutrients through their waxy leaves and aerial roots; (iii) ability to store water in specialized aquiferous leaf tissue; (iv) multicellular trichomes functioning as water valvulae and reflecting radiation; (v) a thick cuticle; (vi) location of stomates in furrows limiting evapotranspiration; and (vii) CAM metabolism. Their root system is not well developed and functions mostly to anchor the plant. In spite of their remarkable diversity, *Bromeliaceae* show a relative conservation regarding chromosome features (size, condensing behaviour, and interphase nuclei) and numbers, with a strong predominance of $2n=50$, representing the diploid level (Gitaí *et al.*, 2014). They would be paleopolyploids with a base chromosome number $x=25$ that probably evolved from hybridization among ancestral species with lower ploidy levels (Brown and Gilmartin, 1989; Brown *et al.*, 1997). The relative cytogenetic homogeneity of *Bromeliaceae* is an important factor in their radiation, facilitating hybridization and genetic interactions between species of the same ploidy. Thus, radiating species have a strong tendency to remain porous for extended periods of time. Indeed, many groups of bromeliads can be hybridized and evidence for natural hybridization has accumulated (Lexer *et al.*, 2016).

The *Bromeliaceae* were long divided into three subfamilies, the *Pitcarnioideae*, the *Tillandsioideae* and the *Bromelioideae*. However, their phylogenetic study has been confounded by high levels of homoplasy for morphological and ecological characters. Most experts agree that certain attributes have arisen several times independently as adaptations to their extreme environment. Early molecular studies gave contradictory results, placing different subfamilies at the base of the *Bromeliaceae* (Clark and Clegg, 1990; Givnish *et al.*, 1990; Ranker *et al.*, 1990; Terry *et al.*, 1997). The subfamilies *Tillandsioideae* and *Bromelioideae* were considered monophyletic. The former include mostly epiphytic species, with smooth leaf margins, flowers usually hypogenous, and dry dehiscent capsules containing many plumose seeds, adapted to wind dispersal. The second is the most numerous. Most *Bromelioideae* are epiphytic, with serrated leaves, covered with multicellular trichome scales, epigynous flowers, petals that are generally appendaged, and fleshy or leathery berries containing naked seeds as an adaptation to dispersion by birds or mammals. With the exception of those of the genus *Greigia*, all the species examined exhibit CAM metabolism, a trait much less frequent in the other subfamilies (Medina, 1990).

The *Pitcarnioideae* were long held to be the most archaic. They are almost terrestrial, with armed leaf margins, hypogenous or epygenous flowers, and dry dehiscent capsules containing naked or appendaged seeds adapted to wind dispersal. Molecular phylogenetic studies, essentially based on plastid DNA markers, have increasingly questioned their monophyly (Ranker *et al.*, 1990; Terry *et al.*, 1997; Horres *et al.*, 2000) and *Pitcairnioideae* were finally split into six subfamilies (*Brocchinioideae*, *Lindmanioideae*, *Hechtioideae*, *Navioideae*, *Pitcairniodideae sensu stricto*, and *Puyoideae*), while confirming the monophyly of *Tillandsioideae* and *Bromelioideae* (Givnish *et al.*, 2007). However, nuclear DNA data have questioned the monophyly of *Bromelioideae* (Jabaily and Sytsma, 2010) and the position of their sister *Puyoideae* (Schütz *et al.*, 2016), suggesting plastid capture events related to hybridization and indicating that the current classification has to be validated with non-plastid markers and a larger species sampling (Palma-Silva *et al.*, 2016). Combined with the bromeliad phytogeography, the new phylogenetic data allowed proposing a chronology for the bromeliad evolution and diversification of the family. According to Givnish *et al.* (2007, 2011), they arose in the Guiana Shield, in humid, nutrient-poor habitats at low elevations, *c.*100 million years ago (Ma); they diverged and expanded into other regions of South and Central America between 15 and 10 Ma, and reached West Africa by long-distance dispersal *c.*9.3 Ma. Modern lineages began to diverge from each other roughly 19 Ma, and *Bromelioideae* would have diverged from *Puyoideae* around 9.1 Ma, in and around the southern Andes, before spreading to lowland areas and the Atlantic forest of eastern and south-eastern Brazil, where their radiation led to their present centre of diversity (Horres *et al.*, 2007; Givnish *et al.*, 2011). Its early divergent representatives include genera *Bromelia* and *Ananas*, which would have dispersed throughout lowland South and Central America (Givnish *et al.*2011). The inter- and intrageneric relationships of the subfamily are poorly understood, particularly in the core group, the *eu-Bomelioideae*, where *Aechmea* and related genera form the largest alliance (Horres *et al.*, 2007; Schulte *et al.*, 2009; Sass and Specht, 2010; Heller *et al.*, 2015). Much of the confusion comes from an insufficient understanding of morphological variability.

> Frequent changes of generic limits reflect the considerable uncertainties concerning the taxonomic value of morphological characters. Furthermore, since the last comprehensive monograph, the number of species has increased by more than one third, and a thorough revision is urgently needed.
>
> (Schulte *et al.*, 2009)

Pineapple is by far the most important economic plant in the *Bromeliaceae*. However, in the same *Bromelioideae* subfamily, some *Aechmea*, *Bromelia* and *Greigia* species also yield edible fruits, including *Aechmea bracteata* (Swartz) Grisebach, *A. corymbosa* (Martius ex Schultes f.) Mez, *A. distichanta* Lem., *A. kuntzeana* Mez, *A. longifolia* (Rudge) L.B. Smith & M.A. Spencer, *A. magdalenae* (André) André ex Baker, *A. nudicaulis* (L.) Grisebach, *A. rubiginosa* Mez, *Bromelia antiacantha* Bertoloni, *B. balansae* Mez, *B. chrysantha* Jacquin, *B. karatas* L., *B. hemisphaerica* Lamarck, *B. nidus-puellae* (André) André ex. Mez, *B. pinguin* L., *B. plumieri* (E. Morren) L.B. Smith, *B. trianae* Mez, and *Greigia sphacelata* Regel (Rios and Khan, 1998; Hornung-Leoni, 2011). The most common are known and consumed locally, under names like *cardo* or *banana-do-mato* (bush banana), *piñuelas* (small pineapple), or *karatas*, *gravatá* and *croata* (derived from Amerindian names given to terrestrial bromeliads). Many other bromeliads are cultivated or gathered as ornamentals, for drinks, vegetables, for fibre extraction or used in traditional medicine (Corrêa, 1952; Purseglove, 1972; Reitz, 1983; Rios and Khan, 1998; Hornung-Leoni, 2011; Negrelle *et al.*, 2012).

Evolution of pineapple classification

From the first observations of the pineapple by European explorers to the present time, pineapple taxonomy has varied considerably. The first botanical description of cultivated pineapples was by Charles Plumier at the end of the 17th century (Plumier, 1755), when he collected plants called '*karatas*' and '*ananas*' on the island of Hispaniola. Following the native classification, he created the genus *Bromelia* for the *karatas*, in honour to the Swedish physician Olaf Bromel, and described the *ananas*, using polynomials such as *Ananas aculeatus fructu ovato, carne albida*. In his *Species Plantarum*, Linnaeus (1753) designated the pineapple as *Bromelia ananas* and *Bromelia comosa*, while Miller (1754, 1768) maintained the name *Ananas*, with six varieties, all cultivated. In the following classifications of the 18th and 19th centuries, as pineapple was mainly known from attractive large-fruited types, these cultivars, named in Latin, were easily confused with species, which resulted in an overwhelming number of different names (Leal *et al.*, 1998). Thus, Lindley (1827) used such names as *Ananassa sativa* for ordinary cultivars, *A. lucida* for smooth-leafed cultivars such as 'Smooth Cayenne' (from the variety *Ananas* (*Lucidus*) of the eighth edition of Miller's Gardener's Dictionary, published in 1768), *A. debilis* for a particular cultivar with undulated leaves, and *A. bracteata* for a crowned pineapple with long bracts. Schultes and Schultes (1830) returned to the original name *Ananas*, with *A. sativus*, *A. debilis*, *A. semiserratus* (instead of *A. lucida*), and *A. bracteatus*.

Linden (1869) described a Colombian smooth-leafed cultivar with the 'piping' character under *A. mordilona*. Morren (1878) gave the first clear description of a distinct pineapple, the *yvira*, characterized by long bracts, the absence of crown, and a propagation by stolons, which he named *A. macrodontes*. In 1889, both Baker and André described a wild pineapple, with a long scape and a small crowned fruit, respectively under the names of *Acanthostachys ananassoides* and *Ananas pancheanus*. In contradiction with this profusion of species descriptions, a few authors, such as Bentham and Hooker (cited by André, 1889), claimed that the genus *Ananas* is monospecific, with multiple wild and cultivated forms. Thus, in his first attempt at simplification, published in the *Flora Brasiliensis*, Mez (1892) recognized only one species, *A. sativus*, with five botanical varieties. The variety *lucidus* included the pineapples with smooth leaves and a large fruit. *A. debilis* was downgraded to a second variety *debilis*, 'only known from European glasshouses'. The variety *bracteatus* included *A. bracteatus* (Lindl.) Schult. f., but also *A. macrodontes* Morren. The variety *muricatus* was made from *A. muricatus* Schult. f., but Mez expressed doubts about its existence. The last variety, *microstachys*, corresponded to the wild pineapple *Acanthostachys ananassoides* Baker.

In 1917, Merrill established the binomial *Ananas comosus*. In 1919, Hassler divided the genus *Ananas* in two sections *Euananas* and *Pseudoananas* (Hassler, 1919). The latter was raised to a distinct monospecific genus by Harms (1930), with *Pseudananas macrodontes* (Morren) Harms. In his second classification, Mez (1934) did not recognize this new genus, and proposed three species: (i) *A. comosus*, including the wild pineapples and the cultivated forms *sativus* (spiny types), *lucidus* (smooth-leafed edible pineapples), and *debilis* (with doubts); (ii) *A. sagenaria*, corresponding to *A. bracteatus* (Lindl.) Schult. f.; and (iii) *A. macrodontes* Morren. Unfortunately, the second simple classification of Mez was no more successful than the first. From 1934 on, pineapple taxonomy has been dominated by the views of L.B Smith and F. Camargo. These authors divided the genus *Ananas*, renaming and multiplying the species in a long series of publications (Camargo, 1939, 1943, 1956; Smith, 1939, 1961, 1962; Camargo and Smith, 1968; Smith and Downs, 1979). The resulting over-classification into two genera and nine species (Smith and Downs, 1979) has been severely criticized on the basis of practicability as well as inconsistency with available data on reproductive behaviour and morphological, biochemical, and molecular diversity (Leal, 1990; Loison-Cabot, 1992; Leal and Coppens d'Eeckenbrugge, 1996; Coppens d'Eeckenbrugge *et al.*, 1997; Leal *et al.*, 1998). Indeed, there are no reproductive barriers between the *Ananas* species defined by Smith and Downs and no differences either in their floral structure and cytology, or in their chromosome number or breeding system. The studies on diversity of isozymes (García, 1988; Aradhya *et al.*, 1994), RFLP and PCR-RFLP markers on ribosomal, total and chloroplast

DNA (Noyer, 1991; Noyer and Lanaud, 1992; Duval *et al.*, 2001, 2003) showed a limited differentiation between the different taxa of *Ananas sensu* Smith and Downs, a strong geographic component of variation, and a close affinity between certain wild pineapples and those cultivated for the fruit or for the fibre. The molecular studies of Duval *et al.* (2001, 2003) confirm the distinction between the generally diploid crowned pineapples propagating by shoots and the tetraploid crownless yvira propagating by stolons, although both taxa show much more morphological and genetic similarity with each other than with other bromeliads, justifying their inclusion within a same genus.

For these reasons, in the first edition of the present book, we returned to a much simpler and consistent classification, taking into account morphological, biochemical, and molecular variation, geographical distribution, and reproductive biology, according to previous reviews (Leal and Coppens d'Eeckenbrugge, 1996; Leal *et al.*, 1998). Recently, Coppens d'Eeckenbrugge and Govaerts (2015) revised synonymies in *Ananas* and corrected a formal error, restoring the epithet *microstachys* at the varietal rank, instead of *ananassoides*.

All pineapples are regrouped into one *Ananas* genus, characterized among *Bromeliaceae* by the unique feature of their inflorescence which is fused into a syncarp.

Ananas *Miller (1754)*

Synonyms: *Bromelia* Linnaeus (1753) – *Ananassa* Lindley (1827).
Description: Leaves densely rosulate, blades linear, scarcely enlarged at base, often spiny. Scape evident. Inflorescence densely strobiliform, flowers hermaphrodite, sessile. Sepals free, obtuse, asymmetric, persistent; petals free, erect, violet or purplish pink. Stamens included, pollen ellipsoid, biporate; ovaries coalescing with each other and with the bracts and axis to form a fleshy syncarpic fruit; epigynous tube short; placentae apical; ovules caudate. Fruit succulent.
Type: *Bromelia ananas* Linnaeus (1753, p. 285).

KEY TO THE SPECIES AND BOTANICAL VARIETIES

1a. Apical coma inconspicuous or lacking, plant propagating by elongate stolons .... *Ananas macrodontes*

1b. Foliaceous coma usually crowning the syncarp. Plant propagating by the crown and shoots from the stem or peduncle (slips) .... *Ananas comosus*

- 2a. Flower bracts conspicuous, imbricate and covering the ovary.................................. *Ananas comosus* var. *bracteatus*
- 2b. Flower bracts relatively inconspicuous, soon exposing the ovary apices
  - 3a. Spiny leaves. Leaf spines both antrorse and retrorse, blades narrowed toward base. . ... *Ananas comosus* var. *parguazensis*
  - 3b. Spiny or smooth leaves. Leaf spines all antrorse, blades not narrowed at base
    - 4a. Dense rosette, wide blades, scape wide, fruit medium to large: .... *Ananas comosus* var. *comosus*
    - 4b. Sparse rosette, blades usually narrow, scape thin and long, usually erect, fruit small to medium
      - 5a. Blades recurved or erect, spiny. Fruit small to medium............................ *Ananas comosus* var. *microstachys*
      - 5b. Blades erect, usually narrow and unarmed with the exception of the apical thorn. Abundant production of shoots .. .. *Ananas comosus* var. *erectifolius*

ANANAS MACRODONTES MORREN

Synonyms: *Ananas microcephalus* Bertoni (1919) – *Ananas microcephalus* var. *major* Bertoni (1919) – *Ananas microcephalus* var. *minor* Bertoni (1919) – *Ananas microcephalus* var. *missionensis* Bertoni (1919) – *Ananas microcephalus* var. *mondayanus* Bertoni (1919) – *Ananas microcephalus* var. *robustus* Bertoni (1919) – *Pseudananas macrodontes* (Morren) Harms (1930) – *Pseudananas sagenarius* (Arruda da Câmara) Camargo (1939) – *Pseudananas sagenarius* var. *macrodontes* (Morren) Camargo (1943).
Refuted synonyms: *Bromelia sagenaria* Arruda. Arruda da Câmara (1810) states that the fruit is similar to that of the edible pineapple but smaller and unpalatable, with very long and imbricate bracts, covering the fruit; in no case does he mention the absence of the leaf crown on the fruit or the vegetative reproduction by long stolons. Thus *B. sagenaria* is a synonym for *A. comosus* var. *bracteatus*, not for *A. macrodontes* – *Ananas sagenaria* (Arruda) Schultes and Schultes. The description of *Ananas sagenaria* was entirely based on that of *Bromelia sagenaria*

Arruda. Schultes and Schultes (1830) themselves questioned the differentiation between *A. sagenaria* and *A. bracteatus* Lindley (1827) – *Bromelia silvestris* Vellozo. Vellozo (1829) clearly mentions the crown on the fruit and hooked spines, which refers to a spiny form of *A. comosus*, probably *A. comosus* var *bracteatus* – *Ananas silvestris* (Vellozo) Müller. Müller (1896) gave the first precise description of *A. fritzmuelleri* (Camargo, 1943), a diploid form with a crown, now considered under *A. comosus* var. *bracteatus*.

Local names: *yvira* (Paraguay), *gravatá de rede*, *gravatá de cerca brava*, *nana caçaba* (Brazil). These names respectively mean fibre, fishing net bromeliad, fierce fence bromeliad, and strong spine pineapple, in reference to the uses of the plant.

Description: Plant propagating by elongate basal stolons, very rarely by axillary shoots, up to 2 m high. Leaves densely rosulate, generally arching decurved, 2–3 m long, blades 2–7 cm wide, coriaceous, lustrous above, densely lepidote beneath, laxly serrate with curved retrorse and antrorse spines 3–10 mm long. Scape erect, 20–50 cm high, 15–25 mm wide, furfuraceous; scape bracts foliaceous, reduced, red or green. Inflorescence simple, strobilate, globose to ovoid, 8–20 cm long and 6–9 cm wide, 100–200 flowered. Floral bracts densely imbricate, wide at base, lanceolate, acuminate, 2–5 cm long, erect, spiny, red, reddish brown or green, cinereous-lepidote, persistent; flowers opening in upward succession. Sepals erect, asymmetric, ovate to lanceolate, narrowed at base, apiculate, 10–15 mm long, thick with membranous margins, green to red orange, sparsely lepidote; petals tubular-erect, 3–5 cm long, the claw oblong, white, the blade lanceolate, purplish; stamens included, adnate to the petals, shorter than the petals. Syncarp up to about 20 cm long and10 in diameter, not growing appreciably after anthesis, succulent, without leafy crown.

Type: *Ananas macrodontes* E. Morren (1878, p. 140, Pl. IV & V).

A. macrodontes lacks a crown at the top of the syncarp and reproduces vegetatively by stolons (see Figs 2.7a and 4.15 in this volume). Exceptional genotypes may also produce shoots directly from the stem (e.g. accession CRF001 at the Venezuelan Centro Nacional de Conservación de Recursos Fitogenéticos), or produce a rudimentary crown (as in specimen US-2622957). Its leaf margins bear strong spines which are retrorse at the leaf base. The fruits are low in acid. *A. macrodontes* has been reported as self-compatible, with $2n=4x=100$ chromosomes (Collins, 1960). Its natural distribution corresponds to coastal and southern Brazil, up to Pernambuco, and to the drainage of the Paraguay and Paraná rivers, from south-eastern Paraguay and north-eastern Argentina up to Mato Grosso (Coppens d'Eeckenbrugge *et al.*, 1997). Its habitat is limited to forest areas, under semi-dense shade. It is subjected to a rainy season during most of the year or even to periods of flooding. Populations of *A. macrodontes* are rare now because of a very strong reduction of its habitat.

Camargo (cited by Reyes-Zumeta, 1967) considered that this species includes at least three botanical varieties, one corresponding to the description by Morren, and another to *A. microcephalus* described by Bertoni (1919) in Paraguay. On the contrary, Baker and Collins observed very little variation in those areas (Smith, 1939). However, the few types observed by Ferreira *et al.* (1992) in Paraguay and South Brazil showed significant variation, which was confirmed by the RFLP study of Duval *et al.* (2001). This nuclear DNA diversity contrasts with the strong uniformity observed on the same sample in the cpDNA study of 2003.

ANANAS COMOSUS (LINNAEUS) MERRILL (1917)

Synonyms: *Bromelia ananas* Linnaeus (1753) – *Bromelia comosa* Linnaeus (1754), based on *Anassa domestica* (Rumphius, 1747) – *Ananas sativus* Schultes and Schultes (1830) *sensu* Mez (1892), in part – See also synonyms under botanical varieties.

Common names: *Nanas* or *ananas* (in many Indian languages of South America, and in French and Portuguese), *piña* (Spanish), pineapple.

Description: Plant not producing stolons but axillary shoots from the stem and/or slips from the scape, up to 1.8 m high. Leaves densely rosulate, scarcely enlarged at base, coriaceous, densely lepidote beneath. Scape evident, erect; scape bracts foliaceous, usually reddish at base. Inflorescence densely strobiliform, usually crowned with one or several rosettes of foliaceous bracts; flowers sessile, highly variable in number, from fewer than 50 to more than 150.

Fig. 2.7. Pineapple species and varieties. Upper row: wild taxa: *A. macrodontes*, after completing anthesis (a); *A. comosus* var. *microstachys*, at anthesis (b); *A. comosus* var. *parguazensis*, flowering head and mature fruit (c). Lower row: botanical varieties domesticated for fibre production: *A. comosus* var. *erectifolius* (d) and *A. comosus* var. *bracteatus*, the common form at early anthesis (e), and the rare form corresponding to *A. fritzmuelleri* Camargo, after completing anthesis (f). Photos: (c) M.-F. Duval; all others G. Coppens d'Eeckenbrugge.

Floral bracts broadly elliptic to lanceolate, acuminate, sublepidote or lepidote. Sepals free, obtuse, slightly asymmetric; petals free, erect, violet or purplish pink, stamens included, pollen biporate and ellipsoid; ovaries coalescing with each other and with the bracts and axis to form a fleshy compound fruit; epigynous tube short; placentae apical; ovules caudate.
Type: *Anassa domestica* Rumphius (1747, p. 230, t81) from Amboina, Moluccas.

Instead of producing stolons, *A. comosus* multiplies by stem shoots (terrestrial and aerial), slips (from the peduncle), and crown. The syncarpic fruit is formed of 50–200 berries. The spines are generally antrorse but some genotypes also exhibit a few retrorse spines. As commonly found in *Bromeliaceae*, *A. comosus* is diploid, with 50 minute and almost spherical chromosomes (Collins and Kerns, 1931; Canpinpin and Rotor, 1937; Marchant, 1967; Sharma and Ghosh, 1971; Lin *et al.*, 1987; Brown and Gilmartin, 1989; Dujardin, 1991). Giant unreduced gametes may appear and produce natural triploids and tetraploids (Collins, 1933, 1960). Most genotypes exhibit reduced self-fertility because of a self-incompatibility system with variation in its expression (Coppens d'Eeckenbrugge *et al.*, 1993). The natural distribution of *Ananas comosus* corresponds to that of its variety *microstachys*, extending east of the Andes (with some presence in the foothills of the Central Andes of Colombia), from northern South America to northern Argentina and Paraguay.

Ananas comosus *var.* comosus Synonyms: *Bromelia ananas* Linnaeus (1753) – *Bromelia semiserrata* Willdenow (1809) – *Bromelia ananas* subvar. *variegata* Noisette (1825) as '*ananas variegata*' – *Ananassa sativa* Lindley (1827) – *Ananas sativus* Schultes and Schultes (1830) – *Ananas comosus* f. *sativus* (Schultes and Schultes) Mez (1934). Spiny cultivars – *Ananassa lucida* (Aiton) Lindley (1827) – *Ananas lucidus* (Aiton) Schultes and Schultes (1830). Smooth-leaved cultivars – *Ananas sativus* var. *lucidus* Baker (1889) – *Ananas comosus* f. *lucidus* Mez. Smooth-leaved cultivars; 'preferred in the European Ananas-houses' (Mez, 1934) – *Ananassa debilis* Lindley (1827). *Ananas debilis* (Lindley) Schultes and Schultes (1830) – *Ananas comosus* f. *debilis* Mez. Cultivar with undulated leaves; 'mihi dubia' (Mez, 1934) – *Ananas semiserratus* (Willd.) Schultes and Schultes (1830). Cultivars with spines at leaf apex – *Ananassa sativa* var. *variegata* (Noisette) Lowe (1868) – *Ananas comosus* var. *variegatus* (Noisette) Moldenke (1938, pl. 38) – *Ananassa mordilona* Linden (1869) – *Ananas mordilona* (Linden) Van Geert (1873) – *Ananas sativus* f. *mordilona* (Linden) Bertoni (1919). Smooth-leaved cultivar from the Colombian Andes, probably cv. Manzana – *Ananassa bracamorensis* Linden (1869) – *Ananas bracamorensis* (Linden) Van Geert (1878). Large-fruited cultivar from Maranhão – *Ananas porteanus* Veitch ex Koch (1871). Ornamental variegated cultivar with fasciated crown, introduced in European glasshouses from the Philippines (see also Morren, 1872) – *Ananas sativus* var. *viridis* Bertoni. Bertoni (1919) relates to this name an inconsistent list of very diverse cultivars – *Ananas sativus* var. *pyramidalis* (Miller) Bertoni (1919) – *Ananas sativus* var. *hispanorum* Bertoni (1919).

Refuted synonyms: *Ananas monstrosus* (Carrière) Smith (1961). Crownless pineapple, an unstable form; the species was invalidated by Leal (1990) – *Ananas penangensis* Baker. Ornamental variegated form, listed among many cultivars by Baker (1889), who only recognized three species in *Ananas* (*A. sativus*, *A. bracteatus*, and *A. macrodontes*) – *Ananas proliferus* Baker. Fruit malformation, caused by a dominant mutation (Collins, 1960) and known as 'knobs' by pineapple growers. Only listed as a horticultural variant by Baker (1889).

Most ancient synonyms of this botanical variety come from horticultural treatises, printed at a time when the fruit was a luxury good grown in European glasshouses, with no direct knowledge about the plant relatives and their habitat. The confusion between botanical forms (species and varieties) and horticultural forms, (cultivars), both designated in Latin, is the most frequent problem in determining synonymies.

Common names: *Ananas*, *nanas*, *piña*, pineapple, *abacaxi* (Brazil), *matzatli* (Aztec).

Description: Leaves numerous and wide, spines antrorse if present. Scape stout, usually 2 cm or more in width and less than 40 cm in length. Inflorescence large, many-flowered. Floral bracts inconspicuous, soon exposing the apices of the ovaries, weakly serrulate or entire. Syncarp grows appreciably after anthesis, to reach a final size several times superior to that of the inflorescence; fruit fleshy; seeds usually rare. The highly variable *A. comosus* var. *comosus* is characterized by its large fruit (up to several kilograms in some cultivars), borne on a wide and strong peduncle. Its leaves are numerous (40–80) and wide (>5 cm). In spiny cultivars, spines are antrorse, and generally denser than in other botanical varieties. Some cultivars are partially spiny or completely smooth.

Ananas comosus *var.* microstachys *(Mez) Smith (1934b)* This is the wild pineapple.

Synonyms: *Acanthostachys ananassoides* Baker (1889) – *Ananas microstachys* Lindman (1891), nom. superfl – *Ananas sativus* Schultes and Schultes var. *microstachys* Mez (1892) – *Ananas ananassoides* (Baker) Smith (1939) – *Ananas comosus* var. *ananassoides* (Baker) Coppens and Leal (2003), nom. superfl – *Ananas guaraniticus* Bertoni (1919) – *Ananas ananassoides* Baker var. *nanus* Smith (1939) – *Ananas microstachys* var. *nanus* (L.B.Sm.) Camargo (1942) – *Ananas nanus* (L.B.Sm.) Smith (1962) – *Ananas genesiolinsii* Reitz (1968). The 'Ananas do Indio' population still exists in Aguas Emendadas, close to Brasilia. Its general morphology relates it clearly to *A. comosus* var. *microstachys*, except for the size of the fruit, which is intermediate between that of wild and domesticated pineapples. This unusual fruit size must be related to its triploid nature (Lin *et al.*, 1987; Dujardin, 1991).

Similar clones are found in the wild and in home gardens, in the Guianas (personal observations).
Local names: *Ananaí* or *nanaí*, *ananas de ramosa* (Brazil, Pará), *curibijul*, *maya piñon*, *piñuela*, *ananas do indio*.
Description: Leaves up to 2 m long and narrow; blades linear, usually less than 4 cm wide, subdensely serrate with wholly antrorse spines. Scape elongate, slender, usually less than 15 mm thick; scape bracts large, subfoliaceous. Inflorescence small to medium, globose to cylindric. Fruit slightly enlarged after anthesis, flesh scant to abundant, fibrous; coma relatively large at maturity.
Type: Regnell III 1261, Caldas, Minas Gerais, Brazil.

Ananas nanus (L.B. Smith) L.B. Smith (see Fig. 4.11b in this volume) is considered here just a dwarf form of *A. comosus* var. *microstachys*. The constancy of the distinctions reported by Smith (1962) from a vegetatively propagated population was not a valid argument for establishing a distinct species.

Ananas comosus var. *microstachys* (see Figs 2.7b and 4.11 in this volume) is the most common wild pineapple and the botanical variety which shows the highest genetic diversity; it is very probably at the origin of the cultivated pineapples (Leal and Coppens d'Eeckenbrugge, 1996; Coppens d'Eeckenbrugge *et al.*, 1997; Duval *et al.*, 2001, 2003). It is found in all tropical areas of South America east of the Andes, generally in savannahs or in low-shaded forest, growing well on soils with limited water holding capacity (sand, rocks), and forming populations of very variable densities. However, a few genotypes thrive in denser rainforest (in the Guianas). Most populations are monoclonal, but some are polyclonal, with variation of recent sexual origin (Duval *et al.*, 1997). The plant has long and generally narrow spiny leaves and bears a small to medium, globular to cylindrical syncarp on a long and thin peduncle. The fruit is often seedy, and its pulp is white or yellow, firm and fibrous, sometimes palatable, with a high sugar and acidity content. Some clones producing a fruit of intermediate size (as compared with the *comosus* variety) are found both wild and cultivated in the Guianas. Some dwarf cultivars are grown as ornamentals.

Ananas comosus *var.* parguazensis *(Camargo and Smith) Coppens and Leal (2003)* Synonyms: – *Ananas parguazensis* Camargo and Smith (1968) – *Ananas pancheanus* André (1889).
Common names: *Piña montañera* (Amazonas, Venezuela), *ananaí* (Roraima, Brazil), *kurupira-nana* (Río Negro, Brazil), *gravatá* (Pará, Brazil).
Description: Leaves up to 2 m long, blades wide (usually more than 3 cm), somewhat contracted toward base, green to red, sinuate-serrate, spines strong, antrorse and retrorse. Scape slender, elongate, often decumbent (depending on fruit size); scape bracts foliaceous, notably dilated above base. Inflorescence few-flowered, subglobose, 4–10 cm long fruit; apical coma large (reaching 30 cm in height), basally more or less retrorse serrate. Fruit succulent, flesh palatable.
Type: Camargo 3 (holotype, COL; isotype, US), Angelita, Bolívar, Venezuela, 16 June 1966.

While the exsiccata of *Ananas pancheanus* that André observed in Kew is too poor on which to offer an opinion, the description brings several elements. Indeed, André (1889) mentions a forest habitat in the Orinoco drainage, wide (6 cm) and convoluted, antrorse and retrorse spines, and a small subspherical syncarp on a long and slender peduncle. The short and thin spines, with a density increasing regularly from the base to the top of the leaf, are not typical of the variety, but intermediate forms with var. *microstachys* are relatively common in the Orinoco and Rio Negro basins (personal observations) as well as in the Guianas (Gouda, 1999).

The wild *Ananas comosus* var. *parguazensis* (see Figs 2.7c and 4.12b in this volume) is very similar to *A. comosus* var. *microstachys* with a difference in the retrorse orientation of some spines and a wider leaf slightly constricted at its base. Its distribution is mostly limited to the Orinoco and Rio Negro basins, with a few observations in eastern Colombia and in the north-eastern Amazon (Coppens d'Eeckenbrugge *et al.*, 1997) and it shows less variability than *A. comosus* var. *microstachys* (Duval *et al.*, 2001, 2003). While *A. comosus* var. *microstachys* shows a higher water use efficiency and grows in both partial shade and full sun, *A. comosus* var. *parguazensis* is restricted to shadier environments (Leal and Medina, 1995). It grows in the lowland forests, under canopies of variable densities, from clearings or riverbanks to dense forest.

Ananas comosus *var.* erectifolius *(Smith) Coppens and Leal (2003)* Synonyms: *Ananas sativus* var. *Duckei* Camargo (1939), without Latin diagnosis – *Ananas erectifolius* Smith (1939) – Synonymy with the *lucidus* variety of Miller (1768) is not founded because the latter was created for smooth-leafed cultivars giving a large fruit.
Common names: *Curagua* (Venezuela), *curauá* or *curaná* (Brazil), *kulaiwat* (Guianas), *pita*; the latter is shared with other bromeliads exploited for fibre, and even agaves.
Description: Leaves stiff and erect, 1 m or longer, blades about 35 mm wide, entire or nearly so, except for the pungent apex. Scape erect, elongate, slender, to 15 mm thick; scape bracts foliaceous, erect. Syncarp small, cylindric or subcylindric; floral bracts small, entire. Fruit slightly larger than inflorescence, coma relatively large, sometimes surrounded by additional crownlets. Flesh scant, fibrous and unpalatable.
Type: Ducke in Baker & Collins 9 (GH), Amazon Basin, Brazil, cultivated 1938.

Ananas comosus var. *erectifolius* (see Figs 2.7d, 4.14 and 10.2 in this volume) is essentially distinguished from *A. comosus* var. *microstachys* by its smooth leaves. It is cultivated by the natives in the Guianas, including the Orinoco basin, and in the North of the Amazon basin for its very strong and long fibres to make hammocks and fishing nets (Leal and Amaya, 1991). The dry fibres constitute 6% of the plant weight (Camargo, 1943). *A. comosus* var. *erectifolius* has never been found in the wild. Plants are medium sized, with abundant shoots and numerous erect leaves, and with a small very fibrous (inedible) fruit. It is also discriminated from *A. comosus* var. *microstachys* by the absence of spines along the leaf margin. However, spiny types have been observed under cultivation or as mutants in germplasm collections. The absence of spines, as well as its erect habit, are probably the product of human selection for high yield of easily extractable fibres from strains of *A. comosus* var. *microstachys*. According to the DNA diversity studies of Duval *et al.* (2001, 2003), different clones of *A. comosus* var. *erectifolius* show more affinity with different clones of *A. comosus* var. *microstachys* than with each other, which suggests that this process may have taken place independently at different times and/or places or some hybridization with other botanical varieties has occurred.

Ananas comosus *var.* bracteatus *(Lindl.) Coppens and Leal (2003)* Synonyms: *Bromelia sagenaria* Arruda da Câmara (1810) – *Ananas sagenaria* (Arruda da Câmara) Schultes and Schultes (1830) – *Ananassa sagenaria* (Arruda da Câmara) Dietrich (1840) – *Bromelia silvestris* Vellozo (1829) – *Ananas silvestris* Müller (1896) – *Ananas fritzmuelleri* (Camargo, 1943) – *Ananassa bracteata* Lindley (1827) – *Ananas bracteatus* (Lindley) Schultes and Schultes (1830) – *Ananas bracteatus* (Lindley) Baker (1889) – *Ananas sativus* var. *bracteatus* (Lindley) Mez (1892) – *Ananas bracteatus* var. *typicus* Smith (1939), nom. inval – *Ananas bracteatus* var. *albus* Smith (1939) – *Ananas bracteatus* var. *hondurensis* Bertoni (1919). With many doubts – *Ananas bracteatus* var. *macrodontes* Bertoni (1919). With many doubts – *Ananas bracteatus* var. *paraguariensis* Bertoni (1919). With many doubts – *Ananas bracteatus* var. *rudis* Bertoni (1919). With many doubts – *Ananas bracteatus* var. *rudis* subvar. *tricolor* Bertoni (1919) – *Ananas bracteatus* var. *tricolor* (Bertoni) Smith (1939) – *Ananas comosus* var. *tricolor* (Bertoni) Camargo (1942) – *Ananas bracteatus* var. *striatus* Foster (1958).
Common names: *Ananas de cerca, ananas bravo, ananas do mato, karaguata-ruha* (southern Brazil).
Description: Leaf blades wide, coarsely and usually laxly spinose-serrate, spines antrorse or antrorse and retrorse. Scape thick and stout. Syncarp medium to large. Floral bracts conspicuous, imbricate and covering the ovaries, coarsely serrate, usually red or pink. Fruit succulent at maturity.
Type: Lindley (1827) pl. 1081.

Ananas comosus var. *bracteatus* (see Figs 2.7e and 4.12a in this volume) has the same original southern distribution area as *Ananas macrodontes*. It is always found cultivated as a living hedge, for fibre and fruit juice, or abandoned in ancient settlements (Baker and Collins, 1939; Duval *et al.*, 1997). The plant is very vigorous with wide and long leaves, large spines and abundant production of shoots. The inflorescence is characterized by its long bracts (longer than the fruitlet) of a bright pink to red colour. The fruit and peduncle are medium-sized. There are two main forms of *Ananas comosus* var. *bracteatus*. The most common corresponds to *A. bracteatus* sensu Smith & Downs, with a spectacular bright red inflorescence and antrorse spines, of which a variegated cultivar has been propagated as a garden ornamental. It has also been used for the cut flower market and a further mutation suppressing spines has been selected. The second form corresponds to *A. fritzmuelleri* Camargo (see Figs 2.7f and 4.13 in this volume), with antrorse and retrorse spines. Both forms show very little variation, suggesting that they are very particular types probably selected by man and dispersed through clonal propagation, and still capable to maintain small populations in disturbed habitats. They share certain morphological, biochemical and DNA characters with *A. macrodontes*. In diversity studies, the common form appears uniform and well grouped (García, 1988; Duval and Coppens d'Eeckenbrugge, 1993; Aradhya *et al.*, 1994; Duval *et al.*, 2001), at some distance from the other botanical varieties of *A. comosus*, but closer to them than to *A. macrodontes*; in the cpDNA study (Duval *et al.*, 2003), it presents a unique haplotype, shared with representatives of the other four botanical varieties of *A. comosus*. In contrast, the rare form corresponding to *A. fritzmuelleri* Camargo appears more closely related to *A. macrodontes* in the total DNA RFLP study (Duval *et al.*, 2001), and its haplotype is closely related to that of the *A. macrodontes* accessions in the cpDNA study (Duval *et al.*, 2003). These results indicate that *A. comosus* var. *bracteatus* originated from an introgression process involving *A. macrodontes*. This hypothesis is consistent with the sympatric distribution of both taxa, in a region where they were long exploited as sources of fibre, and with their morphological similarity, particularly the long pink bracts covering the ovary. *Ananas comosus* var. *bracteatus* hybridizes easily with other varieties of *A. comosus*.

Pineapple taxonomy and evolution under domestication

The *Ananas* classification is clearly based on the biological species concept, as it recognizes two major morphotypes, distinguishing *A. comosus*, reproducing vegetatively by its crown, slips, and suckers, from *A. macrodontes*, lacking these propagules and reproducing vegetatively by stolons. The former is essentially diploid and its botanical varieties are differentiated by quantitative traits, qualitative traits that are governed by allelic series, such as leaf spininess, or traits that vary in the entire genus, as the presence of retrorse spines. The latter is a tetraploid species, and it seems to lack the self-incompatibility system that operates in *A. comosus*. Their ploidy difference constitutes a reproductive barrier, albeit a relative one. Indeed, *A. macrodontes* may produce 5–10% viable seeds when pollinated by *A. comosus*. These seeds are usually tetraploid (see Chapter 4 in this volume).

Despite recognizing that the low genetic distances between pineapple varieties were consistent with the *Ananas* classification, Matuszak *et al.* (2015) proposed to consider 'monophyly as an essential criterion for classification.' This approach is not practicable in *Ananas*, as one of the botanical varieties (*A. comosus* var. *bracteatus*) resulted from an introgression between the two species (Duval *et al.*, 2001, 2003). Thus, we have to admit that paraphyly exists within the genus *Ananas*. This is certainly not exceptional in *Bromeliaceae*, as interspecific barriers are weak in the family and spontaneous hybridization relatively common (Lexer *et al.*, 2016; Palma-Silva *et al.*, 2016), so models involving reticulate evolution should be better adapted than the classical 'tree' approach.

Within *A. comosus*, there is no absolute division among botanical varieties, and some specimens may be difficult to classify at that level. For example, in the drainages of the Rio Negro and Rio Orinoco, where varieties *microstachys* and *parguazensis* coexist, many wild clones present small spines, very few of them retrorse, mostly at the leaf base, without the leaf base constriction that is characteristic of var. *parguazensis*. Similar observations of intermediate specimens have been reported by Gouda (1999) for the Guianas. Intermediate phenotypes also exist between wild and cultivated botanical varieties: Guianese clones that are morphologically similar to var. *microstachys*, or clones with wider leaves, reminiscent of var. *comosus*, produce fruits of intermediate size on relatively thin and long peduncles, that cannot be clearly classified in one or the other variety. This is not unexpected between a cultivated botanical variety and its wild progenitor. This is a second important characteristic of the classification: botanical varieties reflect the evolution of the species, which is an evolution under domestication. Three of the five botanical varieties are cultigens, i.e. they only exist under cultivation, or as escapes from cultivation. They result from different selective processes, which explain why the differentiation among them is much clearer than the differentiation between their wild relatives or the differentiation between themselves and these wild forms. Current morphological, biogeographical and genetic data indicate that the varieties *comosus* and *erectifolius* were domesticated north of the Amazon, respectively for the fruit and for fibre, while var. *bracteatus* was domesticated in the south-east, initially as a fibre crop.

References

André, E. (1889) *Description et histoire des Broméliacées récoltées dans la Colombie, l'Equateur et le Vénézuéla.* Paris: G. Masson.

Aradhya, M., Zee, F. and Manshardt, R.M. (1994) Isozyme variation in cultivated and wild pineapple. *Euphytica* 79, 87–99.

Arruda da Câmara, M. (1810) *Dissertação sobre as plantas do Brazil que podem dar linhos proprios para muitos usos da sociedade, e suprir a falta de canhamo.* Rio de Janeiro, Brazil: Impressão Regia, pp. 3–7, 13–22.

Baker, J.G. (1889) *Handbook of the Bromeliaceae.* London.

Baker, K. and Collins, J.L. (1939) Notes on the distribution and ecology of *Ananas* and *Pseudananas* in South America. *American Journal of Botany* 26, 697–702.

Benzing, D.H. (1980) *The Biology of the Bromeliads.* Eureka, CA: Mad River Press Inc.

Bertoni, M.S. (1919) Contributions à l'étude botanique des plantes cultivées. I. Essai d'une monographie du genre *Ananas. Anales Científicos Paraguayos* (Ser. II) 4, 250–322.

Brown, G.K. and Gilmartin, A.J. (1989) Chromosome numbers in Bromeliaceae. *American Journal of Botany* 76, 657–665.

Brown, G.K., Palaci, C.A. and Luther, H.E. (1997) Chromosome numbers in Bromeliaceae. *Selbyana* 18, 85–88.

Camargo, F. (1939) Ananás e abacaxí. *Revista de Agricultura de Piracicaba* 14, 321–338.

Camargo, F.C. (1942) Pesquisas taxonômicas sôbre os gêneros *Pseudananas e Ananas. Proceedings of the Eighth American Scientific Congress* 3, 178–192.

Camargo, F. (1943) Vida e utilidade das Bromeliáceas. *Boletim Técnico Instituto Agronômico do Norte,* Belem, Brazil.

Camargo, F. (1956) Nota previa. *Ananas lyman-smithii* n. sp. *Arquivos do Jardim Botánico do Rio de Janeiro* 14, 281–289.

Camargo, F. and Smith, L.B. (1968) A new species of *Ananas* from Venezuela. *Phytologia* 16, 464–465.

Canpinpin, J.H. and Rotor, G.B. (1937) A cytological and morphogenetic study of some pineapple varieties and their mutant and hybrid derivatives. *The Philippines Agriculturist* 26, 139–158.

Clark, L.G. and Clegg, M.T. (1990) Phylogenetic comparisons among rbcL sequences in the Bromeliaceae. *American Journal of Botany* 77, 115.

Collins, J.L. (1933) Morphological and cytological characteristics of triploid pineapples. *Cytologia* 4, 248–256.

Collins, J.L. (1960) *The Pineapple, Botany, Utilisation, Cultivation.* London: Leonard Hill Ltd.

Collins, J.L. and Kerns, K.R. (1931) Genetic studies of the pineapple. I. A preliminary report upon the chromosome number and meiosis in seven pineapple varieties (*Ananas sativus* Lindl.) and in *Bromelia pinguin* L. *Journal of Heredity* 22, 139–142.

Coppens d'Eeckenbrugge, G. and Leal, F. (2003) Morphology, anatomy and taxonomy. In: Bartholomew, D.P., Paull, R.E. and Rohrbach, K.G. (ed.) *The Pineapple: Botany, Production and Uses.* CAB International, Wallingford, Oxfordshire, pp. 13–32.

Coppens d'Eeckenbrugge, G., Duval, M.F. and Van Miegroet, F. (1993) Fertility and self-incompatibility in the genus *Ananas. Acta Horticulturae* 334, 45–51.

Coppens d'Eeckenbrugge, G., Leal, F. and Duval, M.F. (1997) Germplasm resources of pineapple. *Horticultural Reviews* 21, 133–175.

Coppens d'Eeckenbrugge, G. and Govaerts, R. (2015) Synonymies in *Ananas* (Bromeliaceae). *Phytotaxa* 293, 273–279.

Corrêa, M.P. (1952) *Diccionário das plantas úteis do Brasil e das exóticas cultivadas*, Vol III. Rio de Janeiro, Brazil: Imprensa Nacional.

Dietrich, D.N.F. (1840) Pineapple. *Synopsis Plantarum* 2, 1087–1088.

Dujardin, M. (1991) Cytogénétique de l'ananas. *Fruits* 46, 376–379.

Duval, M.F. and Coppens d'Eeckenbrugge, G. (1993) Genetic variability in the genus *Ananas*. *Acta Horiculturae* 334, 27–32.

Duval, M.F., Coppens d'Eeckenbrugge, G., Ferreira, F.R., Cabral, J.R.S. and Bianchetti, L. de B. (1997) First results from joint EMBRAPA-CIRAD *Ananas* germplasm collecting in Brazil and French Guyana. *Acta Horticulturae* 425, 137–144.

Duval, M.F., Noyer, J.L., Perrier, X., Coppens d'Eeckenbrugge, G. and Hamon, P. (2001) Molecular diversity in pineapple assessed by RFLP markers. *Theoretical and Applied Genetics* 102, 83–90. DOI: 10.1007/s001220051621.

Duval, M.F., Buso, G.C., Ferreira, F.R., Bianchetti, L.B., Coppens d'Eeckenbrugge, G., Hamon, P., and Ferreira, M.E. (2003) Relationships in *Ananas* and other related genera using chloroplast DNA restriction site variation. *Genome* 46, 990–1004. DOI: 10.1139/g03-074.

Ferreira, F.R., Giacometti, D.C., Bianchetti, L. de B. and Cabral, J.R.S. (1992) Coleta de germoplasma de abacaxizeiros (*Ananas comosus* (L.) Merrill) e espécies afins. *Revista Brasileira de Fruticultura* 14, 5–11.

Foster, M.B. (1958) A new variety of pineapple. *The Bromeliad Society Bulletin* 8, 97.

García, M.L. (1988) Etude taxinomique du genre *Ananas*. Utilisation de la variabilité enzymatique. Doctoral thesis, USTL, Montpellier, France.

Gilmartin, A.J. and Brown, G.K. (1987) Bromeliales, related monocots, and resolution of relationships among Bromeliaceae subfamilies. *Systematic Botany* 12, 493–500.

Gitaí, J., Paule, J., Zizka, G., Schulte and K., Benko-Iseppon, A.M. (2014) Chromosome numbers and DNA content in Bromeliaceae: additional data and critical review. *Botanical Journal of the Linnean Society* 176, 349–368. DOI: 10.1111/boj.12211.

Givnish, T.J., Sytsma, K.J., and Smith, J. F. (1990) A re-examination of phylogenetic relationships among bromeliad subfamiles using cpDNA restrictionsite variation. *American Journal of Botany* 77, 133.

Givnish, T.J, Millam, K.C, Berry, P.E, Sytsma, K.J. (2007) Phylogeny, adaptive radiation, and historical biogeography of Bromeliaceae inferred from ndhF sequence data. *Aliso* 23, 1–26.

Givnish, T.J., Barfuss, M.H.J., Van Ee, B., Riina, R., Schulte, K., Horres, R., Gonsiska, P.A., Jabaily, R.S., Crayn, D.M., Smith, J.A.C., Winter, K., Brown, G.K., Evans, T.M., Holst, B.K., Luther, H., Till, W., Zizka, G., Berry, P.E. and Sytsma, K.J. (2011) Adaptive radiation and diversification in Bromeliaceae: insights from a 7-locus plastid phylogeny. *American Journal of Botany* 98, 872–895.

Gouda, E.J. (1999) Studies on the Flora of the Guianas No. 90: Checklist of Bromeliaceae of the Guianas with notes on critical species. *Selbyana* 20, 30–39.

Harms, H. (1930) Pflanzenfamilien. Bromeliaceae. *Engler Prantl*, 65–159.

Hassler, E. (1919) Bromeliacearum paraguariensium conspectus. *Annuaire du Conservatoire et du Jardin Botanique de Genève*, 268–341.

Heller, S., Leme, E.M.C., Schulte, K., Benko-Iseppon, A.M. and Zizka, G. (2015) Elucidating phylogenetic relationships in the *Aechmea* alliance: AFLP analysis of *Portea* and the *Gravisia* Complex (Bromeliaceae, Bromelioideae). *Systematic Botany* 40, 716–725.

Hornung-Leoni, C.T. (2011) Bromeliads: traditional plant food in Latin America since prehispanic times. *Polibotánica* 32, 219–229.

Horres, R., Zizka, G., Kahl, G. and Weising, K. (2000) Molecular phylogenetics of Bromeliaceae: evidence from trnL (UAA) intron sequences of the chloroplast genome. *Plant Biology* 2, 306–315.

Horres, R., Schulte, K. and Zizka, G. (2007) Systematics of Bromelioideae: evidence from molecular and anatomical studies. *Aliso* 23, 27–43.

Jabaily, R.S. and Sytsma, K.J. (2010) Phylogenetics of *Puya* (Bromeliaceae): placement, major lineages, and evolution of Chilean species. *American Journal of Botany* 97, 337–356.

Kerns, K.R., Collins, J.L. and Kim, H. (1936) Developmental studies of the pineapple *Ananas comosus* (L.) Merr. *The New Phytologist* 35, 305–317.

Koch, K. (1871) Wochenschrift. *Vereines zur Beförderung des Gartenbaues Königilch Preussischen Staaten* 14, 130.

Krauss, B.H. (1948) Anatomy of the vegetative organs of the pineapple *Ananas comosus* (L.) Merr. I. Introduction, organography, the stem and the lateral branch or axillary buds. *Botanical Gazette* 110, 159–217.
Krauss, B.H. (1949a) Anatomy of the vegetative organs of the pineapple, *Ananas comosus* (L.) Merr. II. The leaf. *Botanical Gazette* 110, 333–404.
Krauss, B.H. (1949b) Anatomy of the vegetative organs of the pineapple, *Ananas comosus* (L.) Merr. Concluded. III. The root and the cork. *Botanical Gazette* 110, 550–587.
Leal, F. (1990) On the validity of *Ananas monstruosus*. *Journal of the Bromeliad Society* 40, 246–249.
Leal, F. and Amaya, L. (1991) The curagua (*Ananas lucidus*, Bromeliaceae) crop in Venezuela. *Economic Botany* 45, 216–224.
Leal, F. and Coppens d'Eeckenbrugge, G. (1996) Pineapple. In: Janick, J. and Moore, J.N. (ed.) *Fruit Breeding*. New York: Wiley Publishing Co., pp. 565–606.
Leal, F. and Medina, E. (1995) Some wild pineapples in Venezuela. *Journal of the Bromeliad Society* 45, 152–158.
Leal, F., Coppens d'Eeckenbrugge, G. and Holst, B.K. (1998) Taxonomy of the genera *Ananas* and *Pseudananas* – a historical review. *Selbyana* 19, 227–235.
Lexer, C., Marthaler, F., Humbert, S., Barbará, T., Harpe, M. de la, Bossolini, E., Paris, M., Martinelli, G. and Versieux, L.M. (2016) Gene flow and diversification in a species complex of *Alcantarea inselberg* bromeliads. *Botanical Journal of the Linnean Society* 181, 505–520.
Lin, B.Y., Ritschel, P.S. and Ferreira, F.R. (1987) Número cromossômico de exemplares da família Bromeliaceae. *Revista Brasileira de Fruticultura* 9, 49–55.
Linden, J.J. (1869) *Catalogue et prix courant pour 1869 de J. Linden*. Brussels, Belgium.
Lindley, J.J. (1827) Billbergia. *Botanical Register* 13, 1068.
Lindman, C.A.M. (1891) Bromeliaceae. Herbarii Regnellianii. I. Bromeliae. Kongl. *Svenska Vetenskaps Academiens Handlingar* 24, 1–49.
Linnaeus, C. (1753) *Species plantarum*. Stockholm, Sweden.
Linnaeus, C. (1754) *Herbarium Amboinense*, vol. 21.
Loison-Cabot, C. (1992) Origin, phylogeny and evolution of pineapple species. *Fruits* 47, 25–32.
Lowe, E.J. (1868) *Beautiful Leaved Plants*. London: Groombridge & Sons.
Marchant, C.J. (1967) Chromosome evolution in the Bromeliaceae. *Kew Bulletin* 21, 161–168.
Matuszak, S., Paule, J., Heller, S., Leme, E., Steinbeisser, G. and Zizka, G. (2015) *Ananas* (Bromelioideae, Bromeliaceae) – Phylogenetic relationships based on nuclear, plastid and AFLP data. In: Benko-Iseppon, A.M., Alves, M. and Louzada, R. (ed.) An overview and abstracts of the First World Congress on Bromeliaceae Evolution. *Rodriguésia* 66(2), A34.
Medina, E. (1990) Eco-fisiología y evolución de las Bromeliaceae. *Boletín de la Academia Nacional de Ciencias, Córdoba* 59, 71–100.
Merrill, E.D. (1917) *An interpretation of Rumphius's Herbarium Amboinense*. Manila, Philippines: Bureau of Science, pp. 46, 133–134. DOI: 10.5962/bhl.title.79163.
Mez, C. (1892) Bromeliaceae; *Ananas*. *Martius, Flora Brasiliensis*, Vol. 3(3). Reprinted 1965 Verlag von J. Cramer, Weinheim, Codicote (Herts). New York: Wheldon & Wesley Ltd, pp. 288–294.
Mez, C. (1934) Das Pflanzenreich. Bromeliaceae. *Ananas* Adans. *Engler Prantl*, 160. Reprinted 1965 Verlag von J. Cramer, Weinheim, pp. 101–104.
Miles Thomas, E.N. and Holmes, L.E. (1930) The development and structure of the seedling and young plants of the pineapple (*Ananas sativus*). *New Phytologist* 29, 199–226.
Miller, P. (1754) *Garden Dictionary*, 4th edition. London: Henrey, Staflen and Cowan.
Miller, P. (1768) *Garden Dictionary*, 8th edition. London: Henrey, Staflen and Cowan.
Moldenke, H.N. (1938) *Cultivated Plants*. New York: New York Botanical Garden.
Morren, E. (1872) Notes sur l'*Ananas porteana* Hort. ou Ananas de Marius Porte. *La Belgique Horticole – Annales d'Horticulture* 22, 193.
Morren, E. (1878) Description de l'*Ananas macrodontes*. sp. nov. Ananas à fortes épines. *Belgique Horticole (Liège)* 28, 140–172.
Müller, F. (1896) Die *Bromelia silvestris* der Flora fluminensis. *Berichte der Deutschen Botanischen Gesellschaft* 14, 3–11.
Negrelle, R.R.B., Mitchell, D. and Anacleto, A. (2012). Bromeliad ornamental species: conservation issues and challenges related to commercialization. *Acta Scientiarum. Biological Sciences* 34, 91–100.
Noisette, L. (1825) *Manuel complet du jardinier maraîcher, pépiniériste, botaniste, fleuriste et paysagiste*, Vol. 2. Paris: Rousselon.

Noyer, J.L. (1991) Intérêt des RFLP pour l'amélioration des variétés cultivées du genre *Ananas*. DES thesis, USTL Montpellier II, France.
Noyer, J.L. and Lanaud, C. (1992) Diversité du genre *Ananas* par étude des RFLP. In: *Complexes d'espèces, flux de gènes et ressources génétiques des plantes*. Paris, France: Lavoisier – Techniques et Documentation, pp. 560–561.
Okimoto, M.C. (1948) Anatomy and histology of the pineapple inflorescence and fruit. *Botanical Gazette* 110, 217–231.
Palma-Silva, C., Leal, B.S.S., Chaves, C.J.N., Fay, M.F. (2016) Advances in and perspectives on evolution in Bromeliaceae. *Botanical Journal of the Linnean Society* 181, 305–322.
Plumier, C. (1755) *Plantarum Americanarum*. Amstelaedami, 262pp.
Purseglove, J.W. (1972) Tropical crops. *Monocotyledons*. London: Longman, pp. 75–91.
Py, C., Lacoeuilhe, J.J. and Teisson, C. (1987) *The Pineapple: Cultivation and Uses*. G.P. Maisonneuve Et Larose, Paris, p. 568.
Ranker, T.A., Soltis, D.E., Soltis, P.S. and Gilmartin, A.J. (1990) Subfamilial phylogenetic relationships of the Bromeliaceae: evidence from chloroplast DNA restriction site variation. *Systematic Botany* 15, 425–434.
Rao, A.N. and Wee, Y.C. (1979) Embryology of the pineapple, *Ananas comosus* (L.) Merr. *The New Phytologist* 83, 485–497.
Reitz, P.R. (1968) A new species of pineapple from Central Brazil. *The Bromeliad Society Bulletin* 18, 109–111.
Reitz, R. (1983) *Bromeliáceas e a malária – Bromélia endêmica*. Santa Catarina, Brazil: Flora Ilustrada Catarinense.
Reyes-Zumeta, H. (1967) Breve nota taxonómica sobre piñas cultivadas *Ananas comosus* (L.) Merr. con mención de dos nuevas variedades silvestres. *Revista de la Facultad de Agronomía L.U.Z. (Maracaibo)* 39, 131–142.
Rios, R. and Khan, B. (1998) List of ethnobotanical use of Bromeliaceae. *Journal of the Bromeliad Society* 48, 75–87.
Rumphius, G.E. (1747) *Herbarium amboinense*, Vol. 5, pp. 227–231.
Sakai, W.S. and Sanford, W.G. (1980) Ultrastructure of the water-absorbing trichomes of pineapple (*Ananas comosus*, Bromeliaceae). *Annals of Botany* 46, 7–11.
Sass, C. and Specht, C.D. (2010) Phylogenetic estimation of the core bromelioids with an emphasis on the genus *Aechmea* (Bromeliaceae). *Molecular Phylogenetics and Evolution* 55, 559–571.
Schulte, K., Barfuss, M.H.J. and Zizka, G. (2009) Phylogeny of Bromelioideae (Bromeliaceae) inferred from nuclear and plastid DNA loci reveals the evolution of the tank habit within the subfamily. *Molecular Phylogenetics and Evolution* 51, 327–339.
Schultes, J.A. and Schultes, J.H. (1830) *Ananas sativus* n.n. *Systema vegetabilis* 7, 1283–1285.
Schütz, N., Krapp, F., Wagner, N. and Weising, K.. (2016) Phylogenetics of Pitcairnioideae s.s. (Bromeliaceae): evidence from nuclear and plastid DNA sequence data. *Botanical Journal of the Linnean Society* 181, 323–342. DOI: 10.1111/boj.12403.
Sharma, A.K. and Ghosh, I. (1971) Cytotaxonomy of the family Bromeliaceae. *Cytologia* 36, 237–247.
Smith, L.B. (1934a) Geographical evidence on the lines of evolution in the Bromeliaceae. *Botanische Jahrbücher* 66, 446–448.
Smith, L.B. (1934b) IV. Studies in the Bromeliaceae. *Contributions from the Gray Herbarium* 104, 72.
Smith, L.B. (1939) Notes on the taxonomy of *Ananas* and *Pseudananas*. *Botanical Museum Leaflet, Harvard* 7, 73–81.
Smith, L.B. (1961) Notes on Bromeliaceae. *Phytologia* 8, 12.
Smith, L.B. (1962) A new look at the species of pineapple. *Bromeliad Society Bulletin* 12, 54–55.
Smith, L.B. and Downs, R.J. (1979) Bromelioidees (Bromeliaceae). *Flora Neotropica, Monograph* 14, pt 3.
Terry, R.G., Brown, G.K. and Olmstead, R.G. (1997) Examination of subfamilial phylogeny in Bromeliaceae using comparative sequencing of the plastid locus ndhF. *American Journal of Botany* 84, 664–670.
Van Geert, A. (1873) *Le Catalogue général de l'établissement d'horticulture d'August van Geert* 70, 3.
Van Geert, A. (1878) *Le Catalogue général de l'établissement d'horticulture d'August van Geert* 76, 2.
Vellozo, J.M. (1829) Pineapple. *Florae Fluminensis* 3, 129–130.
Wee, Y.C. and Rao, A.N. (1979) *Ananas* pollen germination. *Grana* 18, 33–39.
Willdenow, C.L. (1809) *Enumeratio plantarum Horti Regii Botanici Berolinensis*, pp. 345–346.

3 Origin and Evolution

Geo Coppens d'Eeckenbrugge,[1] Marie-France Duval[2] and Freddy Leal[3]

[1]CIRAD, UMR AGAP, Montpellier, France; [2]CIRAD, UMR AGAP, Le Lamentin Cedex, Martinique; [3]Universidad Central de Venezuela, Aragua, Venezuela

Alphonse de Candolle (1886) is famous for proposing a modern interdisciplinary method to study the origin of cultivated plants, combining the contributions of botanical geography, particularly for the ecology, distributions and diversity of wild relatives, feral forms and primitive domesticates, paleontology, archaeology, history, and historical linguistics. After integrating the significant contribution of genetics, this method is still the basis for current research on the subject.

The Pineapple Wild Relatives: Taxonomical Status, Ecology and Distributions

For the pineapple, de Candolle only presented potential elements on the distribution of wild populations in Mexico (with doubts), Panama (in fact, plants escaped from cultivation, as they were classified as *Ananassa sativa* Lindl. by Seemann (1852), the Upper Orinoco drainage (based on Von Humboldt's report), Guiana, and the Brazilian state of Bahia (according to a letter received from Martius). He was not acquainted with two earlier reports in the Orinoco region, one by Jacinto de Carvajal who observed in 1647, the existence of 'small wild pineapples, white and very green, with black stones or seeds' (de Carvajal, 1892), and a second one by Father Gumilla (1741); however the first one was only published in 1892, and the second one was largely ignored. De Candolle provided no elements of description. Beer (1857) observed a dried specimen that had been collected in the lower Amazon and concluded that the only difference between the edible and the wild pineapple consisted in the hypertrophy of the fruit and peduncle, which he attributed entirely to cultivation. It was Baker who published the first botanical description of a wild pineapple in 1889, based on two dried specimens from Central Brazil, but he did not recognize it as a pineapple and classified it as *Acanthostachys ananassoides*, noting only that it 'has entirely the habit of a small *Ananas*, but the bract is quite free from the ovary'. Its direct relation with the cultivated pineapple was recognized by Lindman (1891) who reclassified it under *Ananas microstachys*. Mez (1892) went further considering it conspecific to the edible pineapple and naming it *Ananas sativus* Schult. & Schult.f. var. *microstachys*, now a synonym of *A. comosus* var. *microstachys*.

In 1966, Camargo visited the area of the Parguaza River, an Orinoco affluent, where Vélez and Badillo (1946) had collected wild pineapples, then classified as *A. ananassoides* (now *A. comosus* var. *microstachys*). On the basis of these two collections, Camargo and Smith (1968) described another wild pineapple taxon, *A. parguazensis* (now *A. comosus* var. *parguazensis*), and the collections of this region had to be reclassified and divided between the two wild botanical varieties of the current classification; however, a good part of this material could not be assigned to either and was left aside.

The natural distribution of *A. comosus* var. *microstachys* covers that of the whole species. It is generally found in savannahs or clear open forests (Brazilian *cerrados*), growing on soils with limited water-holding capacity (sand dunes or 'campinas', inselbergs and 'rock savannahs') and forming populations of variable densities. In the Guianas, it can also be found, although rarely, in secondary rain forest. In contrast, it is absent from the seasonally flooded lands along the Amazon and its main southern tributaries, which seem to act as a barrier dividing its distribution into two main areas, whose populations appear genetically differentiated (Duval *et al.*, 2003). The northern area corresponds to the Guiana shield, Orinoco basin, and drainage of Rio Negro (i.e. from the Brazilian state of Amapá, north of the lower Amazon, to eastern Colombia). The southern area roughly corresponds to the Brazilian shield, north-eastern Brazil, and the basins of the upper Paraná and Paraguay rivers (from Bolivia and the Brazilian states of Acre, Mato Grosso over to Pernambuco and down to Paraguay and northern Argentina).

In the southern part of its distribution, in drier conditions, *A. comosus* var. *microstachys* presents a very small (generally less than 8 cm), globose to cylindrical, inflorescence. Its growth after anthesis is limited, so the fruit has little flesh. The pulp is white or cream, very firm and fibrous, with high sugar content and acidity, and numerous seeds. In the northern area, wild pineapple habitats appear more variable (Leal and Medina, 1995), and a higher morphological diversity is observed. Some forms present traits that are more reminiscent of var. *parguazensis*, such as wider leaves and/or a few retrorse spines, or var. *comosus*, such as wider leaves and larger fleshy fruits (up to 12–15 cm long); however, they remain smaller than in the edible pineapple, even under cultivation. Such fruits were consumed in the Orinoco (Patiño, 2002) and are still occasionally consumed in the Guianas. Some of them are occasionally found in patches in secondary forest and savannahs in French Guiana, indicating an ancient settlement, or cultivated in gardens (Fig. 3.1).

A. comosus var. *parguazensis* is sympatric with *A. comosus* var. *microstachys* in most of its distribution north of the Amazon, but it appears more centred on the drainage basins of the Orinoco, its area of greatest diversity, and upper Rio Negro, with few observations in the Colombian Amazonia and in the Guianas (Coppens d'Eeckenbrugge *et al.*, 1997). It grows in lowland forests, under canopies of variable densities, from clearings or riverbanks to dense forest. As compared to specimens of *A. comosus* var. *microstachys* growing in close proximity, it seems restricted to shadier environments, because of lower water use efficiency (Leal and Medina, 1995). Typical representatives of the var. *parguazensis* have wide leaves; their small inflorescences grow after anthesis to produce a globose fruit.

The Domesticated Pineapples and Their Pre-Columbian Distribution

The three domesticated botanical varieties are *A. comosus* var. *comosus*, the pantropical pineapple cultivated for its large fruit, *A. comosus* var. *erectifolius*, the curagua, a small-fruited pineapple cultivated for its fibre, and *A. comosus* var. *bracteatus*, a very robust, spiny plant initially cultivated for fibre (Arruda da Câmara, 1810) and more recently for other uses, involving its medium-sized fruit for juice and its spiny leaves for fences (Müller, 1896; Bertoni, 1919). The two latter botanical varieties are now cultivated as ornamentals.

The inflorescence of the edible pineapple grows very significantly after anthesis, so the fruit is large and fleshy, with many fruitlets, borne on a wide and strong peduncle. Seeds are rare in the fruits, because of reduced fertility, conjugated with stronger self-incompatibility and monoclonal cultivation (Coppens d'Eeckenbrugge *et al.*, 1993). The plant has numerous wide leaves (40–80), with antrorse spines; these marginal spines are generally smaller and denser than in wild varieties. The very wide Neotropical distribution of the edible pineapple, its strong differentiation from its wild relatives, and the high diversity of cultivars and uses, all point to an ancient domestication.

The curagua has long been cultivated by the Native Americans all along the Amazon, up to Peru, and north of the river, in the Orinoco basin, the Guianas, and east to the Brazilian state of Maranhão, for the production of a strong and silky white fibre used to make loin

Fig. 3.1. Diversity of pineapple forms intermediate between *A. comosus* var. *comosus* and *A. comosus* var. *microstachys* for fruit size, collected in the wild or in secondary vegetation, except for picture (c) (home garden). Photos: (d) M.-F. Duval; all others G. Coppens d'Eeckenbrugge.

cloths, hammocks, strings and ropes, fishing lines and nets (Leal and Amaya, 1991). There are early reports of its cultivation in southern and eastern Venezuela (Caulin, 1779; Gilij, 1782), and in Maranhão (d'Evreux, 1615). Now, while suffering competition from synthetic fibres, the curagua has found a new economic use in the production of cut flowers, and more recently for the reinforcement of polymer composites for the automotive industry (Zah *et al.*, 2007; Monteiro *et al.*, 2013).

Ananas comosus var. *bracteatus* is an assemblage of two cultivated forms that show the same geographic distribution as *Ananas macrodontes* in south-eastern Brazil and Paraguay, and that are morphologically and genetically intermediate between *A. comosus* and *A. macrodontes*. Morphological and genetic variations are very limited, being comparable to within-cultivar variations (Duval *et al*., 2001, 2003) and suggesting very narrow origins, probably a single genotype. The second form corresponds to *A. fritzmuelleri* Camargo, which is now very rare, so its variation is not known. The most common form was first described by Arruda da Câmara (1810) as a plant from Pernambuco, under the binomial *Bromelia sagenaria* (from the Latin *sagena* for 'net'). He mentioned the vernacular names of *crauatá/caroatá de rede* (net bromeliad) or *ananás bravo* (fierce pineapple), as opposed to the *ananás manso* (meek pineapple) for *A. comosus* var. *comosus*. Its fibre, rougher than that of the curagua and inferior in quality to that of the *ananás manso*, was also used to make ropes for the sugarcane industry. Bertoni (1919) mentioned that *A. comosus* var. *bracteatus* was used as a living hedge and harvested for fibre and fruit juice, or for traditional medicine, in southern Brazil and Paraguay. The rare form described by Fritz Müller (Müller, 1896) was also used in living fences. Obviously, this use necessarily followed the introduction of cattle and the development of private property with the European colonization. The dense, large leaves of the *ananás bravo*, have strong spines and form impenetrable barriers. It is very robust and still thrives today in abandoned farms, but it seems unable to naturalize and colonize new habitats. Several variegated somatic mutations have been selected as garden ornamentals. A private flower producer selected a smooth-leaved mutant in Ivory Coast for the European cut flower market.

Domestication Syndrome in the Cultivated Forms of *Ananas comosus*

A comparison of the edible pineapple and its wild relative shows the direct effects of human selection and the associated evolution for phenological and reproductive traits. Larger fruits resulted from the combination of larger individual fruitlets and an increase in their number, which, in some cases, modified their phyllotaxy. A larger number of wider, and generally shorter, leaves provided more energy, while a stouter and longer stem increased the starch storage capacity during the vegetative phase. The latter has been extended by reduced susceptibility to natural flowering initiation. In many cultivars, seed production has been reduced through the combination of lower sexual fertility and stronger self-incompatibility (Coppens d'Eeckenbrugge *et al*., 1993; Leal and Coppens d'Eeckenbrugge, 1996).

Thus, the domestication syndrome includes not only the partial loss of sexuality in *A. comosus* var. *comosus*, but also those changes in the vegetative organs and the vegetative cycle, which result in a lack of adaptation to the wild. While pineapple plants from rustic cultivars can survive when their cultivation is abandoned, they do not propagate efficiently to form subspontaneous populations. This is the likely cost of a lengthened phenological cycle, a high harvest index, possibly greater predation and the loss of dispersal capacity for the large fruit and crown.

A. comosus var. *comosus*, *A. comosus* var. *bracteatus* and *A. comosus* var. *erectifolius* show no capacity for spontaneous colonization in the wild. The relatively large fruit size of var. *bracteatus* must be limiting to its natural dispersal and the absence of leaf spines in the small-fruited var. *erectifolius*, likely increases its susceptibility to herbivory. For those cultivars of var. *erectifolius* that rarely produce a fruit, there is an additional restriction on sexual recombination, and hence on the plant adaptive potential. On the other hand, we must keep in mind that the absence of spines is also the main characteristic that distinguishes var. *erectifolius* from var. *microstachys*. Thus, when smooth-leaved clones of var. *erectifolius* mutate back to the spiny condition, as has been observed under cultivation, these reverse mutants display the essential characteristics of var. *microstachys*.

Pineapple Diversity and Genetic Relations Among Wild and Cultivated Forms

Duval *et al*. (2001) gathered a diverse *Ananas* sample for a wide genetic study, with 301 accessions representing all taxa and most of the South American distribution of the genus. In this RFLP (restriction fragment length polymorphism) study, *A. comosus* and *A. macrodontes* displayed a high polymorphism, but a limited interspecific differentiation, the latter sharing 59% of its bands with the former.

Within *A. comosus*, the wild varieties *microstachys* and *parguazensis* are the most variable, with 89% and 84% polymorphic bands for 57 and 19 accessions respectively. They show a geographic structure, with a group of *microstachys* accessions from the south of the Amazon basin (Brazilian state of Mato Grosso) and a group of *parguazensis* accessions from the Rio Negro and Rio Orinoco drainages. Their Guianese accessions contribute to other groups, which also involve representatives of varieties *comosus* and *erectifolius*.

Despite its wide morphological variation, var. *comosus* (75% polymorphism for 167 accessions) appeared relatively homogeneous in the factorial analysis, but not genetically distinct, as it showed affinities with representatives of var. *microstachys*, var. *parguazensis*, as well as intermediate, medium-fruited, forms (79% polymorphism; 28 accessions), most of them of Guianese origin. Cultivars sharing the dominant monogenic leaf trait known as 'piping' are grouped with accessions of the same West Amazonian/Andean geographical origin, confirming the intervention of sexual reproduction to disseminate this character.

The highly diverse var. *erectifolius* (69% polymorphism; seven accessions) appears much more scattered in the graphical representations of diversity, with strong affinities with different groups of var. *microstachys*, some of them involving also var. *comosus* and medium-fruited forms. The third domesticate, var. *bracteatus*, is uniform and genetically intermediate between the two *Ananas* species, the common form (28% polymorphism; 18 accessions) remaining close to *A. comosus*, while the rare form is closer to *A. macrodontes*. The general picture reflects the important role of sexual reproduction, without reproductive barriers among botanical varieties, the much higher genetic diversity in the Guiana shield, and the relationship between two domesticates, varieties *comosus* and *erectifolius*, and wild pineapples from this region, essentially representatives of var. *microstachys*.

A second study, using polymerase chain reaction RFLP to analyse chloroplast DNA (cpDNA) diversity, was conducted on a 99-accession subsample (Duval *et al.*, 2003). Phenetic and cladistic analyses identified three strongly supported groups, two of which are geographically consistent: a northern group of var. *parguazensis* accessions restricted to the Rio Negro and Orinoco basins; a southern group made of *A. macrodontes* and the rare form of var. *bracteatus* (anciently *A. fritzmuelleri*); the third and largest one, including all remaining botanical varieties plus the Guianese accessions of var. *parguazensis* and intermediate phenotypes. The wild var. *microstachys* is dominant in this third group and is highly variable (11 haplotypes). The domesticates were less variable, and all haplotypes detected in var. *comosus*, *bracteatus*, and *erectifolius* were shared with var. *microstachys*, and some were also shared with var. *parguazensis* from the eastern Guiana region. Accessions with intermediate phenotypes were represented by four haplotypes, of which three were common with var. *comosus*. This confirms the central role of var. *microstachys* as the wild ancestor of the domesticated pineapples. The comparison of the cpDNA and genomic DNA data point to the hybrid origin of var. *bracteatus*. The distribution of haplotypes shows a maximal diversity in the Guianas, for both wild and domesticated forms. The resulting geographical structuration is consistent with the higher morphological diversity of var. *microstachys* north of the Amazon River and the east–west contrast observed for morphological diversity of the edible pineapple (Duval *et al.*, 1997). A microsatellite (SSR) study consistently showed a strong contrast between Guianese cultivars of the edible pineapple and the western Amazonian ones, with a higher heterozygosity in the former group (Blanc *et al.*, 2004).

Domestication Scenarios

Genetic data are consistent with morphological data, indicating that *A. comosus* var. *comosus* was domesticated from var. *microstachys* in the Guianas, a region of high diversity, where these two botanical varieties, as well as intermediate forms, present genetic and morphological affinities indicative of significant gene flow among them. The potential intervention of var. *parguazensis* seems much less important. On one hand, although it is less common in the Guianas, some of its local representatives also present genetic affinity with domesticated pineapples, and this variety may have contributed to wider leaves and more globulous fruits, as compared with the smaller, often cylindrical fruits of var. *microstachys*. On the other hand, no pineapple cultivar has been reported to present retrorse spines or the typical constriction of leaves at their base.

The most likely scenario is that intermediate materials were collected from 'rock savannahs', sand dunes and similar places, and planted in village gardens. Nowadays, inhabitants of the Guianese forests still collect materials from the wild to incorporate into their cultivated plots. This explains why some cultivars are found both under cultivation and in patches of secondary vegetation in the wild, marking likely sites of ancient cultivation. Such practices constitute a basis for a process involving 'domestication cycles'. In these cycles, pineapples are sampled in the wild, put in polyclonal cultivation, semi-abandoned, and re-sampled for cultivation with possible genetic recombination and selection at each step. Indeed, fields and home gardens are never completely abandoned and forgotten, as they are located near pathways and remain useful (Vélez, 1998). They also serve as stocks of useful planting materials. The most interesting genotypes are thus progressively concentrated (Gnecco, 2003; Miller and Nair, 2006).

Sexual reproduction can contribute to the exploitable diversity by the creation of new cultivars. However, wild types may be more fertile than semi-domesticated materials (Coppens d'Eeckenbrugge *et al*., 1993), such that wild genes are transmitted more efficiently, counteracting the effects of selection. Nevertheless, they may confer robustness so growers may want to maintain them. The result is the multiclonal production system still observed in the Guianas, which maintains a balance between genotypes at different stages of domestication. Such regressive processes can delay the evolution to more highly selected forms of the crop (Galinat, 1974).

In the case of pineapple, a second phase of evolution probably occurred in western Amazonia (along the upper Amazon, close to the triple frontier between Colombia, Peru and Brazil, and along the lower Rio Negro), where we can observe a high diversity of cultivars, except that wild forms are rare or absent (Bello and Julca, 1993; Duval *et al*., 1997). There the prevalence of large, seasonally flooded areas and the rarity of elevated sites with good drainage seem to limit the extension of *A. comosus* var. *microstachys*. Once established in this area, the cultivated pineapple could evolve and diversify in completely artificial conditions, in a dynamic process combining sexual recombination restricted to domesticated germplasm, somatic mutations and clonal selection. This area is an important centre of domestication and diversification for many other fruits (Clement 1989, 1999). Peoples like the Tikunas and the Huitotos still value and maintain a wide diversity of pineapple cultivars and other fruits. In the course of their collecting trips, Duval *et al*. (1997) observed as many as twelve cultivars in a small plot maintained by a single Tikuna family. Schultes (1991) gives similar numbers for the pineapple cultivars named by the Huitotos. The species is culturally very important for peoples of the area. For example, the Yukunas celebrate nine fruit festivals yearly, five of them being pineapple festivals (Jacopin, 1988). Cristancho (2001; cited in Cristancho and Vining, 2004) ranked pineapple among the three primary cultural keystone species (CKS) of the Letuama people (CKS are 'species whose existence and symbolic value are essential to the stability of a cultural group over time'), along with tobacco and coca.

The existence of two centres for the diversification of *A. comosus* var. *comosus*, a primary one in the Guianas, with diversification involving clones at different stages of domestication, and a secondary one in the western Amazon/Andean foothills, with sexual recombination between large-fruited cultivars, is also suggested by the distribution of particular leaf margin types (Duval *et al*., 1997). Thus, the dominant S gene that partially suppresses spines is particularly frequent in the Guianas, while in the western Amazon and in the Andes (from Peru to Colombia), leaf smoothness is determined by the dominant 'piping' gene (Coppens d'Eeckenbrugge and Sanewski, 2011). The existence of homozygotes for the 'piping' gene (Cabral *et al*., 1997) indicates sexual recombination among cultivars within this western pool. Another indication of sexual reproduction in the cultivar diversification process is the advent of triploid pineapples which result from the fertilization of unreduced ovules (Collins, 1960). Triploids include the Brazilian cultivar Gigante de Tarauacá, selected for its huge fruit (up to 15 kg) (Scherer *et al.*, 2015), as well as landraces from Ecuador and Brazil (Heilborn, 1921; Lin *et al.*, 1987).

The high morphological and genetic affinity of *A. comosus* var. *erectifolius* with var. *microstachys* points to a domestication of the former in the Guiana region. The high genetic diversity of the curagua indicates several parallel events of domestication, which is easily explained by the fragile difference between the crop and its wild relative. Cultivation probably started with fibrous plants selected from the wild, and any smooth-leaved mutant or seedling from such materials would provide a new clone or cultivar.

South of the Amazon, fibre production was ensured from the exploitation of other bromeliads, including the yvira, *A. macrodontes*, whose extended and dense spontaneous populations provide a good quantity of raw materials. Much longer leaves constitute another advantage justifying a minimal management of these populations. Introducing the edible pineapple, farmers may have favoured unconsciously the first hybridization with an yvira clone (or an ancestral diploid form of it). Such a scenario implies a recent hybridization, which would be consistent with: (i) the survival of the product (retained for cultivation, protected from competition in the wild); (ii) a limited number of backcross generations; and thence (iii) the very low diversity of var. *bracteatus*. An alternative scenario, following an ancient hybridization in the wild between *A. macrodontes* and *A. comosus* var. *microstachys* would raise many questions on the encounter between species with different habitats, the survival of the successive progenies until their adoption by farmers and their extinction in the wild. The first scenario corresponds to the classical introgression of established crops with wild relatives, and transforms the evolution of var. *bracteatus* into a side-product of that of var. *comosus*. The second scenario would correspond to the *de novo* domestication of a very rare hybrid.

Pineapple Prehistory

Because of very poor conservation conditions in the rainforest and/or the lack of archaeobotanical research in Amazonia, no ancient pineapple remains had been observed in the two putative centres of diversification/domestication until recently when pineapple phytoliths were identified in relatively deep levels (0.80 m) of anthropic soils, indicating cultivation of the species around 3000 BP in western Amazonia (Morcote-Ríos *et al.*, 2013). Larger pineapple remains have only been conserved under arid conditions, and identified in archaeological layers dated from 3200 to 2800 BP on the Peruvian Coast (Pearsall, 1992), while seeds and bracts were found in coprolites from the Tehuacán Valley caves (Mexico) dated between 2200 and 1300 BP (Callen, 1967). Another indication of the early presence and the importance of pineapple in Ancient Mesoamerica is the stone block known as the Cascajal Block, which is the oldest written document known in the Americas. It was discovered near the archaeological site of San Lorenzo, in the Mexican state of Veracruz, associated with the Olmec civilization, and dated 2800–3000 years BP. The pineapple is represented four times on the block, among several natural organisms or products (Coppens d'Eeckenbrugge *et al.*, 2011).

The application of historical linguistics to the names of the pineapple in ancient Mesoamerica (glottochronology) also gives us a minimal estimation of the antiquity of pineapple domestication. Consistently, glottochronological data indicate that the crop was highly significant to Mesoamerican people more than 2500 years BP (Brown, 2010). Thus, domesticated pineapple was traded and adopted as an important fruit crop at the continental scale more than 3000 years BP. Such an early extension of its cultivation area implies the preliminary development of cultivars specifically adapted to the high latitudes and/or altitudes of Peru and Mexico. Indeed, highland cultivars from Andean countries show specific adaptations and tend to perform poorly in lowland conditions. Given the relative rarity of reproduction through seeds in

A. comosus var. *comosus*, the development of environmentally specialized cultivars was necessarily a long and slow process, taking place *in situ*, after the arrival of the domesticated pineapple in these particular environments. In conclusion, a likely time frame for pineapple domestication lies between 6000 and 10,000 years BP, which is consistent with current hypotheses for other major American crops. For example, cassava, another crop combining sexual and vegetative reproductions, was domesticated in south-western Amazonia (Clement *et al.*, 2010), probably before 10,000 BP, as it arrived in northern Peru before 8500 BP (Dillehay, 2011) and in Mexico around 5800 BP (Pope *et al.*, 2001).

References

Arruda da Câmara, M. (1810) *Dissertação sobre as plantas do Brazil que podem dar linhos proprios para muitos usos da sociedade, e suprir a falta de canhamo*, pp. 3–7, 13–22. Rio de Janeiro, Brazil: Impressão Regia.

Baker, J.G. (1889) *Handbook of the Bromeliaceae.* London: George Bell & Sons, pp. 22–25.

Beer, J.G. (1857) *Die familie der Bromeliaceen.* Vienna: Tendler and Comp.

Bello, S. and Julca, A. (1993) Colección y evaluación de ecotipos de piña *Ananas comosus* de la Amazonía peruana. *Memorias Primer Simposio Latinoamericano de Piñicultura*, Cali, Colombia, 25–29 May, 1–13.

Bertoni, M.S. (1919) Contribution à l'étude botanique des plantes cultivées. I. Essai d'une monographie du genre *Ananas. Anales Científicos Paraguayos* (Serie II) 4, 250–322.

Blanc, J., Noyer, J.L., Horry, J.P. and Duval, M.F. (2004) Characterization and diversity of pineapple varieties using microsatellite markers. *Pineapple News* 11, 30–32.

Brown, C.H. (2010) Development of agriculture in prehistoric Mesoamerica: the linguistic evidence. In: Staller, J.E. and Carrasco, M.D. (ed.) *Pre-Columbian Foodways: Interdisciplinary Approaches to Food, Culture, and Markets in Mesoamerica.* New York: Springer Verlag, pp. 71–107.

Cabral, J.R.S., de Matos, A.P. and Coppens d'Eeckenbrugge, G. (1997) Segregation for resistance to fusariose, leaf colour and leaf margin type from the EMBRAPA pineapple hybridization programme. *Acta Horticulturae* 425, 193–200.

Callen, E.O. (1967) Analysis of the Tehuacan coprolites. In: Byers, D.S. (ed.) *The Prehistory of the Tehuacan Valley*, Vol. 1. Austin: the University of Texas Press, pp. 261–289.

Camargo, F.C. and Smith, L.B. (1968) A new species of *Ananas* from Venezuela. *Phytologia* 16, 464–465.

Caulin, A. (1779) *Historia corográfica, natural y evangélica de la Nueva Andalucía.* Biblioteca de la Academia Nacional de la Historia, Vol. 81. Caracas, Venezuela. Reprinted in 1966.

Clement, C.R. (1989) A center of crop genetic diversity in western Amazonia. *BioScience* 39, 624–631.

Clement, C.R. (1999) 1492 and the loss of Amazonian crop genetic resources. II. Crop biogeography at Contact. *Economic Botany* 53, 203–16.

Clement, C.R., Cristo-Araújo, M. de, Coppens d'Eeckenbrugge G., Pereira A.A. and Picanço-Rodrigues, D. (2010) Origin and domestication of native Amazonian crops. *Diversity* 2, 72–106.

Collins, J.L. (1960) *The Pineapple. Botany, Cultivation, and Utilization.* New York: Interscience Publishers Inc.

Coppens d'Eeckenbrugge, G. and Sanewski, G.M. (2011) Leaf margin in pineapple. *Pineapple News* 18, 32–37.

Coppens d'Eeckenbrugge, G., Duval, M.-F. and Van Miegroet, F. (1993) Fertility and self-incompatibility in the genus *Ananas. Acta Horticulturae* 334, 45–51.

Coppens d'Eeckenbrugge, G., Leal, F. and Duval, M.-F. (1997) Germplasm resources of pineapple. *Horticultural Reviews* 2, 133–175.

Coppens d'Eeckenbrugge, G. Uriza Avila, D.E., Rebolledo Martínez, A. and Rebolledo Martínez, L. (2011) The Cascajal Block: another testimony of the antiquity of pineapple in Mexico? *Pineapple News* 18, 47–48.

Cristancho, S. and Vining, J. (2004) Culturally defined keystone species. *Research in Human Ecology* 11, 153–164.

de Candolle, A. (1886) *Origine des plantes cultivées*, 3rd edn, Paris: Félix Alcan.

de Carvajal, J. (1892) *Relación del descubrimiento del Río Apure hasta su ingreso en el Orinoco.* Diputación Provincial de León.

d'Evreux, Y. (1615) *Suitte de l'histoire des choses plus memorables advenuës en Maragnan és années 1613 & 1614.* Paris: Huby.

Dillehay, T.D. (2011) *From Foraging to Farming in the Andes.* Cambridge: Cambridge University Press.

Duval, M.-F., Coppens d'Eeckenbrugge, G., Ferreira, F.R., Cabral, J.R.S. and Bianchetti, L. de B. (1997) First results from joint EMBRAPA-CIRAD *Ananas* germplasm collecting in Brazil and French Guyana. *Acta Horticulturae* 425, 137–144.

Duval, M.-F., Noyer, J.L., Perrier, X., Coppens d'Eeckenbrugge, G. and Hamon, P. (2001) Molecular diversity in pineapple assessed by RFLP markers. *Theoretical and Applied Genetics* 102, 83–90.

Duval, M.-F., Buso, G.C., Ferreira, F.R., Bianchetti, L.B., Coppens d'Eeckenbrugge, G., Hamon, P. and Ferreira, M.E. (2003) Relationships in *Ananas* and other related genera using chloroplast DNA restriction site variation. *Genome* 46, 990–1004.

Galinat, W.C. (1974) The domestication and genetic erosion of maize. *Economic Botany* 18, 31–37.

Gilij, F.S. (1782) *Ensayo de historia americana.* Biblioteca de la Academia National de la Historia, Vol. 17. Caracas, Venezuela. Reprinted in 1965.

Gnecco, C. (2003) Against ecological reductionism: Late Pleistocene hunter-gatherers in the tropical forests of northern South America. *Quaternary International* 109–110, 13–21.

Gumilla, J. (1741) *El Orinoco ilustrado y defendido.* Biblioteca de la Academia Nacional de la Historia, Caracas, Venezuela.

Heilborn, O (1921) Notes on the cytology of *Ananas sativus* Lindl. and the origin of its parthenocarpy. *Arkiv för Botanik* 17, 1–7.

Jacopin, P.Y. (1988) Anthropological dialectics: Yukuna ritual as defensive strategy. *Schweizerische Amerikanisten-Gesellschaft Bulletin* 52, 35–46.

Leal, F. and Amaya, L. (1991) The curagua (*Ananas lucidus*, Bromeliaceae) crop in Venezuela. *Economic Botany* 45, 216–24.

Leal, F. and Coppens d'Eeckenbrugge, G. (1996) Pineapple. In: Janick, J. and Moore, J.N. (ed.) *Fruit Breeding*, Vol. 1: *Tree and Tropical Fruits*, pp. 515–557. New York: Wiley and Sons.

Leal, F. and Medina, E. (1995) Some wild pineapples in Venezuela. *Journal of the Bromeliad Society* 45,152–158.

Lin, B.Y., Ritschel, P.S. and Ferreira, F.R. (1987) Número cromossômico de exemplares da família Bromeliaceae. *Revista Brasileira de Fruticultura* 9, 49–55.

Lindman, C.A.M. (1891) Bromeliaceae. Herbarii Regnellianii. I. Bromeliae. Kongl. *Svenska Vetenskaps Academiens Handlingar* 24(8), 1–49.

Mez, C. (1892) Bromeliaceae; *Ananas.* Martius, *Flora Brasiliensis* 3(3). Reprinted 1965 Verlag von J. Cramer, Weinheim, Codicote (Hertfordshire). New York: Wheldon & Wesley, pp. 288–294.

Miller, R.P. and Nair, P.K.R. (2006) Indigenous agroforestry systems in Amazonia: from prehistory to today. *Agroforestry Systems* 66, 151–164.

Monteiro, S.N., Lopes, F.P.D., Nascimento, D.C.O., Ferreira, A. da S. and Satyanarayanac, K.G. (2013) Processing and properties of continuous and aligned curaua fibers incorporated polyester composites. *Journal of Materials Research and Technology* 2, 2–9.

Morcote-Ríos, G., Lauren Raz, D., Giraldo-Cañas, C.E.F. and León Sicard, T. (2013) Terras pretas de Índio of the Caquetá-Japurá river (Colombian Amazonia). *Tipití* 11, 30–39.

Müller, F. (1896) Die *Bromelia silvestris* der Flora fluminensis. *Berichte der Deutschen Botanischen Gesellschaft* 14, 3–11.

Patiño, V.M. (2002) *Historia y dispersión de los frutales nativos del Neotrópico.* Cali, Colombia: CIAT.

Pearsall, D.M. (1992) The origins of plant cultivation in South America. In: Cowan, C.W. and Watson, P.J. (ed.) *The Origins of Agriculture. An International Perspective.* Smithsonian Series in Archaeological Inquiries. Washington, DC: Smithsonian Institution Press, pp. 173–205.

Pope, K.O., Pohl, M.E.D., Jones, J.G., Lentz, D.L., von Nagy, C., Vega, F.J. and Quitmyer, I.R. (2001) Origin and environmental setting of ancient agriculture in the lowlands of Mesoamerica. *Science* 292, 1370–1373.

Scherer, R.F., Olkoski, D., Souza, F.V.D., Nodari, R.O. and Guerra. M.P. (2015) Gigante de Tarauacá: a triploid pineapple from Brazilian Amazonia. *Scientia Horticulturae* 181, 1–3.

Schultes, R.E. (1991) Ethnobotanical conservation and plant diversity in the Northwest Amazon. *Diversity* 7, 69–72.
Seemann, B. (1852) *The Botany of the Voyage of H.M.S. Herald.* London: Lovel Reeve.
Vélez, G.A. (1998) The chagra: collective patrimony of the indigenous Amazonian communities. *Beyond Law* 6, 121–142.
Vélez, I. and Badillo, V.M. (1946) Comentarios sobre los piñales silvestres y cultivados de la región del Parguasa, Guayana Venezolana. *Revista de Agricultura de Puerto Rico (San Juan)* 37, 127–130.
Zah, R., Hischier, R., Leão, A.L. and Braun, I. (2007) Curauá fibers in the automobile industry – a sustainability assessment. *Journal of Cleaner Production* 15, 1032–1040.

4 Varieties and Breeding*

Garth M. Sanewski,[1] Geo Coppens d'Eeckenbrugge[2] and Davi T. Junghans[3]

[1]Department of Agriculture and Fisheries, Nambour, Australia; [2]CIRAD, UMR AGAP, Montpellier, France; [3]Embrapa Mandioca e Fruticultura, Cruz das Almas, Brazil

Ananas comosus is the most economically important species in the *Bromeliaceae* family, with three of the five botanical varieties domesticated for fruit consumption or fibre use and, more recently, exploited for ornamentation.

Early European pineapple experts recognized as many as 52–70 pineapple cultivars, including those obtained from seeds in European glasshouses, and classified them on the presence of leaf spines, fruit shape and flower colour (Munro, 1835; Beer, 1857). Most of these cultivars disappeared, but three, 'Smooth Cayenne', 'Queen' and 'Singapore Spanish' spread throughout the world. Others became or remained economically important in their region of origin. Some of these are 'Pérola' in Brazil, 'Española Roja' in the Caribbean, and 'Perolera' and 'Manzana' in the Andes of Colombia and Venezuela.

Germplasm movement in post-Columbian times has been accompanied by clonal differentiation and/or selection and renaming giving the impression of more diversity worldwide than actual. Many cultivars are in fact somaclonal variants and not new cultivars. New cultivars generally arise by sexual recombination. To take into account clonal variation and regional nomenclature, and to avoid confusion over genotype identity, Coppens d'Eeckenbrugge *et al.* (1997) proposed to mention local names or specific clones together with the cultivar name, such as 'Queen', clone Victoria; 'Pérola', clone Sugarloaf; 'Smooth Cayenne', clone Champaka F-153. Such a system is essential to enable information sharing among growers and researchers, facilitate germplasm exchanges and allow comparisons of research in different countries (Bartholomew, 2016).

Main Cultivars

There are very few cultivars grown on a large commercial scale and exported to world markets but there are many local cultivars grown for local consumption throughout the world. The most planted pineapple cultivars worldwide are 'MD-2', 'CO-2', 'Smooth Cayenne', 'Singapore Canning', 'Queen', 'Española Roja', 'Pérola', 'Perolera' and 'Manzana'. In Latin American countries there are many local cultivars and wild genotypes, some of which might be suitable for wider use or used in breeding programmes once properly characterized.

'Smooth Cayenne' ('Cayenne Lisse') is the most planted cultivar worldwide. It is the leading variety for canning because of its high yields, good fruit shape, taste and sufficient fibre for firm slices and cubes. This pre-Columbian cultivar has become the standard variety for canning and has been the backbone of the global pineapple industry for more than a century (Chan

* A revision of the chapter by Y.K. Chan, G. Coppens d'Eeckenbrugge and G.M. Sanewski.

et al., 2003). 'Smooth Cayenne' was first collected in French Guiana and sent to Versailles by the botanist Samuel Perrottet in 1820 (Perrottet, 1825). It was sent from England to Florida, Jamaica and Australia in the late 19th century. From there it was sent to Hawaii, where it became predominant and formed the basis of large-scale, modern pineapple culture (Collins, 1951). 'Smooth Cayenne' is still common in the Guianas, under names such as 'Maipuri' or 'the true pineapple' in local language. It has been renamed 'Kew' in some former British colonies, 'Sarawak' in Malaysia, 'Delika Subang' in Indonesia and 'Pattavia' in Thailand. Some of its other clones in Thailand are 'Sriracha', 'Lampang', and 'Nanglae' (Rattanathawornkiti *et al.*, 2016). The plant is vigorous, with 60–80 spiny tip leaves with some anthocyanin. The ovoid, medium to large fruit (1.5–2.5 kg, but up to 4.0 kg) is held on a relatively short and strong peduncle (Fig. 4.1). It has flat fruitlets and orange-yellow skin. The crown is relatively small and the plant produces few slips depending on the clone. It ripens progressively, turning yellow from the base to the top, which is reflected in a strong internal maturity gradient. The flesh is close-textured, pale yellow to yellow colour at maturity, soft and juicy, with variation in soluble solids content (13–19%) and acidity depending on environmental conditions, and low ascorbic acid content. Despite the moderate to high sugar content of 'Smooth Cayenne', its acidity is often considered excessive among subtropical consumers, especially when harvested during winter. The juice is not high quality, because of poor colour, high sugar and turbidity. The production cycle for 'Smooth Cayenne' is longer than most other cultivars and this is exacerbated in cool climates. 'Smooth Cayenne' is sensitive to many known pests (fruit borers, mites, symphillids, nematodes) and diseases (mealybug wilt, fusariosis, fruitlet core rot, butt rot), and to internal browning (Rohrbach and Schmitt, 1994; Bello *et al.*, 1997a,b). However, it is considered resistant to fruit collapse, caused by *Erwinia chrysanthemi* (Lim and Lowings, 1979).

Fig. 4.1. *A. comosus* var. *comosus* cv. Smooth Cayenne, clone F180. Photo G. Sanewski.

'Singapore Canning' is a pre-Columbian variety cultivated in South Asia, particularly Malaysia. Other names and somaclonal variants are 'Singapore Spanish', 'Ruby', 'Red Pine', 'Nanas Merah', 'Nangka', 'Gandul', 'Betek', and 'Masmerah' (Wee, 1972). 'Selangor Green' ('Green Pine', 'Green Spanish', 'Nanas Hijau' and 'Selassie') is a mutation without anthocyanins. Clones in Thailand include 'Intra Chit Dang' (with anthocyanin) and 'Intra Chit Khaw' (without anthocyanin). The plant is medium-sized with 35–70 leaves usually with abundant anthocyanin. Leaf margin varies from smooth to part spiny to complete spiny depending on the clone. The bracts of the peduncle and inflorescence are deep bright red. The plant produces 2–6 slips and several suckers. Fruits are small (1 kg), cylindrical, dark purple turning copper/orange (Fig. 4.2(a)). The anthocyanless mutant exhibits green leaves and bracts, pale flowers, and green fruits turning yellow. Sugar and acidity are low (10–12%) and the taste is poor, however, juice is of good colour and quality and the fruit is a popular canned product in Malaysia. Multiple crowns are frequent. It shows severe chlorosis when exposed to high soil manganese. It is also sensitive to bacterial fruit collapse (*Dickeya chrysanthemi*) in Malaysia (Lim and Lowings, 1979), and to nematodes (Collins and Hagan, 1932).

Fig. 4.2. (a) *A. comosus* var. *comosus* cv. Singapore Canning clone Ruby. Photo G. Sanewski. (b) *A. comosus* var. *comosus* cv. Queen. Photo D. Junghans.

The cultivar Queen is widely distributed, grown principally for the fresh market but also a small volume for processing. 'Queen' is a pre-Columbian cultivar and the oldest one named, having been mentioned by de Rochefort in 1658 (Miller, 1954). There are many clonal variants or synonyms including 'Victoria', 'Ripley Queen', 'Mauritius', 'Formosa', 'Malacca', 'Red Ceylon', 'Buitenzorg', 'Alexandra, 'Moris' (Malaysia); 'McGregor' (Australia); 'Comte de Paris' (China); 'Palembang' and 'Mahkota Bogor' (Indonesia); and 'Phuket', 'Sawee', 'Phulae', 'Trad Sithong', and 'Phetburi' (Thailand) (Rattanathawornkiti *et al.*, 2016; Chan and Lee, 1995; Li *et al.*, 2010). The plant is small (60–80 cm), with short, spiny, silvery leaves with some anthocyanin, a small, trapezoidal fruit (0.5–1 kg), with a full yellow skin and small prominent eyes (Fig. 4.2(b)). The golden yellow pulp is dry, crisp, porous, low acid and sweet (14–18%), with a good flavour and long shelf life. The production of slips and shoots varies widely between clones. 'Queen' is a robust cultivar, showing more general tolerance to stress, pests and diseases than 'Smooth Cayenne'. On the other hand, it is susceptible to *Phytophthora* and fruit collapse, and highly susceptible to internal browning, fruitlet core rot and butt rot (Lim, 1971; Winks *et al.*, 1985; Swarts, 1990; Lutchmeah, 1992; Sanewski *et al.*, 2016).

'Española Roja' or 'Red Spanish' is a pre-Columbian cultivar widely cultivated in Venezuela, the Caribbean basin and the Canary Islands under names such as 'Black Spanish', 'Key Largo', Havannah', 'Habana', 'Cubana', 'Cowboy', 'Bull Head', 'Cumanesa', and 'Native Philippine Red'. The plant is vigorous, usually with spiny or part-spiny leaf margins and abundant anthocyanin. Smooth clones have been selected. Floral bracts are an intense bright red colour, hence the name. The plant regularly produces 1–3 slips and several suckers. The fruit is medium (1.2–2 kg), dark orange, with a barrel shape (Fig. 4.3). The flesh is firm, white to pale yellow, aromatic, and sweet, with moderate sugar content (12%) and some acidity. 'Española Roja' is vigorous and tolerant to high temperature, drought, internal browning, butt rot, wilt and *Phytophthora* (Py *et al.*, 1987; Rohrbach and Schmitt, 1994), but not to high soil manganese

Fig. 4.3. *A. comosus* var. *comosus* cv. Española Roja clone 'sin picos'. Photo G. Sanewski.

or nematodes (Ayala *et al.*, 1969). It is highly susceptible to the South American lepidopter *Strymon basilides* (Geyer).

'Pérola' is the main cultivar in the Brazilian fresh market, with more than 85% of the commercial area. It appears well adapted to a wide range of soils and climate in that country. It has long been cultivated as 'Sugarloaf' ('Pain de Sucre') in Western African countries, 'Selangor Sweet' in Malaysia, 'Nanas Yankee' in Thailand and 'Eleuthera' in some other regions. It is prone to natural flowering and so is less planted than 'Smooth Cayenne' in the Southern Brazil region that has a colder climate. 'Jupi' is a Brazilian clone with a more cylindrical fruit shape. The plant is medium-sized and vigorous, with erect, spiny leaves containing anthocyanin. It has few plant or ground suckers. Long and erect basal slips (5–15) surround the fruit and it has a long peduncle (27–35 cm). The fruit is small to medium (0.9–1.6 kg), very conical or bottle shape, and green with a little yellow in the centre of the mature eyes, and a long, erect crown (Fig. 4.4(a)). The flesh is soft, white to pale yellow and juicy with a good aroma. Its sugar (13–16%) and ascorbic acid contents are moderately high, but with low to medium titratable acidity (0.4–0.6%). As with most conical fruits, it shows a strong gradient of maturation from base to top. Despite its good organoleptic characteristics, the fruit shape, green peel at maturity and a very high level of translucency, limit its suitability for export. 'Pérola' is a very robust cultivar, showing some moderate resistance to *Phytophthora* (Collins, 1960; Sanewski *et al.*, 2016) and

Fig. 4.4. (a) *A. comosus* var. *comosus* cv. Pérola. Photo G. Sanewski. (b) cv. Perolera. Photo D. Junghans.

tolerance to drought, mealybug wilt and nematodes (Collins and Hagan, 1932; Sipes and Schmitt, 1994; Sarah *et al.*, 1997). However, it is in decline because of its high susceptibility to fusariosis that severely affects the Brazilian production.

'Perolera', also named 'Lebrija', 'Motilona', 'Capachera', or 'Tachirense' is an important pre-Columbian cultivar of the north-eastern Andes of Colombia and Venezuela, where it is well adapted to high altitude (800–1400 m). The leaves are green with some anthocyanin and a piping margin. The suckers are numerous (often >6). The fruit can be large (1.5–3 kg), yellow to orange, with an irregular cylindrical shape (Fig. 4.4(b)). It is borne on a long peduncle, which makes it susceptible to fruit lodging and sunburn, particularly in hot lowlands. The fruit is considered resistant to transportation, perhaps by the thicker peel. The flesh is pale yellow to yellow and firm with moderate sweetness (12%). Numerous crownlets protrude from the base of the crown and the upper eyes. Slips are also numerous (4–11). 'Perolera' is susceptible to fruitlet core rot and to the fly *Melanoloma canopilosum* Hendel. It, and its progeny, have been used in the Brazilian breeding programme because of its resistance to fusariosis.

The cultivar Manzana, also called 'Bumanguesa', is cultivated in the same zone as 'Perolera', as well as in the Western Andes of Colombia (Valle del Cauca). It is said to be a somaclonal variant of 'Perolera', from which it differs in the more regular globular to cylindrical shape and larger flat fruitlets. The fruit is dark to bright red, especially when produced in tropical highlands (Fig. 4.5). It has numerous crownlets protruding from the base of the crown. It is less tolerant to handling and transport than 'Perolera'. It is susceptible to the nematode *Pratylenchus neglectus* but tolerant to *Meloidogyne incognita* Kofoid & White (Redondo-Echeverri and Varón, 1992). It is very susceptible to *Phytophthora cinnamomi* (Sanewski *et al.*, 2016).

Fig. 4.5. Main photo shows an immature *A. comosus* var. *comosus* cv. Manzana and inset shows mature, highly coloured fruit. Photos D. Junghans.

The cultivar MD-2, Golden Ripe, or Gold Extra Sweet resulted from a cross between '58-1184' (♀) and '59-443' made in 1970, by Pineapple Research Institute of Hawaii (PRI). Originally selected in 1973 as '73-114', the parentage of 'MD-2' includes 'Smooth Cayenne', 'Smooth Guatemalan', 'Ruby', 'Queen' and 'Pérola' (Williams and Fleisch, 1993). 'MD-2' was first marketed on a large scale by Del Monte Inc. in 1996 and within ten years revolutionized the pineapple fresh fruit market (Bartholomew, 2009). As indicative of its huge success, 'MD-2' has become the market standard for the fresh market. Both the cultivar and its supply chain management were strategic to achieve such success (Vagneron *et al.*, 2009). 'MD-2' represents more than 80% of the fresh pineapples traded worldwide. It produces a medium to large (1.3–2.5 kg) cylindrical, square-shouldered fruit, with large flat fruitlets, and glossy, orange-yellow skin (Fig. 4.6(a)). The yellow pulp is sweet, compact, and slightly fibrous. It is marginally high in sugar (15–17%) and very high in ascorbic acid but lower in acidity than 'Smooth Cayenne'. The leaves have a spiny tip and are solid green with no anthocyanin. 'MD-2' can be as productive as 'Smooth Cayenne'. It is highly resistant to internal browning, but highly susceptible to bacterial heart rot (*Dickeya chrysanthemi*), *Phytophthora cinnamomi* (Sanewski *et al.*, 2016) and natural initiation of flowering (NI).

Fig. 4.6. (a) *A. comosus* var. *comosus* cv. MD-2. (b) Cv. CO-2. Photos G. Sanewski.

The cultivar CO-2 (73-50 in Australia, Maspine in Malaysia, 'Hawaiian Gold' in Hawaii) is a sibling to 'MD-2' arising from the same PRI programme (Oda and Williams, 1994). 'CO-2' is lower in acidity than 'MD-2' and hence more suited to subtropical climates. It produces a medium to large fruit with noticeably large, flat fruitlets (Fig. 4.6(b)). The flesh is yellow with intense yellow ovarian tissue. It is sweet, low-acid, high in ascorbic acid and exhibits a distinct floral aroma. The plant is moderately vigorous with narrow, piping leaves containing some anthocyanin (Taniguchi *et al.*, 2008). It has good resistance to internal browning and bacterial heart rot but is susceptible to *Phytophthora cinnamomi* and NI. 'CO-2' is grown for the fresh market in Australia, Malaysia, South Africa and Hawaii.

Minor Cultivars

Despite the existing diversity in pineapple genotypes, few have been commercialized on a large scale outside their region of origin. However, many minor cultivars that are grown on a small scale for local markets might be worth wider utilization. Table 4.1 lists 38 minor cultivars. A few of these may be synonyms for better-known cultivars. Figures 4.7 to 4.10 show some minor cultivars.

Other Species and Botanical Varieties

The wild pineapple, *A. comosus* var. *microstachys* (Fig. 4.11) and *A. comosus* var. *parguazensis* (Fig. 4.12(b)) constitute the main reservoir of genetic diversity. The former is generally considered the most drought-resistant pineapple. The small fruit is borne on a very long, thin peduncle. The plant is usually highly vegetative, producing many plant and ground suckers with spiny leaves. Its fruit is sometimes seedy, with white to yellow, hard and fibrous pulp and is usually not palatable. Crownlets can be common. Sugar and acid contents are variable with good to poor flavour. The core is narrow. *A. comosus* var. *microstachys* is a potential source of resistance or tolerance to wilt, nematodes (*Meloidogyne javanica*, *Rotylenchulus reniformis*), and crown and root rot (Collins and Hagan, 1932; Hagan and Collins, 1932; Collins, 1960; Ayala, 1961; Ayala *et al.*, 1969; Sipes and

Table 4.1. Minor pre-Columbian cultivars.

Cultivar	Origin	Plant characters	Fruit characters	Other characters	References
'Gigante de Tarauacá' ('Arroba Tarauacá')	Grown in Tarauacá city, Acre state, Brazil	Spreading, green with purple toned leaves, spines, long peduncle, 1–2 suckers	Conical shape, 4.5–15.0 kg, yellow skin, yellow juicy flesh, SSC 11%, AsA 22 mg/100 g	Triploid cultivar	Aldrigue and Lima, 1988 Ritzinger, 1992 Pereira, 2006 Scherer *et al.*, 2015
'Rio Branco' ('RBR-1')	Grown in Acre state, Brazil	Semi-erect habit; smooth, green leaves, long peduncle; one sucker, 8 slips	Cylindrical shape, 1.7 kg, yellow flesh, SSC 14%, low acidity	Susceptible to fusariosis	Ritzinger, 1992, 2000b
'Cabeça de Onça' ('RBR-2')	Grown in Acre state, Brazil	Dark green, piping leaves with red central band, long peduncle, no suckers, many slips	Small erect crown, short cylindrical shape, 3 kg, yellow flesh, SSC 12–13%, very low acidity	Resistant to fusariosis	Ritzinger, 1992, 2000a
'Cabeça de Onça' (Upper Amazon)	Grown in Upper Amazon (Amazonas state of Brazil)	Dark green spiny leaves, short strong peduncle, many slips	Small erect crown, short cylindrical shape, narrow core, sweet yellow flesh, low acidity	Different to 'RBR-2' above	Personal observation
'Quinari' ('SNG-2')	Grown in Acre state, Brazil	Erect habit, green spiny leaves, long peduncle, no suckers but 12 slips	Cylindrical shape, 1.8 kg, pale yellow flesh, SSC 13%, medium acidity	Susceptible to fusariosis	Ritzinger, 1992
'RBR-3'	Grown in Acre state, Brazil	Light green spiny leaves, 1–2 suckers	Small cyclindrical fruit, *c.*1 kg, yellow skin, white flesh, acid	Susceptible to fruitlet core rot	Personal observation
'Roxo de Tefé'	From Upper Amazon, Amazonas state, Brazil	Semi-upright habit, with dark purple, spiny leaves	Small cylindrical fruit, dark orange to red skin, pointed eyes, pale yellow flesh, low acidity	A single dominant gene determines the red skin colour. Tolerant to fusariosis under natural conditions	Cabral *et al.*, 1997
'Primavera'	Collected in Brazilian Amazon	Green, piping leaves, medium length peduncle, early suckering (1–2) and 7–10 slips	Cylindrical, 1.3 kg, yellow skin and white flesh, 13% SSC, medium acidity	Does not tolerate water deficit well. Resistant to fusariosis	Cabral, 1999

'Turiaçu'	Grown in Maranhão and Amazonas states, in Brazil	Erect plant, spiny, dark green leaves with reddish tones, reddish spines, long peduncle, 11 slips, few or no suckers	Cylindrical to conical, weighs 1.6 kg, small to medium crown, full yellow skin, small prominent eyes, yellow flesh, SSC 14–18%, TA 0.38%	Worth twice the price of 'Pérola' due to superior eating quality. Susceptible to heart rot and mealybug wilt. Tolerant to fusariosis under natural conditions. Resistant to *Meloidogyne* sp.	Araújo *et al.*, 2012 Garcia *et al.*, 2013 Silva *et al.*, 2012
'Tchahí'	Upper Amazon, Amazonas state of Brazil	Dark green piping leaves, with anthocyanins, many slips, 1–2 suckers	Large, ovoid to cylindrical fruit, with dark skin turning reddish orange, narrow core, yellow, fibrous and juicy flesh, sugar low to medium, low acidity	Susceptible to fruitlet core rot	Personal observation
'Abacaxi Roxo'	Upper Amazon, Amazonas state of Brazil	Dark green spiny leaves, slips, very erect habit	Long ovoid to conical fruit on a long and strong peduncle, dark skin turning reddish orange, flesh white, low acid, medium sugar, sweet		Personal observation
'Zolhudo'	Upper Amazon, Amazonas state of Brazil	Dark spiny leaves, many slips	Medium to large fruit with large prominent eyes, on a strong peduncle, orange skin, yellow juicy flesh, low in sugar and acid		Personal observation
'Maipouri rouge' (Red Maipuri)	French Guiana	Dark red spiny leaves, slips	Long and conical fruit		Personal observation
'Nana Woli' ('Oli', 'Wori', 'Vori', sometimes called 'Maipuri')	French Guiana	Vigorous plants, wide and spiny green leaves, reddish in the centre, no sucker, many slips	Large fruit with a bottleneck, long and strong peduncle, large flat eyes, yellow skin, yellow flesh, juicy, often translucent, low sugar and acid, flat taste		Personal observation
'Nana Iakín'	French Guiana	Spiny tip to half-spiny leaves, suckers and many slips	Medium, ovoid to cylindrical fruit on medium and strong peduncle, yellow orange skin, pale yellow flesh, low in acid		Personal observation

continued

Table 4.1. *continued.*

Cultivar	Origin	Plant characters	Fruit characters	Other characters	References
'Valera Amarilla'	Grown in Lara and Trujillo states, Venezuela	Medium-sized, large, spiny leaves, long peduncle (prone to sunburn)	Barrel shape, 1.5–3.0 kg, large crown, yellow skin, 72–88 deep eyes, pale yellow sweet flesh, juicy	'Red Spanish' strain according to Morton (1987)	Montilla *et al.*, 1997
'Branco' ('Green Maipouri' in French Guiana)	The most common in French Guiana, also present in Brazilian state of Amapá	Most commonly spiny tip, but also half- or completely spiny, green leaves, without anthocyanins, many slips	Medium to large (>3 kg), ovoid to conical fruit, with a bottle neck, green turning yellow shell, strong scape; wide core, fibrous and tasty flesh	Susceptible to fruitlet core rot	Personal observation
'Valera Roja'	Grown in states of Trujillo, Lara and Merida, Venezuela	Small to medium-sized, narrow, reddish to purple spiny leaves	Cylindrical fruit, 0.6–1.0 kg, red skin, pale yellow flesh, 90–100 flat but deep fruitlets		Leal *et al.*, 2010 Morton, 1987
'Monte Oscuro' ('Pilon')	Grown in Monagas state and in south of Sucre state, Venezuela	Large, with green, wide, spiny leaves	Large with barrel shape, >3 kg, 160–180 fruitlets, strong peduncle, orange-yellow skin, deep yellow flesh, high fibre and SSC (>18%)	Grown in Moriche palm wetlands, triploid	Montilla *et al.*, 1997 Leal *et al.*, 2010
'Cacho de Venado'	Grown to a small extent in Monagas and Sucre	Medium plants with green leaves and small pink spines	Conical fruits, 0.8–1 kg, small crown, yellow skin and flesh, 190–210 deep and prominent eyes		Leal *et al.*, 2010
'Injerta'	Grown to a small extent in Monagas and Sucre	Medium plants with dark green leaves, with reddish base, and soft spines	Conical fruits, 4–5 kg, reddish orange skin, pale yellow flesh, narrow core, 140–150 pronounced eyes		Leal *et al.*, 2010
'Panare'	Grown in small areas of Bolívar state, Venezuela	Medium-sized with long, spiny, green leaves with reddish tinge	Short peduncle, fruit oblong with bottleneck, 0.45–0.7 kg, orange skin, bright yellow flesh, 95–100 prominent fruitlets which are very small at crown base, low fibre content, small core	Related to Panare people	Leal *et al.*, 2010 Morton, 1987

'Brecheche'	Grown to a limited extent in Bolivar state, Venezuela	Small plants with narrow, smooth (not piping), olive green leaves	Cylindrical, 0.7–1.0 kg, small crown, 100–105 flat deep eyes, yellow skin and flesh, little fibre, small core, fragrant, juicy		Leal *et al.*, 2010 Morton 1987
'Morada'	Grown in state of Monagas, Venezuela	Large erect plants, long narrow purplish-reddish leaves with spaced spines	Barrel shape fruit, 0.9–1.5 kg, 90–100 flat fruitlets, purple skin, white flesh	Possible mutation from 'Española Roja'; susceptible to internal browning	Leal *et al.*, 2010
'Maipure'	Grown in Amazonas state, Venezuela	Large plants, long green piping leaves with reddish tinge	Cylindrical, 2.5–3.0 kg, pale yellow flesh, 170–190 flat eyes, bracts longer than eyes, yellow skin	Slash and burn cultivation, related to Maipure people	Leal *et al.*, 2010 Morton 1987
'Mayanez'	Grown in Meta Department, Colombian Orinoquia	Long, spiny, heavy leaves with distal half reddish-green	Barrel shape, 1.3 kg, long fruitlet bracts, yellow flesh, low acid, high SSC	Commercial cultivar with good organoleptic qualities	Delgado-Huertas and Arango-Wiesner, 2015
'Mitú'	Grown in Vaupés Department, Colombian Amazon	Long, wide, piping leaves with dark green margin and purple centre, long, thin peduncle causing fruit lodging	Long shape, 1.3 kg, SSC 14%, TA 0.6%		Delgado-Huertas and Arango-Wiesner, 2015
'Clavo' ('Clavo Rojo')	Grown in Huila Department, Colombia	Small plant with spiny or half-spiny leaves, 8–10 slips, 2–3 suckers	Conical, 1.0 kg, dense flesh, deep red skin, high SSC	Very resistant to transport, cultivated at high altitude (1000 to 1450 m)	Delgado-Huertas and Arango-Wiesner, 2015
'Piamba' ('Castilla', 'Blanca Castilla')	Grown in Tolima Department, Colombia	Vigorous plant with smooth, erect leaves	Similar to 'Perolera', but with white flesh, high SSC and acidity		Salazar, 1944
'De Agua' ('Hartona', 'Cambray')	Valle del Cauca, close to the Pacific Coast of Colombia	Vigorous plant with wide spiny leaves, slips	Conical shape, up to 4.5 kg, pale yellow flesh, low SSC and acidity, flat taste, high juice and low fibre content		Rojas and Aguilera, 1999
'Monte Lirio', ('Cambray', 'Milagreña', 'Smooth Guatemala')	Grown from Mexico to Peru	Medium-sized, reddish-green, piping leaves. 5–8 slips	Conical to globose fruit, with big eyes, 2.5 kg, yellow-green skin, white juicy pulp, SSC 13%, TA 1.0%, 27 mg/100 g AsA	Susceptible to internal browning. Cited in 1789 by the presbyter Juan de Velasco. Popular in Nicaragua	Arriola *et al.*, 1976 Velasco, 1927

continued

Table 4.1. *continued.*

Cultivar	Origin	Plant characters	Fruit characters	Other characters	References
'Morada' ('India', 'Crespa')	Grown in Florencia, Caquetá Department, Andean foothill of Colombia	Green leaves with red to purple from the middle part to the tip; hard red spines, many slips	Globular to elliptic, 1.5–2.0 kg, white to pale yellow flesh, SSC 15%, TA 0.46%		Pulido, 2000 Morales *et al.*, 2001 Hernández *et al.*, 2004
'Montserrat'	Main cultivar in Guyana	Erect habit, long spiny leaves with a brownish tint along leaf centre, 7–15 slips near base of fruit, 2–3 suckers	Conical to cylindrical, yellow to orange, 2.0 kg, SSC 17%, TA 6.6 meq/100 ml, AsA 3.4 meq/100 ml	Very soft flesh	Rouffiange, 1993
'Black Antigua'	West Indies	Spiny	Small to medium fruit, dark green skin turning copper read at its base, firm golden yellow pulp.		Personal observation
'Cabezona' ('Puerto Rico', 'Piña de Agua')	Puerto Rico	Tall and vigorous plant with spiny leaves	Conical shape, yellow to orange skin, >3 kg, white fibrous pulp, little commercial value due to poor consistency which makes postharvest handling and transportation difficult	Triploid cultivar. It was cultivated in the Caribbean Basin	Collins, 1933b Isidrón *et al.*, 2003 Leal *et al.*, 2010 Garcia *et al.*, 2006
'Samba de Chanchamayo'	Grown in 'Selva Central' region, Peru. Main Peruvian cultivar	Rustic and medium-sized, long, smooth leaves with abundant anthocyanins, long, peduncle prone to lodging	Dark red cylindrical fruits, 1.5 kg, yellowish-white flesh, SSC 11%, low acidity	Numerous crownlets from base of crown, tolerant to *Strymon basilides*	Bello *et al.*, 1997b Cáceres, 2009
'Roja Trujillana'	Grown in Department of La Libertad, Peru	Medium-sized, with smooth dark green leaves with some red, short peduncle, many slips, few ground suckers	Medium-sized cylindrical, black skin turning red, flat eyes, white-cream flesh, SSC (12%) and slightly acid	Fruits have good resistance to transport and long shelf life, resistant to *Strymon basilides*	Bello *et al.*, 1997b Pariona, 2008
'Pucallpa'	Grown in Peru and Bolivia	Tall plant (up to 1.5 m high), with smooth reddish green leaves	Large sized conical fruit with reddish-orange skin, deep eyes and pale yellow flesh, 13% SSC	Introduced to Bolivia	

AsA, ascorbic acid; SSC, soluble solids contents; TA, titratable acid.

Fig. 4.7. (a) *A. comosus* var. *comosus* cv. Cabeça de Onça FRF1010. (b) Cv. Piña de Agua. (Photos G. Coppens d'Eeckenbrugge.

Fig. 4.8. (a) Mature fruit of *A. comosus* var. *comosus* cv. Roxo de Tefé. (b) Cv. Turiaçu. Photos D. Junghans.

Schmitt, 1994; Sanewski *et al.*, 2016). *A. comosus* var. *parguazensis* is less variable. The plant appears less vigorous with spiny leaves. It has a globular fruit with relatively wide flat eyes, and a white and fibrous pulp. Resistance to fusariosis is variable in *A. comosus* var. *microstachys* and *A. comosus* var. *parguazensis* (Matos *et al.*, 1991).

Fig. 4.9. (a) *A. comosus* var. *comosus* cv. Quinari. (b) Cv. Primavera. Photos D. Junghans.

Fig. 4.10. (a) *A. comosus* var. *comosus* cv. Gigante de Tarauacá. Photo D. Junghans. (b) Cv. Tapiracanga. Photo G. Sanewski.

Ananas comosus var. *bracteatus* (Fig. 4.12(a)), first cultivated for the fruit and as a living fence, is now widely used as a tropical garden ornamental. Its variability is very limited. It grows well on poor and dry soil and it is somewhat adapted to cool conditions and altitude. The plant is spiny (antrorse) and highly vigorous. The syncarp bracts are red at anthesis,

Fig. 4.11. (a) *A. comosus* var. *microstachys* accession FRF223. (b) Cv. Nanus. (c) Wild *A. comosus* var. *microstachys* in Rondônia, Brazil. Photos (a) and (b) G. Sanewski. Photo (c) Everton Hilo Souza.

producing a spectacular inflorescence. The barely edible fruit is small to medium (0.5–1.0 kg), with white fibrous flesh, a small core, and an average sugar and acid content. It is resistant to heart rot, root rot and fusariosis (Matos *et al.*, 1991; Sanewski *et al.*, 2016). Studies on nematode resistance produced contradictory results (Collins, 1960; Sipes and Schmitt, 1994). Crosses with *A. comosus* var. *comosus* yield hybrids with fruits of good size and weight and a wide variation in flavour (Collins, 1960).

Ananas friztmuelleri Camargo is now considered a form of *A. comosus* var. *bracteatus* (Coppens

Fig. 4.12. (a) *A. comosus* var. *bracteatus* accession FRF414: main photo shows mature fruit; inset shows flowering syncarp. (b) *A. comosus* var. *parguazensis* accession FRF870. Photos G. Sanewski.

d'Eeckenbrugge *et al.*, 2009), genetically closer to *A. macrodontes*, but its exact relationship is unclear. It's very long and erect leaves have regularly spaced spines, retrorse at the leaf base and the syncarp bracts are pale pink.

A. comosus var. *erectifolius* is quite similar to the wild *A. comosus* var. *microstachys*, except for its smooth leaves (Collins, 1960) and its erect habit, which is much less common in the latter. The curagua plant is smaller, with abundant and early shoots, frequent crownlets at the base of the main crown, numerous erect, very fibrous leaves and a small, very fibrous, inedible fruit borne on a long and slender peduncle (Fig. 4.14). In some cultivars, the fruit appears to be rare. Curagua has been cultivated for its long and strong fibres (Leal and Amaya, 1991). A dark red/brown cultivar (Selvagem 6) has also been exploited as an ornamental for the European cut-flower market (Fig. 4.14). The anthocyanin trait is controlled by a single dominant gene (Cabral *et al.*, 1997). *A. comosus* var. *erectifolius* is resistant to heart rot and root rot (Collins, 1960; Sanewski *et al.*, 2016), and relatively tolerant to nematodes (Sipes and Schmitt, 1994).

Ananas macrodontes (Fig. 4.15), a wild tetraploid species, has been mainly used as a source

Fig. 4.13. Foliage of *A. comosus* var. *bracteatus* form *fritzmuelleri* accession BRA–011983. Photo G. Sanewski.

of fibre. *A. macrodontes* differs morphologically from other *Ananas* in that it does not produce a crown or substantially fleshy fruit but does produce underground stolons which can be 50 cm or so long. As with other pineapples, it is resistant to drought but it grows under humid forest. Furthermore, it tolerates temporary flooding. It is resistant to heart rot, root rot, wilt and fusariosis (Collins, 1960; Matos *et al.*, 1991). When selfed, *A. macrodontes* shows little variation, indicating that it is relatively homozygous (Collins,

Fig. 4.14. Mature fruit of *A. comosus* var. *erectifolius* accession 'Selvagem 6'. Photo G. Sanewski.

Fig. 4.15. *A. macrodontes* accession I.26–803. Main photo shows long stolon of immature plant and insets show flowering syncarp and leaf margin spines. Photos G. Sanewski.

1960). Hybridization with *A. comosus* var. *comosus* produces 5–10% fertile seeds, most of which are tetraploid. These tetraploid progenies are self-fertile.

Genetics and Breeding System

The breeding system of *A. comosus* combines efficient vegetative reproduction with functional allogamous sexual reproduction. Pineapple allogamy, as well as its consequent heterozygosity, is related to the existence of self-incompatibility. A high chromosome number also favours genetic recombination. The species *A. macrodontes* is self-compatible.

Cytogenetics

Chromosome number

All the botanical varieties of *A. comosus* have a diploid number of 50 minute, almost spherical, chromosomes with a length of 0.5–1.7 µm. Among three pineapple accessions, Sharma and Ghosh (1971) identified four chromosome types and observed a slight variation in their distribution and in chromatin content. They also observed secondary constrictions in 2–5 pairs of chromosomes. Arumuganathan and Earle (1991) reported values of 1.09 and 0.92 pg (corresponding to *c.*526 and 444 Mbp) for the DNA content of 2C nuclei of *A. comosus* var. *comosus* and *A. comosus* var. *bracteatus* respectively. Triploids (e.g. cultivars Cabezona, Caicara, Monte Oscuro), tetraploids, and even heteroploids (60 chromosomes) may also occur (Collins and Kerns, 1931; Canpinpin and Rotor, 1937; Collins, 1960; Antoni, 1983; Lin *et al.*, 1987; Dujardin, 1991). Collins and Kerns (1931) reported six cases of triploidy out of 8000 plants resulting from a cross between 'Smooth Cayenne' and a clone of *A. comosus* var.*microstachys*. Collins (1933b) demonstrated that these triploids had arisen from unreduced female egg cells fertilized by normal haploid pollen.

Tetraploids could be obtained from unreduced male and female gametes, although their occurrence is rarer than triploids. Tetraploids produce about 90% viable pollen, of a larger size than pollen cells of diploid plants. By crossing them with diploids, hybrid tetraploid seedlings were obtained, which grew faster and larger than diploids. The species *A. macrodontes* is a natural tetraploid with 100 chromosomes (Collins, 1960; Dujardin, 1991).

Ploidy level and morphology

Collins (1933b), Kerns and Collins (1947) and Collins (1960) compared triploid and diploid hybrids and diploid and autotetraploid plants of 'Smooth Cayenne'. The polyploids were taller, with wider leaves, placed at a greater internodal length, so the total leaf area does not differ much. The cross-section of the tetraploid leaves is more closed. Their larger fibres increase leaf rigidity. The size of cells, trichomes and stomata also increases with the ploidy level, while the stomata density is reduced. Water content is higher in polyploids. Triploid hybrids and tetraploid 'Smooth Cayenne' have longer vegetative growth periods, mature more slowly and have a lower fruit sugar content. Tetraploid fruits are often smaller, with fewer but larger fruitlets.

Micro- and megagametogenesis and fertility

Meiosis is primarily regular in diploid and tetraploid *A. comosus*, as well as in *A. macrodontes*, with only rare irregularities such as lagging chromosomes and some abnormal tetrads (Collins and Kerns, 1931; Canpinpin and Rotor, 1937; Collins, 1960). However, some irregular gametes are produced. The proportion of giant pollen cells may reach 6.5% and that of giant ovules 1%. Bhowmik (1977) observed large variations in size and fertility of pollen grains in 'Kew' ('Smooth Cayenne') and 'Queen' (35–81 μm and 36–68 μm respectively) and correlated them with the frequencies of univalents and secondary associations of up to three or seven bivalents in these cultivars. Pollen size and fertility also varies between cultivars and clones (Collins, 1960; Ramírez, 1966; Wee and Rao, 1979; Nayar *et al.*, 1981; Antoni, 1983; Coppens d'Eeckenbrugge *et al.*, 1993). Triploid genotypes are sterile because of irregular meiosis. Tetraploids produce 90% of good pollen grains that are uniform but larger than pollen of diploids (Collins, 1960).

The number of ovules varies widely in *A. comosus*. While *A. comosus* var. *bracteatus* may have 40–70 ovules per flower, this number generally ranges from 14–30 in the wild botanical varieties and in *A. comosus* var. *erectifolius*, and from 16–71 in *A. comosus* var. *comosus* (Coppens d'Eeckenbrugge *et al.*, 1993). There is also variation in all the varieties in the relative frequencies of anatropous and orthotropous ovules. The embryo sac develops according to the *Polygonum* type. Shrinking embryo sacs have been commonly observed at the eight-nucleate stage in 'Masmerah' (Rao and Wee, 1979) in the six main cultivars of *A. comosus* var. *comosus*, in *A. comosus* var. *microstachys*, *A. comosus* var. *erectifolius*, and *A. comosus* var. *bracteatus* (F. Van Miegroet, unpublished results). The proportion of ovules with normal embryo sacs is correlated with the proportion of stained pollen and with fertility (F. Van Miegroet and Coppens d'Eeckenbrugge, unpublished results).

Fertility rate (i.e. the percentage of ovules producing a seed) is generally lower in *A. comosus* var. *comosus*, than in the other botanical varieties. In the latter, fertility ranged from 6% (0.85 seeds/flower) in *A. comosus* var. *microstachys* to 35% (18.4 seeds/flower) in *A. comosus* var. *bracteatus*. Within *A. comosus* var. *comosus*, the cultivars with piping leaves often exhibit a higher fertility (4–11%; 2–5 seeds/flower) than the common cultivars Smooth Cayenne, Española Roja, Singapore Spanish, Pérola, and Queen (less than 5%; 0–2 seeds/ flower). However, the highest fertility rate (29%; 9.45 seeds/ flower), which is comparable to that of wild types, was observed in a spiny landrace. These fertility rates were estimated under an open pollination situation and higher values can be expected under assisted pollination. Fertility is correlated with ovule fertility, pollen stainability and the amount of pollen produced per flower. It is not correlated with ovule number probably due to crowding or competition effects between fertilized ovules. Many genotypes of *A. comosus* var. *comosus* and *A. comosus* var. *bracteatus* can only develop small proportion of their numerous ovules into seeds (Coppens d'Eeckenbrugge *et al.*, 1993). Fertility may be affected by abnormal anthesis as in certain clones of 'Española Roja' and 'Singapore Spanish', where a significant proportion of flowers do not open, hampering pollination.

Self-incompatibility

In *A. comosus*, the self-incompatibility is brought about by the inhibition of pollen tube growth in the upper third of the style (Kerns, 1932; Majumder *et al.*, 1964). It is gametophytically controlled by a single locus with multiple alleles

(Brewbaker and Gorrez, 1967), and generally stronger in *A. comosus* var. *comosus* than in the other botanical varieties. Later work, however, suggested that more than one gene might be involved (D.D.F. Williams, Colorado, 2001, personal communication). However, many cultivars exhibit pseudo-self-compatibility, expressed in the variable production of self-seeds, although the resulting self-fertility is always lower than cross-fertility. For instance, pseudo-self-compatibility is common in 'Singapore Spanish', 'Perolera', 'Manzana', 'Primavera', 'Rondon', 'Cambray', 'Amarelo de Uaupés', 'Roxo de Tefé', 'Morada', 'Alto Turi', and in some clones of 'Queen' (DeWald *et al.*, 1992; Coppens d'Eeckenbrugge *et al.*, 1993, Muller, 1994; Coppens d'Eeckenbrugge *et al.*, 1997). 'MD-2' also is known for its production of a few seeds per fruit under monoclonal cultivation. In a sample of 84 clones from 51 cultivars and primitive landraces studied by Muller (1994), 55 appeared strongly self-incompatible, 21 pseudo-self-compatible, and eight exhibited a self-fertility comparable to their fertility, which could be due to self-compatibility or a very weak self-incompatibility. A dominant mutation suppresses the incompatibility reaction in the pollen of 'Smooth Cayenne' and results in production of selfed-seeds (Collins, 1960). The self-incompatibility of *A. comosus* var. *comosus* does not appear to be affected by tetraploidy (Kerns and Collins, 1947).

Pseudo-self-compatibility is more frequent in wild genotypes. In Muller's 1994 study, five out of 11 *A. comosus* var. *microstachys* clones and four out of seven *A. comosus* var. *parguazensis* clones exhibited pseudo-self-compatibility. One *A. comosus* var. *parguazensis* clone appeared self-compatible. In *A. comosus* var. *bracteatus*, three clones that were phenotypically identical appeared to be self-compatible. A variegated clone of the same species was self-incompatible, while another was pseudo-self-compatible. *A. macrodontes* is highly self-fertile and produces uniform progenies, suggesting that this species is autogamous.

Hybridization between botanical varieties and between species

There is no difficulty in crossing cultivars of *A. comosus* var. *comosus* with cultivars of other botanical varieties. Instead, these crosses often show high fertility, as a result of higher gamete viability of the latter, and heterosis. When *A. comosus* is crossed with *A. macrodontes* (Fig. 4.16), a few fertile seeds are produced that usually yield vigorous tetraploids and sometimes self-sterile triploids. The tetraploid hybrids are highly self-fertile (Collins, 1960).

Breeding and Selection

The pineapple industry of the world is dominated by 'Smooth Cayenne' that is used both for fresh fruit and processing. This near-total reliance on 'Smooth Cayenne' has been so for more than a century. The 'Smooth Cayenne' monopoly is undoubtedly due to its high yield, adaptability and good characteristics for canning. Further, highly specialized systems of production and processing protocols have been

Fig. 4.16. Mature fruit of *A. macrodontes* accession I.26–803 (top left) and 'Smooth Cayenne' (top right) and their tetraploid progeny (bottom row). The inset shows selfed seed of a progeny. Photos G. Sanewski.

developed almost exclusively for this cultivar. These factors, together with high costs and attendant inconveniences in changing to a new cultivar, partly account for the difficulty in replacing ‘Smooth Cayenne’. ‘Smooth Cayenne’ is still a dominant cultivar grown throughout the world despite it probably being developed before the arrival of Christopher Columbus in the West Indies in 1492.

Pineapple is a clonally propagated, principally out-crossing, heterozygous species. Since its commercialization using large, monoclonal production systems, there have been many efforts to improve the genetics of pineapple using a range of techniques from basic to highly advanced.

Spontaneous mutations

Pineapple is known to produce many types of somaclonal mutations. The incidence of mutations is likely a function of the large number of plants propagated and an unstable genome (Collins and Kerns, 1938; Ming *et al.*, 2015).

At least 30 major mutations have been recorded in ‘Smooth Cayenne’ pineapple since the early 1920s. These include but are not restricted to white flowers (as opposed to purple), foliar floret proliferation, multiple sepals and bracts, enlarged fruit segments, protruding fruit segments, increased or decreased trichome density, seediness, multiple crowns, a lower harvest index, misshapen fruits, smaller diameter fruits, leaf spininess, reduced chlorophyll or anthocyanin levels, dwarfed habit, translucent fruits, collar-of-slips, increased or decreased slip number and fruits with an increased incidence of basal knobs (Anonymous, 1925; Kerns, 1928; Collins and Kerns, 1938; Groszmann, 1939; Anonymous, 1940; Dalldorf, 1975; Sanewski *et al.*, 1992). Other measurable clonal differences likely caused by mutations include sucker number, length of the fruit development period, and peduncle length and width (Sanewski, 1997). Only two mutations, non-translucent fruit and resistance to mealybug wilt, could be considered substantially progressive (Collins and Carter, 1954; Obrero, 1995). Mutations can also be induced with a range of techniques including tissue culture (Williams and Fleisch, 1993), exposure of *in vitro* leaf cultures to Co gamma rays (Cisneros *et al.*, 1998), and the treatment of seed and plants with chemical mutagens (Singh and Iyer, 1997).

Clonal Selection and Rogueing

Many commercial pineapple cultivars are improved through selecting desirable genetic variants. Clonal plantings form the basis of modern pineapple culture and many different clones are used throughout the world. All ‘Smooth Cayenne’ clones in Hawaii and Australia were developed through clonal selection (Groszmann, 1982; Py *et al.*, 1987). These are usually high-yielding types with few slips or basal knobs and one strong sucker. Several clones of the cultivar Queen have also been established in Australia, including Ripley Queen, Alexander and McGregor (Lewcock, 1940). In Malaysia, an improved cultivar of ‘Singapore Spanish’, called ‘Masmerah’, was obtained by clonal selection (Wee, 1974). Clonal selection has also been used in the Philippines to develop a new clone of ‘Smooth Cayenne’ less prone to the ‘green-ripe’ disorder (Obrero, 1995) and in India to develop the more vigorous ‘Queen’ clone PGM-1 (Table 4.2). Clonal selection aims to identify superior individual plants that are eventually multiplied as a new clone. This commences with a small number of putative superior plants.

Rogueing is a form of clonal maintenance where plants carrying regressive mutations are removed from fields to avoid replanting. This is performed at the grower operational level and does not produce superior clones per se.

Rogueing

Most mutations are regressive and if allowed to proliferate will lead to a gradual loss of productivity or efficiency. Rogueing can rapidly reduce the percentage of most easily identifiable off-types. Most major pineapple producers routinely remove inferior mutants to maintain productivity and efficiency.

Rogueing aims to minimize the number of undesirable variants in a field by not replanting material from those plants. Rogueing is by

Table 4.2. Cultivars from breeding programmes.

Cultivar	Origin	Plant characters	Fruit characters	Other characters	References
'Johor' or 'Nanas Johor'	A cross between 'Smooth Cayenne' and 'Singapore Spanish', by Malaysian Agricultural Research & Development Institute (MARDI)	High yield, wide adaptability	Good canning characteristics	Susceptible to marble disease	Chan and Lee, 1985 Chan, 2006
'Josapine'	Cross between 'Johor' ('Singapore Spanish'×'Smooth Cayenne') and 'Sarawak' ('Smooth Cayenne'). Released by MARDI 1996	Vigorous, 2–3 suckers, spiny tip leaves are lightly purple-tinged	Cylindrical, 1.1–1.3 kg, dark purple to orange-red skin, deep yellow flesh, strong aroma, SSC 17–22%, good storage-life and resistant to internal browning	For fresh market, ripens early, grown on annual cycle in Malaysia, susceptible to bacterial heart rot (*Dickyea* sp.)	Chan and Lee, 1996
'N36'	A cross between 'Gandul' ('Singapore Spanish') and 'Sarawak' by MARDI	Spiny	Cylindrical, 1.0–2.3 kg, yellow flesh, SSC 16–17%, TA 0.5–1.2%, AsA 26 mg/100 g	Mainly used for processing in Malaysia, resistant to internal browning	Chan (2010)
'Tainung 4' (or 'Sugar-apple Pineapple', 'Easy Peeler', 'Phetchaburi' in Thailand, 'IAC Gomo de Mel' in Brazil)	Cross between Yellow Mauritius ('Queen') × 'Smooth Cayenne', Taiwan	Small plant, spiny leaves	Cylindrical shape, 1.2 kg, bulging fruitlets, dense flesh, little fibre, crisp, fragrant, yellow colour, low juice content, SSC 17–20%, TA 0.43 %	The fruit can be eaten by cutting the pineapple in half or quarters, then pulling apart the fruitlets one by one, without having to peel	Zheng *et al.*, 2012 Lin *et al.*, 2015 Feng *et al.*, 2013 Usberti Filho *et al.*, 1999
'Tainung 5'	Cross between 'Yellow Mauritius' ('Queen') and 'Smooth Cayenne', Taiwan				

continued

Table 4.2. *continued.*

Cultivar	Origin	Plant characters	Fruit characters	Other characters	References
'Tainung 6' ('Apple pineapple')	Cross between 'Yellow Mauritius' ('Queen') and 'Smooth Cayenne', Taiwan	Spiny, green leaves with some red	Squat, ball shape, 1.3 kg, thin skin, yellow flesh with fine texture, crisp, juicy, SSC 13–15%, TA 0.34%, good aroma	Popular for fresh fruit market	Lin and Chang, 2000 Lu *et al.*, 2014 Feng *et al.*, 2013
'Tainung 11' (or 'Perfum')	Cross between Smooth Cayenne' and 'Queen', Taiwan	Erect habit, spiny tip leaves	Cylindrical, 1.0 kg, fruitlets slightly bulging, pale yellow flesh, fine texture, juicy, SSC 15% and TA 0.57%	Low yield, good resistance to transport, popular in the Japanese fresh market	Lin and Chang, 2000 Lu *et al.*, 2014 Lin *et al.*, 2015
'Tainung 13' ('Winter Honey')	Taiwanese cultivar	Spiny tip, green leaves with a purple-red streak down the centre	Slightly conical, 1.2 kg, fruitlets slightly prominent, deep yellow golden flesh, somewhat fibrous, SSC 16–20%, TA 0.27%	Strong pineapple fragrance	Lin and Chang, 2000 Lu *et al.*, 2014
'Tainung 16' ('Sweet Honey')	Cross between Smooth Cayenne' and 'Rough', ('Queen'), Taiwan. Released 1995	Spiny tip, leaf centre is light purple-red	Elongated cone or football shaped, 1.3 kg, fruitlets slightly prominent, pale yellow flesh, low fibrosity, SSC 18 %, TA 0.47%, good aroma		Lin and Chang, 2000 Lu *et al.*, 2014 Feng *et al.*, 2013 Lin *et al.*, 2015
'Tainung 17' ('Taiwan Diamond' or 'Golden Diamond')	Cross between Smooth Cayenne' and 'Rough' ('Queen'), Taiwan. Released 1998	Moderate vigour, spiny tip, leaves slightly rust coloured with green at the end	Cylindrical, 1.4 kg, thin skin, yellow to deep yellow, delicate flesh, large edible core, SSC 14%, TA 0.28, juicy	High yield, susceptible to NI, moderately susceptible to mealybug wilt, most popular cultivar in Taiwan, 80% of fresh pineapple market, main variety for export	Lin and Chang, 2000 Bartholomew *et al.*, 2010 Lu *et al.*, 2014 Feng *et al.*, 2013 Lin *et al.*, 2015
'Tainung 18' ('Golden Osmanthus')	Cross between Smooth Cayenne' and 'Queen', Taiwan. Released 2000	Small plant, spiny tip, leaves green with slight dark red in the centre	Conical, 1.0 kg, fruitlets slightly prominent, thin skin, yellow golden flesh, slight fibre, SSC 14–17%, TA 0.39–0.62%, AsA 7.99 mg/100 g		Lu *et al.*, 2014

'Tainung 19' ('Honey Treasure')	Cross between Smooth Cayenne' and 'Rough' ('Queen'). Taiwan. Released 2001	Small plant, spiny tip, deep green leaves	Cylindrical, 1.0–1.6 kg, many small fruitlets, thin skin, golden flesh, low fibrosity, SSC 17–19%, TA 0.46–0.58%, AsA: 16.93 mg/100 g		Feng *et al.*, 2013 Lu *et al.*, 2014 Lin *et al.*, 2015
'Tainung 20' ('Milky Pineapple')	Imported cultivar, possibly 'Monte Lirio'. Taiwan. Released 2004	Large, tall plant, spineless, deep green leaves	Rounded to cylindrical, 0.55–1.3 kg, many small fruitlets, thin skin, white flesh, low fibrosity, SSC 14%, TA 0.57, AsA: 6.15 mg/100 g. Very fragrant	Next popular in Taiwan after 'Tainung 17'	Lu *et al.*, 2014
'Tainung 21' ('Golden')	Second generation cross with 5/8 'Smooth Cayenne', 1/4 'Queen' and 1/8 'Singapore Spanish'. Taiwan. Released 2005	Short height, strong vigour, short, green leaves, few slips and plant suckers, medium number ground suckers	Cylindrical shape, 1.3 kg, yellow skin, soft pulp, close texture, juicy, yellow to intense yellow pulp, SSC 19%, moderate to high acidity	Free from the problems of stem-end splitting, pineapple core rupture, and cracking peel	Tang *et al.*, 2014
'Tainung 22' ('Honey Fragrance')	Cross between 'Smooth Cayenne' and 'Queen'. Taiwan. Released 2012		Cylindrical shape, 1.8 kg, yellow flesh, orange-red peel, medium texture, juicy, moderate acidity, SSC 18%		Tang *et al.*, 2015
'N67-10'	A vegetative selection from 'Hawaiian Smooth Cayenne', released by Okinawa Prefectural Agricultural Research Center (OPARC) 1985	Vigorous, few slips and suckers	Oblong shape, yellow skin, 1.4 kg, yellowish white flesh, SSC 14%, TA 0.7%	For processing, ripens late August in Okinawa	Ogata *et al.*, 2016
'Soft Touch' ('Okinawa No. 1')	Cross between 'Hawaiian Smooth Cayenne' and 'I-43-880 (from Brazil)'. Japan. Released 1999	Moderate plant vigour, 6.6 slips and 0.4 suckers	Medium ovate shape; reddish orange skin, <1 kg, creamy white flesh, SSC 16%, TA 0.60%	Ripens in mid-July in Okinawa, peach-like aroma	Ogata *et al.*, 2016

continued

Table 4.2. *continued.*

Cultivar	Origin	Plant characters	Fruit characters	Other characters	References
'Bogor'	Cross between 'Smooth Cayenne' and 'Singapore Spanish'. Japan				Ogata *et al.*, 2016
'Gold Barrel'	Cross between 'Cream Pineapple' and 'McGregor T–1'. Japan. Released 2009	Moderate plant vigour with few slips and suckers	Oblong shape, yellowish orange skin, 1.4 kg, yellow flesh, SSC 16%, TA 0.5%, good eating quality	Matures mid-July, high fruit quality, good sensitivity to ethephon makes crop control easy	Ogata *et al.*, 2016
'Julio Star'	Cross between 'N67–10' and 'Cream Pineapple'. Japan, Released 2009	Moderate plant vigour, 5 slips and few suckers	Oblong shape, orange skin, 1.2 kg, pale yellow flesh, SSC 16%, TA 0.6%	Matures late July in Okinawa, resistant to fruitlet core rot, long shelf life	Ogata *et al.*, 2016
'Okinou P17'	Cross between 'Yugafu' and 'Summer Gold'. Japan. Released 2015	Moderate plant vigour, 3.2 slips and 3.0 suckers.	Oblong shape, yellowish orange skin, 1.1 kg, yellowish white flesh, SSC 19%, TA 0.6%.	Ripens late August. Good eating quality, resistant to fruitlet core rot, and tolerant of typhoon winds	Ogata *et al.*, 2016
'Honey Bright'	Cross between 'Mitsubishi Smooth Cayenne' and 'I-43-908'. Japan. Released 1999		Oblong shape, yellowish orange skin, <1 kg, light yellow flesh. SSC 17%, TA 0.67%	Matures mid-August	Ogata *et al.*, 2016
'Summer Gold'	Cross between 'Cream Pineapple' and 'McGregor ST-1' Japan. Released 2004		Oblong shape, yellowish orange skin, >1 kg, light yellow flesh, SSC 16%, TA 0.52%	Matures early-August	Ogata *et al.*, 2016
'Yugafu'	Cross between 'Cream Pineapple' and 'HI101'. Japan. Released 2004		Oblong shape, yellow skin, 1 kg, white flesh, SSC 16%, TA 0.67%	Matures mid-August	Ogata *et al.*, 2016

'PQM–1' ('Pineapple Queen Mutant–1')	A clonal selection of 'Queen'. India	Plant more vigorous than Queen and Mauritius, spiny, reddish leaves, similar suckers but fewer slips than 'Queen'	Cylindrical, 1.3 kg, slightly heavier than 'Queen', fruitlets fewer but larger than 'Queen', SSC 17%, TA 0.6%	Flowering time and fruit maturity late compared to Queen	Prakash *et al.*, 2009
'Amritha'	Cross between 'Kew' (syn. 'Smooth Cayenne') and 'Ripley Queen'. India	Spiny, 1.8 slips	Cylindrical, 1.5–2.0 kg, crown 80–100 g, yellow skin, firm, non-fibrous, pale yellow flesh, good aroma, good taste, high SSC and low acidity		Nisha and Radhamany, 2016
'Aus-Jubilee'	Parents 'Smooth Cayenne' and 'CO-2', Department of Agriculture and Fisheries (DAF), Australia	Dark green piping leaves, medium peduncle, few ground suckers; medium plants suckers, 0.3 slips	Ovoid, golden yellow skin 1.6 kg, pale yellow to yellow flesh, SSC 18%, moderate acidity, AsA ≥25 mg/100 ml juice, good flavour	Fresh market. Some resistant to *Phytophthora cinnamomi*, NI and translucency. Plant Variety Protection grant	Sanewski, 2007b
'Aus-Carnival'	Parents 'CO-2' and 'PRI 53-116', developed by DAF, Australia	Semi-erect habit. Long leaves with piping margin, anthocyanins present, mainly on margins and towards the base, none or few ground suckers, 1.2 large plant suckers, 0.4 slips	Ovoid, 1.6 kg, large crown, yellow skin, SSC 17%, low acidity, good flavour, yellow flesh and small core	Fresh market. High ratoon yield. Plant Variety Protection grant	Sanewski, 2007b Bartholomew *et al.*, 2010
'Aus-Festival'	Parents 'Smooth Cayenne' and 'CO-2'. Developed by DAF, Australia	Vigorous with short piping leaves, some anthocyanin at distal end, 1.2 slips/plant, none or few plant/ground suckers	Elliptic to globular shape, 1.6 kg, medium crown size, yellow to orange skin, pale yellow flesh, low incidence of translucency, low fibrosity, SSC 16%; low acidity, flavour sweet with coconut overtones	For fresh market. Moderate susceptibility to *P. cinnamomi* and NI. Plant Variety Protection grant	Anonymous, 2015

continued

Table 4.2. *continued.*

Cultivar	Origin	Plant characters	Fruit characters	Other characters	References
'IAC Fantástico'	Open pollinated cross between a 'Tapiracanga'× 'Smooth Cayenne' (female parental) and an unknown male parental, released by IAC, Brazil, 2010	Vigorous plant, leaves dark green slightly purple on the tips, spiny tip, 5.6 slips and 1.8 suckers	Barrel to oblong shape; 1.5 kg, yellow to orange skin, yellow flesh, SSC >16%, TA 0.52%, AsA 21.6 mg/100 g	Fresh fruit and processing. Resistant to fusariosis	Spironello *et al.*, 2010
'BRS Imperial'	Cross between 'Perolera' and 'Smooth Cayenne'. EMBRAPA. Released 2003	Medium-sized, slow growth, compact habit, piping, dark green leaves with anthocyanin in centres. peduncle 20×3 cm, 3–5 slips close to fruit base; 1 sucker at harvest	Cylindrical, 0.90–1.60 kg, yellow-orange skin, prominent eyes, firm, yellow flesh, SSC 19%, TA 0.29–0.70%, AsA 29 mg/100 g	First Brazilian hybrid resistant to fusariosis. Disease resistance favours organic production. Resistant to transport damage and internal browning	Cabral and Matos, 2009
'BRS Vitória'	Cross between 'Primavera' and 'Smooth Cayenne', EMBRAPA. Released 2006	Light-green, piping leaves, peduncle 20 mm wide, 4.2 slips	Barrel to cylindrical, bright yellow skin, flat eyes, 1.5 kg, small core (12 mm); firm, white flesh, good flavour, SSC 15.8%, TA 0.8%	Resistant to fusariosis	Ventura *et al.*, 2009
'BRS Ajubá'	Cross between 'Perolera' and 'Smooth Cayenne', EMBRAPA. Released 2009	Long, green, piping leaves, centre and upper half reddish, 1 sucker, 4 slips	Weighs 1.3–1.6 kg, cylindrical shape, yellow skin, yellow pulp, SSC 14.5%, TA 0.6%	Resistant to fusariosis. Well adapted to the subtropical conditions of Southern Brazil. Low incidence of NI	Reinhardt *et al.*, 2012
'Florhan 41', initially named 'Scarlett'	Cross between 'Smooth Cayenne' and 'Manzana' by CIRAD-FLHOR	Compact plant with erect, piping leaves. Long peduncle	Cylindrical, 1.4–2 kg, orange to red skin, medium to large flat eyes, firm, golden yellow flesh, SSC 15–18%, high ascorbic acid content, thin core, low fibre flesh	Susceptible to fruitlet core rot	Coppens d'Eeckenbrugge and Marie, 2000

'Honey Gold'	Developed by Del Monte Fresh Produce International Inc., using clonal selection of Tainung 11 ('Perfum Pineapple')	Spiny tip leaves from yellowish green to reddish green, peduncle 21 cm, 0–3 slips	Cylindrical to slightly conical, <1.0 kg, medium core, intense yellow skin, prominent fruitlets, yellow flesh, SSC 16%, strong aroma and good taste	Fresh market. Very resistant to sunburn, but susceptible to internal browning. Long shelf life	Morales *et al.*, 2006
'Dole 14'	Cross between 'Mayan Gold 03' ('MG-03', 'Tropical Gold') and 'P-1972', by Dole. 'MG-03' is similar to 'MD-2'	Long, piping, pale to greyish green leaves, few suckers, 0–2.0 slips	Conical to cylindrical, yellow skin, deep golden coloured flesh due to a high carotenoid content, medium acidity, low fibre, improved aroma, distinct fruit flavour	High tolerance to NI and low susceptibility to skin cracking. Granted Plant Variety Protection Rights in Indonesia on 13 August 2013	Young and Gonzales, 2010
'P1972' ('75–80')	Cross between PRI varieties '64-337' and '59-443'	Open foliage attitude, long (106–130 cm) piping leaves, pale to greyish green (lighter than MD-2)	Oval shaped, very thin skin, yellow to golden-yellow skin, 1.9–2.4 kg, orange-yellow to golden-yellow flesh, core diameter 2.7 cm, higher carotene, SSC 15%, TA 0.53%, mango-like aroma, AsA: 22.8–27.5 mg/100 ml	Low incidence of fruitlet core rot and marbling. Granted Plant Variety Protection Rights in Indonesia on August 13, 2013	Gonzales and Vriesenga, 2006
'Meli Kalima' ('Honey Cream' in English) or 'Franklynn Rose'	Cross between 'Dry Sweet' and 'Hilo White', developed by Frankie's Nursery, Hawaii	Small plant (53 cm from the ground to the base of fruit), piping green leaves with purple coloration, peduncle 21 cm, 3.9 slips	Cylindrical, medium-sized crown, greyed-orange to dark orange skin, pale yellow to cream flesh, SSC 21–26%	Fruit sold without crown	Sekiya, 2012
Rosé	A genetically modified pineapple cultivar, developed by Del Monte Fresh Produce Company in 2011, using 'MD-2'	Plant characteristics similar to 'MD-2', leaves dark green, spiny tip	Similar to 'MD-2', except for pink or light red flesh colour (with orange-yellow to golden yellow spots) due to accumulation of lycopene, cylindrical shape, 1.5 kg, core 2.7 cm, skin turns a 'Tiger' colour (green, yellow, orange and red pattern) on the shoulder of each fruitlet due to carotenoid accumulations, SSC 15%, AsA 35 mg/100 g	'Rosé' is genetically-modified 'MD-2'. Flesh and skin colour (lycopene) results from over-expression of tangerine phytoene synthase gene under transcriptional control of the pineapple bromelain inhibitor gene. Also contains anti-sense ACC synthase gene for resistance to NI	Firoozabady and Young, 2015

necessity an ongoing process and is best-performed on well-grown plant crops. To identify the inferior plants the grower walks through the field before harvest and marks the crown, slips or plant itself using a can of spray paint. The exact process will depend on what type of planting material is used. In Australia, it is performed on summer plant crops where the expression of defects such as collar-of-slips is greatest (Groszmann, 1945).

Somaclonal mutation rate should not be more than about 5% where plants have been multiplied by *in vitro* culture but can be almost 100% if protocols are not optimized. As an example, cultivars expressing the spiny tip leaf margin, such as 'Smooth Cayenne' and 'MD-2', can mutate to the spiny leaf state. In such cases, rogueing should be used early in the multiplication process to ensure that only true-to-type plants are continued.

Clonal selection

Clonal selection is a slow, more intensive process of starting a new clone from a single or small number of plants. It has been demonstrated that apparently normal plants can carry some undesirable traits such as multiple crowns. If enough plants of that genotype are observed over several environments, these traits will eventually be expressed and that line or clone can be eliminated. Clonal selection is therefore more successful than rogueing against defects such as multiple crowns that are strongly influenced by environment but is more tedious.

The initial population in a clonal selection programme is established usually by walking through fields of maturing pineapples and selecting the crown from superior types or marking the plant with paint and a stake to later collect the entire plant. These selections are then assessed and multiplied over several crops until only the most consistent clones remain. When sufficient plants are available, replicated trials can be established in different environments. The method of obtaining the initial population will vary depending on the specific characters in which the breeder is most interested. Successful clonal selection relies on good data collection, robust statistical designs and careful separation of almost identical clones.

The main characteristics targeted in rogueing or clonal selection are described below.

Fruit weight

Fruit weight is strongly affected by environment and cultural operations. Also, subjective assessment of fruit weight in the field is difficult even where the differences are as much as 10%. For these reasons, rogueing against decreased fruit weight is not easy (Kerns, 1928). There is however, a tendency for the very heavy and very light types within a field to be genetically different. Rogueing can therefore be used to discriminate against very small-fruited types (Anonymous, 1925). Where more objective methods of assessment of fruit weight are used, gains in fruit weight are possible provided measures are taken to avoid inadvertent selection for more translucent types. As an example, a 10–15% increase in fruit weight was achieved in the Australian clonal selection programme (Wassman, 1982).

Multiple crowns

Multiple or fasciated crowns can be heritable but are also strongly influenced by environment and cultural operations (Dalldorf, 1975). Excessive nitrogen fertilization and high temperature during flower cone development are known to increase the incidence of multiple crowns (Anonymous, 1925; Linford and Spiegelberg, 1933). Linford and Spiegelberg (1933) observed that the condition can start at the top of the fruit or as evidenced by the change in eye pattern, at a position in the fruit itself. Re-planting only single crowns from vigorously growing plants that flowered during the hotter periods of the year should reduce its incidence (Dalldorf, 1975). Rogueing cannot completely eliminate this disorder.

Collar-of-slips and knobbiness

Collar-of-slips is a dominant mutation where several slips emerge from the base of the fruit (Collins and Kerns, 1938). The incidence of knobs on the base of the fruit is closely associated with collar-of-slips. Collar-of-slips is highly heritable but influenced by environment. In Australia, collar-of-slips is usually expressed in fruit maturing in summer but not winter (Groszmann, 1940). Also, plants propagated from a collar-of-slips plant will occasionally produce about ten normal types (Collins, 1933a; Groszmann, 1945). Fruit from collar-of-slips types is usually smaller and slip number is more than double that of normal plants (Collins, 1933a; Py *et al.*, 1987). Collar-of-slips is one of the main characteristics targeted in a rogueing programme (Py *et al.*, 1987).

The number of slips produced lower down on the peduncle is not considered the same as collar-of-slips as it is a quantitative trait whose expression is more strongly influenced by plant vigour and environment (Collins and Kerns, 1950). Vigorous plants on the outer edges of fields commonly produce more slips. Rogueing against slip number might not make much difference because of the large environmental and cultural influences.

Fruit conformity

There are several distinct mutations that affect fruit shape or appearance. 'Long Tom' or 'long, slender fruits' is a heritable condition where the fruits are long with a small diameter. Other common mutations in fruit conformity include 'dry fruit' and 'bottle neck'. With these off-types, the flowers are absent from either the upper portion of the fruit or the entire fruit. The effect is a small malformed fruit with obvious bracts but poorly developed eyes. Rogueing against these off-types should eventually eliminate them.

Spiny leaves

Genetic studies have shown that the spiny leaf margin is recessive to the spiny tip margin and 'Smooth Cayenne' and 'MD-2' carry the gene for complete spiny in a recessive condition (Collins and Kerns, 1946). The mutation of the dominant gene for spiny tip to the recessive gene for spines can occur at any time in the growth of the plant. Depending on the developmental stage of the plant when the mutation occurs and the position in the plant of the mutating cell, it can result in a chimera with partially spiny leaves or a completely spiny plant (Collins, 1936).

The mutation to spiny leaves should not be confused with the tendency of 'Smooth Cayenne' to develop spines as a result of stressful conditions and the cessation of leaf growth. As an example, a high night temperature increases the development of spines along the leaf (Friend, 1981). A genetic change is not thought to have occurred in these instances and new leaves develop normally once the growing conditions are favourable. Rogueing is very effective in removing spiny-leaf off-types.

Sexual recombination

While 'Smooth Cayenne' is ideal for processing, the fresh pineapple markets in the world are more diversified. The *Ananas* genepool is fortunately diverse providing great scope for genetic improvement.

In 1914, the Pineapple Growers Association of Hawaii started one of the earliest efforts in modern pineapple improvement. The experimental station of this association later became the Pineapple Research Institute of Hawaii (PRI). One of the main objectives of the PRI programme was to develop a 'Smooth Cayenne'-type variety with a few improved traits. Eventually, genotypes were developed which had better resistance to *Phytophthora*, mealybug wilt, nematodes, pink disease and internal browning. Some selections also had higher vitamins A and/or C, less acid in winter ripened fruits, better harvesting peak, higher yield and plant vigour and improved cannery recovery. Many varieties were selected which were better than 'Smooth Cayenne' in certain individual characteristics, but eventually none could replace 'Smooth Cayenne' because after extensive evaluation, each would reveal a fatal flaw (Williams and Fleisch, 1993). The PRI closed in 1975 and evaluation of progenies was completed at the Maui Pineapple Company in 1985. From this programme, two cultivars, MD-2 and its sibling CO-2, both fifth-generation crosses selected in 1973, have been commercialized (PRI breeding records).

The main PRI programme which started in the 1920s used a system of out-crossing between the five cultivars Smooth Cayenne, Smooth Guatemala (syn. 'Monte Lirio'), Pérola, Queen and Ruby (syn. 'Singapore Spanish') and then inter-crossing the F1 selections to maintain a high component of 'Smooth Cayenne' in progeny. Other programmes started in a similar fashion with the F1 population composed of crosses between 'Smooth Cayenne' and other pre-Columbian cultivars. Examples are 'Smooth Cayenne' and 'Queen' in the Thailand programme (Sripaoraya, 2009), 'Smooth Cayenne' and 'Perolera' in the French programme (Cabot, 1987), 'Smooth Cayenne' or 'Pérola' and 'Primavera' or 'Perolera' in the Brazilian programme (Cabral *et al.*, 1993), and 'Singapore Spanish', 'Smooth Cayenne' and 'Queen' in the Malaysian programme (Chan, 1989, 2006). Most current breeding programmes are focused on the development of cultivars for the fresh market.

Most cultivars of pineapple (*Ananas comosus* var. *comosus*) are considered highly heterozygous (Chan, 1986). The difficulty in finding a superior recombinant among the widely varying progenies obtained through crossing of highly heterozygous pineapple cultivars is the main problem experienced by breeding programmes (Collins, 1960; Williams and Fleisch, 1993; Coppens d'Eeckenbrugge and Duval, 1995).

Heterozygosity is the breeder's source of diversity and it is essential for crop improvement. However, a high level of heterozygosity means it is necessary to develop large populations and apply selection pressure for a large number of traits and incorporate multiple generations unless a carefully chosen breeding strategy is used.

Breeding strategies

Selection of breeding strategy is critical to ensure efficiency, progress and long-term viability of the programme. The best breeding system will usually be the one that maintains genetic variability in the breeding population, either within or between genomes, but allows the frequency of desirable genes to increase.

Inbreeding depression

Inbreeding depression is seen in pineapple (Collins, 1960; Cabral *et al.*, 2003; Sanewski, 2009) but there are no genetic studies reported. Inbreeding depression is an inclusive term that encompasses aspects of seed development post fertilization, germination and plant growth. Coppens d'Eeckenbrugge *et al.* (1993) observed similar seed abortion rates after self- and open-pollination of 71 cultivated and wild clones, which indicates no inbreeding depression effect at the zygotic or embryonic level. Also, no difference was observed in germination of self- and cross-seeds, or in the early development of the resulting seedlings from 'Española Roja', 'Primavera', two clones of *A. comosus* var. *microstachys*, and two clones of *A. comosus* var. *bracteatus*. When transplanted to the field, these seedlings showed about 50% size reduction as compared with progenies from open pollination. In contrast, a strong growth depression was observed in the early development of seedlings of self-progenies from 'Manzana,' and from a clone of *A. comosus* var. *microstachys*, causing their death in the nursery.

Inbreeding depression can be minimal with lesser forms of inbreeding (Sanewski, 2009). Depending on the parent varieties, a level of inbreeding might be a worthwhile strategy to include in a pineapple breeding programme.

Outcrossing

Most pineapple breeding programmes start with heterozygotic pre-Columbian varieties such as 'Smooth Cayenne' and follow outcrossing with simple recurrent selection. In heterozygotic species such as pineapple, out-crossing with simple recurrent selection should shift the mean for characters under selection in the direction of selection. Heterozygosity, particularly between genomes, will remain relatively high with this strategy if there are sufficiently diverse foundation genotypes. Several generations of large seedling populations will be needed to improve or maintain many traits simultaneously. Because of the need to progress through several generations, the foundation population needs to include several unrelated genotypes to avoid problems with inbreeding or incompatibilities in subsequent generations. There should also be an effort to improve the breeding pool. The number of foundation varieties needed will depend on the expected longevity of the programme and number of generations anticipated but they should contain all the genes and hence traits required for the programme to be successful (Sanewski *et al.*, 2011).

Selfing

Selfing has also been proposed as a strategy to develop homozygous varieties of pineapple for use as parents (Dujardin, 1991; Cabral *et al.*, 2000). This strategy is sound if the plants are homozygous for desirable traits. Fifty per cent of alleles will be homozygous with each cycle of selfing (Allard, 1960) and seven cycles of selfing are required to achieve full homozygosity. The immediate bottle-neck is the genetic self-incompatibility (SI) that exists in most domesticated pineapple. Although they are distinct phenomena, the effects of pollen–pistil incompatibility (SI) will be aggravated by inbreeding depression. Pollen–pistil incompatibility can be strong in some cultivars but subject to environment, possibly temperature, in other varieties. The latter condition is often termed 'pseudo-compatibility' (Majumder *et al.*, 1964). Strong incompatibility or limited pseudo-compatibility are evident in most pineapple cultivars likely to be used in pineapple breeding. None are truly self-compatible. A low level of compatibility in some pineapple varieties has been attributed to the existence of a low percentage of mutated pollen grains which are able to fertilize an ovule (Majumder *et al.*, 1964).

There are several reports of attempts at selfing with variable success. Selfing of 'Smooth Cayenne' was investigated by the PRI in the late 1930s to early 1940s to identify desirable recessive alleles. Collins (1960) experimented with selfing a self-compatible mutant of 'Smooth Cayenne' and achieved only two generations before inbreeding depression was so severe that only a few plants produced fruits and none of them set seed. Despite considerable effort, no desirable types were ever produced. Coppens d'Eeckenbrugge *et al.* (1993) selfed several pre-Columbian clones with variable results. Following on from this, Coppens d'Eeckenbrugge and Duval (1995) proposed to introduce one cycle of selfing before the final hybridization to produce inbred genotypes from the parental cultivars. This was seen as a way to lower the variability of the hybrid progeny and identify recessive alleles. The final hybridization between the best inbreds from two different cultivars would restore vigour. This follows a reciprocal recurrent selection (RRS) strategy of breeding. Sanewski (2009) attempted selfing in four modern fresh market cultivars with effectively no success.

To obtain useful parents requires the production of a reasonable number of seedlings, possibly thousands, to allow screening for recessive undesirable alleles. This applies to progeny from selfing more so than outcrosses because the selfed progeny can be homozygous for undesirable traits. Choice of parents should not be based on their propensity to produce selfed seed. Methods to overcome self-incompatibility are therefore essential. Attempts have been made to overcome self-incompatibility with X-rays (Marr, 1964) or gamma rays (Subramanian *et al.*, 1981). Bhowmik and Bhagabati (1975), tested bud pollination, style-grafting and NAA application but only the last technique produced a few self-seeds in 'Queen'. The cut-style technique of pollination has also been unsuccessful (Sanewski, 2008). Anecdotal evidence suggests self-incompatibility in some varieties of pineapple might be at least partially overcome by high temperature (Kerns *et al.*, 1968; Williams, 1970b).

Sib and half-sib crosses

While most programmes have commenced by out-crossing between unrelated varieties, subsequent generations usually encounter some mild in-breeding, usually in the form of sib or half-sib crosses because of the use of 'Smooth Cayenne' in most F1 crosses.

Mild forms of in-breeding such as half-sibcrosses which produce large populations might allow a greater chance of recombination of favourable alleles compared to selfing which produces few progeny (Allard, 1960). The more gradual increase in inbreeding depression that accompanies milder inbreeding might also facilitate the gradual selection against inbreeding depression (Porcher and Lande, 2016).

Sibcrosses are often incompatible or encounter excessive inbreeding depression (Sanewski, 2009). Half-sib crosses are often compatible (perhaps in one direction only), but might be accompanied by mild in-breeding depression so will require selection pressure for vigour. Germination of seed and early growth of seedlings should be optimal to allow identification and discard of the weaker types (Williams, 1970a).

Mild inbreeding generally does not appear very deleterious in pineapple (Sanewski, 2008, 2009). The leading fresh market pineapple variety, MD-2, has an inbreeding coefficient of slightly more than a half-sib cross (Sanewski *et al.*, 2011; PRI Breeding Records).

Mild inbreeding in the form of half-sib crosses and back crosses has been used in the Australian pineapple breeding programme over three generations (Sanewski and De Faveri, 2017). Half-sib selection is equivalent to selecting for general combining ability (Viana *et al.*, 2014) so when followed over one cycle only shouldn't replace additive genetic effects.

Back-crossing

'Smooth Cayenne' was often used as the recurrent parent in back-crosses (Anonymous, PRI breeding records) in an attempt to introgress desirable traits into a 'Smooth Cayenne'-like cultivar. Back-crossing to an unimproved variety, like 'Smooth Cayenne', is likely to decrease the frequency of desirable alleles in breeding stocks of an advanced programme. It will almost always be better to use current generation varieties as recurrent parents. Back-crossing is often feasible, at least in one direction, seed yield can be good and inbreeding depression mild but acceptable.

Breeding for hybrid vigour

Hybrid vigour (heterosis) usually arises where hybridization between two different genotypes gives rise to new and unique allele combinations in the progeny. From a functional perspective, heterosis can be considered as the opposite of inbreeding depression (Allard, 1960).

Heterosis can be an important contributor to the stability of a phenotype in different environments (Fridman, 2015) and is highly desirable in crop plants if it is expressed in commercially valuable traits. Progeny of crosses between pre-Columbian pineapple cultivars can display transgressive recombination suggesting a level of heterosis. Marked heterosis has been shown to exist in pineapple in crosses between wild and domesticated botanical varieties (Sanewski *et al.*, 2011) but there are no examples in the literature of pineapple breeding programmes which deliberately target heterosis although it was noted by Collins (1960) in some crosses.

Reciprocal recurrent selection (RRS) might be a worthwhile breeding strategy to actively exploit heterosis in clonal crops (Ceballos *et al.*, 2015) like pineapple and at the same time minimize heterozygosity. RRS entails the development of several unrelated, inbred lines which are then combined to exploit heterosis. RRS is thus dependent on a level of homozygosity in parents but with different alleles. The process of mild inbreeding through back-crossing and/or half-sib crosses over perhaps two generations would be necessary to develop the inbred lines needed. The first steps are to better understand the genetic basis of heterosis in pineapple and to identify heterotic patterns between such inbred materials.

Polyploid breeding

Manipulation of ploidy was first explored in the PRI programme in the 1940s. Tetraploids, in particular tetraploid 'Smooth Cayenne', it was discovered, were vigorous but produced poor fruit quality (Kerns and Collins, 1947). This work also revealed several natural triploids among seedling populations. These triploids were generally of a good type sparking interest in the generation of triploids (Collins, 1933b). It was believed therefore that tetraploids of elite varieties, when used in combination with diploid parents, would produce triploid progeny very similar to the tetraploid parent with only a few traits inherited from the diploid. Considerable promise was placed in the production of tetraploids of elite varieties using colchicine on growing points (Kerns and Collins, 1947; Collins, 1960). In the years 1947–1949 approximately 59,000 seedlings were produced with strategies to increase ploidy (Collins, 1950). After much effort with few useful triploids produced, together with the fact that triploids were mostly sterile and hence, couldn't easily be used for further breeding, this approach was abandoned.

Japanese researchers (Nagatomi *et al.*, 1997) produced tetraploid pineapple of the cultivar Taino No. 4 (possibly 'Tainung 4') using colchicine on callus. The polyploid plants produced larger fruits with more and larger fruitlets but a lower SSC compared to diploids. They concluded that significant potential existed to produce genotypes with enlargements in some characters.

Considerations in hand pollination

Technique of cross-pollination

Synchrony in flowering of parents is an important consideration in cross-pollination. The time taken from induction to anthesis can be different in pineapple cultivars. Depending on temperatures the difference might be in the order of 10–14 days. Where the time to anthesis is unknown for each parent cultivar, a percentage of plants of each cultivar should be induced at say, five-day intervals.

Time of day for pollination

Dawn to 9 am is probably the best time for pollination (Cabral *et al.*, 1993; Sanewski, 1998). However, pollination may be done during the entire length of the day, although seed set may vary depending on the time of day and appears to be temperature related. In work by Chan (1986) higher seed set was obtained when crossing was done during the early morning or cooler evening and significantly less so during the hotter midday or early afternoon.

Segregation in seedling populations

Evaluation of seedling populations indicates that there is wide segregation for most quantitative characters (Chan, 1989; Loison-Cabot, 1990; Chan, 1991; Sanewski, 1998). For soluble solids content (SSC) for example, the range of values showed transgressive segregation i.e. the minimum and maximum values of the progeny exceeded the lowest and highest values of both parents. Sanewski (1998) reported that the range of SSC from progeny of the 'Queen'× 'Smooth Cayenne' cross was 9–24% while the parents had a narrower range from 14% to 19%.

Trait heritability

There are few detailed genetic studies in pineapple but the genetic determinism of some characters is known. Quantitative characters with a potentially high heritability under standard field conditions include sugar content, fruitlet number, fruitlet size, fruit weight, peduncle length and width, crown weight, core diameter and number of basal knobs. Characters with high heritability are fruit shape and the fruit weight: plant weight ratio (Mathew *et al.*, 1979; Loison-Cabot, 1990). Crown fasciation has a low heritability. Leaf margin type is determined by genes P, determining piping leaves, and S, with a multiple allelic series, from smooth-leaved to partly spiny to spiny (Collins, 1960).

Fruit weight

Fruit weight is composed of the quantitative characters fruitlet number and fruitlet weight. In well-grown populations, narrow sense heritability can be 0.60, indicating that good progress in improving fruit weight is possible. Fruitlet number has been shown to be even more highly heritable (0.77) and is positively correlated to fruit weight (0.63). The inclusion of parents with a large number of fruitlets should add another dimension to improving fruit size (Sanewski *et al.*, 2011).

SSC

Chan (1989) suggests SSC in pineapple is a quantitative trait controlled by multiple additive genes. SSC displays a moderately high heritability (0.52). There is however a weak negative genetic correlation (−0.22) between fruit weight and SSC. SSC can therefore be difficult to increase substantially and quickly without some slight reduction in fruit weight (Sanewski *et al.*, 2011).

Peduncle length

Peduncle length appears quantitative and displays a maternal effect (Cabot, 1987). Collins (1960) indicates it is controlled by several genes with a long peduncle being dominant. Heritability can be high (0.63) indicating significant response to selection is possible (Sanewski *et al.*, 2011).

Slip number

Slip number can have a bimodal expression with populations segregating into zero or several slips. Slip number has a low heritability (0.33) with a strong environmental influence. A separate condition where a large number of slips are produced around the base of the fruit is sometimes called 'collar-of-slips'. Collins (1960) indicates collar-of-slips is a dominant, multi-gene character difficult to remove from populations. The number of slips and plant suckers can be negatively correlated (–0.36) (Sanewski *et al.*, 2011).

Plant sucker number

In the reciprocal crosses between 'Smooth Cayenne' and 'Manzana' Loison-Cabot (1990) detected a maternal effect. On a wider sample of parent combinations Sanewski *et al.* (2011) found low heritability (0.05).

Root rot resistance

Resistance to root rot was considered quantitative and additive (Smith, 1966). Varieties could differ in their susceptibility to both *P. cinnamomi* and *P. nicotianae* var. *parasitica* (Johannessen and Kerns, 1964). While quantitative, resistance to *P. cinnamomi* appears dominant and has been associated with thirteen molecular markers positioned in one broad locus on chromosome five (Sanewski *et al.*, 2017).

Fusariosis resistance

Resistance to fusariosis was shown to follow a simple, dominant inheritance in F1 progeny of 'Perolera' and 'Primavera' (Junghans *et al.*, 2005).

Choosing parents

In hybridization for improvement of pineapple, the choice of parents has usually been based on the strength of one in complementing the weakness of the other. In practice, breeders are always faced with selection of a set of traits and not just single characters. Chan (1991) computed breeding values for parents based on consideration of a set of important characters. The values were computed by obtaining the product of the percentage of progenies that qualified for selection in each of the characters. Sanewski (1998) used a similar computation and found that advanced cultivars such as CO-2 developed by PRI, had a high breeding value. Best Linear Unbiased Predictions (BLUPS) for selected traits for parents can also be calculated based on progeny performance in small experimental trials (De Faveri and Sanewski, 2014). These calculations will also give estimates of broad and narrow sense heritability for each trait. Breeding values or BLUPS are helpful for guidance in the choice of parents, estimating population size and in ascertaining the probability of obtaining useful progenies.

Fertility, compatibility and direction of cross

Along with the heritability of the traits sought and their allelic state in parental varieties, fertility and compatibility are important considerations in deciding the parents to be used, the direction of the cross (which is the pollen parent) and the number of flowers to be pollinated. Chan (1986) found that crosses between 'Smooth Cayenne', 'Queen', 'Singapore Spanish' and 'Johor' were compatible, but there was differential seed set between the crosses and their reciprocals.

In the Brazilian programme, 'Smooth Cayenne' has been shown to be a poor female parent bearing only about 200 seeds per fruit in a cross with 'Perolera'. When a reciprocal cross was made there were more than 2000 seeds per fruit (Cabral *et al.*, 1993).

The direction of the cross was reported by Loison-Cabot and Lacoeuilhe (1990) to have an influence on the performance of the progeny. In a cross between 'Smooth Cayenne' and 'Manzana', the latter used as the female parent improved the yield of the progeny compared with the reciprocal. In the case of soluble solids content (SSC), 'Smooth Cayenne' as the female parent generated progenies with higher SSC.

While crosses need to be sufficiently fertile to generate useful seed number, this should not guide the programme. Heritability of traits and parental breeding value, including maternal effects, should be the primary considerations.

Suitable seedling population size

When the breeding values of parents are known, the population size required can be estimated (Chan, 1991). The number of selections in the second cycle that the breeder wants to evaluate must first be ascertained. At MARDI, it was estimated a seedling population of 23,000–45,000 progenies was needed to obtain 300 selections that all contained the set of eight chosen traits. Loison-Cabot (1987), using 'Smooth Cayenne' and 'Perolera' parents, recommended 30,000 progenies for each annual selection cycle while Cabral *et al.* (1993) and Benega *et al.* (1998) maintained 23,000 and 34,000 progenies respectively to obtain 120–169 first-cycle selections. The PRI breeding programme has been the most extensive to date and by 1970 was producing 400,000 seed per year (Williams, 1970a). Up to 100,000 seedlings were developed from some of the better parent combinations (Williams and Fleisch, 1993). Sanewski (1998) using some of these PRI selections, obtained 300 selections from populations of 5000–10,000 seedlings because the parents (fifth-generation selections) were calculated to have high breeding values.

Collins (1960) indicated that where crosses with wild germplasm (var. *microstachys*) were performed, only 1000 progenies were necessary in the first generation but 15,000–25,000 per back-cross were required for each of four back-cross generations.

Estimating seed number

It is critical that the breeder is aware of expected seed and resulting seedling numbers with preferred parent combinations and compensates where necessary with extra pollinations otherwise the population might be increasingly composed of more compatible crosses rather than preferred combinations. The number of seed obtained is the result of a combination of many factors including fruitlet number, ovule number, ovule viability, pollen abundance and viability, ovule crowding as well as genetic incompatibility.

The pineapple fruit is a fusion of up to approximately 200 fruitlets but usually 75–150 depending on the cultivar (Chan, 1986). It is uncommon to obtain more than 20 seeds per fruitlet in cultivars, even in the most fertile crosses (Sanewski, 2009). Depending on the parent combination, from 3–20 seeds per flower can be expected (Sanewski, 2009). If fruits have a large number of fruitlets and all flowers are pollinated, 200–1000 seeds can be expected per fruit in compatible cultivars with moderate to good fertility. Highly fertile out-crosses can yield 2000 seeds per fruit, whereas compatible half-sib crosses might yield around 500 seeds. The direction of the cross can have a major impact on compatibility and seed number.

Self-compatibility (pollen–pistil compatibility) can be checked using the fluorescence microscopy and aniline blue method (Currier, 1957; Majumder *et al.*, 1964). A few flowers are pollinated and the growth of the pollen through the pistil checked the following morning. Compatible pollen will have most pollen tubes beyond one-third the length of the style. These show as thin, bright lines down the tube with highly fluorescing callose plugs.

Pollination technique

Pollen is collected from flowers at anthesis early in the morning. The flowers are removed from the inflorescence by cutting at the base of the flower using a scalpel. An alternative, slightly more laborious method is to make three deep incisions in a triangle at the base of the flower and then gouge out the whole intact flower. This method of collecting pollen from excised flowers is sometimes better than removing only the stamens. Pollen in excised flowers should stay fresh longer and dehydrate less rapidly. There is no conclusive data on storage of pollen but Kerns (1931) indicated that 'Smooth Cayenne' pollen could be stored in cool, dry air for 15 days with minimal loss of viability.

During hand-pollination, if anthers are loose, an anther is picked up with forceps and gently brushed onto the stigma of the flower to be pollinated. Where flowers are intact, a single flower can be picked up with fingers and used to pollinate. In most cases, there is no necessity to emasculate the flowers before pollination because most cultivars are self-incompatible but new or cleansed forceps should be used for each parent combination. In cases where fidelity of the pollen source is vital, as in genetic studies, or where varieties are planted in plots in close proximity with overlapping flowering and pollinators are present, syncarps should be protected with a fine mosquito netting sleeve or paper bag tied at either end (Leal and Coppens d'Eeckenbrugge, 1996). Usually about three to seven flowers open daily over about 2–3 weeks but this will be dependent on the variety and environment (Chan, 1986).

Some varieties can also display a genetic condition called 'closed flower'. While this is uncommon, it can occur to varying degrees and appears to be modified by the environment. An application of 10 ml of 400 ppm gibberellic acid just before the first petal opens is reported to overcome closed flower (Williams and Yamane, 1970).

Seed handling, germination and nursery management

Seeds develop in the ovarian cavities just under the skin. To remove them, thinly peel the fruit and scrape out the seeds or blend the pulp using a low-speed blender. Most non-viable seeds are empty and remain afloat and these can be eliminated with the rinse water. Clean seeds are air-dried then disinfected with 2–3% sodium hypochlorite and wetting agent for 15 min, dried and treated with a suitable fungicide before storage (Sanewski, 2007a). The use of a fungicide is critical. Storage of seeds with silica gel in sealed plastic bags in a refrigerator at 4–5°C can maintain viability for two years (Leal and Coppens d'Eeckenbrugge, 1996).

Seed pre-treatment with concentrated sulphuric acid for 30 s improves uniformity in germination (Collins, 1960), but is not essential, especially if already treated with sodium hypochlorite. In the tropics, good germination can be obtained by spreading seeds evenly in shallow trays filled with fine river-sand. It is important not to bury the seed (Sanewski, 2007a). The trays can be covered with clear polyethylene sheets to retain high humidity (Chan, 1986) where misting is not available. In countries with cooler temperatures, germination in temperature-controlled glasshouses or in heated mist-beds inside a shade-house may be required (Sanewski, 1998). A range of different temperatures, from 24°C to 35°C, have been used to germinate seed (Iyer *et al.*, 1978; Sanewski. 2007a) and 28°C usually works well. Seeds can also be sown on sterile agar medium in culture rooms (Loison-Cabot, 1987; Benega *et al.*, 1997; Sanewski, 2007a). Germination starts after two weeks but can proceed over several weeks. The germination percentage of healthy seed from out-crossing is generally 80% or more.

Seedlings can be carefully replanted into trays, pots or peat beds in the nursery once they reach the 4–6-leaf stage. This occurs after 2–3 months after germination. During transplanting plants with poor vigour, dwarfism, albinism or spiny leaves are discarded. The later the transplanting is done, the more easily they can be screened. The seedlings are kept in the nursery for 6–12 months before they are transplanted in the field for evaluation or subjected to disease screening. Most disease screening requires plants to be sun-hardened. In the Brazilian programme, resistance to fusariosis (*Fusarium guttiforme*) is evaluated at the nursery stage during winter by inoculation of seedlings with the pathogen (Matos and Souto, 1984).

The simplest approach to field planting is to plant in blocks coded by the respective parent combinations. About 2% of the population are planted with check plants, usually the parents or popular industry clones, interspersed randomly within the block (Sanewski, 1998). More complex designs can be used where parent combinations are being compared. Numbered field pegs can be used throughout the field to allow identification of plants. At harvest, plant number can be written on the crown using ink or paint markers.

Evaluation trials

The extent of evaluation will depend on the size of the population, resources available and objectives. For small populations (<5000 plants), evaluation is carried out on each plant. Here, statistical procedures that can handle large numbers of treatments or unbalanced designs are used. As an example, seedling families (parent combinations) can be compared in this way using BLUPs (Sanewski and De Faveri,

2017). Usually minimal data covering mostly qualitative characters or easily measured quantitative traits are given emphasis. Fruit traits evaluated might include fruit weight, skin colour, crown weight, core diameter, flesh colour, flesh porosity, translucency, and incidence of disease, acidity and soluble solids content. Plant traits might include peduncle length, and slip and sucker numbers. Selection can be based on weighted trait scores or application of a stringent level of acceptability for each trait. The individually selected plants are usually propagated to some extent depending on the resources available. The leaf bud technique can be used to propagate from the crown, suckers, slips and stem of each selection. It should be possible to obtain up to 100 plants if all materials are utilized. A less intensive approach is to replant the crown, slips and suckers and section the mother plant.

Where large seedling plantings are conducted (> approximately 10,000 plants), data are only collected on visually acceptable fruit. This is achieved by walking through the field and harvesting only fruit from plants with easily observed desirable traits such as short peduncles, the correct number of slips and suckers and large attractive fruit. A quick SSC measurement and visual assessment of internal quality can also be made at the time of harvest.

The crown, slips and suckers from the seedling cycle are then used for the second round of selection. Molecular markers for important traits might be used for further selection before the remaining genotypes are evaluated in more highly replicated trials. For replicated trials, linear mixed models (row:column designs) can be used if more than 10–12 genotypes are involved. Usually ANOVA-based randomized block designs are used if fewer than, say, 12 genotypes. More agronomic characters such as ascorbic acid, fruit titratable acid, incidence of diseases and flavour are recorded. Selected genotypes might then be carried through to genotype×environment trials (regional trials). Meristem and leaf-bud culture can be used to great advantage at this stage (Chan, 1997). Stem sectioning is used throughout to multiply plants.

In Malaysia, the hybridization programme took about 12 years before release of the cultivar Josapine (Chan, 1993; Chan and Lee, 1996). Loison-Cabot (1987) estimated it would require 15 years to develop cultivars with the majority of time spent on evaluation and *in vitro* propagation. These estimates are based on single generations. The Hawaiian cultivar, MD-2, the result of five generations, took approximately 34 years of breeding and multiplication before release (Sanewski *et al.*, 2011). Collins (1960) estimated that 20–25 years would be required for the development of commercially acceptable cultivars involving wild germplasm because of the need to incorporate about four additional back-crosses.

References

Aldrigue, M.L and Lima, H.C. (1988) Caracterização física, química e físico-química do abacaxi [*Ananas comosus* (L.) Merril variedade Gigante-de-Tarauacá]. Série B – Ciência e Tecnologia. Rio Branco, Brazil: UFAC, pp. 81–88.

Allard, R.W. (1960) *Principles of Plant Breeding*. John Wiley and Sons, New York.

Anonymous (1925) Mass selection of pineapple planting material. *Experiment Station of the Association of Hawaiian Pineapple Canners Bulletin* No. 1. Honolulu, HI: University of Hawaii.

Anonymous (1940) Improvement of the pineapple by breeding and selection. *Queensland Agricultural Journal* 53, 14–44.

Anonymous (2015) Grant Application No. 2012/149 – 'Aus-Festival'. *Plant Varieties Journal* 28(3), 282.

Antoni, M.G. (1983) Taxonomy and cytogenetics of pineapples. MSc thesis, University of Florida, Gainesville, FL.

Araújo, J.R.G., Aguiar Jr, R.A., Chaves, A.M.S., Reis, F.O. and Martins, M.R. (2012) Abacaxi 'Turiaçu': cultivar tradicional nativa do Maranhão. *Revista Brasileira de Fruticultura* 34(4), 1270–1276.

Arriola, M.C. de, Menchú, J.F. and Rolz, C. (1976) Caracterización, manejo y almacenamiento de piña. *Informe Técnico ICAITI*, 76–105.

Arumuganathan, K. and Earle, E.D. (1991) Nuclear DNA content of some important plant species. *Plant Molecular Biology Reporter* 9, 208–218.

Ayala, A. (1961) An analysis of quantitative and qualitative composition of nematodes populations in pineapple fields in Puerto Rico. *Journal of Agriculture of the University of Puerto Rico* 45, 285–299.

Ayala, A., González-Tejera, E. and Irizarry, H. (1969) Pineapple nematodes and their control. In: Peachey, J.E. (ed.) *Nematodes of Tropical Crops.* Wallingford, Oxfordshire: CAB International, pp. 210–224.

Bartholomew, D.P. (2009) 'MD-2' pineapple transforms the world's pineapple fresh fruit export industry. *Pineapple News* 16, 2–5.

Bartholomew, D.P. (2016) Proper naming of pineapple cultivars: problems and recommendations for improvement. *Pineapple News* 23, 3–5.

Bartholomew, D.P., Coppens d'Eeckenbrugge, G. and Chen, C.C. (2010) Register of new fruit and nut cultivar list 45 (Pineapple, pp. 740–742). *HortScience* 45, 716–756.

Beer, J.G. (1857) Die Familie der Bromeliaceen. In: Monographie des Ananas. *Belgique Horticole (Liège)* 28, 144–172.

Bello, S., Julca, A. and Villachica, H. (1997a) Mancha de la fruta de piña tipo galerías asociada a *Melanoloma canopilosum* Hendel. *Acta Horticulturae* 425, 493–500.

Bello, S., León, H.V. and Otiniano, A.J. (1997b) Resistencia de cultivares de piña a la 'broca de la fruta' (*Thecla basilides* Geyer) en Chanchamayo – Perú. *Acta Horticulturae* 425, 187–192.

Benega, R., Isidrón, M., Arias, E., Cisneros, A., Martínez, J., Torres, I., Hidalgo, M. and Borroto, C. (1997) In vitro germination and callus formation in pineapple seeds (*Ananas comosus.* (L.) Merr.). *Acta Horticulturae* 425, 243–246.

Benega, R., Cisneros, A, Hidalgo, M., Martínez, J., Arias, E., Arzola, M., Carvajal, C. and Isidrón, M. (1998) Hybridization in pineapple results and strategies to save time for obtaining and releasing new hybrid varieties for growers. In: *Abstracts of the Third International Pineapple Symposium, 17–20 November 1998.* Pattaya, Thailand.

Bhowmik, G. (1977) Meiosis in two varieties of pineapple. *Indian Journal of Genetics and Plant Breeding* 37, 1–4.

Bhowmik, G. and Bhagabati, A. (1975) Self-incompatibility studies in pineapple (*Ananas comosus* L.). *Indian Agriculture* 19, 259–265.

Brewbaker, J.L. and Gorrez, D.D. (1967) Genetics of self-incompatibility in the monocot genera, *Ananas* (pineapple) and *Gasteria. American Journal of Botany* 54, 611–616.

Cabot C. (1987) Practice of pineapple breeding. *Acta Horticulturae* 196, 25–36.

Cabral, J.R.S. (1999) *Cultivares de abacaxi.* Circular Técnica, 331. Cruz das Almas, Brazil: Embrapa Mandioca e Fruticultura.

Cabral, J.R.S. and Matos, A.P. (2009) Imperial, a new pineapple cultivar resistant to fusariosis. *Acta Horticulturae* 822, 47–50.

Cabral, J.R.S., Matos, A.P. and Cunha, G.A.P. (1993) Selection of pineapple cultivars resistant to fusariose. *Acta Horticulturae* 334, 53–58.

Cabral, J.R.S., Matos, A.P. and Coppens d'Eeckenbrugge, G. (1997) Segregation for resistance to fusariose, leaf colour and leaf margin type from the EMBRAPA pineapple hybridization programme. *Acta Horticulturae* 425, 193–200.

Cabral, J.R.S., Coppens d' Eeckenbrugge, G. and Matos, A.P. (2000) Introduction of selfing in pineapple breeding. *Acta Horticulturae* 529, 165–168.

Cabral, J.R.S., Souza, A. da S., Matos, A.P. and Caldas, R.C. (2003) Effects of self-pollination in pineapple cultivars. *Revista Brasileira de Fruticultura* 25(1), 165–168.

Cáceres, E. (2009) Manual de Piña (Proyecto especial Pichis Palcazu – Mejoramiento de la producción del cultivo de la piña mediante sistemas agroflorestales en el distrito de Perené – Chanchamayo). Available at: http://www.pepp.gob.pe/archivos/transparencia/2906217203442manual_de_la_pina.pdf (accessed 27 July 2018).

Canpinpin, J.H. and Rotor, G.B. (1937) A cytological and morphogenetic study of some pineapple varieties and their mutant and hybrid derivatives. *The Philippines Agriculturist* 26, 139–158.

Ceballos, H., Kawuki, R.S., Gracen, V.E., Yencho, G.C. and Hershey, C.H. (2015). Conventional breeding, marker-assisted selection, genomic selection and inbreeding in clonally propagated crops: a case study for cassava. *Theoretical and Applied Genetics* 128, 1647–1667.

Chan, Y.K. (1986) Hybridization and selection in F_1 as a methodology for improvement of pineapple (*Ananas comosus* (Merr.). In: Chan, Y.K., Raveedranathan, P. and Mahmood, Z. (ed.) *Prosiding Simposium Buahbuahan Kebangsaan.* Serdang, Malaysia, pp. 307–314.

Chan, Y.K. (1989) F_1 variation from hybridization of two 'Spanish' pineapple cultivars. *MARDI Research Journal* 17, 172–177.

Chan, Y.K. (1991) Evaluation of F1 populations from a 4x4 diallel in pineapple and estimation of breeding values of parents. *MARDI Research Journal* 19, 159–168.

Chan, Y.K. (1993) Recent advancements in hybridization and selection of pineapple in Malaysia. *Acta Horticulturae* 334, 33–44.

Chan, Y.K. (1997) Performance and stability of pineapple hybrids in G×E trials in Malaysia. *Acta Horticulturae* 425, 201–211.

Chan, Y.K. (2006) Hybridization and selection in pineapple improvement: the experience in Malaysia. *Acta Horticulturae* 702, 87–92.

Chan, Y.K. (2010) Pineapple (*Ananas comosus*). In: Chan, Y.K, Tan, S.L and Jamaluddin, S.H. (ed.) *Breeding Horticultural Crops @ MARDI.* Serdang, Malaysia: Malaysian Agricultural Research and Development Institute (MARDI).

Chan, Y.K. and Lee, C.K. (1985) The Hybrid 1 pineapple: a new canning variety developed at MARDI. *Teknologi Buah-buahan* 1, 24–30.

Chan, Y.K. and Lee, H.K. (1995) Evaluation of performance and stability of six genotypes of Queen pineapple. *MARDI Research Journal* 23, 1–9.

Chan, Y.K. and Lee, H.K. (1996) 'Josapine': a new pineapple hybrid developed at MARDI. In: Osman, M., Clyde, M.M. and Zamrod, Z. (ed.) *The Second National Congress on Genetics.* Bangi, Malaysia: Genetics Society of Malaysia, UKM, pp. 217–220.

Chan, Y.K., Coppens d'Eeckenbrugge, G. and Sanewski, G.M. (2003) Breeding and variety improvement. In: Bartholomew, D.P., Paull, R.E. and Rohrbach, K.G. (eds) *The Pineapple; Botany, Production and Uses.* Wallingford, Oxfordshire: CAB International.

Cisneros, A., Benega, R., Borras, O., Martínez, J., Arzola, M., Arias, E., Hidalgo, M. and Isidrón, M. (1998) Mutation induction in pineapple using Co Gamma rays. In: *Abstracts of the Third International Pineapple Symposium, 17–20 November 1998.* Pattaya, Thailand.

Collins, J.L. (1933a) Studies of genetic variations in Cayenne. *Pineapple Quarterly* 3, 48–55.

Collins, J.L. (1933b) Morphological and cytological characteristics of triploid pineapples. *Cytologia* 4, 248–256.

Collins, J.L. (1936) A frequently mutating gene in the pineapple *Ananas comosus* (L.) Merr. *American Naturalist* 70, 467–476.

Collins, J.L. (1950) Review of current breeding work. *PRI Abstract of Monthly Seminar Review.*

Collins, J.L. (1951) Notes on the origin, history, and genetic nature of the Cayenne pineapple. *Pacific Science* 5(1), 3–17.

Collins, J.L. (1960) *The Pineapple.* New York: Interscience Publishers Inc.

Collins, J.L. and Carter, W. (1954) Wilt resistant mutations in the cayenne variety of pineapple. *Phytopathology* 44, 662–666.

Collins, J.L. and Hagan, H.R. (1932) Nematode resistance of pineapples. Varietal resistance of pineapple roots to the nematode *Heterodera radicicola* (Greef) Muller. *Journal of Heredity* 23, 459–465, 503–511.

Collins, J.L. and Kerns, K.R. (1931) Genetic studies of the pineapple. I. A preliminary report upon the chromosome number and meiosis in seven pineapple varieties (*Ananas sativus* Lindl.) and in *Bromelia pinguin* L. *Journal of Heredity* 22, 139–142.

Collins, J.L. and Kerns, K. (1938) Mutations in pineapples. A study of thirty inherited abnormalities in the cayenne variety. *Journal of Heredity* 29, 169–173.

Collins, J.L. and Kerns, K.R. (1946) Inheritance of three leaf types in the pineapple. *Journal of Heredity* 37, 123–128.

Collins, J.L. and Kerns, K.R. (1950) The effect of variation in number of slips per mother plant and variation in plant spacing on yield of slips, suckers and fruit in the progeny in a mass-selection experiment. *Pineapple Research Institute Report* No. 20.

Coppens d'Eeckenbrugge, G. and Duval, M.-F. (1995) Bases genéticas para definir una estrategia de mejoramiento de la piña. *Revista de la Facultad de Agronomía (Maracay)* 21, 95–118.

Coppens d'Eeckenbrugge, G. and Marie, F. (2000) Pineapple breeding at CIRAD. II. Evaluation of 'Scarlett', a new hybrid for the fresh fruit market, as compared to 'Smooth Cayenne'. *Acta Horticulturae* 529, 155–163.

Coppens d'Eeckenbrugge, G., Duval, M.F. and Van Miegroet, F. (1993) Fertility and self-incompatibility in the genus *Ananas. Acta Horticulturae* 334, 45–51.

Coppens d'Eeckenbrugge, G., Leal, F. and Duval, M.F. (1997) Germplasm resources of pineapple. *Horticultural Reviews* 21, 133–175.

Coppens d'Eeckenbrugge, G., Sanewski, G.M., Smith, M.K., Duval, M.-F. and Leal, F. (2009) Ananas. In: Kole, C. (ed) *Wild Crop Relatives: Genomic and Breeding Resources, Tropical and Subtropical Fruits.* Berlin/Heidelberg: Springer-Verlag. DOI: 10.1007/978-3-642-20447-0_2.

Currier, H.B. (1957) Callose substance in plant cells. *American Journal of Botany* 44, 478–488. DOI: 10.2307/2438916.

Dalldorf, E.R. (1975) Plant selection: multiple tops in 'Smooth Cayenne' pineapples. *Farming in South Africa, Pineapple Series* No. D1.

De Faveri, J. and Sanewski, G. (2014) The use of best linear unbiased predictions (BLUPs) in pineapple breeding. *Pineapple News* 21, 12–14.

Delgado-Huertas, H. and Arango-Wiesner, L. (2015) Caracterización morfoagronómica de genotipos de piña (*Ananas* spp.) en un suelo de terraza alta de Villavicencio. *Orinoquia* 19, 153–165.

DeWald, M.G., Moore, G.A. and Sherman, W.B. (1992) Isozymes in *Ananas* (pineapple): genetics and usefulness in taxonomy. *Journal of the American Society for Horticultural Science* 117, 491–496.

Dujardin, M. (1991) Cytogénétique de l'ananas. *Fruits* 46, 376–379.

Feng, S., Tong H., Chen, Y., Wang, J., Chen, Y., Sun, G., He, J. and Wu, Y. (2013) Development of pineapple microsatellite markers and germplasm genetic diversity analysis. *BioMed Research International, v. 2013,* Article ID 317912. DOI: 10.1155/2013/317912.

Firoozabady, E. and Young, T.R. (2015) Pineapple plant named Rosé. *U.S. Plant Patent No. 25,763 (P3).* Washington, DC: US Patent and Trademark Office.

Fridman, E. (2015) Consequences of hybridization and heterozygosity on plant vigor and phenotypic stability. *Plant Science* 232, 35–40.

Friend, D.J.C. (1981) Effect of night temperature on flowering and fruit size in pineapple (*Ananas comosus* [L.] Merrill). *Botanical Gazette* 142, 188–190.

García, M., Salcedo, F., Barrios, R. and Carrera, L. (2006) La piña 'Pilón' o cabezona, cultivada en las laderas de los estados Monagas y Sucre. *INIA Divulga* 7, 27–29.

Garcia, M.V.B., Garcia, T.B., Matos, A.P., Junghans, D.T. and Cabral, J.R.S. (2013) Situação e perspectivas da abacaxicultura no Amazonas. In: *Simpósio Brasileiro da Cultura do Abacaxi*, 5, 2013, Palmas (CD-ROM). Available at: http://www.alice.cnptia.embrapa.br/handle/doc/981108 (accessed 2 March 2018).

Gonzales, J. and Vriesenga, J. (2006) Pineapple plant named 'P-1972'. *U.S. Plant Patent No. 16396 (P3).*

Groszmann, H.M. (1939) Pineapple plant selection – with special reference to the elimination of inferior types. *Queensland Agricultural Journal* 52, 27–42.

Groszmann, H.M. (1940) Improvement of the pineapple by breeding and selection. *Queensland Agricultural Journal* 53, 14–44.

Groszmann, H.M. (1945) Pineapple plant selection with special reference to the elimination of inferior types. *Queensland Agricultural Journal* 6, 203–215.

Groszmann, H.M. (1982) Pineapple plant selection with special reference to clonal selection in Queensland. In: *RNA Field Day Notes.* Fortitude Valley, Qld: Queensland Fruit and Vegetable Growers.

Hagan, H.R. and Collins, J.L. (1932) Studies of varietal resistance of pineapple plants. II. Plant resistance to *Heterodera radicicola* (Greef) Müller. *Pineapple Quarterly* 2, 4–20.

Hernández, M.S, Barrera, J.A., Páez, D., Ardilla, A.O. and Rubio, H.R. (2004) Aspectos generales de la piña. In: *Aspectos biológicos y conservación de frutas promisorias de la Amazonia Colombiana.* Bogota, Colombia: SINCHI, pp. 61–84. Available at: https://www.sinchi.org.co/files/publicaciones/publicaciones/pdf/Aspectos%20biologicos.pdf (accessed 20 March 2018).

Isidrón, M., Rosales, Y., Pifferrer, A., Cisneros, A. Benega, R. and Carvajal, C. (2003) Caracterización del germoplasma de piña colectado en Cuba mediante prospección nacional: I. Localización, diversidad genética y situación actual. *Cultivos Tropicales* 24(1), 65–71.

Iyer, C.P.A., Singh, R. and Subramanyam, M.D. (1978) A simple method for rapid germination of pineapple seeds. *Scientia Horticulturae* 8, 39–41.

Johannessen, G.A. and Kerns, K.R. (1964) Screening and classifying new varieties for resistance to *P. cinnamomi* and *P. parasitica. PRI Research Report* No. 111, 96–100.

Junghans, D.T., Brasileiro, H.S., Santos, V.J., Cabral, J.R.S. and Matos, A.P. (2005) Herança da resistência a fusariose em abacaxizeiro. *Fitopatologia Brasileira* 30, S123–S123, Supplement.
Kerns, K.R. (1928) Persistence of characters in the 'Smooth Cayenne' pineapple. In: *Experiment Station of the Association of Hawaiian Pineapple Canners*, Bulletin No. 11. Honolulu, HI: University of Hawaii.
Kerns, K.R. (1931) Studies on the pollen of Cayenne. *Pineapple Quarterly* 1, 145–150.
Kerns, K.R. (1932) Concerning the growth of pollen tubes in pistils of Cayenne flowers. *Pineapple Quarterly* 1, 133–137.
Kerns, K.R. and Collins, J.L. (1947) Chimeras in pineapple. Colchicine-induced tetraploids and diploid-tetraploids in the Cayenne variety. *Journal of Heredity* 38, 322–330.
Kerns, K.R., Williams, D.F. and Smith, J.B. (1968) A review of seediness in pineapple. *PRI Research Report* No. 24.
Leal, F. and Amaya, L. (1991) The Curagua (*Ananas lucidus*, Bromeliaceae) crop in Venezuela. *Economic Botany* 45(2), 216–224.
Leal, F. and Coppens d'Eeckenbrugge, G. (1996) Pineapple. In: Janick, J. and Moore, J.N. (ed.) *Fruit Breeding, Vol. I. Tree and Tropical Fruits.* New York: Wiley Publishing Co., pp. 565–606.
Leal, F., Coppens d'Eeckenbrugge, G., Avilán Rovira, L. and Medina, E. (2010) *La piña de América o ananás.* Consejo de Desarrollo Científico y Humanístico. Caracas, Venezuela: Universidad Central de Venezuela.
Lewcock, H.K. (1940) Pineapple culture in Queensland. *Queensland Agricultural Journal* 53, 6–13.
Li, Y.-H., Zhang, Z. and Sun, G.-M. (2010) Changes in cell number and cell size during pineapple (*Ananas comosus*) fruit development and their relationship with fruit size. *Australian Journal of Botany* 58, 673–678.
Lim, W.H. (1971) Evaluating the susceptibility of pineapple cultivars to bacterial heart rot. *Malaysian Pineapple* 1, 23–27.
Lim, W.H. and Lowings, P.H. (1979) Pineapple fruit collapse in Peninsular Malaysia: symptom and varietal susceptibility. *Plant Disease Reporter* 63, 170–174.
Lin, C.H. and Chang, C.C. (2000) Pineapple production and industry in Taiwan. *Acta Horticulturae* 529, 93–97.
Lin, B.Y., Ritschel, P.S. and Ferreira, F.R. (1987) Número cromossômico de exemplares da família Bromeliaceae. *Revista Brasileira de Fruticultura* 9, 49–55.
Lin, Y.S., Kuan, C.S., Weng, I.S. and Tsai, C.C. (2015) Cultivar identification and genetic relationship of pineapple (*Ananas comosus*) cultivars using SSR markers. *Genetics and Molecular Research* 14(4), 15035–15043.
Linford, M.B. and Spiegelberg, C.H. (1933) Illustrated list of pineapple fruit diseases, blemishes and malformities. *Pineapple Quarterly* 3, 135–178.
Loison-Cabot, C. (1987) Practice of pineapple breeding. *Acta Horticulturae* 196, 25–36.
Loison-Cabot, C. (1990) Pineapple genetics: heritability of certain characters and their stability during vegetative cycles. *Fruits* 45, 447–455.
Loison-Cabot, C. and Lacoeuilhe, J.J. (1990) A genetic hybridization programme for improving pineapple quality. *Acta Horticulturae* 275, 395–400.
Lu, X.H., Sun, D.Q., Wu, Q.S., Liu, S.H. and Sun, G.M. (2014) Physico-chemical properties, antioxidant activity and mineral contents of pineapple genotypes grown in China. *Molecules* 19, 8518–8532.
Lutchmeah, R.S. (1992) Common disorders and disease of pineapple fruit cv. Victoria in Mauritius. *Revue Agricole et Sucrière* 7, 27–31.
Majumder, S.K., Kerns, K.R., Brewbaker, J.L. and Johannessen, G.A. (1964) Assessing self-incompatibility by a pollen fluorescence technique. *Proceedings of the American Society for Horticultural Science* 84, 217–223.
Marr, G.S. (1964) Study of self-incompatibility in pineapple. *Agricultural Research, Pretoria*, 2–6, 561–652.
Mathew, V, Lyla, K.R. and Nayar, N.K. (1979) Estimation of genetic variability in pineapple for quantitative and qualitative traits. *Indian Journal of Agricultural Science* 49 (11), 855–857.
Matos, A.P. and Souto, G.F. (1984) Avaliação de técnicas de inoculação para detectar resistência à fusariose em plântulas de abacaxi. *Revista Brasileira de Fruticultura* 6, 39–42.
Matos, A.P., Mourichon, X. and Lapeyre, F. (1991) Reaction of pineapple accessions on inoculations with *Fusarium monoliforme* var. *subglutinans*. *Fruits* 46, 647–652.
Miller, E.V. (1954) The natural origins of some popular varieties of fruit. *Economic Botany* 8(4), 337–348.
Ming, R., VanBuren, R., Wai, C.M., Tang, H.B., Schatz, M.C., *et al.* (2015) The pineapple genome and the evolution of CAM photosynthesis. *Nature Genetics* 47, 1435.

Montilla, I., Fernández, S., Alcalá, D. and Gallardo, M. (1997) El cultivo de la piña en Venezuela. Centro de Investigaciones Agropecuarias del Estado de Lara. IICA/CReA/PROCIANDINO/FRUTHEX. Maracay, Venezuela: Fondo Nacional de Investigaciones Agropecuarias.

Morales, M., Hernández, M.S., Cabezas, M., Barrera, J. and Martínez, O. (2001) Caracterización de la maduración del fruto de piña nativa *(Ananas comosus* L. Merrill) cv. India. *Agronomía Colombiana* 18(1–2), 7–13.

Morales, J.L., Sauter, H. and Young, T.R. (2006) Pineapple plant named 'Honey Gold'. *U.S. Plant Patent No. 16328 (P3).* Washington, DC: US Patent and Trademark Office.

Morton, J. (1987). Pineapple. In: Morton (ed.) *Fruits of Warm Climates.* Miami, FL: Julia F. Morton, pp. 18–28.

Muller, A. (1994) Contribution à l'étude de la fertilité et de l'autofertilité dans le genre *Ananas.* Thesis, ISTOM, College of International Agro-Development, Montpellier, France.

Munro, D. (1835) Classification of pineapple varieties. *Transactions of the London Horticultural Society (Ser. 2)* 1, 1–34.

Nagatomi, S., Degi, K. and Ikemiya, H. (1997) Development of polyploid induction and discrimination methods in pineapple. Institute of Radiation Breeding. *Technical News* No. 59.

Nayar, N.K., Valsamma, M. and Lyla, K.R. (1981) Varietal variations on pollen size and fertility in pineapple (*Ananas comosus*, L. Merr.). *Progressive Horticulture* 13, 85–87.

Nisha, A.P. and Radhamany, P.M. (2016) Fruit morphological variability in different commercial cultivars and local varieties of *Ananas comosus* (l.) Merr. from Kerala. *International Journal of Botany and Research* 6(5), 1–8.

Obrero, F. (1995) Information about a new pineapple clone. *Pineapple News* 1, 11–12.

Oda, C. and Williams, D.D.F. (1994) Pineapple plant named 'CO-2'. *U.S. Plant Patent No. 8863.* Washington, DC: US Patent and Trademark Office.

Ogata, T., Yamanaka, S., Shoda, M., Urasaki, N. and Yamamoto, T. (2016) Current status of tropical fruit breeding and genetics for three tropical fruit species cultivated in Japan: pineapple, mango, and papaya. *Breeding Science* 66, 69–81.

Pariona, P.R (2008) *Diagnóstico del cultivo de piña en la provincia de Chanchamayo.* Unidad de Promoción Agraria de la Agencia Agraria de Chanchamayo. Dirección Regional de Agricultura Junín. Available at: http://es.scribd.com/doc/97110641/diagnostico-del-cultivo-de-pina (accessed 2 March 2018).

Pereira, P.F. (2006) Os solos-sedimentos da região central do Estado do Acre (Feijó-Tarauacá), sua aptidão ao cultivo de abacaxi e a relação com os sedimentos fluviais atuais. Master's thesis, Programa de Pós Graduação em Geologia e Geoquímica, Universidade Federal do Pará, Belém, Brazil.

Perrottet, S. (1825) Catalogue raisonné des plantes introduites dans les colonies françaises de Mascareigne et de Cayenne, et de celles rapportées vivantes des mers d'Asie et de la Guyane, au Jardin des Plantes de Paris. *Mémoires de la Société Linnéenne* 3, 89–151.

Porcher, E. and Lande, R. (2016) Inbreeding depression under mixed outcrossing, self-fertilization and sib-mating. *BMC Evolutionary Biology* 16, 105.

Prakash, J., Bhattacharyya, D., Chattopadhyay, K., Roy, S., Das, S.P. and Singh, N.P. (2009) PQM-1: A newly developed superior clone of pineapple for northeastern India as evident through phenotype, fruit quality and DNA polymorphism. *Scientia Horticulturae* 120, 288–291.

Pulido, P. (2000) Desarrollo reproductivo de la piña en el piedemonte amazónico colombiano y su respuesta a la inducción con etileno. Thesis, Facultad de Ciencias Biológicas, Departamento de Biología, Universidad de los Andes, Bogotá, Colombia.

Py, C., Lacoeuilhe, J.J. and Teisson, C. (1987) *The Pineapple: Cultivation and Uses.* Paris: G.P. Maisonneuve et Larose.

Ramírez, O.D. (1966) Pollen fertility as a means of selecting male parent plants in pineapple. *Journal of Agriculture of the University of Puerto Rico* 50, 231–240.

Rao, A.N. and Wee, Y.C. (1979) Embryology of the pineapple, *Ananas comosus* (L.) Merr. *New Phytologist* 83, 485–497.

Rattanathawornkiti, K., Kanchanaketu, T., Suwanagul, A. and Hongtrakul, V. (2016) Genetic relationship assessment of pineapple germplasm in Thailand revealed by AFLP markers. *Genomics and Genetics* 9(2), 1–10.

Redondo-Echeverri, E. and Varón, F. (1992) Efecto de los nemátodos en el cultivo de la piña *Ananas comosus* L. (Merr.). *Fitopatología Colombiana* 16, 180–191.

Reinhardt, D.H., Cabral, J.R.S., Matos, A.P. and Junghans, D.T. (2012) 'BRS Ajubá', a new pineapple cultivar resistant to fusariosis and adapted to subtropical conditions. *Acta Horticulturae* 928, 75–79.

Ritzinger, R. (1992) Avaliação e caracterização de cultivares de abacaxi no Acre. Rio Branco, Brazil: EMBRAPA-CPAF/Acre.
Ritzinger, R. (2000a) Abacaxi 'Cabeça-de-Onça'. In: Luiz Carlos Donadio (ed.) *Novas variedades brasileiras de frutas*. Jaboticabal, Brazil: Sociedade Brasileira de Fruticultura, pp. 6–7.
Ritzinger, R (2000b) Abacaxi 'Rio Branco'. In: Luiz Carlos Donadio (ed.) *Novas variedades brasileiras de frutas*. Jaboticabal, Brazil: Sociedade Brasileira de Fruticultura, pp. 14–15.
Rohrbach, K.G. and Schmitt, D.P. (1994) Pineapple. In: Ploetz, R.C., Zentmyer, G.A., Nishiyima, W.T. and Rohrbach, K.G. (ed.) *Compendium of Tropical Fruit Diseases*. St Paul, MI: APS Press, pp. 45–55.
Rojas, O.E. and Aguilera, E.H.A. (1999) Respuesta del cultivo de la piña (*Ananas comosus* (L.) Merril) a diferentes dosis de N, P, K, en un suelo Oxisol del departamento del Meta. Thesis, Universidad de los Llanos, Villavicencio, Colombia.
Rouffiange, N. (1993) Diagnosis of the pineapple industry in Guyana: major constraints and perspectives. Port-of-Spain, Trinidad and Tobago: IICA.
Salazar, C. R. (1944) Situación del cultivo de la piña en Colombia. Memorias del curso regional de actualización en frutas tropicales. Programa de frutas tropicales C.1. Nataima. Espinal-Tolima. Mayo.
Sanewski, G.M. (1997) Evaluation of four pineapple cultivars on a winter plant crop cycle in south-east Queensland for the fresh market. *Maroochy Research Station Report* No. 7, 87–89.
Sanewski, G.M. (1998) The Australian pineapple fresh market breeding programme. In: *Abstracts of Third International Pineapple Symposium, 17–20 November 1998.* Pattaya, Thailand.
Sanewski, G. (2007a) Pineapple seed germination. *Pineapple News* 14, 5–7.
Sanewski, G.M. (2007b) 'Aus-Jubilee' and 'Aus-Carnival' pineapple description. *Plant Varieties Journal* 20(4), 228–238.
Sanewski, G. (2008) The effect of cut style pollination on fruit set in selfed pineapple. *Pineapple News* No. 15, 7–9.
Sanewski, G.M. (2009) The effect of different levels of inbreeding on self-incompatibility and inbreeding depression in pineapple. *Acta Horticulturae* 822, 63–70.
Sanewski, G. and De Faveri, J. (2017) The Australian fresh market pineapple breeding programme. *Acta Horticulturae* 1166, 41–46.
Sanewski, G.M., Simpson, D.R., Winks, C.W. and Sinclair, E.R. (1992) Selection of 'Smooth Cayenne' clones. In: *Annual Pineapple Field Day Notes*. Fortitude Valley, Qld: Queensland Fruit and Vegetable Growers.
Sanewski, G.M., Smith, M.K., Pepper, P.M and Giles, J.E (2011) Review of genetic improvement of pineapple. *Acta Horticulturae* 902, 95–108.
Sanewski, G.M., Ko, H.L., De Faveri, J. and Kilian, A. (2016) Genetic resistance to the root rot pathogen *Phytophthora cinnamomi* in *Ananas. Acta Horticulturae* IIII: 281–286.
Sanewski, G., Ko, L., Innes, D., Kilian, A., Carling, J. and Song, J. (2017) DArTseq molecular markers for resistance to *Phytophthora cinnamomi* in pineapple (*Ananas comosus* L.) *Australasian Plant Pathology* 46(5), 499–509. DOI: 10.10007/s13313-017-0512-1.
Sarah, J.L., Mesnildrey, L., Marguerite, E. and Boisseau, M. (1997) Laboratory screening of pineapple germplasm for resistance to the lesion nematode *Pratylenchus brachyurus. Acta Horticulturae* 425, 179–186.
Scherer, R.F., Olkoski, D., Souza, F.V.D., Nodari, R.O. and Guerra, M.P. (2015) Gigante de Tarauaca: A triploid pineapple from Brazilian Amazonia. *Scientia Horticulturae* 181, 1–3.
Sekiya, F.F. (2012) Pineapple plant named 'Franklynn'. *U.S. Plant Patent No. 23,193 (P3).* Washington, DC: US Patent and Trademark Office.
Sharma, A.K. and Ghosh, I. (1971) Cytotaxonomy of the family Bromeliaceae. *Cytologia* 36, 237–247.
Silva, G.S., Krasucki, A.I.S., Rozário, I.L.M. and Leite, R.R. (2012) Resistência do abacaxizeiro 'Turiaçu' a *Meloidogyne arenaria, M. enterolobii, M. incognita* e *M. javanica. Summa Phytopathologica* 41(3), 227–228.
Singh, R. and Iyer, C.P.A. (1997) Standardisation of dosimetry and techniques for inducing mutations in pineapple. In: Nijjar, G.S. (ed.) *Fruit Breeding in India.* Delhi, India: Oxford & IBH Publishing Co.
Sipes, B.S. and Schmitt, D.P. (1994) Evaluation of pineapple, *Ananas comosus*, for host-plant resistance and tolerance to *Rotylenchulus reniformis* and *Meloidogyne javanica. Nematropica* 24, 113–121.
Smith, J.B. (1966) II – Breeding for resistance. In: Review of the root rot and heart rot problem in pineapple. *PRI Research Report* No. 118.

Spironello, A., Siqueira, W.J., Usberti Filho, J.A., Teófilo Sobrinho, J., Carvalho, C.R.L. *et al.* (2010) Cultivar de abacaxizeiro IAC Fantastico. Campinas: Instituto Agronômico de Campinas. Available at: http://www.iac.sp.gov.br/cultivares/inicio/Folders/Abacaxi/IACFant%C3%A1stico.htm (accessed 2 March 2018).
Sripaoraya, S. (2009) Pineapple hybridization and selection in Thailand. *Acta Horticulturae* 822, 57–62.
Subramanian, N., Iyer, C.P.A. and Singh, R. (1981) Surmounting self-incompatibility in pineapple (*Ananas comosus* L.) with pollen irradiation. *Indian Journal of Horticulture* 38, 162–164.
Swarts, D.H. (1990) The effects of pre-storage conditions and storage temperatures on the keeping quality of Queen pineapples under simulated sea freight conditions. *Subtropica* 11, 15–17.
Tang, C.H., Tsai, M.H., Tsai, H.W. and Kuan, C.S. (2014) A new variety of pineapple [*Ananas comosus* (L.) Merr.] 'TN21 Golden'. *Journal of.Taiwan Agricultural Research* 63(2), 167–177.
Tang, C.H., Tsai, M.H., Tsai, H.W. and Kuan, C.S. (2015) A new variety of pineapple (*Ananas comosus* (L.) Merr.) 'Tainung No.22 (Honey Fragrance)'. *Pineapple News* 22, 49–50.
Taniguchi, G., Sanewski, G.M., Bartholomew, D.P. and Paull, R.E. (2008) Characteristics of the Pineapple Research Institute of Hawaii hybrids 73-50 and 73-114. *Pineapple News* 15, 27–33.
Usberti Filho, J.A., Siqueira, W.J., Spironello, A., Tanaka, M.A.S., Sigrist, J.M.M. *et al.* (1999) Abacaxi IAC Gomo-de-mel. Campinas: Instituto Agronômico de Campinas. Available at: http://www.iac.sp.gov.br/cultivares/inicio/Folders/Abacaxi/cIACGomo-de-Mel.htm (accessed 2 March 2018).
Vagneron, I., Faure, G. and Loeillet, D. (2009) Is there a pilot in the chain? Identifying the key drivers of change in the fresh pineapple sector. *Food Policy* 34, 437–446.
Velasco, J. (1927) Historia del Reino de Quito, en la América meridional. Escrita por el Presbítero Don Juan de Velasco, nativo del mismo reino. Año de 1789. Quito. Imprenta Nacional. Tomo I and Part 1a. Que contiene la Historia Natural.
Ventura, J.A., Costa, H., Cabral, J.R.S. and Matos, A.P. (2009) 'Vitória': new pineapple cultivar resistant to fusariosis. *Acta Horticulturae* 822, 51–56.
Viana, J.M.S., Mundin, G.B., Lima, R.O., Silva, F.F.E. and Resende, M.D.V. (2014) Best linear unbiased prediction for genetic evaluation in reciprocal recurrent selection with popcorn populations. *Journal of Agricultural Science* 152(3), 428–438.
Wassman, R.C. (1982) The importance of selected clones in pineapple production. In: *Annual Pineapple Field Day Notes.* Fortitude Valley, Qld: Queensland Fruit and Vegetable Growers.
Wee, Y.C. (1972) Some common pineapple cultivars of West Malaysia. *Malaysian Pineapple* 2, 7–13.
Wee, Y.C. (1974) The Masmerah pineapple: a new cultivar for the Malaysian pineapple industry. *World Crops* March/April, 64–67.
Wee, Y.C. and Rao, A.N. (1979) *Ananas* pollen germination. *Grana* 18, 33–39.
Williams, D.F. (1970a) Production, propagation and testing of new varieties. *PRI News* 18(1).
Williams, D.D.F. (1970b) Seediness in pineapple. *PRI News* 18(2), 51–60.
Williams, D.D.F. and Fleisch, H. (1993) Historical review of pineapple breeding in Hawaii. *Acta Horticulturae* 334, 67–76.
Williams, D.D.F and Yamane, G.M. (1970) Open flowers induced on closed-flower varieties by gibberellin. *PRI News* 18(1), 11–12.
Winks, C.W., Glennie, J.D. and Lanham, T. (1985) Progress report: pineapple breeding. *Maroochy Horicultural Research Station Report 1984–1985* 4, 175–177.
Young, R. and Gonzales, J. (2010) Pineapple plant named 'Dole–14'. *U.S. Plant Patent No. 20,885 (P3).* Washington, DC: US Patent and Trademark Office.
Zheng, L.Y., Sun, G.M., Liu, Y.G., Lv, L.L., Yang, W.X., Zhao, W.F. and Wei, C.B. (2012) Aroma volatile compounds from two fresh pineapple varieties in China. *International Journal of Molecular Sciences* 13, 7383–7392.

5 Biotechnology*

José Ramon Botella,[1] Sharon Hamill,[2] Hian-Lien Ko,[2] Tracie Matsumoto[3] and Fernanda Vidigal Duarte Souza[4]

[1]University of Queensland, St Lucia, Australia; [2]Queensland Department of Agriculture and Fisheries, Nambour, Australia; [3]ARS Daniel K. Inouye US Pacific Agricultural Research Center, Hilo, HI, USA; [4]Embrapa Mandioca e Fruticultura, Cruz das Almas, Brazil

---

Biotechnology is a broad field of several disciplines and their applications including cell and meristem culture, *in vitro* manipulation of tissues, genetic studies and genetic manipulation. It is a rapidly expanding and advancing field of study and there has been much activity in pineapple in recent years.

Tissue Culture

Applications for tissue culture include micropropagation via axillary bud proliferation, adventitious bud proliferation and regeneration from callus cultures, somatic embryogenesis, use of bioreactors and temporary immersion systems, *in vitro* germplasm conservation, and culture of excised protoplasts, ovules or anthers.

Micropropagation

In practice, micropropagation is mostly used for the establishment of multiplication blocks which then provide conventional planting material for larger production blocks. The initial limited use of micropropagation in the multiplication of new cultivars allows screening of variants on fruit characteristics before conventional methods of multiplication are used.

The most common explant used to initiate cultures is the axillary bud dissected from crown leaves. Fitchet (1990) also suggests the crowns of some cultivars, such as 'Smooth Cayenne', should be first desiccated for a short period to break bud dormancy. Even though contamination of the initial cultures is common, enough buds can be obtained to ensure successful establishment of some explants following surface disinfestation. 'Clean' buds are most commonly grown on a Murashige and Skoog (1962) solid medium (MS) supplemented with a cytokinin, usually benzyladenine (BA) at between 1.75 and 5 mg l^{-1} (Hamad and Taha, 2008a). Using these concentrations with a cultivar such as 'Smooth Cayenne', 10–15 plants per month can be produced. Sucrose is the common carbon source added at 3 g l^{-1} for pineapple (Mengesha *et al.*, 2013; Kunze *et al.*, 2014) although Nelson *et al.* (2015) found that 5 g l^{-1} sucrose to be optimum. Hamad and Taha (2008b) found that pineapple multiplication decreased during time in culture and that the optimum subculture interval was 60 days.

A micropropagation method developed by Kiss *et al.* (1995) involved the induction of

---

* A revision of the chapter by Mike K. Smith, Hian-Lien Ko, Sharon D. Hamill, Garth M. Sanewski and Mick W. Graham.

 The Pineapple, 2nd Edition: Botany, Production and Uses (eds G.M. Sanewski *et al.*)

etiolated shoots with subsequent multiplication of shoots along the nodal segments when placed horizontally on the culture medium.

A double phase culture system for pineapple was developed by Scherwinski-Pereira *et al.* (2012) to improve multiplication and reduce labour costs.

Once shoots have been produced and multiplied they are usually transferred to an MS semi-solid medium containing an auxin, such as Indole-3-butyric acid (IBA) or 1-Naphthaeneacetic acid (NAA) at between 0.5–2.0 mg l^{-1} (Hamad *et al.*, 2013) or to a growth regulator-free medium for root development (Kunze *et al.*, 2014).

A summary of some available pineapple micropropagation methods is presented in Table 5.1.

Plants derived from tissue culture display similar chemical and nutrient profiles, maturation periods and physical properties as traditionally grown pineapple (Jackson *et al.*, 2016).

Organogenesis and somatic embryogenesis

Plants can also be regenerated from cell or callus cultures. When the cytokinin is supplemented with an auxin such as naphthalene acetic acid (NAA) or 4-amino-3,5,6-trichcloropicolinic acid (picloram), the development of 'lumpy' tissue with protocorm-like bodies or callus tissue can result (Mathews and Rangan, 1979, 1981; Wakasa, 1989; Devi *et al.*, 1997; Yapo *et al.*, 2011). Proliferation of these types of tissue results in cultures with an enhanced regeneration capacity via either organogenesis (Fitchet, 1990) or somatic embryogenesis (Daquinta *et al.*, 1996; Cisneros *et al.*, 1998; Yapo *et al.*, 2011). Regeneration of plants from cell and callus cultures is the basis of the genetic transformation of pineapples. However, cell and callus culture can contribute to an increased level of somaclonal variation (Scowcroft, 1984).

Liquid culture systems

Liquid systems have now been developed that improve the efficiency of pineapple micropropagation. Conventional semi-solid media and different types of liquid media (temporary immersion and continuous immersion with aeration) have been compared and temporary immersion was found to be the most efficient method of propagation (Silva *et al.*, 2007; Rizzal *et al.*, 2014). Immersion systems are now widely used for mass propagation of pineapple and improved systems have been described by several researchers and include temporary immersion (Escalona *et al.*, 1998, 1999; Ayenew *et al.*, 2013; Scherer *et al.*, 2013, 2015b) and immersion by bubbles (Scheidt *et al.*, 2009).

In vitro production of metabolites

Bromelain is a valuable mixture of proteolytic enzymes found in several parts of the pineapple plant and fruit and is in high demand (Norsyalieza and Illias, 2017). Tissue culture has been evaluated as a way to both produce bromelain using *in vitro* plants and to increase levels of bromelain in plants (Neta *et al.*, 2012). The level of bromelain activity varied depending on culture conditions (Leite *et al.*, 2012).

Cultivar improvement

Pineapple tissue culture has been used to develop new cultivars by selection processes or mutation breeding. A clone with improved fruit quality and resistance to bacterial heart rot (*Erwinia chrysanthemi*) was developed from susceptible 'Josapine' pineapple (Ibrahim *et al.*, 2009) by exposing micropropagated plants to gamma irradiation. There have been attempts to develop resistant pineapple plants by exposing callus to filtrates of pathogens and selecting callus that survives exposure. Borras *et al.* (2001) used culture filtrate of *Fusarium subglutinans* (fusariose), to select plants with improved tolerance and Valencia *et al.* (2014) screened callus using filtrate from *Phytopthora cinnamomi* (Phytopthora heart rot). Although these methods offer potential for cultivar improvement, commercial resistant cultivars have not yet been reported.

Beneficial microorganisms

Several naturally occurring, endophytic, potentially symbiotic bacteria have been identified in pineapple (Gonzales *et al.*, 1996; Guillemin *et al.*,

Table 5.1. Recent methods of micropropagation of pineapple.

Stage I	Stage II	Stage III	Stage IV	Reference
Explant: Apical bud from crown Medium: MS with 1.0 mgl^{-1} BA and 0.1 mgl^{-1} NAA	Medium: as for Stage I, except 2.0 mgl^{-1} BA 'regeneration from proliferating callus at base of plantlets' 56 plantlets/month	MS with 2.0 mgl^{-1} IBA	Not stated	Devi *et al.* (1997)
Explant: axillary bud shoot from crown, Medium: MS with 2.0 mgl^{-1} BA and 0.25 mgl^{-1} NAA and 30 g l^{-1} sucrose	Medium: MS with 2.0 mgl^{-1} BA and 0.25 mgl^{-1} NAA and 30 g l^{-1} sucrose semi-solid media with liquid top layer same ingredients DP after four cycles over 5 months starting with five explants; cultivars 'Rio Branco', 'Senador Guiomard' and 'Quinari' produced 431, 363 and 362 plantlets respectively.	MS full or ½ strength with 0.5–2.5 μM IBA	Soil, vermiculite and cow manure 3:1:1	Scherwinski-Pereira *et al.* (2012)
Explant: apical shoot from slip, TIS Medium: MS no growth regulators and 20 g l^{-1} sucrose	Medium: MS with 2.0 mgl^{-1} BA and 30 g l^{-1} sucrose, TIS. 13.17 shoots after 6 weeks	MS with 3.0 mgl^{-1} IBA and 40 g l^{-1} sucrose, TIS	Soil, sand and decomposed coffee husk 2:1:1	Ayenew *et al.* (2013)
Explant: block with apical meristem 2×4 cm cut into four vertical sections Medium: MS with 02.5 mgl^{-1} BA and 0.1 mgl^{-1} NAA, 1.0 mgl^{-1} TDZ and 30 gl^{-1} sucrose	Medium: MS with 2.5 mgl^{-1} BA, 0.1 mgl^{-1} NAA and 30 g l^{-1} sucrose in TIS, liquid, semisolid media in order of highest multiplication rate 10.7, 10.5, 5.5 plantlets after 60 days (8.5 weeks)	Not stated	Not stated	Rizzal *et al.* (2014)
Explant: apical bud from crown Medium: MS with 0.1 mgl^{-1} BA and 1 mgl^{-1} NAA	Medium: MS with 6.0 mgl^{-1} BA and 30 g l^{-1} sucrose, nine shoots per bud after 8 weeks	Not stated	Not stated	Nelson *et al.* (2015)

Abbreviations: IAA, indole acetic acid; IBA, indole butyric acid; BA, benzyladenine; KIN, Kinetin; NAA, naphthalene acetic acid; TDZ, Thidiazuron; MS, Murashige and Skoog (1962); MT, Murashige and Tucker (1969); N6, Chu (1978); Nitsch (1951) TIS Temporary Immersion System, DP Double-phase culture system

1996; Matos *et al.*, 1996; Tapia-Hernandez *et al.*, 2000; Weber *et al.*, 2013; Esposito-Polesi *et al.*, 2017) and some have now been shown to improve growth and establishment during the acclimatization phase of micropropagated pineapple (Baldotto *et al.*, 2010, 2011; Gonzalez-Rodriguez *et al.*, 2013; Aviz Silva *et al.*, 2016).

Somaclonal variation

Spiny leaves are often encountered during conventional propagation of spiny-tip cultivars such as 'MD-2' and 'Smooth Cayenne' and it is no surprise that spininess is also commonly reported in micropropagated spiny-tip pineapple cultivars (Wakasa, 1979, 1989; Liu, 1987; Smith and Drew, 1990). Other somaclonal variants recorded in micropropagated pineapple include variations to leaf colour, leaf shape, wax, foliage density and phyllotaxy (Wakasa, 1989).

Banana is micropropagated using similar culture media as pineapple and there is evidence in banana that higher levels of growth regulators are linked to an increased incidence of off-types (Bairu *et al.*, 2006). Some research suggests that off-types more commonly occur in adventitious buds compared to auxillary buds and that low levels of cytokinins should be used to reduce adventitious budding (Damasco *et al.*, 1996). While there has been some conflicting evidence of the role that plant growth regulators play in somaclonal variation, there have been several studies demonstrating that off-type events increase the longer plants are in tissue culture, and the more times the plants are subcultured (Damasco *et al.*, 1996; Borse *et al.*, 2011). A similar effect has been found in pineapple which showed that long-term culture was the main contributor to somaclonal variation (Roostika *et al.*, 2015).

The type of plant growth regulator and method of micropropagation is suggested to influence somaclonal variation. Feuser *et al.* (2003) described higher (although not significantly so) levels of variation in pineapple culture with paclobutrazol and lower levels of variation in temporary immersion systems compared to stationary systems.

To mimimize off-types in pineapple micropropagation, the media should contain low levels of growth regulators, multiplication of plants from each bud should be limited to eight cycles or 500 plantlets, cultures should be established from many buds from many crowns and cultures re-initiated annually (Hamill, 2016).

Roguing of off-types, particularly during nursery establishment, is essential to reduce the percentage of off-types. Many of the somaclonal variants recorded by Wakasa (1989) can be easily identified while the plants are small. Other less obvious variants, such as those that produce small, malformed or cracked fruit, require a fruiting cycle for identification.

In vitro germplasm conservation

Among the various conservation strategies, *in vitro* preservation of micropropagated plants has been used for many species, notably those that are propagated asexually (Souza *et al.*, 2006; Silva *et al.*, 2016; Sugimoto *et al.*, 1991). The advantages of *in vitro* conservation include the possibility of maintaining a large number of accessions in a small physical space, free of intemperate weather and other risks to which fields are exposed, and the ease and security of germplasm exchange.

In vitro conservation involves the maintenance of cultured plants in active but slow growth with periodic subculturing to fresh medium. The factors used to achieve this range from reduced temperature and luminosity to alterations in the formulation of the culture medium (Zee and Munekata, 1992; Canto *et al.*, 2004; Souza *et al.*, 2006; Silva *et al.*, 2016). Prolonged conservation and repeated subculturing can however generate somaclonal variation or cause fungal contamination and loss of plants. Consideration should also be given to the effect of long culture times on the vegetative rescue capacity and multiplication rate of the conserved accessions (Silva *et al.*, 2016).

Three official pineapple germplasm collections exist in the world, one in Hawaii with 181 accessions preserved as potted plants and *in vitro*; that of CIRAD in Martinique, with over 600 accessions kept in greenhouses; and the Active Germplasm Bank (AGB) of the Embrapa Cassava and Fruits research unit (Embrapa Mandioca e Fruticultura) in Brazil, containing approximately 700 accessions preserved in a field, 60% of which are also conserved *in vitro*.

Other smaller collections are reported, such as one in Cuba, with accessions preserved *in vitro* and by cryopreservation.

The pineapple collection in the US National Plant Germplasm System (NPGS) is housed at the USDA ARS Daniel K. Inouye US Pacific Agricultural Research Center in Hilo, Hawaii. A total of 239 accessions of pineapple comprise the collection with 237 accessions grown in the greenhouse and 222 accessions in tissue culture. This storage protocol was further optimized by changing culture vessels to 14 mm diameter test tubes to restrict lateral growth of the plants where plants are transferred every 15–18 months (C. Arakawa, personal communication). Cultures are maintained under two 40-watt cool white fluorescent bulbs located 46 cm above the test tubes for 16 hours each day (8 hours dark) at 24°C.

The Embrapa Cassava and Fruits AGB was established in 2003 using a slow-growth regime (Canto *et al.*, 2004; Souza *et al.*, 2006; Silva *et al.*, 2016). Each accession includes six plants preserved in test tubes containing MS culture medium with a salt concentration of 1/3, supplemented with 30 g l^{-1} of sucrose, 1.8 g l^{-1} of Phytagel ®, no growth regulators, and with pH 5.8. The plants are kept at 21 ± 1°C, with a light intensity of 300 $Em^{-2}\,s^{-1}$ and photoperiod of 12 hours. Plants are first indexed by RT-PCR for pineapple mealybug wilt associated viruses (PMWaVs). If the presence of any viruses is detected, the buds of that accession undergo a cleaning process, consisting of culturing a 0.5 mm long shoot tip in a medium for removal of the viral complex. Cryotherapy can be a complementary technique.

Silva *et al.* (2016) performed a study with 66 pineapple accessions, of all botanical varieties, that had been preserved *in vitro* for ten years under slow-growth regime in the Embrapa conditions described. The results indicated it was possible to perform subculturing at 24-month intervals as with *in vitro* conservation, and still achieve resumed growth of all accessions.

Cryopreservation

Modern cryopreservation techniques have made it possible to conserve collections for long periods at ultra-low temperatures (−196°C) without the need for periodic interventions (Fig. 5.1). The nearly total interruption of the plant's metabolism, while maintaining cellular integrity, makes this form of conservation effective and inexpensive (Benson, 2008). The main cost is the investment to establish the basic storage structure, consisting of cryogenic tanks and a nitrogen flow system (Reed, 2008; Panis, 2009).

Souza *et al.* (2015) developed a protocol based on the droplet vitrification technique for the cryopreservation of wild pineapple cultivars. Freezing of cultured 0.5 mm shoot tips gave success rates ranging from 44% to 86% among the different accessions.

The droplet vitrification protocol includes a pre-culturing step lasting 48 hours in 0.3 M sucrose, followed by vitrification in a PVS2 solution (Sakai *et al.*, 1990) for 45–60 minutes, depending on the variety, and thawing in a sucrose-rich solution (1.2 M) for 20 minutes. The recovery and development of the tips is achieved by culturing in an establishment medium (MS with 0.05 mg l^{-1} of BAP) for approximately 40 days. Healthy conditions of the initial explant, with tissue having dense cells, and good dexterity in excising the shoot tip, are fundamental for the success of this process.

Other slightly more laborious techniques, such as encapsulation and vitrification have also obtained good results (Gonzalez-Arnao *et al.*, 1998, 2000; Martinez-Montero *et al.*, 2005, 2012).

Anther and Ovule Culture

There has been interest in the development of haploid cultivars as a first step to circumvent the heterozygosity observed in segregating seedling populations. To date, limited success has been achieved with ovule culture but anther culture appears more difficult (Benega *et al.*, 1996; Arias Valdes *et al.*, 1998).

Protoplast Isolation and Fusion

Like anther and ovule culture, there is some doubt whether the development of techniques

Fig. 5.1. (a) *In vitro* plants for removal of shoot apices; (b) root apices in regeneration after freezing; (c) acclimated cryopreserved plants and (d) cultivation of cryopreserved accessions for evaluation. Photos F. Vidigal Duarte Souza

for protoplast isolation and fusion are likely to be worthwhile in pineapple. Protoplasts have been used in some monocot species for transformation, but there are inherent problems associated with the use of protoplasts. Protoplast fusion is sometimes used where parent incompatibility prevents hybridization by conventional means. This approach will however result in the development of tetraploid lines. Tetraploid pineapple cultivars studied to date have been inferior types (Collins and Kerns, 1947). Nevertheless, protoplasts of the cultivar 'Perolera' have been successfully isolated (Guedes *et al.*, 1996) but plant regeneration was not achieved.

Zhao *et al.* (2011) also described a method of protoplast isolation.

Genes and Genetic Resources

Expression analyses, including EST, microarray and next generation sequencing studies, have provided insights into the function of many newly discovered genes in pineapple (Koia *et al.*, 2012; Ong *et al.*, 2012b; Moyle and Botella, 2014).

Moyle *et al.* (2005a, 2006) created PineappleDB, an EST sequencing library constructed by identifying expressed gene sequences associated with fruit ripening and nematode-plant interaction. The library contains over 5600 ESTs and 3383 consensus sequences and is accessible online (PineappleDB, 2007).

Seven per cent of pineapple ESTs have been found to contain SSRs, which is relatively high compared to 2–5% in other species. Furthermore, 85.3% of the published pineapple ESTs has been annotated to known proteins (Ong *et al.*, 2012a). By 2011, approximately 6000 pineapple ESTs were entered in the NCBI dbEST database (GenBank, 2017; see also Ong *et al.*, 2012a,b; Zhou *et al.*, 2015).

Another information database that might eventually be useful for pineapple has been created by CIRAD, France (Carlier *et al.*, 2007). This database manages genomic, genetic and phenotypic information about a range of tropical crops

(TropGeneDB, n.d.) but currently does not include pineapple.

Finally, bioinformatics is being increasingly used to provide functional data on genes involved in important physiological and biochemical processes such as CAM photosynthesis, flowering, diurnal cycling (cis-regulatory elements and transcription factors), carbohydrate metabolism and carbon flux, proteolytic enzymes, disease resistance, genes with alternative splicing events, and micro RNAs (Cazzonelli *et al.*, 1998; Yusuf *et al.*, 2015; Ming *et al.*, 2016; Paull *et al.*, 2016; Wai *et al.*, 2016a,b, 2017; Zhang *et al.*, 2016; Zheng *et al.*, 2016; Sanewski *et al.*, 2017; Sharma *et al.*, 2017).

Transcriptomic approaches are being increasingly used to identify differentially expressed genes (DEGs). Li *et al.* (2016a) identified 56 DEGs and Hanqing *et al.* (2013) found 60 DEGs associated with flower initiation in *Aechmea fasciata*. Liu and Fan (2016) found 13 DEGs associated with flower initiation in pineapple and Li *et al.* (2016b) identified ethylene receptors associated with flowering in pineapple. Li *et al.* (2017b) used RNA-seq to study the DEGs associated with the albino tissues in a variegated clone of *A. comosus* var. *bracteatus*. Li *et al.* (2017a) used qPCR to identify stable 'house-keeping' genes (18s, EF1 and alpha-tubulin) in *A. comosus* var. *bracteatus* that would be suitable as reference genes for qRT-PCR studies. Ma *et al.* (2015) used high-throughput transcriptome sequencing of *A. comosus* var. *bracteatus* to identify genes coding for terpenoid and phenylpropanoid biosynthesis. Ma *et al.* (2014) identified three somatic embryogenesis receptor-like kinase genes in 'Shenwan' ('Queen') pineapple. Moyle *et al.* (2005b) performed a transcriptome analysis of genes expressed during fruit development.

Pineapple Genomes

Nuclear genome

Pineapple biotechnology has experienced a quantum leap with the sequencing of the genomes of the pineapple cultivars 'Smooth Cayenne' (clone 'F153') and 'MD–2' as well as the primitive pineapple, *Ananas comosus* var. *bracteatus* (Ming *et al.*, 2015; Redwan *et al.*, 2016).

The first partial genetic map was constructed and deposited in the NCBI database (NCBI, n.d.) by Carlier *et al.* (2004). This initiative used random amplified polymorphic DNA (RAPD), amplified fragment length polymorphism (AFLP) and inter-simple sequence repeat (ISSR) markers in a pseudo-test cross strategy, resulting in a 31.6% and 57.2% coverage of the total genome length, respectively, for *Ananas comosus* var. *comosus* and *Ananas comosus* var. *bracteatus*. Further additions were made by Carlier *et al.* (2006, 2007, 2012) using an F2 population and a combination of random and sequence specific markers, SSRs, EST-SSRs, SCAR and CAPS markers.

Wai (2014) also used an F2 mapping population to assemble a pineapple genome, but engaged a different strategy of whole genome random shotgun genome sequencing. This technique identified restriction sites associated DNA (RAD) markers, from which high-quality SNP markers were obtained and mapped into 30 linkage groups. Nashima *et al.* (2016) employed the same approach to identify candidate SSR sequences or short tandem repeats (STR), from the existing pineapple sequence genome, as according to these authors only 298 *Ananas* SSRs were registered in the NCBI database by 2014.

De Sousa *et al.* (2013) presented an improved version of the genomic map, which was an integration of F1- and F2-based maps of *A. comosus* var. *comosus*×*A. comosus* var. *bracteatus* progeny and the inclusion of 41 new sequence specific markers, assembling 741 loci in 28 integrated linkage groups. This study reported that approximately 1300 genomic sequences had been uploaded to the main NCBI genome database, of which 1163 belonged to *A. comosus* var. *comosus* and 105 belonged to *A. comosus* var. *bracteatus*. Other research groups have also contributed to populating the integrated pineapple genome map (Moyle *et al.*, 2005a, 2006; Kinsuat, 2007; Kinsuat and Kumar, 2007; Botella and Smith, 2008; Wöhrmann and Weising, 2011; de Sousa *et al.*, 2013).

The most recent uploads for 'Smooth Cayenne' by Ming *et al.* (2015), now cover 86% of the assembled genome.

The 'MD-2' genome is 508 Mb and includes 27,017 predicted protein-coding genes (Redwan *et al.*, 2016).

Chloroplast genome

Aside from nuclear genes, the pineapple chloroplast genome has been fully sequenced by two independent groups (Nashima *et al.*, 2015; Redwan *et al.*, 2015). The complete chloroplast genome sequence for *Ananas comosus* has a length of 159,636 bp and a total of 132 genes, of which 113 are unique genes, consisting of 59.8% coding and 40.2% non-coding regions (Nashima *et al.*, 2015).

Genetic Engineering

Given the high heterozygosity present in domestic pineapple, genetic engineering seems especially appropriate for pineapple improvement as it can introduce single genes or highly precise genetic changes without changing other agronomic and quality attributes.

The first genetically modified pineapple to be approved for commercialization in the USA, 'Rosé', has been produced by interfering with the carotenoid synthesis pathway in the cultivar 'MD-2'. Overexpression of the phytoene synthase gene from *Citrus reticulata*, was combined with suppression of two lycopene cyclases to allow lycopene accumulation in the fruit, resulting in a distinctive pink colour (Firoozabady and Young, 2013).

Promoters

Initial work showed that promoters from different plant species can be used in pineapple although it is usually preferable to use promoters from the same species as promoter activity requires coordination with endogenous transcription factors (Patent Lens, n.d.).

The maize ubi-1, enhanced CaMV 35S, SCSV4 (subclover stunt circovirus), *arabidopsis ubq3*, sugarcane *ubi9*, banana ethylene forming enzyme (EFE) promoter, pineapple ubiquitin, pineapple bromelain inhibitor promoter, pineapple acac2 promoter, *arabidopsis* modified mannopine synthase promoter (smas), and rice 35s have been used in pineapple (Firoozabady and Gutterson, 1998; Graham *et al.*, 1998; Espinosa *et al.*, 2002; Wei *et al.*, 2003; Firoozabady *et al.*, 2006; Yabor *et al.*, 2006; Mhatre *et al.*, 2009, 2013; Wang *et al.*, 2009; Ma *et al.*, 2012; Firoozabady and Young, 2013).

Inclusion of an intron in the 5′ untranslated region of ubiquitin promoters usually improves overexpression of transgenes in monocots and this approach has been used in pineapple (Wang and Paull, 2018).

Pineapple promoters with constitutive expression such as the translation factor SUI1 and the ribosomal protein L36 have been cloned and their expression patterns studied in *Arabidopsis thaliana* (Koia *et al.*, 2013).

The AcMADS1 transcription factor which is strongly induced during pineapple fruit ripening and its promoter have been shown to direct strong expression in fruits and flowers of *arabidopsis* and tomato (Koia *et al.*, 2012; Moyle *et al.*, 2014).

Transformation

Transformation techniques include infection with *Agrobacterium tumefaciens* or biolistics. Both techniques have proven to be effective in pineapple (Ko *et al.*, 2005).

Nan *et al.* (1996) first reported the use of biolistics on liquid embryogenic protocorm cultures, obtaining low levels of transgenic material. Leaf bases and calli have been much more efficient (Sripaoraya *et al.*, 2001, 2006; Ko *et al.*, 2006). Transgenic pineapple plants using the *Agrobacterium*-mediated technique on embryogenic callus cultures have been successfully reported by several groups (Firoozabady and Gutterson, 1998; Isidron *et al.*, 1998; Graham *et al.*, 2000; Sripaoraya *et al.*, 2001; Espinosa *et al.*, 2002; Firoozabady *et al.*, 2006; Mhatre *et al.*, 2009, 2013; Ma *et al.*, 2012; Firoozabady and Young, 2013). Aside from conventional tissue culture techniques, temporary immersion bioreactors have been used to improve the transformation efficiency (Espinosa *et al.*, 2002).

Traits that have been targeted in pineapple using genetic engineering include nematode resistance, pineapple mealybug wilt virus resistance, flowering and fruit ripening control, blackheart resistance, disease resistance (*Phytophthora nicotianae*), enhanced iron and zinc content, improved lycopene content and herbicide tolerance (Botella *et al.*, 1998; Graham *et al.*, 1998; Rohrbach *et al.*, 1998; Sanewski *et al.*, 1999; Sripaoraya *et al.*, 2001; Trusov and Botella, 2006; Mhatre *et al.*, 2009, 2011, 2013; Firoozabady and Young, 2013; Ko *et al.*, 2013; Tu *et al.*, 2017).

In vitro selection procedures

For pineapple transformation, a number of selectable markers have been used including *SurB*, *nptII*, *hph*, *Hpt*, *GUS* and *GFP* (Sripaoraya *et al.*, 2001; Espinosa *et al.*, 2002; Firoozabady *et al.*, 2006; Ko *et al.*, 2006; Yabor *et al.*, 2006; Wang *et al.*, 2009).

Molecular Markers

Molecular markers can be used to enhance conventional breeding and genetic engineering, and to determine genetic diversity and phylogenetic relationships. Genetic classification in the genus *Ananas* was originally based on quantitative, morphological (Smith and Downs, 1979) and qualitative variables, followed by the use of enzymatic markers (isozyme analysis) (Aradhya *et al.*, 1994) and most recently using DNA-based marker techniques.

The earliest report of polymorphisms in pineapples was by biochemical detection using isozyme markers, which was based only on phenotypic traits and produced low numbers of useful markers. It also revealed inconsistencies with the *Ananas comosus* group classification (Paz *et al.*, 2005, 2012; Botella and Smith, 2008; Rodríguez *et al.*, 2013; Zhang *et al.*, 2014). Since then, a number of molecular markers have been employed for genetic diversity studies and population genetics, generating high-density linkage maps to develop fingerprints for IP protection, marker-assisted selection, comparative genomics, and detection of somaclonal variation for conservation purposes.

Restriction fragment length polymorphism (RFLP)

The RFLP technique has been used to detect a high level of polymorphism and is robust, but is relatively expensive and time-consuming (Noyer *et al.*, 1996; Duval *et al.*, 1998; Paz *et al.*, 2005, 2012). On the other hand, Kato *et al.* (2004) found that RFLP was not highly polymorphic and not very informative for distinguishing among accessions.

Randomly amplified polymorphic DNA (RAPD)

Several groups have applied RAPD markers (Ruas *et al.*, 1995; Kinsuat, 2007; Popluechai *et al.*, 2007; Ines *et al.*, 2009; Burhooa and Ranghoo-Sanmukhiya, 2012) to reveal genetic relationships among Thai, Malaysian, Philippino and Mauritian pineapple cultivars, respectively. Findings reported were a variable 8–85% of the primers, producing useful polymorphic differentiation, groupings consistent with the morphological classification, and the detection of duplications in some of the collections. Kinsuat (2007) exploited RAPD markers to sequence polymorphic fragments, which were deposited into GenBank.

Amplified fragment length polyporphism (AFLP)

AFLPs have been used for genetic diversity analysis and germplasm evaluation in pineapple (Kato *et al.*, 2004; Paz *et al.*, 2005, 2012; Tapia *et al.*, 2005; Rattanathawornkiti *et al.*, 2016). Rattanathawornkiti *et al.* (2016), Tapia *et al.* (2005) and Paz *et al.* (2005) applied AFLP on Thai, Mexican and Cuban germplasm collections, respectively. The research groups reported that the cluster classifications produced by the analysis were in agreement with the morphological characterization of the germplasm collection. Kato *et al.* (2004) however, employed the method to a broad germplasm collection and found that the groupings did not support the morphological classification or RFLP data.

Simple sequence repeats (SSR)

SSRs offer a higher level of polymorphism due to their abundant presence in coding and non-coding regions, its co-dominant mode of inheritance, its multi-allelic nature and genome specificity. The copy number repeats are highly polymorphic, even among closely related genotypes, making it a reliable method for quantifying and comparing levels of interspecific genetic variations in pineapple (Shoda *et al.*, 2012; Rodríguez *et al.*, 2013; Lin *et al.*, 2015; Chaudhary *et al.*, 2016; Fang *et al.*, 2016; Rattanathawornkiti *et al.*, 2016), as well as for genetic mapping and population genetics (Wöhrmann and Weising, 2011; Nashima *et al.*, 2016). SSR markers make a powerful tool for species comparisons within the same genus, as they are often conserved across different species and genera of the same family (Rodríguez *et al.*, 2013). SSRs have been used widely in pineapple in the last 4–5 years with much effort gone into identifying, characterizing and developing them. Fang *et al.* (2016) and Feng *et al.* (2013) studied SSRs in pineapple and found a frequency of 1 SSR/ 3.73 kb with mono, di and tri-nucleotide repeats accounting for 95.6% of SSRs and 93% found in the intergenic regions (introns).

Kinsuat (2007) and Kinsuat and Kumar (2007) isolated microsatellite loci using the 5′-anchored PCR technique, of which 57 were sufficiently acceptable to contribute to the pineapple genome map. Tong *et al.* (2010) developed genomic SSR markers from a Chinese pineapple cultivar using the selectively amplified microsatellite (SAM) method. Polymorphic levels produced by these SSR markers indicated they were functional primers.

Mining of the existing genomic libraries (Shoda *et al.*, 2012; Feng *et al.*, 2013; Chaudhary *et al.*, 2016; Fang *et al.*, 2016; Nashima *et al.*, 2016) creates opportunities to identify novel SSR markers for genetic diversity studies.

SSR markers have been applied by several research groups to determine genetic relationships in Malaysian (Kinsuat and Kumar, 2007), Japanese (Shoda *et al.*, 2012) and Taiwanese (Lin *et al.*, 2015) accessions, and for differentiation of bract colour in Brazilian ornamental *Ananas comosus* var. *erectifolius* plants (Segeren *et al.*, 2013). Kinsuat and Kumar (2007) found the SSR markers they have employed to produce 2–4 alleles per locus with a very high heterozygosity level, suggesting great opportunities for future improvement of the crop. Similarly, the SSR markers Shoda *et al.* (2012) developed proved to be highly polymorphic resulting in 3–7 alleles per locus, sufficient to confirm parent–offspring relationships and of great value in establishing effective breeding strategies and PBR protection. The Taiwanese cultivars subjected to SSR marker analysis were clustered into three main groups with a high degree of genetic variation, and 2–6 alleles per marker (Lin *et al.*, 2015). When Feng *et al.* (2013) applied the designed primer sets to their pineapple collection, the primers turned out to be highly informative and reproducible, grouping the accessions according to the historic classification. Nashima *et al.* (2016) identified novel repeated sequences in the genomic DNA of cultivar 'N67-10' ('Smooth Cayenne' clone), which made up 8.84% of the total sequence length. Chaudhary *et al.* (2016) tapped into the complete pineapple genome using Next Generation Sequencing and found SSRs that were highly specific and unique to pineapple.

Expressed sequence tags – simple sequence repeats (EST-SSRs)

EST markers have been described as more conserved than genomic SSRs and have been found to be associated with functional genes (Fang *et al.*, 2016). They are therefore expected to be more transferable across taxonomic boundaries, enabling genome evolution and comparative mapping studies. Even though EST-SSRs are expected to be less polymorphic than genomic SSRs, markers that were tested showed a reasonable polymorphism with 2–6 alleles among *A. comosus* accessions (Wöhrmann and Weising, 2011), being in agreement with the new proposed taxonomy. Functional markers such as EST-SSRs will likely play a pivotal role in marker-assisted selection and genetic mapping of pineapple. Tong *et al.* (2011) used EST-SSR markers to build a fingerprint profile for each accession in the Chinese pineapple germplasm collection, showing ample polymorphism between the accessions.

Single nucleotide polymorphism (SNP) and diversity array technique (DArT)

SNP and DArT markers utilize the detection of polymorphisms through hybridization of specific probes or to sequences on an array. SNP and DArT markers are extremely abundant in the genome and lend themselves to automated high throughput assay formats (Zhou *et al.*, 2015; Kilian *et al.*, 2016). While both marker types are a cost-effective fingerprinting tool to study genetic diversity, Chaudhary *et al.* (2016) expressed a concern that for assessing genetic relatedness, SNPs are more likely to display the population structure better if the population is bigger. Zhou *et al.* (2015) designed polymorphic SNPs for genotyping a pineapple germplasm collection and discovered a high level of duplicates. The DArTseq technology, which is a combination of the DArT complexity reduction approach and the NGS technology, was applied to pineapple to study genetic diversity in the *Ananas* metagenome (Kilian *et al.*, 2016) and to identify markers associated with *Phythophthora cinamomi* resistance (Sanewski *et al.*, 2017).

Other markers

Other nucleotide marker systems which have been employed for pineapples include inter-simple sequence repeat (ISSR) (Prakash *et al.*, 2009; Hidayat *et al.*, 2012; Vanijajiva, 2012), restriction-site-associated DNA sequencing (RAD-seq) (Urasaki *et al.*, 2015), start codon targeted markers (SCoT) (Chen *et al.*, 2012), and chloroplast markers (Botella and Smith, 2008; Hamdan *et al.*, 2013; Redwan *et al.*, 2015; Nashima *et al.*, 2015). Several of these methods have been used to evaluate genetic diversity and relationships between accessions in Malaysian (Hidayat *et al.*, 2012), Thai (Vanijajiva, 2012) and Chinese (Chen *et al.*, 2012) collections, respectively, displaying a relatively high level of polymorphism of 50–65%. Others were used as tools to identify markers which can be converted into specific SNP primers to validate the distinctness of a superior mutated clone (Prakash *et al.*, 2009) or into CAPS and SSR markers (Urasaki *et al.*, 2015) linked to specific traits.

Combinations of different marker types have also been reported. Thien (2006) used a combination of RAPD-PCR and direct amplified length polymorphism (DALP) to study sugar content in fruit. Neither of these techniques was able to differentiate between high and low sugar levels in pineapple fruit. Kamaludin (2014) used a combination of ITS primers in a first round of fingerprinting with RAPD-PCR in the second round to create a more specific marker range between certain fragment sizes, with the aim of developing more specific genetic probes to aid selection and identification of different genotypes.

Chloroplasts are maternally inherited; therefore, chloroplast markers are potentially valuable in ecological and evolutionary studies. The large subunit of the ribulose-bisphosphate carboxylase (rbcL) gene is reported to be highly conserved in *Ananas comosus*, therefore having a slower evolution rate than nuclear genes (Botella and Smith, 2008; Hamdan *et al.*, 2013; Redwan *et al.*, 2015). While there are practical constraints to using these markers, Hamdan *et al.* (2013) used them to infer phylogenetic relationships between Malaysian cultivars.

Markers for somaclonal variation

For the purpose of *in vitro* conservation, molecular marker techniques have been employed to reveal polymorphism between the source plant and the somaclones (Feuser *et al.*, 2003; Pérez *et al.*, 2009; Ma *et al.*, 2012; Schrerer *et al.*, 2015a; da Silva *et al.*, 2016). Different *in vitro* culture methods were found to result in different levels of genotypic fidelity, shown with a combination of isozyme and RAPD markers (Feuser *et al.*, 2003). The result is in accordance with that found by Pérez *et al.* (2009), who performed AFLP analysis to demonstrate somaclonal variants from *in vitro* callus explants, while Scherer *et al.* (2015a) found homogeneity along all marker loci of regenerated plants from the same system. Similarly, da Silva *et al.* (2016) using ISSR markers, could not detect any somaclonal variation in *A. comosus* var. *comosus* and *A. comosus* var. *ananassoides* accessions which have been conserved *in vitro* for ten years. Ma *et al.* (2012) proposed that the somatic embryogenesis receptor-like kinase gene AcSERK1, may be useful as a marker to detect embryonic competence of callus tissue. The application would benefit the establishment of a more effective regeneration system and embryo development.

References

Aradhya, K.M., Zee, F. and Manshardt, R.M. (1994) Isozyme variation in cultivated and wild pineapple. *Euphytica* 79, 87–99.

Arias Valdes, E., Benega Garcia, R., Nieves, N., Rodriguez, Cisneros, Pena, A., and Isidrón Perez, M. (1998) Determination of haploids plants in pineapple and alternative methods in the evaluation of ploidy levels. In: *Abstracts of Third International Pineapple Symposium.* Pattaya, Thailand.

Aviz Silva, A. de, Carvalho, A.J.C. de, Freitas F.P. de, Oliveira Pessanha, P.G. de, Santos, P.C. dos *et al.* (2016) Diazotrophic bacteria and nitrogen fertilization on the growth of micropropagated pineapple plantlets during acclimatisation. *Ciencia Rural, Santa Maria* 46(11), 1952–1958.

Ayenew, B., Tadesse, T., Gebremariam, E., Mengesha, A. and Tefera, W. (2013) Efficient use of temporary immersion bioreactor (TIB) on pineapple (*Ananas comosus* L.) multiplication and rooting ability. *Journal of Microbiology, Biotechnology and Food Sciences* 2(4), 2456–2465.

Bairu, M.W., Fennell, C.W. and van Staden, J. (2006) The effect of plant growth regulators on somaclonal variation in Cavendish banana (*Musa* AAA cv. '*Zelig*'). *Scientia Horticulturae* 108, 347–351.

Baldotto L.E.B, Baldotto, M.A., Canella, L.P., Bressan-Smith, R. and Olivares, F.L. (2010) Growth promotion of pineapple 'Vitoria' by humic acid and *Burkholderia* spp. during acclimatization. *Revista Brasileira de Ciência do Solo* 34, 1593–1600.

Baldotto, L.E.B., Olivares, F.L. and Bressan-Smith, R. (2011) Structural interaction between GFP-labelled diazotrophic bacterium *Herbaspirillum seropedicae* RAM10 and pineapple plantlets 'Vitoria'. *Brazilian Journal of Microbiology* 42, 114–125.

Benega, R., Isidron, M., Arias, E., Cisneros, Martinez, J., Torres, I., Hidalgo, M. and Borroto, C. (1996) *In vitro* germination and callus formation in pineapple seeds (*Ananas comosus* (L.) Merr.). *Acta Horticulturae* 425, 243–246.

Benson, E.E. (2008) Cryopreservation of phytodiversity: a critical appraisal of theory and practice. *Critical Reviews in Plant Sciences* 27, 141–219.

Borras, O., Santos, R., Matos, A.P., Cabral, R.S. and Arzola, M. (2001) A first attempt to use a *Fusarium subglutinans* culture filtrate for the selection of pineapple cultivars resistant to fusariose disease. *Plant Breeding* 120, 435–438.

Borse, N., Chimote, V.P. and Jadhav A.S. (2011) Stability of micropropagated *Musa acuminate* cv. Grande Naine over clonal generations; a molecular assessment. *Scientia Horticulturae* 129, 390–395.

Botella, J.R. and Smith, M. (2008) Chapter 18: genomics of pineapple, crowning the king of tropical fruits. In: Moore, P.H. and Ming, R. (ed.) *Genomics of Tropical Crop Plants*, Springer-Verlag, New York, pp. 441–451.

Botella, J.R., Cavallaro, A.S. and Cazzonelli, C.I. (1998) Towards the production of transgenic pineapple to control flowering and ripening. In: *Abstracts of Third International Pineapple Symposium*, Pattaya, Thailand.

Burhooa, K. and Ranghoo-Sanmukhiya, V.M (2012) Evaluation of different pineapple (*Ananas comosus* Merr L.) varieties using morphological and genetic markers in Mauritius. *Biotechnology* 11(5), 272–279.

Canto, A.M.M.E., Souza, F.V.D., Costa, M.A.P.C., Souza, A.S., Ledo, C.A.S. and Cabral, J.R.S. (2004) Conservação in vitro de germoplasma de abacaxi tratado com paclobutrazol. *Pesquisa Agropecuária Brasileira* 39, 717–721.

Carlier, J.D., Reis, A., Duval, M.F., Coppens d'Eeckenbrugge, G. and Leitão, J.M (2004) Genetic maps of RAPD, AFLP and ISSR markers in *Ananas bracteatus* and *A. comosus* using the pseudo-testcross strategy. *Plant Breeding* 123, 186–192.

Carlier, J.D., Nancheva, D., Leitão J.M. and Coppens d'Eeckenbrugge, G. (2006) Genetic mapping of DNA markers in pineapple. *Acta Horticulturae* 702, 79–86.

Carlier, J.D., Coppens d'Eeckenbrugge, G. and Leitão, J.M. (2007) Pineapple. In: Kole, C. (ed.) *Genome Mapping and Molecular Breeding in Plants*, Vol. 4: *Fruits and Nuts*, pp. 331–342.

Carlier, J.D., Sousa, N.H., Santo, T.E., Coppens d'Eeckenbrugge, G. and Leitão, J.M. (2012) A genetic map of pineapple (*Ananas comosus* (L.) Merrill) including SCAR, CAPS, SSR and EST-SSR markers. *Molecular Breeding* 29, 245–260.

Cazzonelli, C.I., Cavallaro, A.S. and Botella, J.R. (1998) Cloning and characterisation of ripening-induced ethylene biosynthetic genes from non-climacteric pineapple (*Ananas comosus*) fruits. *Australian Journal of Plant Physiology* 25, 513–518.

Chaudhary, S., Kumar Mishra, B., Vivek, T., Magadum, S. and Khan Yasin, J. (2016) PineElm SSRdb: a microsatellite marker database indentified from genomic, chloroplast, mitochondrial and EST sequences of pineapple (*Ananas comosus* (L.) Merrill). *Hereditas* 153, 16–21.

Chen, X.L., Su, W.Q., Liu, Y.Q., Ren, H. and Lu, Y.Y. (2012) Analysis on genetic diversity of 36 pineapple collections by SCoT markers. *Southwest China Journal of Agricultural Science* 2, 625–629.

Chu, C.C. (1978) The N6 medium and its application to anther culture of cereal crops (in Chinese). In: *Proceedings of Symposium on Plant Tissue Culture.* Science Press, Beijing, pp. 43–50.

Cisneros, A., Benega, R., Martinez, J., Arias, E. and Isidron, M. (1998) Effect of different plant growth regulators on the subculture of pineapple embryogenic calli. In: *Abstracts of Third International Pineapple Symposium*, Pattaya, Thailand, p. 69.

Collins, J.L. and Kerns, K. (1947) Chimeras in the pineapple – Colchicine-induced tetraploids and diploid tetraploids in the variety. *Journal of Heredity* 38, 322–330.

da Silva, R.L., Ferreira, C.F., da Silva Ledo, C.A., de Souza, E.H., da Silva, P.H., de Carvalho Costa, M.A.P and Souza, F.V.D. (2016) Viability and genetic stability of pineapple germplasm after 10 years of in vitro conservation. *Plant Cell Tissue and Organ Culture* 127, 123–133.

Damasco, O.P., Graham, G.C., Henry R.J., Adkins S.W., Smith M.K. and Godwin, I.D. (1996) Random amplified polymorphic DNA (RAPD) detection of dwarf off-types in micropropagated Cavendish (*Musa* spp. AAA) bananas. *Plant Cell Reports* 16, 118–123.

Daquinta, M., Cisneros, A., Rodriquez, Y., Escalona, M., Perez *et al.* (1996) Somatic embryogenesis in pineapple (*Ananas comosus* (L.) Merr). In: *Pineapple Working Group Newsletter. International Society for Horticultural Science*, pp. 5–6.

de Sousa, N., Carlier, J., Santo, T. and Leitão, J. (2013) An integrated genetic map of pineapple (*Ananas comosus* (L.) Merr.). *Scientia Horticulturae* 157, 113–118.

Devi, Y.S., Mujib, A. and Kundu, S.C. (1997) Efficient regenerative potential from long term culture of pineapple. *Phytomorphology* 47, 255–259.

Duval, M.F., Noyer, J.L., Hamon, P., and Coppens d'Eeckenbrugge, G. (1998) Study of variability in the genus *Ananas* and *Pseudananas* using RFLP. In: *Abstracts of Third International Pineapple Symposium*, Pattaya, Thailand, p. 12.

Escalona, M., Lorenzo, J.C., Gonzalez, B., Daquinta, M., Fundora, Z., Borroto, C.G., Espinosa, D., Arias, E., and Aspiolea, E. (1998) New system for in vitro propagation of pineapple [*Ananas comosus* (L.) Merr]. *Tropical Fruits Newsletter* 29, 3–5.

Escalona, M., Lorenzo, J.C., Gonzalez, B., Daquinta, M., Gonzalez, J.L., Desjardins, Y. and Borroto, C.G. (1999) Pineapple (*Ananas comosus* L. Merr) micropropagation in temporary immersion systems. *Plant Cell Reports* 18, 743–748.

Espinosa, P., Lorenzo, J.C., Iglesias, A., Yabor, L., Menendez, E. *et al.* (2002) Production of pineapple transgenic plants assisted by temporary immersion bioreactors. *Plant Cell Reports* 21, 136–140.

Esposito-Polesi, N.P., Abreu-Tarazi, M.F. de, Almeida, C.V. de, Tsai, S.M. and Almeida, M. de (2017) Investigation of endophytic bacterial community in supposedly axenic cultures of pineapple and orchids with evidence of abundant intracellular bacteria. *Current Microbiology* 74, 103–113.

Fang, J., Miao, C., Chen, R. and Ming, R. (2016) Genome-wide comparative analysis of microsatellites in pineapple. *Tropical Plant Biology* 9, 117–135.

Feng, S., Tong, H., Chen, Y., Wang, J., Chen, Y. *et al.* (2013) Development of pineapple microsatellite markers and germplasm genetic diversity analysis. *BioMed Research International*, Article ID 317912.

Feuser, S., Meler, K., Daquinta, M., Guerra, M.P. and Nodari, R.O. (2003) Genotypic fidelity of micropropagated pineapple (*Ananas comosus*) plantlets assessed by isozyme and RAPD markers. *Plant Cell, Tissue and Organ Culture* 72, 221–227.

Firoozabady, E. and Gutterson, N. (1998) Genetically transformed pineapple plants and methods for their production. International Patent Application Number PCT/US98/03681.

Firoozabady, E. and Young, T.R. (2013) Pineapple plant named Rosé (EF2–114). https://www.google.com/patents/US20130326768.

Firoozabady, E., Heckert, M. and Gutterson, N. (2006) Transformation and regeneration of pineapple. *Plant Cell Tissue and Organ Culture* 84, 1–16.

Fitchet, M. (1990) Clonal propagation of Queen and Smooth Cayenne pineapples. *Acta Horticulturae* 275, 261–266.

GenBank (2017) NCBI dbEST database. Available at: http://www.ncbi.nlm.nih.gov/projects/dbEST (accessed 5 March 2018).

Gonzales, R., Dominguez, Q., Expositio, L., Jorge, L., Martinez, T. and Hidalgo, M. (1996) Effectiveness of 8 strains of Azobacter on the adaption of tissue cultured plantlets of pineapple (*Ananas comosus* Merr.) cv. 'Smooth Cayenne'. *Acta Horticulturae* 425, 277–284.

Gonzales-Rodriguez, R.M., Serrato, R., Molina, J., Aragon, C.E., Olalde V. *et al.* (2013) Biochemical and physiological changes produced by *Azotobacter chroococcum* (INIFAT5 strain) on pineapple in vitro-plantlets during acclimatisation. *Acta Physiologiae Plantarum* 35, 3483–3487.

González-Arnao, M.T., Ravelo, M.M., Urra, C., Martinez-Montero, M.E. and Engelmann, F. (1998) Cryopreservation of pineapple (*Ananas comosus*) apices. *CryoLetters* 19, 375–382.

González-Arnao, M.T., Ravelo, M.M., Urra, C., Martinez-Montero, M.E. and Engelmann, F. (2000) Cryopreservation of pineapple (*Ananas comosus*) apices by vitrification. In: Engelmann, F. and Takagi, H. (eds) *Cryopreservation of Tropical Plant Germplasm.* Japan/Italy: JIRCAS/IPGRI, pp. 390–392.

Graham, M., Ko, L., Hardy, V., Robinson, S., Sawyer, B. *et al.* (1998) The development of blackheart resistant pineapples through genetic engineering. In: *Abstracts of Third International Pineapple Symposium*, Pattaya, Thailand, p. 11.

Graham, M., Ko, L., Hardy, V., Smith, M., Robinson, S. *et al.* (2000) The development of blackheart resistant pineapple through genetic engineering. *Acta Horticulturae* 529, 133–136.

Guedes, N.M.P., Maria, J., Zambolin, L. and Ventura, J.A. (1996) Protoplast isolation of *Ananas comosus* (L.) Merr. CV 'Perolera'. *Acta Horticulturae* 425, 259–265.

Guillemin, J.P., Gianinazzi, S. and Gianinazzi-Pearson (1996) Endomycorrhiza biotechnology and micropropagated pineapple (*Ananas comosus* (L.) Merr.). *Acta Horticulturae* 425, 267–275.

Hamad, A.M. and Taha, R.M. (2008a) Effect of Benzylamniopurine (BAP) on *in vitro* proliferation and growth of pineapple (*Ananas comosus* L. Merr.) cv. Smooth Cayenne. *Journal of Applied Sciences* 8(22), 4180–4185.

Hamad, A.M. and Taha, R.M. (2008b) Effect of sequential subcultures on *in vitro* proliferation capacity and shoot formations pattern of pineapple (*Ananas comosus* L. Merr.) over different incubation periods. *Scientia Horticulturae* 17, 329–334.

Hamad, A.M., Taha, R.M. and Mohajer, S. (2013) *In vitro* induction and proliferation of adventitious roots in pineapple (*Ananas comosus* L.) cultivars of Smooth Cayenne and morris. *Australian Journal of Crop Science* 7(7), 1038–1045.

Hamdan, N., Abd Samad, A., Hidayat, T. and Mohd Salleh, F. (2013) Phylogenetic analysis of eight Malaysian pineapple cultivars using a chloroplastic marker (rbc*L* gene). *Jurnal Teknologi* 64(2), 29–33.

Hamill, S.D. (2016). Processes, costs and traits of plants produced in tissue culture must be considered to develop effective crop production systems. *Acta Horticulturae* 1113, 85–92, Doi: 10.17660/ActaHortic.2016.1113.12. Available at: https://doi.org/10.17660/ActaHortic.2016.1113.12 (accessed 29 March 2018).

Hanqing, C., Zhiying, L. and Li, X. (2013) Characterizing developmental and inducible differentiation between juvenile and adult plants of *Aechmea fasciata* treated with ethylene by transcriptomic analysis. *Plant Growth Regulation* 69, 247–257.

Hidayat, T., Izana Abdullah, F., Kuppusamy, C., Abdul Samad, A. and Wagiran, A. (2012) Molecular identification of Malaysian pineapple cultivar based on internal transcribed spacer region. *APCBEE Procedia* 4, 146–151.

Ibrahim, R., Hamzah, A., Jan Jam, Z., Bahagia, M. and Joyo, M. (2009) Gamma irradiation induced mutation for the improvement of Josapine pineapple against bacterial heart rot disease and improved fruit quality. *Proceedings of an International Joint FAO/IAEA Symposium*, 2008, pp. 276–278.

Ines, M.B.C., Magdalita, P.M., Viña, C.D.D., Cruz Jr, F.S.D and Villegas, V.N. (2009) Randomly amplified polymorphic DNA (RAPD) analysis to reveal genetic relationships among pineapple (*Ananas comosus* (L.) Merr.) genotypes. *Philippine Journal of Crop Science* 34(3), 1–10.

Isidron, P.M., Benega, G.R., Cisneros, P.A., Arias, V.E. *et al.* (1998) Application of biotechnological and traditional methods in Cuban pineapple breeding program. In: *Abstract of Third International Pineapple Symposium*, Pattaya, Thailand, p. 45.

Jackson, D., Williams, S., Newby, D.-M., Hall, S., Higgins *et al.* (2016) Tissue cultured verses traditionally grown pineapples: growth and nutrient profile. *Journal of Biotechnology and Biomaterials* 6(3). Doi: 10.4172/2155-952X.1000237.

Kamaludin, M.E.A. (2014) Fingerprinting analysis of Malaysian pineapple cultivars based on ITS-RAPD method. Master's thesis, Universiti Teknologi Malaysia, Johor, Malaysia.

Kato, C.Y., Nagai, C., Moore, P.H., Zee, F., Kim, M.S. *et al.* (2004) Intra specific DNA Polymorphism in pineapple (*Ananas comosus* (L.) Merr.) assessed by AFLP markers. *Genetic Resources and Crop Evolution* 51, 815– 825.

Kilian, A., Sanewski, G. and Ko, L. (2016) The application of DArTseq technology to pineapple. *Acta Horticulturae* 1111, 181–187.

Kinsuat, M.J. (2007) Development of molecular markers for characterization of local pineapple (*Ananas comosus* var. comosus) cultivars. Master's thesis, Universiti Malaysia Sabah, Biotechnology Research Institute, Sabah, Malaysia.

Kinsuat, M.J. and Kumar, S.V. (2007) Polymorphic microsatellite and cryptic simple repeat sequence markers in pineapples (*Ananas comosus* var. *comosus*). *Molecular Ecology Notes* 7(6), 1032–1035.

Kiss, E., Kiss, J., Gyulai, G. and Heszky, L.E. (1995) A novel method for rapid microprogation of pineapple. *HortScience* 30, 127–129.

Ko, L., Hardy, V., Jobin-Décor, M., Campbell, P., Eccleston, K. *et al.* (2005) The introduction of transgenes to control blackheart in pineapple: biolistics vs Agrobacterium transformation. *Proceedings of IAPTC&B Conference*, 21–24 September, Perth, Australia.

Ko, H.L., Campbell, P.R., Jobin-Decor, M.P., Eccleston, K.L., Graham, M.W. and Smith, M.K. (2006) The introduction of transgenes to control blackheart in pineapple (*Ananas comosus* L.) cv. Smooth cayenne by microprojectile bombardment. *Euphytica* 150, 387–395.

Ko, L., Eccleston, K., O'Hare, T., Wong, L., Giles, J. and Smith, M. (2013) Field evaluation of transgenic pineapple (*Ananas comosus* (L.) Merr.) cv. 'Smooth Cayenne' for resistance to blackheart under subtropical conditions. *Scientia Horticulturae* 159, 103–108.

Koia, J.H., Moyle, R.L. and Botella, J.R. (2012) Microarray analysis of gene expression profiles in ripening pineapple fruits. *BMC Plant Biology* 12, 240.

Koia, J., Moyle, R., Hendry, C., Lim, L. and Botella, J.R. (2013) Pineapple translation factor SUI1 and ribosomal protein L36 promoters drive constitutive transgene expression patterns in *Arabidopsis thaliana*. *Plant Molecular Biology* 81, 327–336.

Kunze, A., Lovato, P.E., Costa, M.D. and Dal Vesco, L.L. (2014) Pineapple (*Ananas comosus)* cv. Perola *ex vitro* growth and mycorrhizal colonization affected by *in vitro* sucrose concentration. *Revista Brasileira de Fruticultura* 36(3), 766–770.

Leite, N.S., Lima, A.A.B de, Santana, J.C.C., Lopes, F.L.G., Silva Leda, A. de *et al.* (2012) Determination of optimal conditions to obtain the bromelain from pineapple plants produced by micropropagation. *Brazilian Archives of Biology and Technology* 55(5), 647–652.

Li, Z., Wang, J., Zhang, X., Lei, M., Fu, Y. *et al.* (2016a) Transcriptome sequencing determined flowering pathway genes in *Aechmea fasciata* treated with ethylene. *Journal of Plant Growth Regulation* 35, 316–329.

Li, Y.H., Wu, Q.S., Huang, X., Liu, S.H., Zhang, H.N. *et al.* (2016b) Molecular cloning and characterization of four genes encoding ethylene receptors associated with pineapple (*Ananas comosus* L.) flowering. *Frontiers in Plant Science* 7, 710. Doi: 10.3389/fpls.2016.00710.

Li, R., Yu, S., Li, X., Xiong, Y., Lin, Z. *et al.* (2017a) Screening of reference genes in *Ananas comosus* var. *bracteatus* for qRT-PCR. *Journal of Tropical and Subtropical Botany* 25(3), 250–256.

Li, X., Kanakala, S., He, Y., Zhong, X., Yu, S. *et al.* (2017b) Physiological characterization and comparative transcriptome analysis of white and green leaves of *Ananas comosus* var. *bracteatus*. *PLoS ONE* 12(1), e0169838-e0169838.

Lin, Y.S., Kuan, C.S., Weng, I.S and Tsai, C.C. (2015) Cultivar identification and genetic relationship of pineapple (*Ananas comosus*) cultivars using SSR markers. *Genetics and Molecular Research* 14(4), 15035–15043.

Liu, L.J. (1987) In vitro propagation of spineless Red Spanish pineapple. *Phytopathology* 77, 1711.

Liu, C.H. and Fan, C. (2016) De novo transcriptome assembly of floral buds of pineapple and identification of differentially expressed genes in response to ethephon induction. *Frontiers in Plant Science* 7, 203. Doi: 10.3389/fpls.2016.00203.

Ma, J., He, Y., Wu, C., Liu, H., Hu, Z. and Sun, G. (2012) Effective Agrobacterium-mediated transformation of pineapple with CYP1A1 by kanamycin selection technique. *African Journal of Biotechnology* 11(10), 2555–2562.

Ma, J., He, Y., Hu, Z., Xu, W., Xia, J. *et al.* (2014) Characterization of the third SERK gene in pineapple (*Ananas comosus*) and analysis of its expression and autophosphorylation activity in vitro *Genetics and Molecular Biology* 37(3), 530–539.

Ma, J., Kanakala, S., He, Y., Zhang, J. and Zhong, X. (2015) Transcriptome sequence analysis of an ornamental plant, *Ananas comosus* var. *bracteatus*, revealed the potential unigenes involved in terpenoid and phenylpropanoid biosynthesis. *PLoS ONE* 10(3). Doi: 10.1371/journal.pone.0119153.

Martinez-Montero, M.E., Martínez, J., Engelmann, F. and Gonzalez-Arnao, M.T. (2005) Cryopreservation of pineapple (*Ananas comosus* (L.) Merr) apices and calluses. *Acta Horticulturae* 666, 127–130.

Martinez-Montero, M.E., González-Arnao, M.T. and Engelmann, F. (2012) Cryopreservation of tropical plant germplasm with vegetative propagation – review of sugarcane (*Saccharum* spp.) and pineapple (*Ananas comusus* (L.) Merrill) cases. In: Katkov, I. (ed.) *Current Frontiers in Cryopreservation.* Rijeka, Croatia: Intech, pp. 359–396.

Mathews, V.H. and Rangan, T.S. (1979) Multiple plantlets in lateral bud and leaf explants *in vitro* cultures of pineapple. *Scientia Horticulturae* 11, 319–328.

Mathews, V.H. and Rangan, T.S. (1981) Growth and regeneration of plantlets in callus cultures of pineapple. *Scientia Horticulturae* 14, 227–234.

Matos, R.M.B., Silva, E.M.P. da, and da Silva, E.M.R. (1996) Effect of inoculation by arbuscular mycorrhizal fungi on the growth of micropropagated pineapple plants. *Fruits* 51(2), 115–119.

Mengesha, A., Ayenew, B. and Tadesse, T. (2013) Energy sources affect *in vitro* propagation and subsequent acclimatisation of *Ananas comosus*, var. smooth cayenne plants. *Journal of Microbiology, Biotechnology and Food Sciences* 2(6), 2372–2376.

Mhatre, M. (2013) Agrobacterium-mediated genetic transformation of pineapple (*Ananas comosus* L., Merr.). *Methods in Molecular Biology* 11013, 435–453.

Mhatre, M., Tilak-Jain, J., De, S. and Devasagayam, T.P.A. (2009) Evaluation of the antioxidant activity of non-transformed and transformed pineapple: A comparative study. *Food and Chemical Toxicology* 47(11), 2696–2702.

Mhatre, M., Srinivas, L. and Ganapathi, T.R. (2011) Enhanced iron and zinc accumulation in genetically engineered pineapple plants using soybean ferritin gene. *Biological Trace Element Research* 144(1–3), 1219–1228.

Ming, R., VanBuren, R., Wai, C.M., Tang, H., Schatz, M.C. *et al.* (2015) The pineapple genome and the evolution of CAM photosynthesis. *Nature Genetics* 47(12), 1435–1445.

Ming, R., Wai, C.M. and Guyot, R. (2016) Pineapple genome: a reference for monocots and CAM photosynthesis. *Trends in Genetics* 32(11), 690–696.

Moyle, R.L. and Botella, J.R. (2014) EST sequencing of *Meloidogyne javanica* infected pineapple root tissues reveals changes in gene expression during root-knot nematode induced gall formation. *Tropical Plant Biology* 7, 43–52.

Moyle, R., Crowe, M., Ripi, J., Fairbairn, D.J. and Botella, J.R. (2005a) PineappleDB: an online pineapple bioinformatics resource. *BMC Plant Biology* 5: 21. Doi:10.1186/1471-2229-5-21.

Moyle, R., Fairbairn, D.J., Ripi, J., Crowe, M. and Botella, J.R. (2005b) Developing pineapple fruit has a small transcriptome dominated by metallothionein. *Journal of Experimental Botany* 56, 101–112.

Moyle, R.L., Ripi, J., Fairbairn, D.J., Botella, J.R and Crowe, M. (2006) The pineapple EST sequencing and microarray project. *Acta Horticulturae* 702, 47–50.

Moyle, R.L., Koia, J.H., Vrebalov, J., Giovannoni, J. and Botella, J.R. (2014) The pineapple AcMADS1 promoter confers high level expression in tomato and Arabidopsis flowering and fruiting tissues, but AcMADS1 does not complement the tomato LeMADS-RIN (rin) mutant. *Plant Molecular Biology* 86(4–5), 395–407. Doi: 10.1007/s11103-014-0236-3.

Murashige, T. and Skoog, F. (1962) A revised method for rapid growth and bioassays with tobacco tissue cultures. *Physiologia Plantarum* 15, 473–497.

Murashige, T. and Tucker, D.P.H. (1969) Growth factor requirements of citrus tissue culture. In: Chapman, H.D. (ed.) *Proceedings of the First International Citrus Symposium.* Riverside, CA: University of California-Riverside, pp. 1155–1161.

Nan, G.-L., Nagai, C., Moore, P.H., Sun, S.S.M., Sipes, B.S., Paull, R.E. and Rohrbach, K. (1996) Tissue culture and genetic transformation studies on pineapple. *In Vitro Cellular and Development Biology* 32, 102A.

Nashima, K., Terakami, S., Nishitani, C., Kunihisa, M., Shoda, M. *et al.* (2015) Complete chloroplast genome sequence of pineapple (*Ananas comosus*). *Tree Genetics and Genomes* 11, 60.

Nashima, K., Hosaka, F., Terakami, S., Nishitani, C., Kunihisa, M. *et al.* (2016) Development of novel SSR markers from the genome sequence of pineapple. *Acta Horticulturae* 1111, 189–194.

NCBI (n.d.) National Center for Biotechnology Information. Available at: https://www.ncbi.nlm.nih.gov (accessed 5 March 2018).

Nelson, B.J., Asare, P.A. and Junior, R.A. (2015) *In vitro* growth and multiplication of pineapple under different duration of sterilisation and different concentrations of Benzylaminopurine and sucrose. *Biotechnology* 14(1), 35–40.

Neta, J.L.V., Silva Ledo, A. da, Lima, A.A.B., Santana, J.C.C., Leite, N.S., Ruzene, D.S., Silva, D.P. and Souza, R.R. de (2012) Bromelain enzyme from pineapple: in vitro activity study under different micropropagation conditions. *Applied Biochemistry and Biotechnology* 168(2), 234–246.

Nitsch, J.P. (1951) Growth and development *in vitro* of excised ovaries. *American Journal of Botany* 38, 566–577.

Norsyalieza, T.A. and Illias, R.M. (2017) Bromelain: from production to commercialisation. *Journal of the Science of Food and Agriculture* 97(5), 1386–1395,

Noyer, J.L., Lanaud, C., Coppens d'Eeckenbrugge, G. and Duval, D.F. (1996) RFLP study on rDNA variability in Ananas genus. *Acta Horticulturae* 425, 153–159.

Ong, W.D., Voo, C.L.Y. and Kumar, S.V. (2012a) Development of ESTs and data mining of pineapple EST-SSRs. *Molecular Biology Reporter* 39, 5889–5896.

Ong, W.D., Voo, L.Y.C. and Kumar, V.S. (2012b) *De novo* assembly, characterization and functional annotation of pineapple fruit transcriptome through massively parallel sequencing. *PLoS ONE* 7(10), e46937.

Panis, B. (2009) *Cryopreservation of Musa Germplasm*, 2nd edn. Technical Guidelines, 9. Montpellier, France: Bioversity International

Patent Lens (n.d.) Types of promoters used to regulate gene expression. Available at: http://www.cambia.org/daisy/promoters/239/1343.html (accessed 5 March 2018).

Paull, R., Chen, N., Ming, R., Wai, C., Shirley, N., Schwerdt, J. and Bulone, V. (2016) Carbon flux and carbohydrate gene families in pineapple. *Tropical Plant Biology* 9, 200–213.

Paz, E.Y., Gil, K., Rebolledo, L., Rebolledo, A., Uriza, D., Martínez, O., Isidrón, M. and Simpson, J. (2005) AFLP characterization of the Mexican pineapple germplasm collection. *Journal of the American Society of Horticultural Science* 130(4), 575–579.

Paz, E.Y., Gil, K., Rebolledo, L., Rebolledo, A., Uriza, D., Martínez, O., Isidrón, M., Díaz, L., Lorenzo, J.C. and Simpson, J. (2012) Genetic diversity of Cuban pineapple germplasm assessed by AFLP markers. *Crop Breeding and Applied Biotechnology* 12, 104–110.

Pérez, G., Yanes, E., Isidrón, M. and Lorenzo, J.C. (2009) Phenotypic amd AFLP characterization of two new pineapple somaclones derived from in vitro culture. *Plant Cell Tissue and Organ Culture* 96, 113–116.

PineappleDB (2007) *PineappleDB: The Online Pineapple Bioinformatics Resource.* Available at: http://genet.imb.uq.edu.au/Pineapple/index.html (accessed 5 March 2018).

Popluechai, S., Onto, S. and Eungwanichayapant, P.D. (2007) Relationships between some Thai cultivars of pineapple (*Ananas comosus*) revealed by RAPD analysis. *Songklanakarin Journal of Science and Technology* 29(6), 1491–1497.

Prakash, J., Bhattacharyya, S., Chattopadhyay, K., Roy, S., Das, S.P. and Singh, N.P. (2009) PQM-1: a newly developed superior clone of pineapple for northeastern India as evident through phenotype, fruit quality and DNA polymorphism. *Scientia Horticulturae* 120, 288–291.

Rattanathawornkiti, K., Kanchanaketu, T., Suwanagul, A. and Hongtrakul, V. (2016). Genetic relationship assessment of pineapple germplasm in Thailand revealed by AFLP markers. *Genomics and Genetics* 9(2), 1–10.

Redwan, R.M., Saidin, A and Kumar, S.V. (2015) Complete chloroplast genome sequence of MD-2 pineapple and its comparative analysis among nine other plants from the subclass Commelinidae. *BMC Plant Biology* 15, 196.

Redwan, R.M., Saidin, A. and Kumar, S.V. (2016) The draft genome of MD-2 pineapple using hybrid error correction and long reads. *DNA Res* 23(5), 427–439.

Reed, B.B.M. (2008) *Plant Cryopreservation: A Practical Guide*. New York: Springer.

Rizzal, W., Janna, O.A., Rusli, I. and Nur Ashikin, P.A. (2014) Comparison in multiplication rate of MD2 pineapple micropropagation with liquid MS, solid MS and temporary immersion system. *Transactions of the Malaysian Society of Plant Physiology* 22, 205–209.

Rodríguez, D., Grajal-Martín, M.J., Isidrón, M., Petit, S. and Hormaza, J.I. (2013) Polymorphic microsatellite markers in pineapple (*Ananas comosus* (L.) Merrill). *Scientia Horticulturae* 156, 127–130.

Rohrbach, K.G., Christopher, D., Hu, J., Paull, R., Sipes, B., Nagai, C., Moore, P., McPherson, M., Atkinson, H., Oda, C., Fleisch, H. and McLean, M. (1998) Management of a multiple goal pineapple genetic engineering program. In: *Abstracts of Third International Pineapple Symposium*, Pattaya, Thailand.

Roostika, I., Khumaida, N. and Ardie, S.W. (2015) Rapid analysis to detect somaclonal variation of pineapple in vitro cultures during micropropagation. *Biotropia* 22(2), 109–119.

Ruas, P.M., Ruas, C.F., Fairbanks, D.J., Andersen, W.R. and Cabral, J.R.S. (1995) Genetic relationship among four varieties of pineapple, *Ananas comosus*, revealed by random amplified polymorphic DNA (RAPD) analysis. *Brazilian Journal of Genetics* 18, 413–416.

Sakai, A., Kobayashi, S. and Oiyama, I. (1990) Cryopreservation of nucellar cells of navel orange (*Citrus sinenesis* Osb. var. *brasiliensis* Tanaka) by vitrification. *Plant Cell Reports* 9, 30–33.

Sanewski, G.M., Botella, J.R., Scott, C. and Sinclair, E. (1999) Development of a pineapple plant not prone to natural flowering using genetic engineering. In: *Final Report – FR424*. Gordon, Australia: Horticultural Research and Development Corporation Publication.

Sanewski, G., Ko, L., Innes, D., Kilian, A., Carling, J. and Song, J. (2017) DArTseq markers for resistance to *Phytophthora cinnamomi* in pineapple (*Ananas comosus* L.). *Australasian Plant Pathology* 46(5), 499–509.

Scheidt, G.N., Arakaki, A.H., Chimilovski, J.S., Portella, A.C.F., Spier, M.R. *et al.* (2009) Utilization of the bioreactor of immersion by bubbles at the micropropagation of *Ananas comosus* L. Merril. *Brazilian Archives of Biology and Technology*, 52 (special number), 37–43.

Scherer, R.F., Garcia, A.C., Freitas Fraga, H.P. de, Vesco, L.L. del, Steinmacher, D.A. and Guerra M.P. (2013) Nodule cluster cultures and temporary immersion bioreactors as a high performance micropropagation strategy in pineapple (*Ananas comosus* var. *comosus*). *Scientia Horticulturae* 151, 38–45.

Scherer, R.F., Freitas Fraga, H.P. de, Klabunde, G.F., da Silva, D.A and Guerra, M.P. (2015a) Global DNA methylation levels during the development of nodule cluster cultures and assessment of genetic fidelity of in vitro-regenerated pineapple plants (*Ananas comosus* var. *comosus*). *Journal of Plant Growth Regulation* 34, 677–683.

Scherer, R.F., Holderbaum, D.F., Garcia, A.C., Silva, D.A. da, Steinmacher, A. and Guerra, M.P. (2015b) Effects of immersion system and gibberellic acid on the growth and acclimatisation of micropropagated pineapple. *Crop Breeding and Applied Biotechnology* 15, 66–71.

Scherwinski-Pereira, J.E., Elequisandra da, C., Araruna L., da Silva, T.L., Antonio, G. *et al.* (2012) Double-phase culture system for large scale production of pineapple. *Plant Cell Tissue and Organ Culture* 109, 263–269.

Scowcroft, W.R. (1984) *Genetic Variability in Tissue Culture; Impact on Germplasm Conservation and Utilization.* International Board of Plant Genetic Resources, Rome.

Segeren, M.I., Rodrigues, D.P., Muca, G., Reis, V. and Gimenes, A. (2013) Genetic diversity of green and pink color of *Ananas comosus* var. *erectifolius* (L.B. Smith). *Acta Horticulturae* 1000, 527–532.

Sharma, A., Wai, C.M., Ming, R. and Yu, Q. (2017) Diurnal cycling transcription factors of pineapple revealed by genome-wide annotation and global transcriptomic analysis. *Genome Biology and Evolution* 9(9), 2170–2190.

Shoda, M., Urasaki, N., Sakiyama, S., Terakami, S., Hosaka, F. *et al.* (2012) DNA profiling of pineapple cultivars in Japan discriminated by SSR markers. *Breeding Science* 62, 352–359.

Silva, A.B. da, Pasqual M., Teixeira, J.B., Arauji, A.G. de (2007) Micropropagation methods of pineapple. *Empresa Brasileira de Pesquisa Agropecuaria* 42(9), 1257–1260.

Silva, R.L., Ferreira, C.F., Ledo, C.A.S., Souza, E.H., Silva, P.H. *et al.* (2016) Viability and genetic stability of pineapple germplasm after 10 years of in vitro conservation. *Plant Cell, Tissue and Organ Culture* 127, 123–133.

Smith, L.B. and Downs, R.J. (1979) *Bromelioidees* (*Bromeliaceae*). Flora Neotropica, Monograph 14. Part 3.

Smith, M.K. and Drew, R.A. (1990) Current applications of tissue culture in plant propagation and improvement. *Australian Journal of Plant Physiology* 17, 267–289.

Souza, F.V.D., Soares, T.L., Cabral, J. R.S., Reinhardt, D.H., Silva, J.L.C. and Benjamim, D.A. (2006) Slow-grow conditions for the in vitro conservation of pineapple germplasm. *Acta Horticulturae* 702, 41–47.

Souza, F.V.D., Kaya, E., Vieira, L.J., Souza, E.H., Amorim, V.B.O. *et al.* (2015) Droplet-vitrification and morphohistological studies of cryopreserved shoot tips of cultivated and wild pineapple genotypes. *Plant Cell, Tissue and Organ Culture* 124, 351–360.

Sripaoraya, S., Marchant, R., Power, J.B. and Davey, M.R. (2001) Herbicide-tolerant transgenic pineapple (*Ananas comosus*) produced by microprojectile bombardment. *Annals of Botany* 88, 597–603.

Sripaoraya, S., Keawsompong, S., Insupa, P., Power, J.B., Davey, M.R. and Srinives, P. (2006) Genetically manipulated pineapple: transgene stability, gene expression and herbicide tolerance under field conditions. *Plant Breeding* 125, 411–413.

Sugimoto, A., Yamaguchi, I., Matsuoka, M., Nakagawa, H., Kato, S. and Nakano, H. (1991) *In vitro* conservation of pineapple genetic resources. *Research Highlights, Tropical Agricultural Center*, 14–16.

Tapia, C.E., Gutiérrez, E.M.A., Guillén, A.H., Warbourton, L.M., Uriza A.D and Rebolledo, M.A. (2005) Characterization of pineapple germplasm (*Ananas* spp) by mean AFLPs. *Acta Horticulturae* 666, 109–114.

Tapia-Hernandez, A., Bustillos-Cristales, M.R., Jimenenz-Salgado, T., Cabellero-Mellado, J. and Fuentes-Ramirez, L.E. (2000) Natural endophytic occurrence of *Acetobacter diazotrophicus* in pineapple. *Microbial Ecology* 39(1), 49–55.

Thien, Y.N. (2006) The isolation and identification of molecular marker linked to sugar production in pineapple (*Ananas comosus* var. *comosus*). Master's thesis, Universiti Malaysia Sabah, Biotechnology Research Institute, Sabah, Malaysia.

Tong, H.L., Feng, S.P., Chen, Y., Chen, Y.Y., Sun, G.M. and Wu, Y.T. (2010) Development of molecular markers from genomic-SSR of pineapple (*Ananas comosus*). *Journal of Fruit Science* 27(4), 551–555.

Tong, H.L., Feng, S.P., He, J.H., Wang, J.Y., Chen, Y.Y., Sun, G.M., Yu F and Wu, YT. (2011) Establishment of fingerprinting for pineapple (*Ananas comosus*) by SSR marker. *Journal of Fruit Science* 28(2) 240–245.

TropGeneDB (n.d.) *TropGene Database*. CIRAD. Available at: http://tropgenedb.cirad.fr (accessed 5 March 2018).

Trusov, Y. and Botella, J.R. (2006) Silencing of the ACC synthase gene *ACACS2* causes delayed flowering in pineapple [*Ananas comosus* (L.) Merr.]. *Journal of Experimental Botany* 57, 3953–3960.

Tu, Y.K. Tseng, C.S., Wu, M.T. and Chen, H.W. (2017) Characteristics evaluation of transgenic delay-flowering pineapple (*Ananas comosus* L. Merr.). *Acta Horticulturae* 1166, 77–85.

Urasaki, N., Goeku, S., Kaneshima, R., Takamine, T., Tarora, K. *et al.* (2015) Leaf margin phenotype-specific restriction-site-associated DNA-derived markers for pineapple (*Ananas comosus* L.). *Breeding Science* 65, 276–284.

Valencia, L.D.C., Pascual, C.B. and Delfin, E.F. (2014) In vitro selection of pineapple cv. 'Queen' with resistance to culture filtrate of *Phytophtora cinnamomi* Rands. *Philippine Journal of Crop Science* 39(1), 58–66.

Vanijajiva, O. (2012) Assessment of genetic diversity and relationships in pineapple cultivars from Thailand using ISSR marker. *Journal of Agricultural Technology* 8(5), 1829–1838.

Wai, C.M. (2014) Development of pineapple (*Ananas comosus* L.) genetic map with restriction site associated DNA (RAD) markers. In: In: *Proceedings of the International Plant and Animal Genome XXII*, San Diego, California, 11–14 January, p. 452.

Wai, C.M., Powell, B., Ming, R. and Min, X.J. (2016a) Genome-wide identification and analysis of genes encoding proteolytic enzymes in pineapple. *Tropical Plant Biology* 9(3), 161–175.

Wai, C.M., Powell, B., Ming, R. and Min, X.J. (2016b) Analysis of alternative splicing landscape in pineapple (*Ananas comosus*). *Tropical Plant Biology* 9(3), 150–160.

Wai, C.M., VanBuren, R., Zhang, J., Huang, L., Miao, W. *et al.* (2017) Temporal and spatial transcriptomic and microRNA dynamics of CAM photosynthesis in pineapple. *Plant Journal* 92(1), 19–30.

Wakasa, K. (1979) Variation in the plants differentiated from the tissue culture of pineapple. *Japanese Journal of Breeding* 29, 13–22.

Wakasa, K. (1989) Pineapples. In: Bajaja, Y.P.S. (ed.) *Biotechnology in Agriculture and Forestry 5: Trees II.* Berlin: Springer-Verlag, pp. 13–29.

Wang, M.L. and Paull, R.E. (2018) Genetic transformation of pineapple. In: *Genetic and Genomics of Pineapple*. Springer. *In Press.*

Wang, M.L., Uruu, G., Xiong, L., He, X., Nagai, C. *et al.* (2009) Production of transgenic pineapple (*Ananas cosmos* (L.) Merr.) plants via adventitious bud regeneration. *In Vitro Cellular & Developmental Biology-Plant* 45(2), 112–121.

Weber, O.B., Videira, S.S. and Araujo, J.L. de (2013) Identification of culturable endophytes in 'Champaka' pineapple grown in an organic system. *African Journal of Agricultural Research* 8(26), 3422–3430.

Wei, H., Wang, M.L., Moore, P.H. and Albert, H.H. (2003) Comparative expression analysis of two sugarcane polyubiquitin promoters and flanking sequences in transgenic plants. *Journal of Plant Physiology* 160, 1241–1251.

Wöhrmann, T. and Weising, K. (2011) In silico mining for simple sequence repeat loci in a pineapple expressed sequence tag database and cross-species amplification of EST-SSR markers across Bromeliaceae. *Theoretical and Applied Genetics* 123, 635–647.

Yabor, L., Arzola, M., Aragon, C., Hernandez, M., Arencibia, A. and Lorenzo, J.C. (2006) Biochemical side effects of genetic transformation of pineapple. *Plant Cell Tissue and Organ Culture* 86, 63–67.

Yapo, E.S., Kouakou, T.H., Kone, M., Kouadio, J.Y., Kouame, P. and Merillon, J.M. (2011) Regeneration of pineapple (*Ananas comosus* L.) plant through somatic embryogenesis. *Journal Plant Biochemistry and Biotechnology* 20(2), 196–204.

Yusuf, N.H.M., Ong, W.D., Redwan, R.M., Latip, M.A. and Kumar, S.V. (2015) Discovery of precursor and mature microRNAs and their putative gene targets using high-throughput sequencing in pineapple (*Ananas comosus* var. comosus). *Gene* 571, 71–80.

Zee, F.T. and Munekata, M. (1992) *In vitro* storage of pineapple (*Ananas* spp.) germplasm. *HortScience* 27, 57–58.

Zhang, J.S., Liu, J. and Ming, R. (2014) Genomic analyses of the CAM plant pineapple. *Journal of Experimental Botany* 65, 3395–3404.

Zhang, X.D., Liang, P.P. and Ming, R. (2016) Genome-wide identification and characterization of nucleotide-binding site (NBS) resistance genes in pineapple. *Tropical Plant Biology* 9, 187–199.

Zhao, W., Yang, W., Wei, C. and Sun, G. (2011) A simple and efficient method for isolation of pineapple protoplasts. *Biotechnology and Biotechnological Equipment* 25(3), 2464–2467.

Zheng, Y., Li, T., Xu, Z.N., Wai, C.M., Chen, K. *et al.* (2016) Identification of microRNAs, phasiRNAs and their targets in pineapple. *Tropical Plant Biology* 9, 176–186.

Zhou, L., Matsumoto, T., Tan, H.W., Meinhardt, L.W., Mischke, S. *et al.* (2015) Developing single nucleotide polymorphism markers for the identification of pineapple (*Ananas comosus*) germplasm. *Horticulture Research* 2, 15056.

6 Crop Environment, Plant Growth and Physiology*

Duane P. Bartholomew
University of Hawaii at Manoa, Honolulu, USA

Pineapple, a crop native to the tropical Americas, is presently grown commercially over latitudes ranging from approximately 30° N in India to 33° 58′ S in South Africa (Bartholomew and Kadzimin, 1977; see also Table 6.1 below) and at elevations ranging from sea level to over 1400m in equatorial Africa (Fig. 6.1). The length of the crop cycle is mainly determined by the prevailing climate and the size of the planting material and seldom requires less than 12 months from planting to harvest in tropical environments and more commonly 14 to as much as 36 months in cool subtropical environments. The information on days from forcing to harvest (Table 6.1) and the growing degree-days (GDD; see also Fig. 6.1 below) provide some indication of crop cycle length in the various environments shown. Dorey *et al.* (2016) reported that 'Queen' fruit required 1400 degree days from flowering to maturity and additional GDD would be required from forcing to flowering.

The induction of flowering of pineapple is almost universally initiated with a growth regulator, which has made large-scale production of pineapple possible in areas where flowering would otherwise be too erratic for commercial production. Where the seasonal variation is large and fruiting is forced in most or all months of the year, it is common for both vegetative and reproductive growth to occur during both warm and cool seasons. Unlike many crops where cultivars are bred to adapt them to fairly specific environments, only four pineapple cultivars are grown widely and only one, 'MD-2', is the result of a modern plant breeding programme.

There are many challenges associated with dealing with a small number of cultivars in such a wide range of environments and the focus of this chapter is on the variable environments where pineapple is grown and the way that physiology and growth of this unique crop interacts with climate changes during vegetative development.

Production at the more extreme latitudes is mainly restricted to areas where the climate is moderated by the ocean, as the crop will not tolerate freezing temperatures. One of the main features of pineapple is its adaptation to areas of low rainfall, although productivity is reduced in drought conditions. This is partly due to the fact that pineapple, as with most *Bromeliacae*, has the crassulacean acid metabolism (CAM) photosynthetic pathway and morphological and anatomical features that minimize transpiration. This specificity of pineapples is shared by several other commercially important crops (Nobel, 2010), though none have the economic importance of pineapple.

High productivity of pineapple depends on successfully managing such controllable factors as water and nutrient supply and pests and diseases. Environmental factors such as temperature, irradiance, and rainfall interacting with the plant's CAM metabolism and unique

* A revision of the chapter by Eric Malézieux, Francois Côte and Duane P. Bartholomew.

 The Pineapple, 2nd Edition: Botany, Production and Uses (eds G.M. Sanewski *et al.*)

Table 6.1. Range of locations where pineapples are grown and, where available, days from forcing (F) to harvest (H).

			Temperature °C				
Location	Latitude	Elev. (m)	Max.	Min.	CV[a]	F to H	Reference
Tenerife	28° 30′ N	50	23.4	16.2	SC[b]	210–270	Galan Sauco *et al.*, 1988
Taichung	24° 09′ N	110	25.8	17.3	Tainung 17		
Kunia	21° 20′ N	216	30.1	14.3	SC	175–223	Fleisch, 1988
Maui	20° 48′ N	792	20.4	14.7	SC	230–255	Fleisch, 1988
Townsville	18° 16′ N	10			SC		
Pattaya	12° 75′ N	73	30.9	27.8	SC		
Foulaya	10° 15′ N	400	30.7	18.7	SC	180	Py *et al.*, 1957
Buenas Aires	9° 10′ N	360–470	32.0	19.9			Romero *et al.*, 1973
Polomolok	6° 15′ N	255–830	30.7	20.7	SC		
Abidjan	5° 23′ N	25	30.9	22.3	SC	156–166	Lacoeuilhe, 1978
Nyombe	4° 33′ N	70			SC		Gaillard, 1970
Tiko	4° 03′ N	30	30.2	22.3	SC	150	Aubert *et al.*, 1973
Molyko	4° 03″ N	550	27.6	17.7	SC	180	
Buea	4° 08′ N	1000	25.0	14.9	SC	210	
Johore	1° 22′ N	5	35.0	18.9	SP		Wee, 1969
Thika	1° 01′ S	1463	26.7	15.2	SC	213–232	Olende, C. (pers. comm.)
Thika	1° 01′ S	1463	26.7	15.2	MD-2	234–258	Olende, C. (pers. comm.)
Lampung	5° 26′ S	10	31.5	20.7	SC	161–166	Suwandi *et al.*, 2016
Sape	7° 5′ S	124	31.2	20.6			Giacomelli and Py, 1982
Coracao de Maria	12° 14′ S	267	30.0	23.6			
Ivoloina	18° 03′ S	20	28.0	19.6		144–221	Moreau and Moreuil, 1976
Nambour	26° 38′ S		25.8	14.0	SC	203–301	Wassman, 1990
Zululand	28° S	125			Queen	210–240	Lacoeuilhe and Sarah, 1990
Osorio	29° 55′ S	38	23.8	19.6	SC or Pérola		
East London	33° 02′ S	55	22.8	14.4	SC		Bouffin, 1991
Port Elizabeth	33° 58′ S	55	21.2	13.3	SC		Tkatchenko, 1947

[a]CV denotes cultivar.
[b]SC indicates 'Smooth Cayenne'.

morphological and anatomical features ultimately determine productivity within a given environment.

Carbon Assimilation and Water Use

Carbon fixation via the CAM photosynthetic pathway

Higher plants have either the C3, C4 or CAM pathway of CO_2 fixation. Pineapple has the CAM pathway and is also characterized as a succulent plant because about one-third of the volume of its leaves is water storage tissue. In unirrigated culture the presence of water storage tissue buffers water availability, which can prolong atmospheric CO_2 assimilation and growth during periods of drought. From a research point of view, having the CAM pathway is a significant disadvantage that for reasons explained below makes the study of photosynthesis and the factors that affect it much more challenging than is the case for C3 and C4 plants.

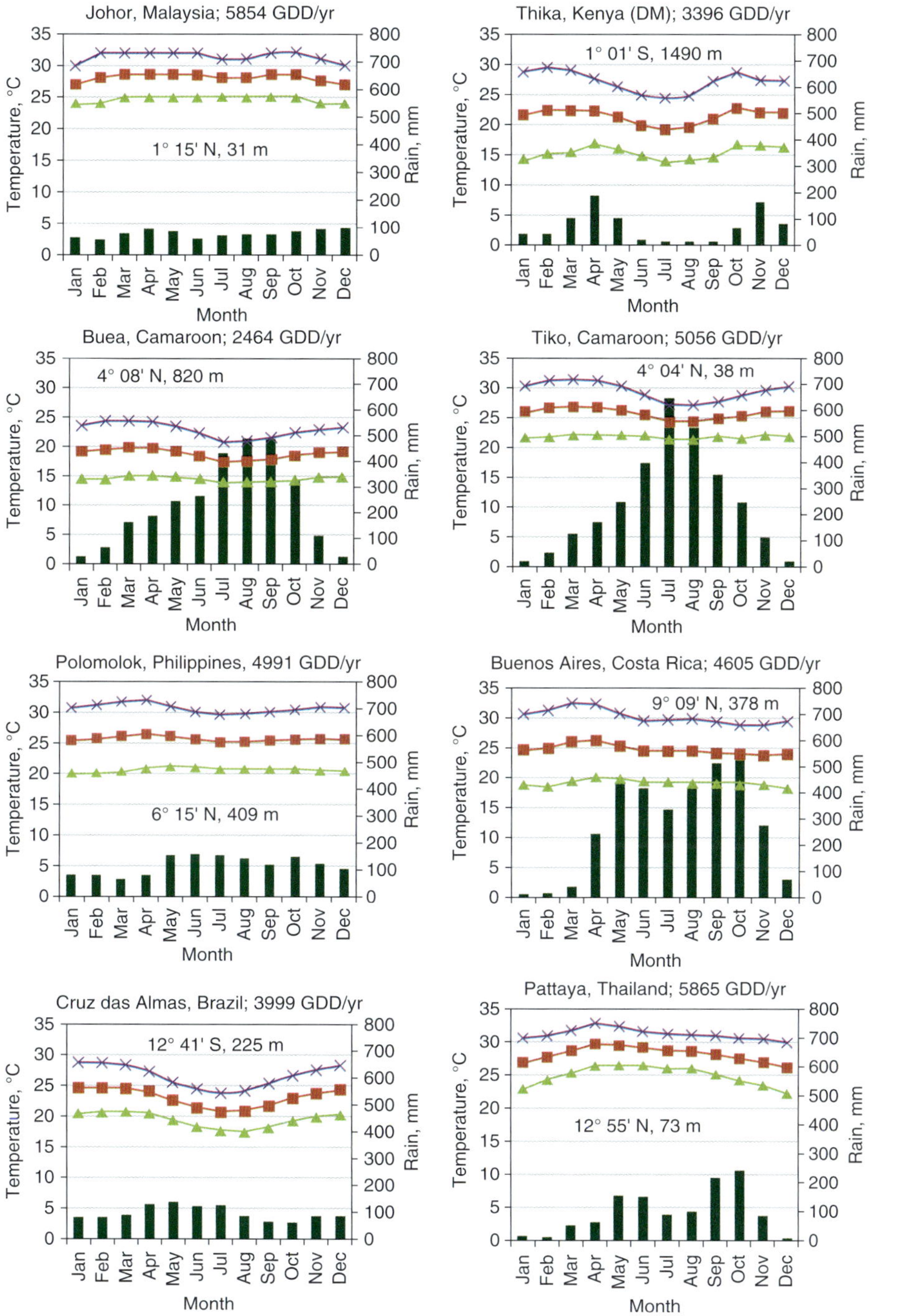

Fig. 6.1. Climates and approximate growing degree days (GDD; base temperature 12°C, for a variety of locations where pineapple is grown. The GDD values show how temperature changes with latitude and elevation could influence plant and fruit growth. No relationship has been established between the displayed GDD values and any specific developmental process. From https://en.climate-data.org/ (Baru, Brazil; Beerwah, Queensland, Australia; Buea, Cameroon; East London, South Africa; Molyko, Cameroon;

continued

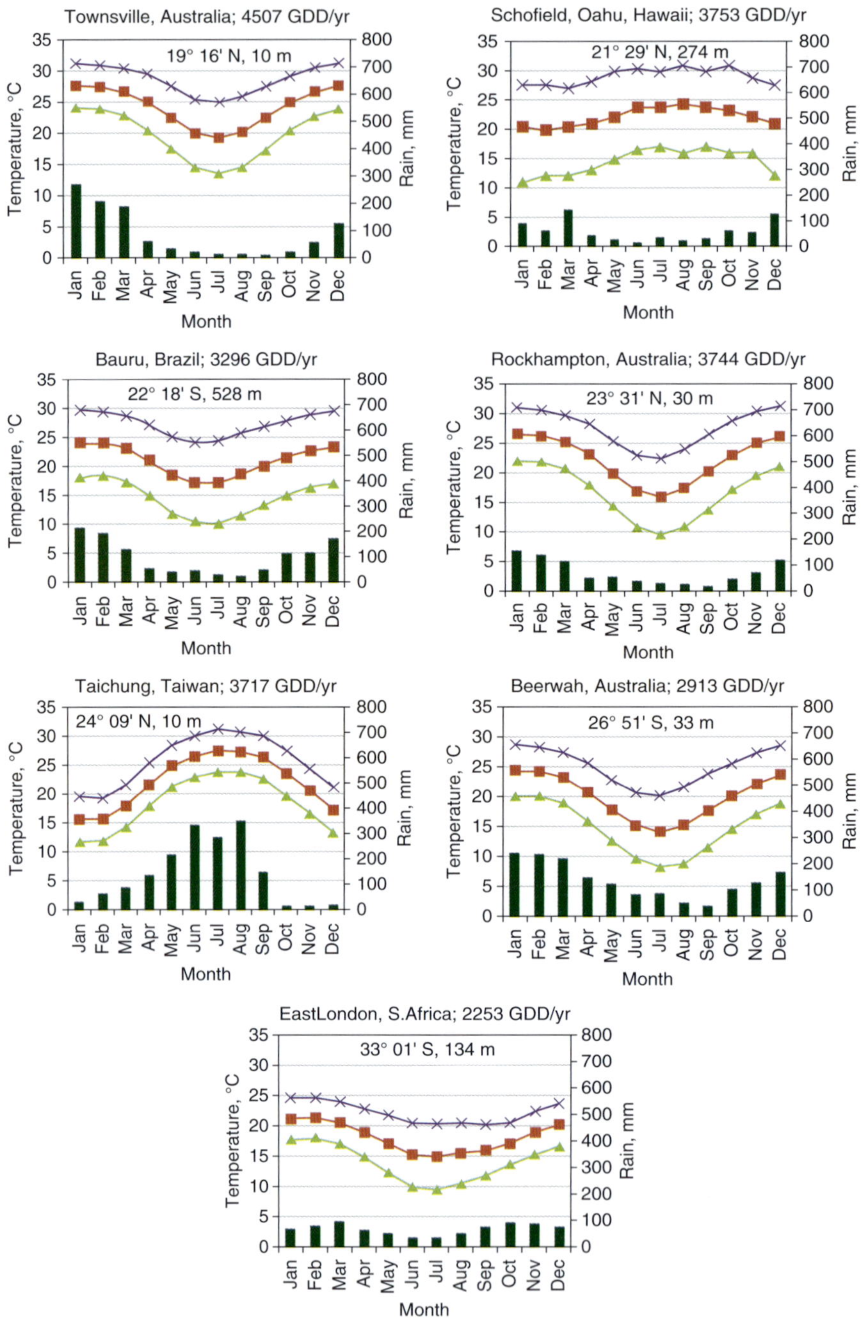

Fig. 6.1. *continued* Pinedeco, Buenos Aires, Costa Rica; Polomolok, Mindanao, Philippines; Rockhampton, Queensland, Australia; Taichung, Taiwan; Thika, Kenya; Tiko, Cameroon); From www.worldweatheronline.com/johor-bahru-weather-averages/johor/my.aspx (Johor Bahru, Malaysia); From us.worldweatheronline.com/pattaya-weather-averages/chon-buri/th.aspx (Pattaya, Thailand); From www.bom.gov.au/climate/data (Townsville, Queensland, Australia); From www.usclimatedata.com/climate/honolulu/hawaii/united-states/ushi0026 (Wahiawa, Oahu, Hawaii). (Accessed 9 August 2017.)

Pineapple is a near-orphan in the crop plant world due to the CAM pathway and because its centre of origin and major centres of cultivation are tropical. This physiological and geographic isolation has greatly limited support for basic research relative to that provided to support research on the important temperate crops. Also, basic research on a crop plant having the C3 or C4 photosynthetic pathway often provided insights into the physiology of other crops that have the same photosynthetic pathway. Unfortunately for researchers of pineapple, there are few commercially important crops with the CAM pathway and none are as far removed from their centres of origin as pineapple, so gains in knowledge of pineapple photosynthesis were slow in coming and hard earned. The unique physiology and morphology of CAM plants in general, and pineapple in particular, has greatly delayed the accumulation of knowledge that provides a clear picture of the functioning of CAM and carbon acquisition and allocation across the diverse environments where pineapple is grown.

An identifying feature of CAM plants is the fixation of a significant percentage of CO_2 at night. From a research point of view, gas exchange data must be measured over 24 hours to get a complete picture of carbon assimilation and its subsequent fixation into carbohydrate for one day. For C3 and C4 plants, data on the effects of multiple light levels on CO_2 fixation can be collected in a few hours at most. This temporal separation between the initial fixation of atmospheric and respiration-generated CO_2 and its subsequent reduction to carbohydrate contributes to a CAM plants high water use efficiency (WUE) (Klug and Ting, 1978; Osmond, 1978; Borland *et al.*, 2011). A summary of the diurnal rhythm of CO_2 uptake and transpiration follows below.

Photosynthetic rhythm in CAM plants and in pineapple

Rhythmic patterns of net CO_2-exchange are well known in CAM plants. In the case of pineapple, the rhythm is reflected in the diurnal changes in leaf acids and soluble sugars with sugars being the primary source of substrate for the production of the organic acids (Fig. 6.2). At the time Sideris *et al.* (1948) discovered this cycling of sugars and organic acids they reported that it strongly indicated that sugars were the substrate used in the formation of organic acids at night in chlorophyllous tissues and that such acids were produced and sugar utilized at much greater rates by rapidly growing plants. Their hypothesis was confirmed in 1989 (Borland and Griffiths, 1989).

While the CAM pattern of CO_2 assimilation is well known, photosynthetic O_2-evolution is less well documented. The diel rhythm of CO_2-fixation and O_2-evolution of an attached pineapple 'D' leaf (Fig. 6.3) displays the typical four-phase pattern of net CO_2 exchange in CAM-plants described by Osmond (1978).

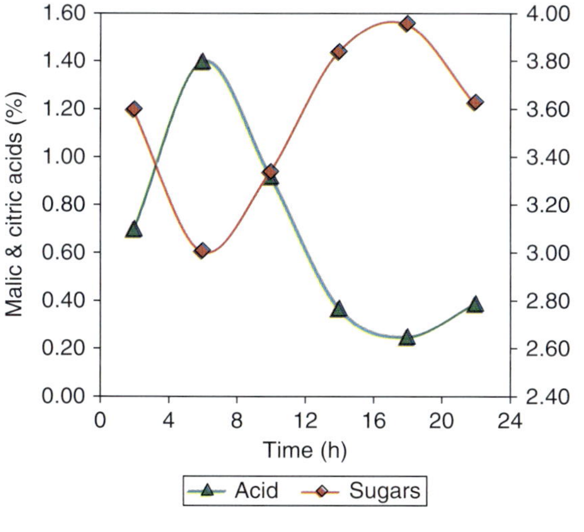

Fig. 6.2. Diurnal changes in sugars and malic-citric acid in the tissue of pineapple D-leaves of plants grown in solution cultures in the greenhouse. Sideris *et al.* (1948). ©American Society of Plant Biologists, reproduced with permission.

Phase I corresponds to nocturnal carboxylation of phosphoenolpyruvate (PEP) catalysed by phosphoenolpyruvate carboxylase (PEPC), a cytosolic enzyme with a high affinity for its bicarbonate substrate, to form oxaloacetate (OAA). OAA is reduced to malate and stored in the chlorenchyma cell vacuoles although some acid may also be stored in the water storage parenchyma (Prigge and Gutiérrez-Soto, 2014). Due to the high affinity of PEPC for bicarbonate, the intercellular CO_2 partial pressure remains very low during this phase and stomatal conductance is high. The general conclusion has been that the intercellular CO_2 level triggers stomatal opening. The substrate for PEP production in pineapple is mainly soluble sugars (Sideris *et al.*, 1948; Borland and Griffiths, 1989; Carnal and Black, 1989). In order for CO_2 dark fixation to occur there must be a reservoir of carbohydrates accumulated during the preceding light period. In terms of balance, respiratory CO_2 is also fixed and contributes to the malate pool. Night O_2 uptake data (Fig. 6.3) indicates that respiratory CO_2 in the D-leaf represents approximately 10% of the total CO_2 fixed into malate at night (Côte, 1988). Night net CO_2-fixation in pineapple generally accounts for at least two thirds of the total net fixation of the day-night period under natural conditions (Neales *et al.*, 1980; Bartholomew, 1982; Nose *et al.*, 1986; Côte, 1988; Zhu *et al.*, 1999). Values for many studies in controlled environments with various photo- and thermo-periods and light levels ranged from 40 to 100% with a mean of 80±17% (Bartholomew and Malézieux, 1994). The proportion of CO_2 fixed nocturnally by pineapple plants propagated *in vitro* increased from 0 to 70% as plant fresh weight increased from 1 to 300 g (Côte *et al.*, 1993). However, Nievola *et al.* (2005) grew *in vitro* plants for 3 months in alternating (AT, 28°C light–15°C dark) and constant (CT, 28°C) temperature and obtained quite different results. In AT, plants weighing 1.2 g fresh weight had high CO_2 dark fixation, thick and rigid leaves and twice the dry weight of plants grown at CT. Acid accumulation in AT plants was accompanied by quite large diurnal changes in abscisic acid (ABA) and IAA while those changes in CT plants were small. Discrimination of $^{13}CO_2$, which ranged from about −11 to −19 in the D-leaf of pineapple plants grown for months at day/night temperatures ranging from 22/18 to 34/30°C (Bartholomew and Malézieux, 1994), is typical of that for constitutive CAM plants. Thus the preponderance of evidence supports the conclusion that pineapple is a constitutive, often called obligate, CAM plant. Recently it was demonstrated that ABA is involved in the CO_2 signalling pathway that controls guard cell responses to the CO_2 concentration (Chater *et al.*, 2015).

Phase II begins after sunrise and begins the transition from net CO_2 fixation catalysed by PEPC to CO_2 fixation via ribulosebisphosphate carboxylase/oxygenase (Rubisco). Rainha *et al.* (2016) reported that about 100 μmol m^{-2} s^{-1} photosynthetically active radiation (PAR) was required to initiate malate transport from the vacuole and begin decarboxylation. As the rate of CO_2 release from malate increases the intracellular CO_2 partial pressure also increases and stomatal conductance decreases. The net

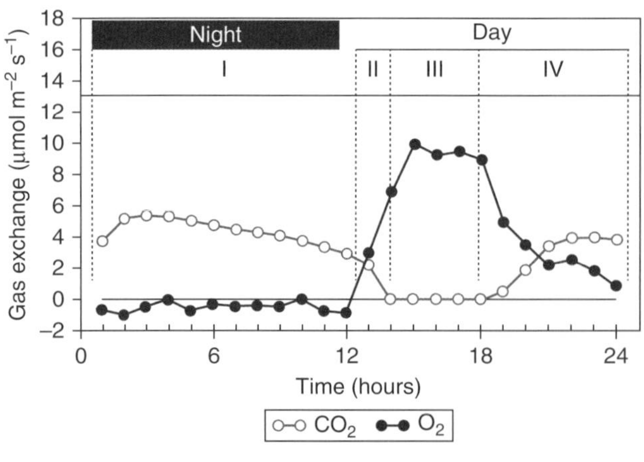

Fig. 6.3. Net CO_2 and O_2 exchange rates of an attached 'Smooth Cayenne' pineapple D-leaf throughout a night/day cycle (Côte, 1988). Fixation of CO_2 illustrates the four phases of CAM while O_2 evolution indicates O_2 uptake at night due to respiration. The photosynthetic photon flux was 400 μmol m^{-2} s^{-1} with a 12 h night/12 h day and a night/day temperature 22/26°C. Data are the average for four consecutive days.

contribution to carbon gain during phase II is only 1–3% of the total net CO_2-uptake by pineapple plants maintained in controlled conditions (Fig. 6.3).

Phase III is a period when net CO_2-fixation is negligible while the CO_2 generated by malate decarboxylation is fixed via Rubisco. Under ideal conditions for CAM, the internal CO_2 partial pressure rises well above atmospheric levels because the rate of malate decarboxylation exceeds the rate of internal CO_2-fixation via Rubisco. This high internal CO_2 partial pressure results in near unmeasurable stomatal conductance, which limits transpiration during this phase and also stimulates photosynthesis by inhibiting the Rubisco oxygenase reaction. Although net CO_2-uptake is nil during phase III, the maximum rate of photosynthetic carbon reduction via Rubisco occurs during this phase. Based on the assumption that 1.0 mole of O_2 is evolved for every mole of CO_2 reduced (Kaplan and Björkman, 1980), the rate of CO_2 fixation via Rubisco during phase III deduced from the rate of O_2 evolution during this phase could be 1.3–6-fold higher than the maximum rate of net CO_2 fixation achieved by pineapple during phase IV (Côte, 1988). In the data of Fig. 6.3, it is 2.4 times higher.

Phase IV begins when the malate pool is depleted, the internal CO_2 partial pressure declines, stomatal conductance increases, and atmospheric CO_2 uptake resumes for the remainder of the day. In pineapple, this fixation generally amounts to 15 to 25% of the total net CO_2-uptake (Fig. 6.3). Following the assumption that 1.0 mole of O_2 is evolved per mole of CO_2 fixed, it was estimated that only 60–80% of the total CO_2 fixed during the night by a pineapple plant was re-assimilated via Rubisco when net-CO_2 uptake resumed in phase IV (Côte *et al.*, 1989). A value of 75% can be deduced from Fig. 6.3. PEPC has been reported to be active during late phase IV in CAM plants (Kluge, 1969; Kluge *et al.*, 1982) and measurement of malate concentration also suggested that CO_2 uptake via PEPC occurs in pineapple during late phase IV (Black *et al.*, 1982; Kenyon *et al.*, 1985). The O_2 and CO_2 gas exchange data suggest that CO_2 fixation via PEPC during phase IV represents 3–15% of the nocturnal net CO_2 fixed by pineapple (Côte, 1988; Côte *et al.*, 1989).

Gas exchange measured under controlled conditions is assumed to only depict in a general way what occurs in the field where stochastic, rather than steady-state, environmental conditions prevail. Diurnal and seasonal changes in weather influence both the duration and intensity of the four phases of CAM as well as daily net CO_2 assimilation. While extrapolation of results obtained in controlled conditions to plants in natural environments must be done carefully, *in situ* measurement of stomatal conductance and leaf titratable acidity confirm the general conclusions derived from studies in controlled environments.

Transpiration rhythm and transpiration rate in pineapple

In pineapple, the diel rhythms of stomatal conductance and transpiration are closely linked to the net CO_2 uptake rhythm (Neales *et al.*, 1968; Nose *et al.*, 1981; Bartholomew, 1982; Côte *et al.*, 1993; Zhu *et al.*, 2005; see also Fig. 6.4 below). The lowest rates of transpiration occur during phases I, II and III, and the highest during phase IV. In phase IV leaf stomata are open, leaf temperature may be higher than air temperature and the vapour pressure gradient between the leaf and air likely is greater than at any other time. Water use efficiency (WUE) is two times greater during phase I than during phase IV and reaches a minimum during phase III (Fig. 6.4). The transpiration rates of pineapple ranged from 0.05 to 0.23 mmol m^{-2} s^{-1} (Ekern, 1965; Joshi *et al.*, 1965; Neales *et al.*, 1968; Nose *et al.*, 1981) while values for sunflower and tobacco, both C3 plants, were at least tenfold higher at about 3.1 mmol m^{-2} s^{-1} (Neales *et al.*, 1968). Transpiration by pineapple in the light in controlled conditions was only 4% of that of sunflower leaves grown in a similar environment (Neales *et al.*, 1968) and only 6% of that for wheat and WUE was 3.3 fold greater (Côte *et al.*, 1993).

Leaf and canopy transpiration rates for pineapple are lower than those for most cultivated crops. At midday, there was no measurable water vapour loss from a full pineapple canopy with a leaf area index (LAI) of 7 (Ekern, 1965). The transpiration ratio (TR, units water lost per

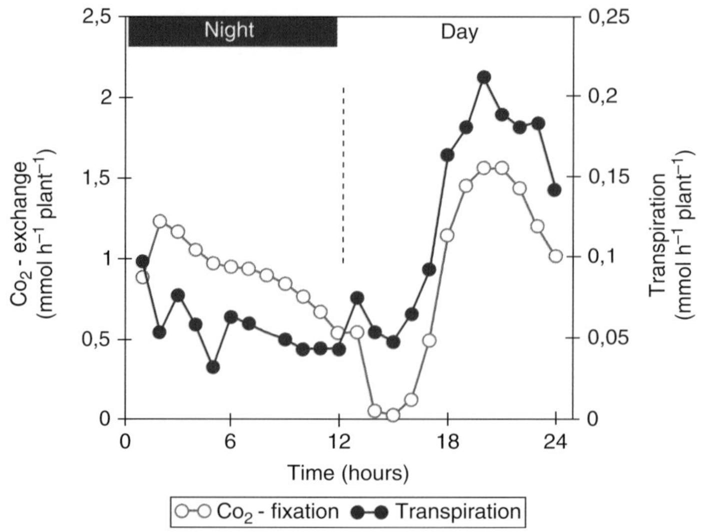

Fig. 6.4. Transpiration and net CO_2 exchange during a night/day cycle by a 'Smooth Cayenne' pineapple plant. Study conditions included a photosynthetic photon flux of 650 µmol m^{-2} s^{-1}, a 12 h night/12 h day photoperiod and a night/day temperature of 22/28°C. The 120 g (fresh weight of aerial parts) plant was obtained from *in vitro* culture. Data are averages for three consecutive days. Reproduced by permission of the International Society of Horticultural Science from Cote, F.X., Folliot, M. and Andre, M. (1993). Photosynthetic crassulacean acid metabolism in pineapple: Diel rhythm of CO_2 fixation, water use, and effect of water stress. *Acta Horticulturae* 334, 113–130. DOI: 10.17660/ActaHortic.1993.334.12 https://doi.org/10.17660/ActaHortic.1993.334.12

unit dry matter gain; kg kg^{-1}), a measure of WUE, for pineapple was about 50 (Sideris and Krauss, 1955; Joshi *et al.*, 1965; Côte, 1988) to 116 (Neales *et al.*, 1968) while the range for C3 crops was 450–950 and that for C4 crops was 250–350 (Klug and Ting, 1978). A consequence of the low rate of water use is that the crop can sustain a LAI of greater than 7 (Malézieux, 1991; Zhang, 1992) over long periods under low rainfall conditions (Ekern, 1965; Aubert, 1973a).

Origins of the high water use efficiency related to CAM

Diffusion of water vapour from and CO_2 into plant leaves can be described by the equations:

$$\Delta W = r\,T \tag{1}$$

$$\Delta CO_2 = 1.6\, r\, P \tag{2}$$

where ΔW is the difference in water vapour pressure between the stomatal cavity and the atmosphere, r the resistance to water vapour diffusion, T the transpiration rate, ΔCO_2 the difference in CO_2 partial pressure between the stomatal cavity and the atmosphere and 1.6 the ratio of the diffusion coefficients for water vapour and CO_2 in air, and P is the rate of net atmospheric CO_2 fixation. Combining equations (1) and (2) gives:

$$T = 1.6 * \Delta W * P/\Delta CO2 \tag{3}$$

The low transpiration rate of CAM plants is related to morphological and physiological adaptations that influence ΔW, P and ΔCO_2 of equation (3) to reduce T:

- Net CO_2 fixation (P) occurs mainly during the night when ΔW is low.
- Stomatal conductance is negligible during the day when ΔW is high, in response to a high internal CO_2 concentration, consequently P and T are low or even zero.
- During phase IV ΔCO_2 is high (Côte *et al.*, 1993) while ΔW is somewhat reduced, which results in reduced rates of P and T.

The highest leaf conductances recorded for pineapple were about 3% of those observed for cotton (Neales *et al.*, 1968). This low conductance was attributed to a thick mesophyll, low stomatal frequency, and sunken stomata located beneath multicellular trichomes (Bartholomew and Kadzimin, 1977).

Effects of Environmental Factors

Temperature effects on photosynthesis

The effects of temperature on photosynthesis of pineapple are complex and relatively few studies have been conducted. While the results of these studies often are not directly comparable, the general trends show that as long as CAM is not limited by low temperature and low PAR (Rainha *et al.*, 2016), the intensity of CAM (phase I) is highest when the light/dark temperature differential is approximately 10°C and the night temperature is cool, about 20°C (Zhu *et al.*, 1999). These results are consistent with other observations that the optimum day/night temperature for growth also is about 30/20°C (Neild and Boshell, 1976; Bartholomew and Malézieux, 1994).

At a given day temperature, an increase in the night temperature decreases night net CO_2 uptake (Connelly, 1972; Neales *et al.*, 1980; Zhu *et al.*, 1999) as well as the proportion of night fixation in a 24 h period (Connelly, 1972; Bartholomew, 1982; Bartholomew and Malézieux, 1994; Zhu *et al.*, 1999). Carbon isotopic fractionation ($\delta^{13}C$) values showed little variation with plant age, generally were lowest (11–15) where night temperatures were 18–22°C, and were highest (18–19) at a day:night temperature of 30°C (Bartholomew and Malézieux, 1994). Neales *et al.* (1980) and Zhu *et al.* (1999) reported a 50% decrease in night uptake as the night temperature was increased from 20 to 25°C at a constant day temperature of 30°C. The decrease could be due in part to increased fixation of respiration-derived CO_2, but leaf titratable acidity also decreases (Min, 1995), indicating less total CO_2 dark fixation. Some studies show little or no change in net CO_2 uptake during the day with an increase in night temperature (Neales *et al.*, 1980; Zhu *et al.*, 1999), but Connelly (1972) reported a dramatic increase when night temperature increased from 15 to 30°C.

At a given night temperature, the fraction of total CO_2 assimilated in a 24-h period that is fixed at night increases with increasing day temperature (Neales *et al.*, 1980; Bartholomew, 1982; Bartholomew and Malézieux, 1994; Zhu *et al.*, 1999). This increase is primarily due to a decrease in net CO_2 uptake in the day and to a lesser extent to an increase in net CO_2 uptake at night (Connelly, 1972; Neales *et al.*, 1980; Zhu *et al.*, 1999). For example, Zhu *et al.* (1999) reported that at a night temperature of 25°C, CO_2 uptake in the day decreased by 50%, from 55.1 to 23.5 mmol CO_2 m^{-2} $period^{-1}$ as the day temperature was increased from 30°C to 35°C. The percentage increase in CO_2 uptake at night was much smaller, however, the absolute increase of 43.1 mmol CO_2 m^{-2} $period^{-1}$ was almost twice the reduction during the day. It is hypothesized that the duration of phase IV is shorter at 35°C than it as at 30°C because more time is required to process the greater accumulation of malate during phase I and possibly also because increased respiratory CO_2 at the 35°C day temperature slows processing of the accumulated malate.

The relationship between net CO_2 fixation and dry matter accumulation under various thermoperiods has not been well studied. In one study, net CO_2 fixation of a single attached leaf of plants grown 8 months at a 30°C day and a 20°C night was 1.6 times higher than that for leaves of comparable plants grown in a 30°C day and 25°C night environment while whole-plant dry matter accumulation was only 1.1 times higher in the cooler environment (Zhu *et al.*, 1997a). However, assimilation was measured only on D-leaves while dry matter accumulation reflects the contribution of leaves of all ages.

Irradiance effects on photosynthesis

Because CO_2 assimilation in CAM plants, including pineapple, occurs via the photosynthetic carbon reduction cycle, it is assumed that photosynthesis will increase with increasing irradiance and the results of Shiroma (1977) and Nose *et al.* (1981) support this assumption. Nose *et al.* (1985) found that light saturation of pineapple plants occurred at a photosynthetic photon flux (PPF) of about 500 $\mu mol\ m^{-2}s^{-1}$. *In vitro* grown plants maintained under continuous illumination also saturated at a PPF of 500 μmol $m^{-2}s^{-1}$ (Côte, 1988). However, variation in irradiance can alter the pattern of CO_2 fixation. For example, high irradiance decreases the duration of phase III and increases the duration of phase IV assimilation relative to what occurs at low irradiance (Nose *et al.*, 1986). This is because the rate of malate-decarboxylation-dependent CO_2 assimilation is more rapid during phase III (Sale and Neales, 1980) so the malate pool is depleted more rapidly.

Significant quantities of soluble sugars are required to provide the phosphoenolpyruvate substrate that is carboxylated to ultimately form the malate that is accumulated in the vacuole during phase I (Sideris *et al.*, 1948). High irradiance levels (perhaps 20 Mj m^{-2} day^{-1}) probably increases the quantity of soluble sugar formed in leaves during phase III and IV and it seems reasonable to assume that a large sugar supply at the beginning of phase I would increase the plant's capacity to accumulate malate in leaf vacuoles during the night. Consistent with this hypothesis, an increase in irradiance during the day increased CO_2 fixation the following night (Shiroma *et al.*, 1977; Nose *et al.*, 1981, 1985, 1986) and leaf titratable acidity also increased with increasing irradiance (Sideris *et al.*, 1948; Connelly, 1969; Aubert, 1971; Sale and Neales, 1980). However, increases in CO_2 fixation in phase I that follow an increase in irradiance generally are smaller than the increase in CO_2 assimilation in the light. Thus, the intensity of CAM (ratio of CO_2 fixed in phase II–IV:phase I) tends to decrease as irradiance increases, at least until saturating levels are attained.

Little is known about the effect of photoperiod on pineapple photosynthesis. Nose *et al.* (1986) reported that as photoperiod increased from 10 to 16 h, total uptake in the light increased while uptake at night decreased. However, expressed on an hourly basis, the increase in CO_2 uptake in the light as daylength increased was small while the decrease at night was relatively large. A 40–50% reduction in CO_2 uptake at night can be deduced from the data of Nose *et al.* (1986) as photoperiod decreased from 16 to 8 h. In natural environments, e.g. Hawaii, the decrease in photoperiod from summer to winter is only 21% while the decrease in irradiance is about 50% so the impact of decreasing photoperiod on CO_2 assimilation probably is not large.

CO_2 concentration

The effect of elevated CO_2 concentration on pineapple photosynthesis has not been well studied, but elevated CO_2 levels do increase CO_2 fixation. At 1700 μmol mol^{-1} CO_2, net CO_2 fixation increased two- to threefold during phase IV (Côte, 1988) and a CO_2 concentration greater than 1000 μmol mol^{-1} was required to saturate photosynthesis in that phase. Saturation at such a high CO_2 level results from the pineapple leaf's low stomatal conductance even under conditions where the stomata are fully open.

Zhu *et al.* (1997b) reported that the dry mass of pineapple plants grown 4 months at 730 μmol mol^{-1} CO_2 was 1.2 times higher than that of plants grown at ambient CO_2 (330 μmol mol^{-1}). In that study, D-leaf titratable acidity at the end of phase I was more than 1.3 times greater for plants grown at elevated than at ambient CO_2. Net CO_2 uptake in phase I also increased, but only at a night temperature of 25°C (Zhu *et al.*, 1999). Uptake of CO_2 during phase IV also increased and the effect of enrichment was greater as the average temperature increased from 25°C to 30°C (Zhu *et al.*, 1999). Regardless of day or night temperatures, stomatal conductance at 700 μmol mol^{-1} CO_2 was approximately half that measured at 350 μmol mol^{-1} CO_2 and water use efficiency (WUE) was at least 30% greater at 700 than at 350 μmol mol^{-1} CO_2 (Zhu *et al.*, 2005). The response of pineapple to increased CO_2 concentration is typical of numerous species that utilize the CAM pathway (Drennan and Nobel, 2000). Based on the limited results obtained with pineapple, the simultaneous increase in CO_2 and temperature associated with global climate change would not have a large negative impact on CO_2 assimilation or on vegetative growth and should increase WUE.

Water stress

When CAM plants are subjected to drought, CO_2 fixation during phase IV eventually ceases; however, CAM plants continue to refix respired CO_2, a phenomenon referred to as CAM idling (Klug and Ting, 1978). Pineapple plants subjected to water deficit exhibit a similar response (Fig. 6.5). When well-watered, CO_2 fixation during the day and night is normal with >80% of the CO_2 being fixed at night (Zhu *et al.*, 1997a). Depending on plant size, soil volume explored by the root system and soil water-holding capacity, drought reduces CO_2 uptake as the soil water potential decreases. For potted plants, CO_2 fixation in the light by attached 'D' leaves ceased after about 5 (Fig. 6.5) or 15 days (see Fig. 6.6; Zhu *et al.*, 2005). Nocturnal CO_2 fixation rate was not affected at the beginning of water stress, but decreased progressively as the duration of stress lengthened (see Fig. 6.6; Zhu *et al.*, 2005). The rate that net CO_2 assimilation declined during drought was related to the day/night temperature regime (Zhu *et al.*, 2005). Fluctuations in leaf titratable acidity indicated significant assimilation, about 35% of the maximum, still occurred after 70 days of drought at 30/20°C while at warmer temperatures (30/25°C and 35/25°C), titratable acidity was 20% or less of the maximum after only 40 days of drought. While pineapple is highly tolerant of drought, assimilation declines fairly rapidly with drought, and warm temperatures hasten the rate of decline. Assimilation of CO_2 in phase I and IV resumed rapidly after re-watering (see Fig. 6.5).

Temperature

At a 35°C day temperature, day transpiration rates for pineapple decreased as the night temperature decreased while transpiration rates at night remained constant (Yoder, 1969). These results are consistent with CO_2 assimilation studies that show that assimilation in the light and WUE were highest when night temperatures were lowest (Zhu, 1996). For example, WUE at night was 1.3 times higher for pineapple plants maintained in a day/night temperature of 30/20°C compared to those in a 30/25°C thermoperiod. In the light, WUE was >1.5 times greater at 30/25°C than at 35/25°C (Zhu, 1996).

Stomatal conductance (g_s) of pineapple typically is lowest during the day and highest at night (see Fig. 6.7). However, in a very uncommon 10/25°C day/night thermoperiod, g_s was much higher during the day (51 mmol m^{-2} s^{-1}) than at night (8.1 mmol m^{-2} s^{-1}) (Neales *et al.*, 1980), reversing the typical inverted pattern of stomatal opening. Assimilation of CO_2 can occur at 10°C (Rainha *et al.*, 2016) though perhaps at too low a rate to provide the sugars required for CO_2 dark fixation.

Few data on g_s are available under field conditions. The dry 'Harmattan' wind in West Africa induced prolonged closing of the stomata, with opening occurring late in the night (Py *et al.*, 1987). The Harmattin wind can be cold, as low as 9°C, or hot and if it is cold, the data of Rainha *et al.* (2016) suggests the low temperature could

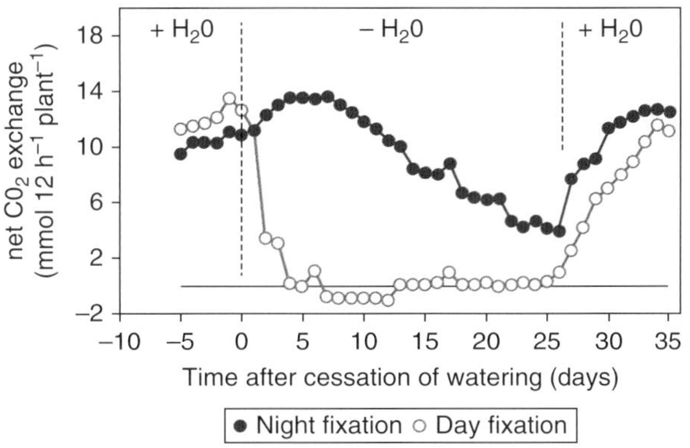

Fig. 6.5. Effect of water deficit on the daily CO_2 exchange by a 'Smooth Cayenne' *in vitro* cultured pineapple plant weighing 100 g. The plant was placed in aluminum silicate and watering was stopped on day 0 and was resumed on day 26. Environmental conditions include: photosynthetic photon flux of 650 μmol m^{-2} s^{-1}, a 12 h night/12 h day photoperiod and a 22/28°C night/day temperature. Reproduced by permission of the International Society of Horticultural Science from Cote, F.X., Folliot, M. and Andre, M. (1993). Photosynthetic crassulacean acid metabolism in pineapple: Diel rhythm of CO_2 fixation, water use, and effect of water stress. *Acta Horticulturae* 334, 113–130. DOI: 10.17660/ActaHortic.1993.334.12. https://doi.org/10.17660/ActaHortic.1993.334.12

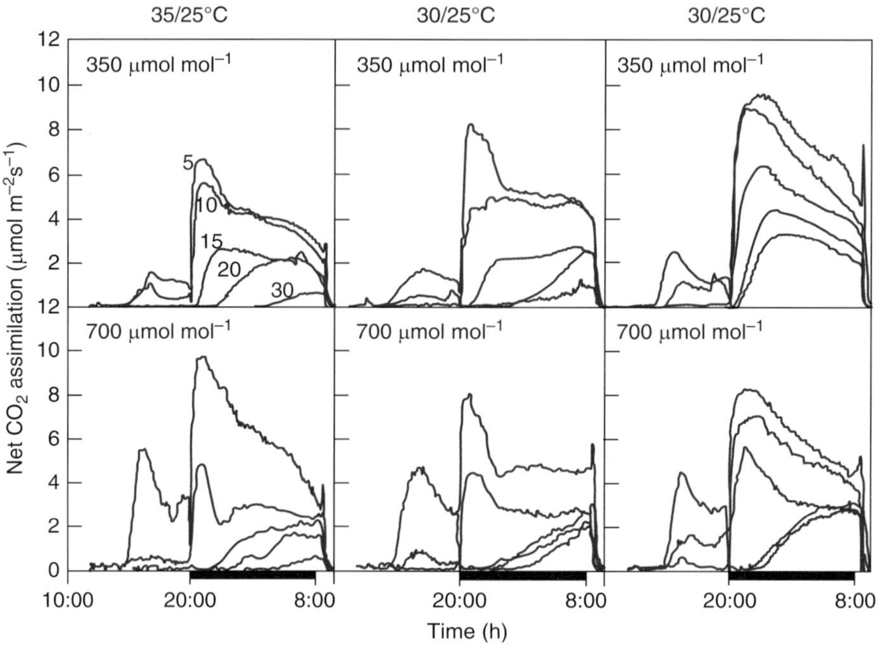

Fig. 6.6. Daily variation in net CO_2 assimilation of leaves of pineapple plants grown at two CO_2 levels and three day-night temperatures on day 5, 10, 15, 20 and 30 after withholding irrigation. Black bars indicate the night the CO_2 levels are indicated on each panel. Reproduced by permission of the International Society of Horticultural Science from Zhu, J., Bartholomew, D.P and Goldstein, G. (2005). Photosynthetic gas exchange and water relations during drought in 'Smooth Cayenne' pineapple (*Ananas comosus* (L.) Merr.) grown under ambient and elevated CO_2 and three day/night temperatures. *Acta Horticulturae* 666, 163–173. DOI: 10.17660/ActaHortic.2005.666.15. https://doi.org/10.17660/ActaHortic.2005.666.15.

be the cause of the inverted g_s. It is concluded that the consumptive use of water by pineapple would be higher in tropical environments with high night temperatures and a small day/night temperature differential than it would be in subtropical environments where larger day/night temperature differentials prevail.

Irradiance

Transpiration of pineapple generally is negligible during the day (Ekern, 1965) so it is expected that irradiance would have a limited effect on transpiration. However, high irradiance could shorten phase III, which could extend the duration of CO_2 fixation during phase IV because the stomata open earlier in the afternoon. This would be expected to increase transpiration during the day and Nose *et al.* (1981) reported an increase in transpiration with increasing irradiance.

CO_2 concentration

Zhu *et al.* (1999) reported a >1.1 fold increase in WUE during the night and a >2.3 fold increase in the day for pineapple plants grown at 700 μmol mol^{-1} CO_2 compared to those grown at 350 μmol mol^{-1} and g_s also was much lower (Fig. 6.7). Depending on the day/night temperature regime, the 24-h WUE was 1.3 to 1.6 times higher at 700 μmol mol^{-1} than at 350 μmol mol^{-1}. Data of Zhu *et al.* (1999) showed that higher WUE was associated with a two- to threefold decrease in stomatal conductance during phases IV and I.

Water stress

The patterns of g_s during the onset of water stress are similar to those of net CO_2 uptake (see Fig. 6.7). Zhu *et al.* (2005) observed a

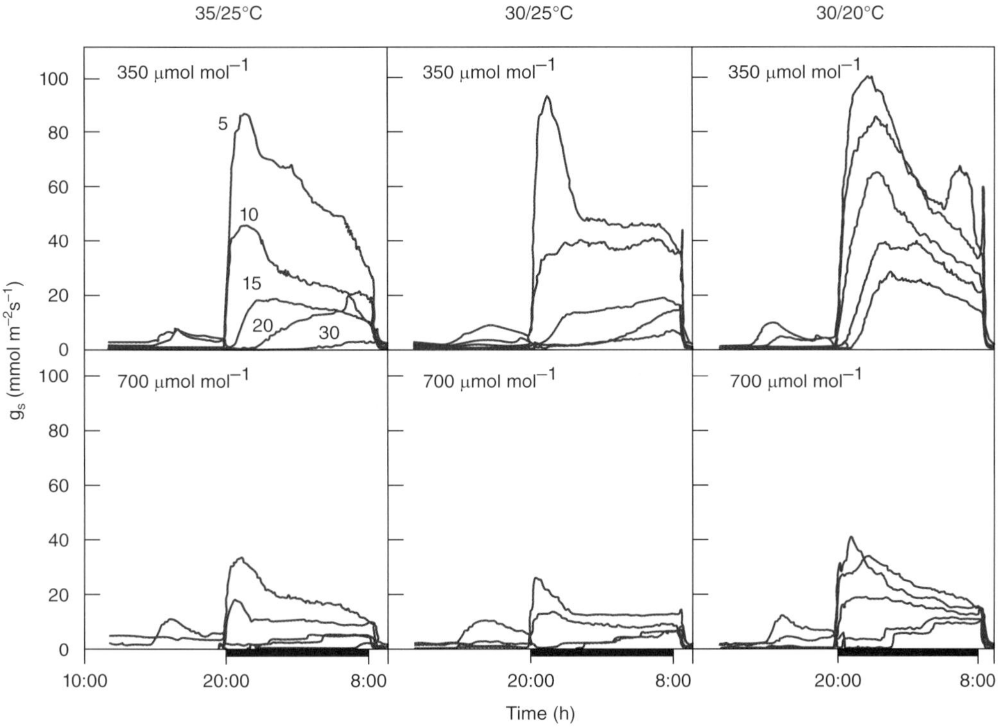

Fig. 6.7 Diurnal stomatal conductance (g_s) of leaves of pineapple plants grown as described above. Black bars indicate the night and the CO_2 levels are indicated on each panel. Reproduced by permission of the International Society of Horticultural Science from Zhu, J., Bartholomew, D.P and Goldstein, G. (2005). Photosynthetic gas exchange and water relations during drought in 'Smooth Cayenne' pineapple (*Ananas comosus* (L.) Merr.) grown under ambient and elevated CO_2 and three day/night temperatures. *Acta Horticulturae* 666, 163–173. DOI:10.17660/ActaHortic.2005.666.15. https://doi.org/10.17660/ActaHortic.2005.666.15.

rapid decrease in g_s during phase IV for pineapple plants subjected to drought and a somewhat slower decrease during phase I. After 15 days of drought, g_s was near zero during phase IV. For stressed plants in phase I, the time when g_s reached a maximum shifted from early in the light period to near its end. In ambient CO_2, WUE decreased progressively as water stress progressed. However, plants maintained at 700 μmol. mol^{-1} CO_2 exhibited a steady increase in WUE (Zhu, 1996). In a field study, Combres (1981, unpublished results) observed that after a rainfall following a period of drought the stomata open progressively earlier in the morning and in the afternoon in the following days.

Plant Growth and the Environment

Main characteristics of vegetative plant growth

Pineapple is grown from a variety of propagules (see Chapter 7) but all develop in a similar manner. If conditions for growth are favourable after planting, root initiation begins and is followed in a month or more by the appearance of new leaves. For MD-2 plants, D-leaf length and weight did not increase for the first 3 months after planting (Van de Poel *et al.*, 2009). Between planting and inflorescence initiation, growth occurs at root, stem, and leaf meristems. Pineapple cultivars Smooth Cayenne, CO-2 and

'MD-2' have strong apical dominance and do not produce shoots from stem axillary buds prior to flower induction. Apical dominance is broken by forcing and shoots begin to grow. However, while more information is needed, shoots only continue to grow in the equivalent of subtropical environments, perhaps those with <5000 or even <4500 GDD (see Fig. 6.1). In environments with more than about 5000 GDD (Fig. 6.1), shoots do not grow until after fruit harvest. Cultivars Victoria (Queen) (Maerere, 1997) and Española Roja have weaker apical dominance and may initiate shoots prior to flower induction even in tropical environments.

Shoots of the same initial mass, whether suckers, slips or crowns, have approximately the same leaf area despite quite large differences in leaf number and size. Crowns have numerous short leaves while suckers have fewer but longer leaves. In an experiment in Côte d'Ivoire, the leaf appearance rate in the month after planting was eight for crowns and three for suckers (Lacoeuilhe, 1976). Beyond 3 months, leaf emergence rate was similar for crowns and suckers. During the first few months of growth, leaves produced by suckers are longer and heavier than those produced by crowns. For planting material of similar size, even if produced by leaf trimming of large suckers to equalize shoot weights, leaf emergence rate and leaf size are similar within 3–5 months after planting (Py *et al.*, 1987).

Leaves

Leaves represent approximately 90% of aerial plant fresh weight during vegetative growth (Py, 1959). Leaves grow from the base, the maximum leaf length is reached several months, at least 8 in the Van de Poel *et al.* (2009) study, after initiation and leaves of a wide range of sizes and ages are present on the plant at the same time (see Fig. 6.8). The time from initiation to full elongation depends on temperature and is approximately 4 months in the nearly equatorial conditions of Côte d'Ivoire, but considerably longer in cooler environments.

The 'D' group of leaves is the longest, has 'succulent-brittle' leaf bases that are not fully mature, and has nutrient contents that reflected current nutrient availability. Over time it became common practice to apply the term 'D' to a single leaf, usually the tallest leaf on the plant. A D-leaf represents an easily identified and easily pulled standard leaf (see Fig. 8.1) that can serve as an index of plant growth and plant nutrient status. The D-leaf can also be a good indicator of plant size variation within a field. If D-leaf weights or lengths in a 20-leaf sample taken from consecutive plants in a bed are not uniform, it is likely that plant sizes also are not uniform and is a good indication that fruit size also will be variable. Interplant variability in size indicates that planting material was not adequately sized before planting.

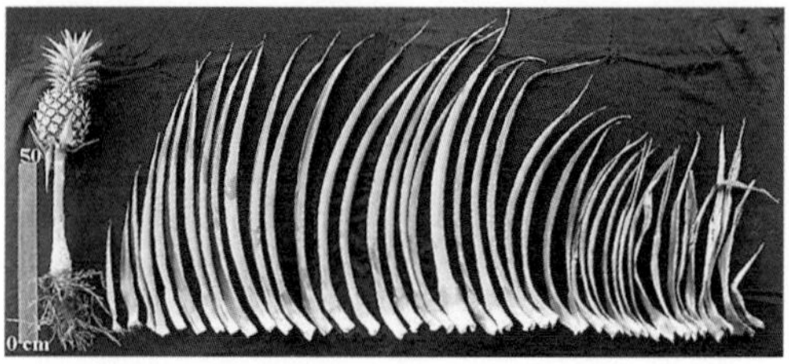

Fig. 6.8. Array of leaves present on a fruiting pineapple plant. The oldest leaves are to the right of the stem and its associated peduncle and fruit. Photo E. Malézieux.

The D-leaf can only serve as a growth index as long as a plant is growing without significant stresses and the D-leaf has not reached its maximum length (Bartholomew, 2008). When growth is rapid, the D-leaf is almost always nearly but not fully expanded (Py *et al.*, 1987). The weight or length, or both, of a D-leaf will vary with the climate where the crop is grown. Therefore, length or weight and plant weight should be established for the general environment where the crop is growing before it can be used to estimate plant weight. The maximum leaf length will be longer in tropical than in cooler climates, would decrease with increasing planting density and may differ among pineapple cultivars.

As leaves grow in a field of pineapple the canopy, defined by the LAI, expands until all soil area is shaded. Without proper management the LAI in a pineapple field can get so large, and interplant competition for light so intense, that photosynthetic efficiency of individual leaves in the canopy actually declines. Lin *et al.* (1994) found the average $\Delta^{13}C$ values in the leaves 8–11 showed less $^{13}CO^2$ discrimination and higher PEPC activity than older leaves. While the leaves had lower CAM activity, no data on net CO_2

uptake with leaf age and exposure was presented or found elsewhere. The LAI and interplant competition can be managed by adjusting plant spacing, by knowing the relationship between plant size at forcing and fruit size at harvest (harvest index) and by forcing when the target plant weight is reached. That information makes it possible to grow and force plants that produce the fruit size desired by the market. So while leaf area per plant can reach 2.2 m^2 for a plant having a fresh weight of 3.6 kg (Py, 1959), at modern planting densities of 70,000 plants ha^{-1}, the LAI would be greater than 15. If the ideal fruit size is a tray count of 7 (1.6–1.8 kg), and the harvest index is 0.75 then the plant size at forcing would be about 2.2 kg. The LAI can't be estimated because there is insufficient information on the relationship between plant weight and leaf area over the range of environments where pineapple is grown. The harvest index is much smaller in the warm tropics, as low as 0.5, so a larger plant is required to maintain fruit size. With a larger plant, the planting density might have to be reduced to obtain the desired fruit size.

Roots

Root morphology is determined by the type of planting material and suckers have few roots, slips have somewhat more roots and some could have elongated while the slip is still attached to the mother plant, and crowns have numerous root primordia that are visible at their base. After planting, root mass increases relatively more rapidly than leaf mass (Py *et al.*, 1987). Though roots of 'Smooth Cayenne' pineapple can theoretically reach a length of 1.8 m and grow to a depth of 85 cm (Sideris and Krauss, 1934), the sensitivity of pineapple roots to soil compaction (Rafaillac *et al.*, 1978; Ikan, 1990) generally confines the root system to the tilled area (Ekern, 1965; Ikan, 1990). This restricted soil volume prospected by the roots limits the amount of water and soil nutrients available to the plant. There is evidence that root growth decreases after flower induction and that maximum root mass is reached at anthesis.

Stem and axillary buds

Stem mass increases progressively after planting with no unique morphological changes until reproductive development begins. Stem dry matter (DM) content increases when starch accumulates in the stem (Fig. 6.9). However, starch accumulation varies with plant size, aerial environment and reproductive development (Table 6.2). Stem DM in the 26/22°C environment began to increase on 16/6/1974 and reached 18.2% on 11/24/1974. A comparison of stem dry weights and DM contents for forced and unforced plants on 6/16, 7/8, and the two values that follow the result for 11/24/1974 show that ethephon forcing suspended stem growth but resulted in increases in stem DM content.

Differences in stem DM in the 34/18°C environment are explained as follows. Stem DM was highest on 7/8/1974 because the plant was forced only 2 months earlier so carbohydrate demand was low. The lower DM contents of plants harvested on 1/5 and 2/14/1975 would result from allocation of starch to sugar accumulation in maturing fruits and perhaps also to support growth of suckers or slips, or both. Plants grown in 30°C night temperatures and treated with ethephon were not forced so there was no change in stem DM. Growth at a 30°C night temperature was not as rapid as in environments with lower night temperatures and it is assumed that low reserves or high temperature stress, or both, rendered them insensitive to ethephon treatment.

No publications were found on the biology of 'Queen' pineapple so it isn't known how intraplant competition between fruits and shoots for resources changes for that cultivar. For the commercially dominant cultivars, when apical dominance is broken by the initiation of reproductive development, resource demand for leaf growth is reduced and early during inflorescence development it is only a minor sink. As a result, dry matter, mainly starch, increases because leaves produce more photosynthate than is required to support the growth of fruit and a sucker or two. Under such conditions, resources are available to simultaneously support the growth of a fruit and an axillary bud or two so any suckers will have reached half the mass required to produce a ratoon fruit by the time the fruit matures. The foregoing is true for relatively large but not for small 'Smooth Cayenne' or 'MD-2' plants in a typical subtropical environment. In tropical environments where the average temperature is near or above 25°C (Fig. 6.1), the physiological

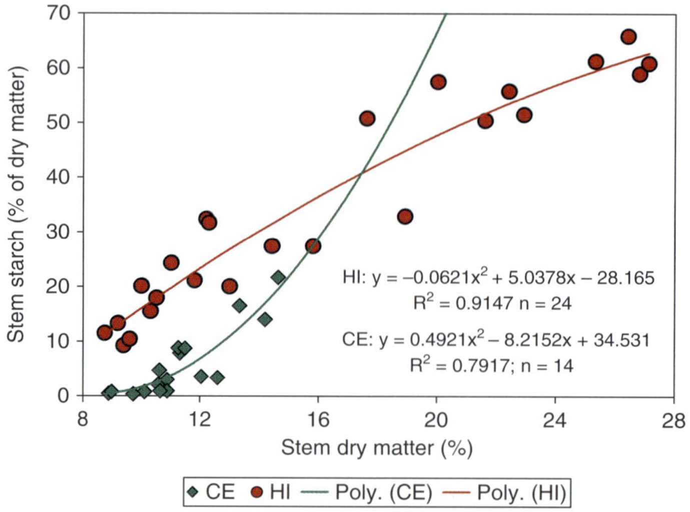

Fig. 6.9. Relationship between stem dry matter and stem starch as a percentage of stem dry matter in Champaka F153 ('Smooth Cayenne') pineapple. Data are: CE, samples from single plants grown in controlled environments at day/night temperatures of 22/22, 26/22, 30/22, 30/30, 34/18 and 34/30 for at least 218 days; and HI, the average of six plants harvested at about monthly intervals from 109 to 661 days after planting. The data were best fit by polynomial functions and extrapolation beyond the data is a software artefact. (Unpublished data of D.P. Bartholomew (CE) and Pineapple Research Institute of Hawaii (HI) reproduced with permission of D. Nellis, of the Pineapple Growers Association of Hawaii).

response of the plant is not quite as poor as that of plants in the environments with a 30°C night environment (Table 6.2) because they at least can be forced to produce a fruit. But the photosynthate supply is not sufficient to simultaneously support the fruit and a sucker and certainly isn't sufficient to build up a starch reserve in the stem. Since the fruit is the strongest sink in a resource-limited plant, axillary buds only begin to increase in size after the fruit has matured. This physiological symphony is a very under-researched area that warrants further study because both irradiance and temperature clearly affect photosynthate supply.

Morphological, anatomical and physiological features and adaptations to environmental conditions

Water use

The trough-shaped pineapple leaves are arranged spirally around the stem in a dense rosette pattern and this shape and orientation distributes radiation uniformly over what eventually becomes a very large leaf area. The leaves also channel light rains and dew to the base of the plant, making a significant contribution to its water economy (Ekern, 1965, 1968). Large trichomes completely cover both ad- and abaxial leaf surfaces (Fig. 6.10) and a highly cutinized upper epidermis with a multicelled hypodermis are two significant morphological and anatomical features that contribute to a pineapple plant's water economy (Krauss, 1930, 1949). There are 30–35 trichomes per mm^2 on the abaxial leaf surface (Aubert, 1973a) that cover strips of stomata located in furrows between ridges devoid of stomates (Fig. 6.10). There are relatively few stomata per unit leaf area (70–85 mm^2) and the stomatal pore is small (Krauss, 1949; Bartholomew and Kadzimin, 1977).

The mature 'Smooth Cayenne' leaf cross-section can be up to 4 mm thick (Krauss, 1949), with one-third to one-half the volume of the leaf occupied by a water storage parenchyma (Fig. 6.11) and the balance of the leaf is chlorenchyma. The water storage parenchyma functions as a reservoir of water that buffers periods of drought. When depleted by drought

Table 6.2. Effects of day and night temperature and forcing on stem dry weight (D wt.) and dry matter content (DM) of plants grown in controlled environments. Plants were potted on 10/10/1973 and grown at 26°C day, 22°C night (12 hour photoperiod) until potted into 25 cm containers and placed into treatment temperatures on 1/19/1974. Results are averages of two plants. Unpublished data of D.P. Bartholomew.

	Day and night temperature, °C									
	34–18		26–22		30–22		30–30		34–30	
Harvest date	D wt. (g)	DM (%)	D wt. (g)	DM (%)	D wt. (g)	DM (%)	D wt. (g)	DM (%)	D wt. (g)	DM (%)
19/1/74			2.9	11.9						
20/1/74			3.5	11.5						
18/2/74	3.1	12.0	3.9	12.4					4.2	12.7
18/3/74	4.4	11.5	5.1	11.7					3.8	10.4
13/4/74	5.4	11.1	5.6	11.2					4.7	10.3
18/5/74	7.4	10.4	7.6	11.7	12.5	11.4	4.4	9.2	4.3	9.4
16/6/74			8.7[1]	10.7						
16/6/74	14.5	12.8	12.6	12.0					6.0	9.3
8/7/74	23.1[1]	13.3	26.5[1]	14.4			12.4[1]	9.4		
8/7/74			49.3	16.2	25.0	12.2	6.8	9.0	10.0	9.5
24/11/74									13.1[1]	9.1
24/11/74			81.4	18.2	126.2	13.0	23.2	8.6	19.6	9.8
4/12/74			29.8[1]	14.1						
5/1/75	30.8[2]	8.3	46.1[2]	11.4						
23/1/75					79.0[2]	11.5				
14/2/75	58.8[2]	10.1								

[1]Plants treated with ethephon on 5/17/1974.
[2]Plants treated with ethephon on 7/11/1974.

(Fig. 6.12) the tissue is replenished after adequate rain (Krauss, 1949; Sanford, 1962). The leaf's anatomy and morphology result in low evapotranspiration (ET) (Ekern, 1965) and render pineapple extremely tolerant to drought. Transpiration values of 1.3 mm day^{-1} on cloudy days and 2.7 mm day^{-1} on fine days, average 2.1 mm day^{-1}, were obtained for 15-to-16-month-old fruiting pineapple plants (Shiroma, 1973) and ET from a 'Pérola' field crop cycle was 1421 mm vs 1615 mm reference ET (Azevedo *et al.*, 2007). Average ET values for 'Smooth Cayenne' pineapple in field experiments in Hawaii were 0.83 mm day^{-1} with a plant-trash mulch and 1.25 mm day^{-1} with plastic mulch (Ekern, 1965). For a leaf canopy formed by small plants (LAI = 4.2) during fruit development, daily maximum ET was 1.3 mm on cloudy days (irradiance = 10.7 MJ m^{-2}) and 2.7 mm on a sunny day (irradiance = 16.6 MJ m^{-2}) with an average of 2.1 mm day^{-1} (Shiroma, 1973). ET in Côte d'Ivoire can reach 4.5 mm day^{-1} when solar radiation is high (Combres and Perrier, 1976) and with irrigation ET averaged 3.0 mm per day over a 2-month period (Combres, 1979).

Drought

Symptoms of drought develop more or less slowly depending on the soil volume accessed by the root system and by the soil water-holding capacity. Despite the high tolerance of pineapple to drought, at best plants survive drought and, at worst drought in a dry season reduces the width and elongation rate of young leaves, the rate of leaf emergence and the length and weight of successive leaves (Py, 1965). In Hawaii, leaf elongation decreased when the gravimetric soil moisture content declined below 30–35% (Ekern, 1964). If drought is prolonged older leaves begin to wilt and with prolonged drought leaf colour changes from dark to pale green,

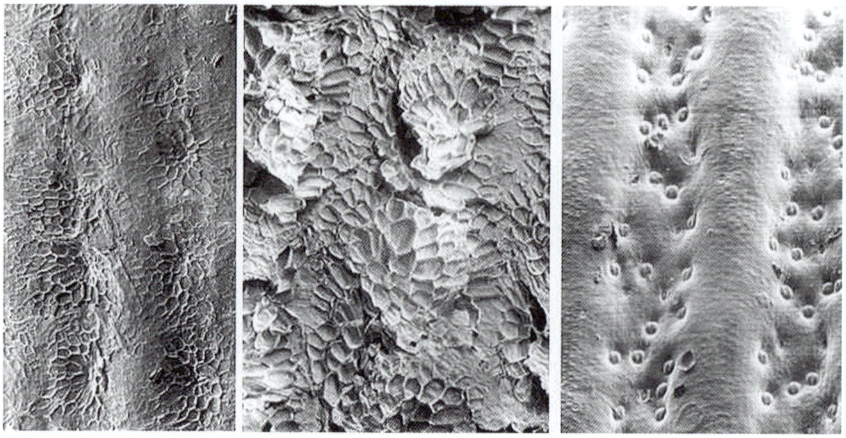

Fig. 6.10. Scanning electron micrographs (SEM) showing the large multicellular trichomes on the adaxial and abaxial surfaces of a 'Smooth Cayenne' pineapple leaf section. Leaf trichomes are more prominent on the abaxial surface and, when removed, show the ridge and valley morphology with stomata located in valleys that parallel the longitudinal axis of the leaf. Photo D.P. Bartholomew.

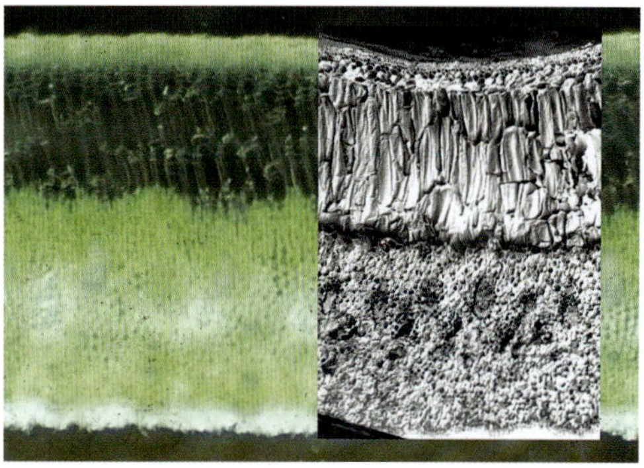

Fig. 6.11. Light and scanning electron microscope cross-sections of a mature pineapple leaf showing the prominent water storage tissue on the adaxial side of the leaf and the chlorenchyma on the abaxial (underside) half of the leaf. Photos D.P. Bartholomew.

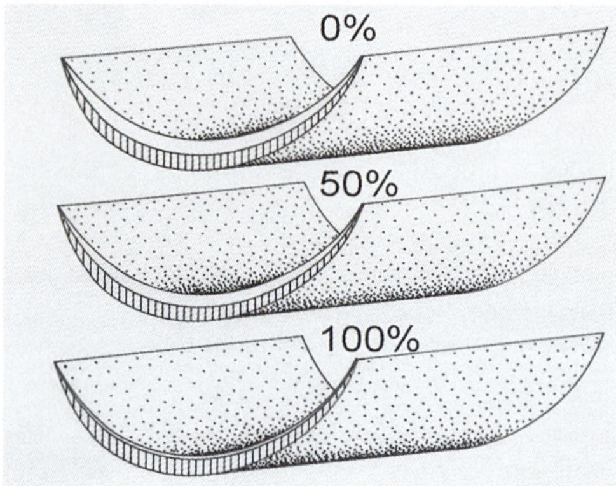

Fig. 6.12. Leaf water storage tissue deficiency, per cent, at inflorescence emergence (red bud) with decreasing availability of water. Modified from Nightingale, 1949.

then to pale yellow, and finally in pigmented cultivars, to red (Swete Kelly and Bartholomew, 1993). At the later stages, leaf margins curl downward, leaves lose their turgidity and become limp and growth stops (Py *et al.*, 1987).

The effects of drought are reversible and when water again becomes available, the leaves rehydrate and normal growth resumes. Leaves not yet fully expanded resume their growth and leaf width rapidly increases, resulting in a constriction at the point where elongation resumed (Py *et al.*, 1987). Leaves of 'Smooth Cayenne' and 'MD-2', often develop spines on the margins at the point where growth was restricted. Symptoms of water stress may appear more rapidly where soil water holding capacity is low, if rooting depth is restricted, or if the root system was damaged by pests or disease (Swete Kelly and Bartholomew, 1993).

Temperature

Leaf and plant temperatures of this relatively non-transpiring crop reach values that are detrimental, perhaps even lethal, to mesophytic crops. The temperature of horizontal leaves of

6-month-old 'Smooth Cayenne' plants reached 48°C between 1 and 3 pm in Hawaii (Aubert and Bartholomew, 1973). The difference between the leaf middle and base reached 18°C and the maximum difference between leaf and air temperatures was 18.6°C at midday. Such extreme leaf temperatures caused no permanent harm, but no data were found regarding the short- and long-term effects of such high temperatures on physiological processes.

Leaf morphology of 'Smooth Cayenne' and 'MD-2' plants varies with the mean temperature. Plants grown with optimum nutrition in environments having a night temperature >25°C and in warm, humid, low altitude climates produce numerous, wide, flaccid and shallowly troughed leaves (see Fig. 6.13 below; Py *et al.*, 1987; Bartholomew and Malézieux, 1994). Leaves are more rigid in dry tropical climates than they are in comparable warm and humid environments (J. Vasquez, 2017, personal observation). Conversely, in environments with cool night temperatures and in cooler subtropical climates, leaves are more erect, stiffer, deeply channelled and are fewer in number (Fig. 6.13; Py *et al.*, 1987; Bartholomew and Malézieux, 1994).

The indices specific leaf area (SLA, leaf area per unit of leaf dry mass, $m^2\ kg^{-1}$) and leaf area ratio (LAR, leaf area per unit of total plant dry mass, $m^2\ kg^{-1}$) allow the effects of temperature on plant morphology to be quantified and make it possible to predict the effects of environment on vegetative growth. Whole vegetative-plant SLAs over time were consistently lower for plants grown at night temperatures of 18, 22 and 26°C than for those grown at 30°C (Fig. 6.14). The lowest SLA was about $3.8\ m^2\ kg^{-1}$ dry mass at a day/night temperature of 22/18°C (Bartholomew, 1982). In a controlled environment SLA declined gradually as plant mass and age increased (Fig. 6.14), especially at the cooler night temperatures, so comparisons of SLA between plants in different environments must be made using plants having approximately the same mass.

Fig. 6.13. Leaf morphology of 'Smooth Cayenne' pineapple plants grown 5 months in the indicated temperatures. At a constant 30°C leaves form a shallow trough and droop while leaves on plants grown at a night temperature of 22°C have a deeper trough and are more erect. Photo D.P. Bartholomew.

Growth slows when night temperatures are cool and a lower SLA at such temperatures will further reduce the rate of increase in LAI and, in conjunction with more erect leaves, will prolong the time required to reach full canopy closure and complete interception of available light. Thus, it is predicted that the rate of leaf area expansion will decrease more rapidly as night temperature decreases than would be predicted just on the basis of the effects of temperature on leaf elongation. Mitigating this effect of cool temperature, at least to some degree, is the fact that plants growing in cooler environments have a higher net assimilation rate, which would partially offset the benefit of a more rapid rate of leaf expansion that would occur in warm environments. Thus, dry matter gain by plants in cooler environments may be as high as that which occurs in warmer environments where leaves expand more rapidly. Unfortunately, simultaneous changes in other climatic factors as temperature, possible differences in the growth rate of 'Smooth Cayenne' clones, and differences in quality of management makes it very difficult to compare productivity across environments having different temperature regimes.

Consistent with the effect of temperature on SLA, LAR values also generally increased with increasing night temperature (Fig. 6.15) and decreased with increasing plant age. An increase in LAR was also observed for plants grown in the field as elevation decreased and average temperature increased (Fleisch, 1988). The SLA and LAR, both on a dry mass basis, were related to average air temperature (T) by the equations:

$$SLA = 1.93T + 3.5,\ R^2 = 0.33,\ \text{and} \quad (4)$$

$$LAR = 1.07T + 64.6,\ R^2 = 0.37 \quad (5)$$

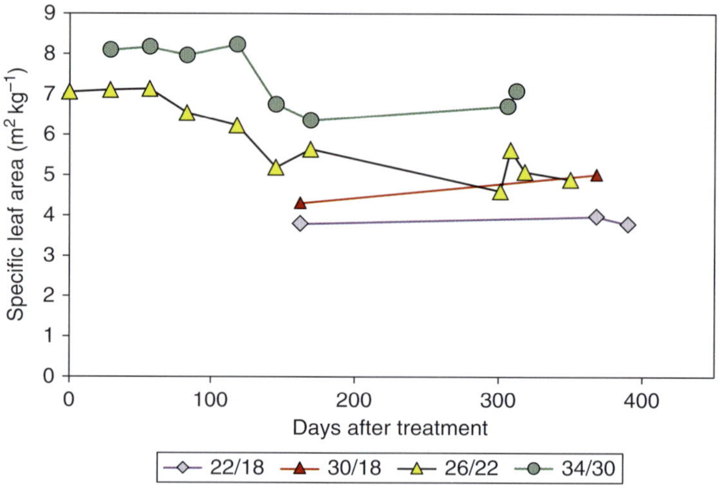

Fig. 6.14. Effects of day and night temperature (°C) on the specific leaf area (SLA, $m^2\ kg^{-1}$ dry weight) of the green leaf tissue of Champaka F153 ('Smooth Cayenne') pineapple grown in controlled environments. Pineapple crowns were started in a 26/22°C day/night temperature environment and transferred to the treatment temperatures 102 days after planting. At a given harvest date, larger plants generally had a lower SLA than did smaller plants. Data points are means of two plants. Redrawn from Bartholomew and Malézieux, 1994.

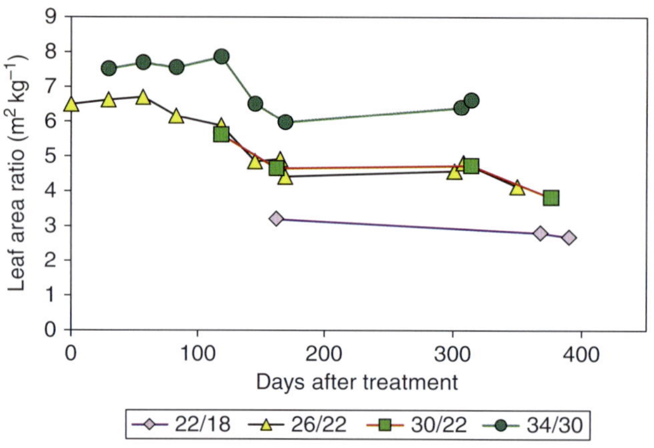

Fig. 6.15. Effects of temperature on the leaf area ratio (m^2 of leaf kg^{-1} dry weight of plant) of Champaka F153 ('Smooth Cayenne') pineapple grown in controlled environments. See Fig. 6.14 for study details. Redrawn from Bartholomew and Malézieux, 1994.

where SLA is expressed in $cm^2\ g^{-1}$ and temperature (T) in °C. The relatively low R^2 values for equations (4) and (5) are assumed to be due, at least in part, to the decrease in SLA and LAR with increasing plant mass and age.

In contrast with many crops (Biscoe and Gallagher, 1977) and regardless of the air temperature, the leaves of the pineapple mother plant persist at least into the first ratoon in Hawaii, a period of more than 24 months. However, the functionality of pineapple leaves of various ages has not been adequately determined.

Both the 'Smooth Cayenne' pineapple leaf and stem are important storage organs and can accumulate large quantities of starch in subtropical environments, but little or no starch, which is indicated by leaf and stem DM of 10% or less, in tropical environments.

The stem of a crown at the time of fruit maturation can contain more than 20% starch on a dry basis. In a Hawaii study, starch and dry matter content of planted crowns initially declined and remained at about 10% until 350 days after planting (see Fig. 6.16 below;

Pineapple Research Institute of Hawaii, unpublished results) at which time both stem DM and starch began to increase. Stem DM and starch reached maximum levels at about 550 days after planting and 100 days after forcing, after which they declined as the maturing fruit and developing suckers and slips drew down storage reserves (data not shown).

Based on the data of Fig. 6.16, at a plant population density of 4.3 plants m^{-2} stem starch accumulation reached about 325 g m^{-2}, of ground area at 458 days after planting when flowering was forced. About 90 days later, starch mass per unit of ground area exceeded 800 g m^{-2}. In environments warmer than Hawaii, there may be no increase in stem dry matter content until vegetative growth is interrupted by forced induction. Data for controlled conditions showed that at night temperatures of 26 or 30°C, plants had lower dry matter contents, about 9%, and little starch in the stem while plants grown at night temperatures of 18 and 22°C had significant amounts of starch (Bartholomew and Paull, 1986). Plant dry matter and starch content are believed to be especially important after floral induction and development (see Chapter 11).

While stem starch levels can exceed 60% of stem dry matter in Hawaii, starch levels do not reach 20% at the time of forcing in Côte d'Ivoire where temperatures are higher. Forcing occurs much sooner after planting in Côte d'Ivoire and other more tropical environments than in Hawaii in part because suckers, typically 400–500 g (Py *et al.*, 1987) are the primary source of planting material. Days from planting to the targeted forcing weight within a given environment are very much dependent on the size of the planted shoot (Py *et al.*, 1987). Because the LAR increases as night or average temperature, or both, increases, large differences exist between statistical models of vegetative growth developed in Hawaii (Fleisch, 1988; Zhang, 1992) and in Côte d'Ivoire (Malézieux, 1991).

Leaf dry matter content also gradually increases with time during vegetative and early reproductive growth and though such changes are relatively small, when total storage per unit of land area is computed, the accumulation is significant. In an unpublished study conducted in Hawaii (Pineapple Research Institute of Hawaii), the starch content of leaves sampled early in the morning averaged about 0.25% for the first 400 days after planting. Starch then gradually increased to 1.27% at 570 days after planting (120 days after forcing). At that time, plants had accumulated an average of 254 g m^{-2}, land area basis, of starch in leaves. Both stem and leaf starch declined sharply after floral

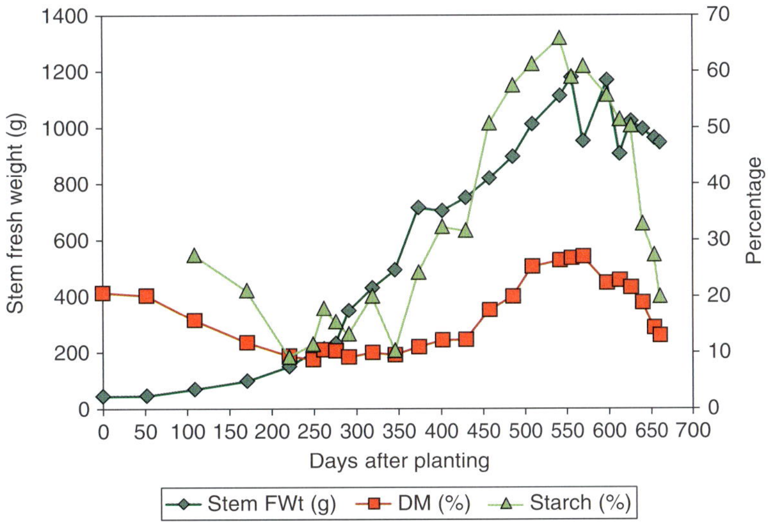

Fig. 6.16. Changes in stem fresh weight, dry matter content and leaf starch of field-grown Champaka F153 ('Smooth Cayenne') pineapple plants from planting of crowns to fruit maturation. Plants forced on day 458 and fruit were ripe on day 640. Unpublished data, Pineapple Research Institute of Hawaii.

induction, presumably as these reserves were drawn upon to support fruit, slip and sucker growth. In hot tropical environments (probably average temperature >25°C), little starch is accumulated in plant tissues by the time of forcing so growth of fruit and propagules such as suckers are dependent on current photosynthesis. In these environments, propagule development is delayed until after the fruit is harvested and management of sucker production (see Chapter 7) becomes almost as important as that of the fruit.

Changes in plant dry matter content and partitioning with changing root temperature at ambient air temperature were similar to the effects of temperature observed in controlled environments. Leaf dry matter content decreased from 16.5 to 12% as the root temperature increased from 15 to 30°C (Ravoof, 1973). Also, as root temperature increased, the percentage of dry matter allocated to leaves increased from 77 to 80% while that allocated to stem decreased from 17 to 13% (Ravoof, 1973). Plant LAR likely increased as root temperature increased.

Light interception

The spectral properties of the adaxial surface of pineapple leaves over the wavelength range from 520 to 750 nm were not significantly different from those of wheat, olive, orange and peach (N'Guessan, 1985). However, pineapple leaves have very low reflectances in the infra-red region with minima at 1440 and 1990 nm (water peaks). Despite reflectances in the visible range comparable to mesophytes, much of the radiant energy falling on a pineapple plant is entrapped by multiple reflections within the rosette leaf array (Ekern, 1965), which results in low canopy total reflectance (Ekern, 1965). As a result, canopy albedos ranged from 0.14 to 0.16 for canopies 1–16 months of age (Shiroma, 1973; Combres, 1983). There was no significant variation in albedo with canopy height (N'Guessan, 1986).

Estimated canopy light extinction coefficients for pineapple are typical of canopies with a relatively erect leaf orientation. Extinction coefficients (k) of 0.56–0.59 were calculated for 'Smooth Cayenne' pineapple in Hawaii (Fleisch, 1988; Zhang, 1992), resulting in 95% light interception at a LAI of about 5.0. Extinction coefficients for mesophytic crops typically range from 0.4 for erectophile canopies to 0.8 or greater for planophile ones (Russell *et al.*, 1989). A significantly lower k of 0.29 was found for pineapple in Côte d'Ivoire by measuring total radiation interception of pineapple crops with a somewhat wider row spacing (Malézieux, 1991). Low k values account in part for the ability of pineapple to sustain very high LAIs and high densities crowd plants so most leaves would be supported in an erect position.

Photoperiod and irradiance

The data on the effects of photoperiod and irradiance are sparse and few useful generalizations can be made from them. Also, changes in these parameters in natural environments are often confounded with temperature. Seedling leaf lengths decreased as the photosynthetic photon flux (PPF) increased from 100 to 325 μmol m^{-2} s^{-1} (Aromose, 1989) and Aragón *et al.* (2013) showed that 260 μmol m^{-2} s^{-1} shifted *ex vitro* plants from C3 to the CAM pathway, produced a more compact plant and significantly increased growth. At a PPF of 400 μmol m^{-2} s^{-1}, the length of the longest leaf of plants grown for 692 days in controlled conditions increased with the photoperiod from 6 to 10 h, was unchanged at 12 h, and declined sharply at 16 h (Friend and Lydon, 1979). The width of the longest leaf increased continuously with increasing photoperiod while leaf thickness (g cm^{-2}, fresh weight basis) declined (Friend and Lydon, 1979).

It is difficult to obtain meaningful data in natural environments because irradiance and temperature are highly correlated and effects attributed to irradiance (Py *et al.*, 1987; Srivastava and Singh, 1971) could also be due to the effects of temperature.

Effects of environment on vegetative growth

Temperature

The plant growth rate of pineapple is strongly influenced by temperature but few studies have been made in controlled environments and the results from field studies are difficult to interpret due to seasonal variation in the weather. Leaf and root elongation rates measured in controlled environments over a range of temperatures (Sanford, 1962) generally correspond well with the results of later studies. The optimum temperature for leaf growth was 29°C and that for roots was 32°C (Sanford, 1962). Growth of both organs was greatly reduced at temperatures below 20°C and ceased at about 10°C.

Data on the long-term effects of temperature on dry matter accumulation are sparse and such data are needed if the vegetative growth responses in the diverse field environments are to be interpreted. The growth of the same cultivars in a wide range of environments makes crop modelling difficult. Pineapple plants grown in controlled environments having night temperatures of 26°C or less had greater weights than those grown at 30°C night temperatures and the differences increased over time (see Fig. 6.17; Bartholomew, 1982). After about 400 days, plants grown at cooler night temperatures accumulated two to three times the dry mass of those grown at a 30°C night temperature (34/30°C and 30/30°C). In environments with warm days (26–34°C) and cool nights (18–22°C) plants also had higher dry matter contents and greater plant relative growth rates and leaf growth rates than did plants grown in warm days with 30°C nights (Bartholomew, 1982). Plant leaf area at the end of the study also was greatest for plants grown at 30°C days and 26 or 22°C night temperatures (Fig. 6.18). The more rapid growth of plants at the cooler night temperatures resulted mainly from higher net assimilation rates.

Net assimilation rates (NAR) ranged from about 1.0 g m^{-2} day^{-1} for plants grown at a 30°C night temperature to 2.0 g m^{-2} day^{-1} for plants grown at night temperatures of 26 or 22°C (Bartholomew, 1982). The NAR for field-grown 'Smooth Cayenne' plants in Hawaii decreased in a curvilinear manner from about 2.7 g m^{-2} day^{-1} at a planting density of 2.16 plants m^{-2} to about 1.9 g m^{-2} day^{-1} at 12.81 plants m^{-2} (Zhang and Bartholomew, 1997). There were small but inconsistent differences in NAR for plantings made June 15, August 15 and October 18, 1989. The field results were most similar to those obtained in the 26–22°C controlled environment.

Under field conditions, the mother plant relative growth rate (RGR) for the period from about 2 months after planting to induction of flowering was reduced by the cool winter temperatures of southern Queensland (Sinclair,

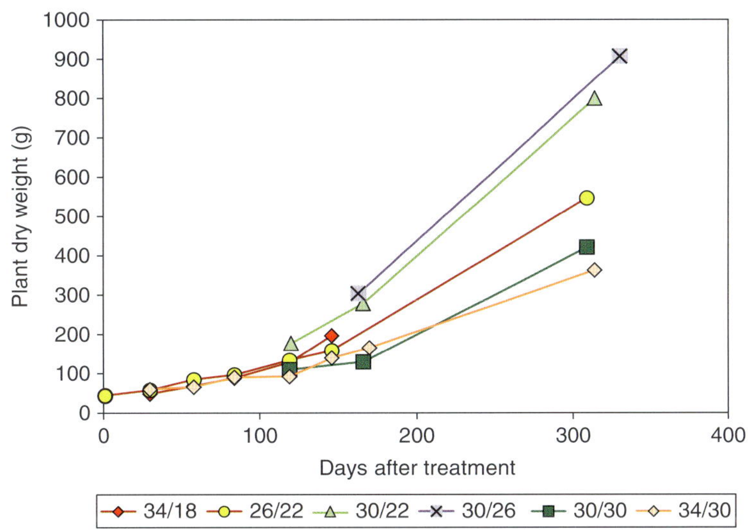

Fig. 6.17. Effect of day and night temperature on growth in dry mass of 'Smooth Cayenne' pineapple plants grown in controlled environments. Pineapple crowns were started in a 26/22°C day/night temperature environment and transferred to the various temperatures 100 days after planting (day 0 in the figure). Each data point is the mean for two plants. Redrawn from Bartholomew and Malézieux, 1994.

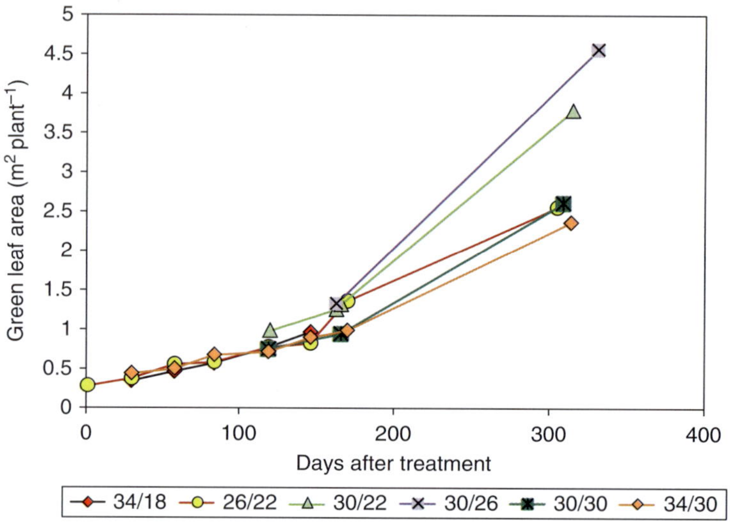

Fig. 6.18. Effect of day and night temperatures on growth in green leaf (white and pale green leaf basal tissue excluded) area per plant of 'Smooth Cayenne' pineapple plants grown in controlled environments. See caption for Fig. 6.17 for methods. Redrawn from Bartholomew and Malézieux, 1994.

1992). In comparing data from Hawaii and Côte d'Ivoire, plants grown from about 200 g crowns in Hawaii have 0.7 m^2 of leaf area 8 months after planting and would be grown another 4–6 months before forcing (Bartholomew and Kadzimin, 1977) while in Côte d'Ivoire, plants of the same age but grown from suckers weighing 400–500 g would have 1.8 m^2 of leaf area and would be large enough to be forced (Malézieux, 1991). Assuming months from planting to forcing in Hawaii requires 12 months, the GDD (Fig. 6.1) required equals 3753. In Côte d'Ivoire, plants reach forcing size in 8 months or 0.67 of a year, which when multiplied by estimated annual GDD for Abidjan of 5440 results in 3645 GDD from planting to forcing, a result similar to that for plants in Hawaii. Much of the difference in months from planting to forcing at the two locations is due to the more rapid expansion of leaf area in tropical Côte d'Ivoire than to the size of the planting material.

In areas where there is considerable seasonal variation in climate, time of planting can also affect months from planting to forcing. Propagules planted during periods of cool weather and low irradiance establish and grow more slowly than do those planted when weather conditions are more favourable for growth while in Thika, Kenya, propagules planted before the rainy season do not grow until the rains arrive. In Hawaii, slips are the main source of planting material and most plantings are 'set' and growth is initiated by about a 5 cm overhead irrigation. Subsequently, plants are drip irrigated.

Theoretically, growth in leaf area depends on the rates of leaf initiation and expansion. Leaf expansion, as indicated by rate of elongation, increased with temperature up to 29°C (Sanford, 1962). Growth in leaf area per plant also was greater when the days were warm, about 30°C, and night temperature was ≤26°C (Fig. 6.18). A relatively sharp temperature optima occurs for leaf elongation rate (Sanford, 1962), but no such sharp optima was observed for dry weight gain (Fig. 6.17), presumably because of morphological and anatomical differences induced by temperature, or the day/night temperature differential, or both.

The SIMPINA model for 'Queen' pineapple in Réunion appears to bypass leaf emergence by using light energy conversion efficiency (E_b) to calculate 'planting to biomass production' and the process appears to be triggered by a threshold sum of degree-days for E_b initiation (Dorey *et al.*, 2015).

Leaf initiation in natural environments, as indicated from rates of leaf appearance, is sensitive to temperature. In Taiwan, the equation

$$NLEAF = 18.125 \cdot \log_{10} T / 11.669 \quad (6)$$

where (T) is mean monthly temperature (°C) and NLEAF is number of leaves to appear in 1 month, indicates the extent to which leaf appearance rate is determined by temperature (Shiroma, 1972). Data on leaf emergence collected from fields at elevations of 91, 305, and 792 m in Hawaii were fit by a similar relationship (J. Zhang, 1988, unpublished data). In Côte d'Ivoire, where the mean air temperature is >26.5 and the diurnal variation low, the number of leaves appearing per 2 months was best fit by a quadratic function with an optimum temperature at about 26–27°C (E. Malézieux, 1992, unpublished results). Similar results showing the existence of an optimum temperature near 28°C were obtained in Hawaii (Zhang, 1992).

Leaf number can be predicted using heat accumulation or daily thermal time (DTT) (Zhang, 1992). An enhanced model was developed that also accounted for the reduction in growth that occurs at temperatures <16°C and >35°C. Such temperatures normally would be reached only during part of the day (Zhang *et al.*, 1997). Daily thermal time (DTT) was calculated for each day by the equation:

$$DTT = TM - BT \quad (7)$$

where TM is the mean temperature and BT the base temperature. A BT of 16°C provided the best fit of the data. For a minimum temperature <BT, or a maximum day temperature >35°C, DTT was calculated using a procedure similar to that reported by Malézieux *et al.* (1994) for the estimation of daily fruit thermal time. The leaf emergence model was calibrated and validated with data from Hawaii and Côte d'Ivoire. Based on Celsius temperature, the phyllochron (interval between the appearance of two successive leaves) was 50 degree days per leaf. Increased planting density increased interplant competition, which significantly reduced leaf number. It was assumed that this effect was due to a reduction in meristem temperature caused by the intense shading that occurs at high plant population densities. This effect was simulated in a model using interplant competition factors calculated from leaves m^{-2} of ground area (Zhang, 1992).

In addition to the effects of root temperature on dry matter partitioning noted above, Ravoof (1973) also showed that the total fresh mass of 40-day-old plants grown from slips increased with increasing root temperature from 15 to 30°C; however, the increase in growth was not proportional to the increase in root temperature. In Okinawa, Japan, a relatively cold climate for pineapple (mean temperature ≈17°C, minimum temperature for more than 1 month ≤6°C) and in Hawaii during the winter (mean temperature ≈22.3°C), growth is enhanced by black polyethylene film mulch (Ekern, 1967; Ogura *et al.*, 1968). The 33% increase in growth in Hawaii resulted from a soil temperature increase of only 1.6°C. Leaf dry weight increases due to plastic mulch at different elevations in West Cameroon were 12% at sea level (avg. T = 26.2°C), 32% at 500 m (avg. T = 22.4°C) and 23% at 1000 m (avg. T = 20°C) (Aubert, 1973b). In the Canary Islands where mean average air temperature is about 20°C, both black polyethylene mulch and a polyethylene cover over the plants increased yields (Galan-Sauco *et al.*, 1988).

Irradiance and plant population density

Few studies have been conducted on the effects of irradiance on vegetative growth. Connelly (1969) found that estimated plant weight of 'Smooth Cayenne' plants grown in pots was unaffected by 25% shade but was reduced 20% under 50% shade. Less nitrogen was required for optimum growth at low than at high light.

In most crops average plant mass decreases with increasing plant population density due to interplant competition for light, but no comparable data were found for pineapple. Interplant competition does occur during vegetative growth of pineapple and growth is reduced at high plant population densities (Zhang, 1992; Malézieux and Sébillotte, 1990a). Results in Hawaii showed that plant mass did not decrease with increasing planting density for the first 200 days after planting, even at 12.35 plants m^{-2}, and light interception did not reach 95% at that density until about 350 days after planting (Zhang, 1992). However, plant mass at forcing, which occurred on the same date for plantings made in June, August and October (respectively 440, 380 and 335 days after planting), decreased significantly with increasing density at all planting dates. Similar results were obtained in Côte d'Ivoire (Malézieux and Sébillotte, 1990a). In the latter experiment, plant mass 8 months after planting increased by 33% with irrigation and by 20% without irrigations, as plant population density decreased from 100,000 to 20,000 plants ha^{-1}. It is clear that competition for light occurs during vegetative growth in pineapple fields even at moderate planting densities. However, it may not be detectable where growth is limited by water or other stresses.

Lest the details obscure the message, it is clear that total biomass and often total yield increases with increasing plant population density. However, studies often show that at densities much above 74,000 plants ha^{-1}, and sometimes well below that, the percentage of fruit recovered as well as the quantity and quality of fruit declines. See Chapter 11, for discussion of the effects of plant population density on fruiting.

Allometric data help to explain the effects of plant population density on vegetative growth. Leaf appearance rate decreased with increasing planting density (Norman, 1977; Malézieux, 1992; Zhang, 1992), and in one study, leaves were longer and narrower at the higher planting densities (Wee, 1969). Specific leaf mass (SLM, dry mass basis) was significantly decreased when planting density was higher than 6 plants m^{-2}. Zhang (1992) reported that SLA at 300 days after planting was related to planting density (PD) by the equation

$$SLA = 50.3 + 0.51PPD \quad (8)$$

and the effect was highly significant.

Developing crowns, slips and suckers require photosynthates at least until they become autotrophic. Competition for light at higher planting densities is assumed to reduce the supply of assimilates available for growth and for storage reserves, because the number of slips and suckers per plant decreases at higher plant populations, probably mainly because the fruit is the dominant sink so forced plants will always develop a fruit, but may not develop suckers until after the fruit matures (Lacoeuilhe, 1974; Py *et al.*, 1987). For clones that produce slips, slip numbers decrease at a faster rate with increasing plant population density than do suckers (Py *et al.*, 1987). Suckers per plant decreased from 1.84 to 1.11 as planting density increased from 46,112 to 80,650 plants ha^{-1} (Scott, 1992). Thus, despite a wide range of planting densities, suckers ha^{-1} ranged from 82,162 to 89,572 (Scott, 1992).

In addition to reducing the number of suckers per plant, sucker development, which in Australia (Sinclair, 1992) and Hawaii is initiated at the time of forcing, usually is delayed at higher planting densities. Circumstantial evidence suggests that when interplant competition is intense, there is insufficient photosynthate for the development of both a fruit and suckers.

In tropical regions such as Côte d'Ivoire, sucker growth, the primary planting material, was increased up to 20% by cutting off the leaves that extend above the developing suckers immediately after fruit harvest (Combres, 1983). Leaf removal may improve sucker exposure, thus promoting their growth; however, the loss of leaf area may decrease the ultimate number of suckers obtained from the field. Sison (2017) reported that slashing 'MD-2' plants to 10–15 cm above ground level reduced suckers ha^{-1} produced between September 2002 and July 2003 from 294,000 to 193,000.

Drought

Despite the xeromorphic characteristics of pineapple, growth and yield are significantly reduced when drought is prolonged. One effect of a water deficit is to reduce the number and length of the roots (Kadzimin, 1975; E. Malézieux and G. Godo, 1991, unpublished results). White root tips visible in the soil, expressed as a percentage, were used as an indicator of root health and water deficit (Sanford, 1962) and root elongation ceases and root tips suberize when the soil moisture approaches −1.5 megapascals (MPa) (Ekern, 1967). Suberization does not protect roots indefinitely as they will die if the soil moisture stress is severe and prolonged (Krauss, 1959). In Côte d'Ivoire, rooting depth and root numbers increase with irrigation or with polyethylene mulch (E. Malézieux and F. Zampatti, 1991, unpublished results).

Most of the quantitative data on the effects of water stress on growth were obtained from plants grown in pots at a time when modern measurement methods were unavailable. Sideris and Krauss (1928) found that plants did not grow in soils containing <5% gravimetric soil moisture but was uniform at soil moisture contents of >20%. Growth in loamy soil in pots continued at a soil moisture content of 10% but almost stopped when soil moisture was <5% (Shiroma, 1971). Kadzimin (1975) grew pineapple in black polyethylene bags for 7 months at average soil water potentials (ψ_{soil}) of −0.1, −0.5, −1.0, −1.5 MPa and with an unirrigated control. Plant total dry mass, relative to the −0.1 MPa treatment, was reduced 27.6% at −0.5 MPa, 32.4% at −1.0 MPa and 46.9% at −1.5 MPa. Only leaf and stem growth were reduced at −1.0 MPa while leaf, stem, and root growth were reduced in the −1.5 MPa and unirrigated treatments. The root system as a proportion of the whole plant decreased significantly in the −1.5 MPa and unirrigated treatments.

The clay soils of Hawaii have a high water holding capacity so when irrigation began when ψ_{soil} reached −0.4 MPa growth was equal to or better than irrigating at a ψ_{soil} of −0.03 or −0.07 MPa (Thorne, 1953). However, growth in sandy soils with low soil water holding capacity was reduced as soon as the ψ_{soil} fell below −0.015 MPa at a 15 cm depth (Combres, 1979). Combres (1983) concluded that a ψ_{soil} of −0.015 could be used as a threshold-value to initiate irrigation.

The SIMPINA model (Dorey *et al.*, 2015) uses a water balance approach to estimate total available soil water. Readily available water in the root zone that can be extracted by the crop is then calculated and presumably is used for the computation of plant biomass.

Plant moisture content slowly decreases as the soil water content decreases and the effect is reflected in the thickness of the leaf water storage tissue (Figs 6.11 and 6.12). However, leaf water content changes slowly and pineapple tolerates long periods of drought with minimal injury. Nose *et al.* (1981) observed only a 0.1% decrease in leaf water content as soil pF decreased from 1.1 (approximate field capacity) to 4.5. Even uprooted plants lose water slowly. Suckers unattached at flowering and kept without water for 3 months on a sunlit glasshouse bench lost only 44.4% of their fresh weight while comparable shaded ones lost 41.6% (Sideris and Krauss, 1955). Despite the high degree of dehydration, significant fruit growth occurred while water loss from the leaves due to translocation to the fruit was estimated at 18% for plants in the sun and 26.8% for plants in shade.

Normal values of pineapple leaf water potential (ψ_L) ranged from about −0.6 MPa at a soil water potential of −0.1 MPa (Kadzimin, 1975) to a pre-dawn level of about −1.2 MPa (Zhu *et al.*, 2005). Leaf ψ_L reached −2.3 MPa at a ψ_{soil} of −1.8 MPa (Kadzimin, 1975). George *et al.* (1984) found that ψ_L of pineapple leaves varied from −0.2 to about −1.8 MPa after 12 weeks without any water while leaf relative water content (RWC) decreased from about 96% to 42% over the same time period. During 70 days of drought pre-dawn leaf RWC decreased from >95% to about 83% at a 30/20°C day/night thermoperiod and to about 67% at a 35/25°C day/night thermoperiod (Zhu *et al.*, 2005). Leaf RWC and ψ_L were linearly correlated with a 10% change in RWC corresponding to a 0.28 MPa change in ψ_L (George *et al.*, 1984). Similarly, Kadzimin (1975) found that as ψ_L decreased from −0.1 Mpa for well-watered plants to −2.2 Mpa for unirrigated plants, leaf RWC decreased from 96 to 70% However, leaf RWC was concluded to be a rather insensitive indicator of water-stress in pineapple (Kadzimin, 1975; Chapman *et al.*, 1983; George *et al.*, 1984).

Pineapple plants subjected to water stress have significantly lower ET values compared to unstressed plants. In central Côte d'Ivoire, ET of complete pineapple leaf canopies averaged 3.8 mm day^{-1} when irrigated and 2.9 mm day^{-1} when unirrigated (Combres and Perrier, 1977). The ET/ET_g ratio, where ET_g is the potential evapotranspiration of a grass cover, is about 0.45 for an irrigated pineapple crop but the ratio decreases during drought (Combres, 1983).

Evapotranspiration of a pineapple crop depends both on the stage of development and climatic conditions. When well-watered, the ET of complete canopies ranged from 0.6 to 0.7 of the potential evaporation of a grass cover (Combres and Perrier, 1977). The crop parameter Kc, in the equation

$$ETR = K_c \cdot ET_o \tag{9}$$

where ETR is actual evapotranspiration, and ET_o is standard evaporation (Priestley and

Taylor, 1972), generally ranges between 0.8 and 1.3 for most crops. An average value of 0.74 was observed for pineapple (Combres, 1983) but K_c can vary significantly over the plant cycle. K_c decreased significantly from 0.93, 5 months after planting (50–60% of soil cover) to 0.41 by 11 months after planting at which time there was 80–90% of soil cover by the canopy (N'Guessan, 1985). Where pineapple was planted through a polyethylene mulch in Hawaii, a 50% decrease in the daily rate of consumptive use of water occurred at 60% canopy closure (Ekern, 1964). A striking feature of the pineapple crop ET rate is that it decreases as the plant canopy cover increases, whether mulched or unmulched (Ekern, 1964). With mulch, 15% of the incident energy was used for ET after the canopy closed. The very low water-use rate of pineapple is further reduced by water stress.

Flooding

Excess water can have the same effect as drought on CAM and RWC (Min and Bartholomew, 2005). Leaf titratable acidity after 14 days of continuous flooding had decreased from >400 to about 130 mmol m^{-2} and took >40 days to recover. Leaf RWC decreased from about 90% to 83% and had a similar recovery time. The effects of excess water in a field environment likely would not be so drastic and would have its greatest impact during root initiation and at fruit filling. As with most crop plants, root growth is restricted by inadequate aeration associated with excess soil moisture. Pineapple grows well in an aerated solution (Iwaoka *et al.*, 1935) and much early research on pineapple nutrition was done in aerated solution cultures (Sideris and Krauss, 1937; Tisseau, 1971). The formation and persistence of root hairs seems to depend upon an oxygen supply to the roots (Py *et al.*, 1987). Root aerenchyma is also increased by a diminished oxygen supply and a comparable effect is seen with an increase in soil compaction (Rafaillac *et al.*, 1978). Excessive water produces a leaf colour change similar to that seen with water stress in that leaves of 'Smooth Cayenne' first become pale yellow, then red, and leaves are reduced in length and are more erect (Py *et al.*, 1987).

General effects

Temperature

Temperature is one of the most important environmental factors determining pineapple distribution and productivity in the world (see Fig. 6.1 and Table 6.1). Pineapple survives in hot, dry environments where other crops would be non-productive, but also is cultivated in the cool subtropics where freezing temperatures may occur (Table 6.1). The lowest annual average temperature where pineapple is grown on a commercial scale appears to be 17.2°C in Port Elizabeth, South Africa. In southern Queensland the mean monthly maximum/minimum temperature ranges from 29/19°C in summer to 20.5/6°C in winter (Wassman, 1990). The plant does not tolerate frost, but temperatures have been reported to drop below 0°C for short periods in pineapple growing areas of southern Queensland, Australia (Swete Kelly and Bartholomew, 1993), South Africa and Sao Paulo, Brazil (Giacomelli and Py, 1982; Py *et al.*, 1987). Exposure to temperatures <0°C can destroy the canopy and lead to the loss of the crop (Giacomelli and Py, 1982; Swete Kelly and Bartholomew, 1993). Overhead irrigation was used to protect against freezes in Florida (J. Tenbruggencate, 1986, personal communication).

Large areas are planted to pineapple in hot, wet intertropical regions and at low elevations, especially along coastal plains where the climate is more moist and hotter than inland regions. In equatorial regions, pineapples are grown where temperatures are continuously cool at elevations of 900 m or more in Cameroon, Kenya, Réunion and Uganda. In equatorial regions the annual range of monthly temperature is small, sometimes not more than ±1°C. In most of the tropics, and especially in hot and wet regions close to the equator the diurnal temperature cycle is often more important than the seasonal cycle (Monteith, 1977).

Optimum day and night temperatures for vegetative growth of pineapple are near 30 and 20°C, respectively, with an optimum mean temperature of 23–24°C (Neild and Boshell, 1976). Plant growth decreases rapidly at mean temperatures below 15°C or above 32°C (Neild and Boshell, 1976). In Hawaii, the slower vegetative growth rates observed from December to April are mainly due to a decrease in the average temperature, especially soil temperature (Ekern, 1967). In Australia, low temperatures in the midwinter period from May to August reduce or even stop plant growth (Glennie, 1981; Wassman, 1986).

The large environmental variation among areas where the crop is grown (Table 6.1) accounts for much of the large variation in days from planting to maturation of the mother plant crop. The rapid expansion of the pineapple fresh fruit industry has caused fruit size to become a critical issue for growers. Within a given environment fruit weight at maturity generally is closely related to plant fresh mass at forcing (Py and Lossois, 1962; Gaillard, 1969; Tan and Wee, 1973; Malézieux, 1988; Malézieux and Sébillotte, 1990a). As a result, vegetative growth is carefully managed so plants of the size required to produce the fruit size demanded by the market place are available on an almost daily basis.

As emphasized previously, with no other stresses, plant growth is determined by temperature. For that reason and also because planting material size helps to determine days from planting to the size for forcing, the interval from planting to forcing varies considerably over the wide range of latitudes and altitudes where pineapple is grown (Table 6.1). In East London (33° S, South Africa), the average temperature is 18.8°C and the average period from planting to forcing is 24 months (Bouffin, 1991) whereas in equatorial West Africa the vegetative growth period is only 6 months.

Seasonal temperature variation in the Canary Islands causes the vegetative cycle to range from 9 months for a spring planting to 14 months when the crop is planted in winter (Galan-Sauco *et al.*, 1988) and similar variations occur in Australia and South Africa. A plant fresh mass of 2.5 kg, a common plant weight for forcing, can be reached within 8 months after planting of 400–500 g suckers in Côte d'Ivoire (lat. 5° N, mean temperature of 26.6°C), within 10–12 months after planting 300 g slips in Hawaii (lat. 21° N, mean temperature of 23°C) and within 13–14 months after planting slips in Queensland, Australia (lat.26° S, mean temperature of 19°C). As a result, the time from planting to harvest ranges from 12 months near the equator (Gaillard, 1969), where the average annual temperature is 26–27°C, to 32 months in Swaziland (26° 30′ S) where mean annual temperature is 16–17°C (Dodson, 1968). There is wide variation in days from planting to forcing for 'Smooth Cayenne' in the pineapple-growing areas of the world and similar variation in the length of the reproductive phase.

An advantage provided by cooler environments is the possibility of producing one or more ratoon crops. Ratooning is done in equatorial climates, but delayed sucker development limits the benefit of a ratoon crop. Since the yield of the ratoon crop is smaller than the plant crop yield, the ratoon profit is relatively small. Also, if fruits are grown for the fresh market suckers would be the main source of planting material so a relatively large area must be reserved for sucker production. When ratoons are grown, the plant and one ratoon crop are harvested in about 2 years in Thailand, 3 years in Hawaii, 3–4 years in Australia (Vuillaume, 1986), and as much as 5 years in Swaziland (Dodson, 1968).

Temperature injury

High temperatures generally are of minor concern during vegetative growth of pineapple. Although leaf and plant temperatures may reach very high values in this non-transpiring crop when there is poor air circulation in the plant canopy, pineapple leaves tolerate high temperatures well (Aubert and Bartholomew, 1973). Leaf sunburn is sometimes seen, but it is not a serious problem in most of the areas where pineapples are grown.

Low temperature injury is the greater concern where freezing temperatures are encountered and the areas affected are relatively small. With mild radiation frost, the upper surface of horizontal leaves develop a red/white flecked, scorched appearance. If the injury is severe, the whole leaf dies and the leaves become papery. Severity determines whether vegetative plants will recover and produce an acceptable crop (Swete Kelly and Bartholomew, 1993). Protection is provided by growing plants under cover, use of fixed fans to mix warmer upper air layers with the cold air at ground level and overhead irrigation.

Irradiance

Irradiance was identified as a factor affecting pineapple yield at least by the mid–1930s in Hawaii (Sideris *et al.*, 1936). However, there is little evidence that irradiance limits pineapple production in most areas where the crop is grown. Sanford (1962) attributed to Sideris an observation that yield decreases about 10% for each 20% decrease in solar radiation. Shiroma *et al.* (1977) concluded that irradiance could limit 'Smooth Cayenne' pineapple growth in Okinawa; however, the conclusion was not based on data obtained *in situ*.

A linear relationship between plant weight 8 months after planting and intercepted radiation by the canopy was established in an experiment in Côte d'Ivoire where plants were planted every month for 6 years (E. Malézieux, 1992, unpublished results). From the data an empirical relationship that associated average irradiance during the entire crop cycle and an indicator of drought from planting to harvest explained 64% of the variation in plant weight at harvest (72 monthly plantings with similar cultural practices). However, interpretation of the results of such studies is very difficult because of the long-term nature of the crop and the fact that changes in irradiance are confounded with simultaneous changes in air temperature. In areas where the variation in temperature is small compared to that for irradiance, such as near the equator, irradiance may be of importance in determining plant growth.

Where both irradiance and temperature change significantly with season, it is very difficult to evaluate the effects of irradiance on pineapple growth. The quality and average size of the available planting material can vary with time of year (Louis and Nightingale, 1937; Zhang, 1992) and these factors are important because they determine the initial plant size. At a given planting density in the same environment, plant mass at a given time after planting is determined by the mass of the propagule (Py *et al.*, 1987). Even modest differences in propagule mass between plantings in different seasons can obscure the effects of seasonal changes. After a 2–3-month establishment period, which can be further prolonged by cool temperatures or lack of rain, vegetative growth integrates irradiance over a period of 6–15 months. Fruit development requires an additional 5–8 months. Since plant mass at forcing is an important determinate of fruit mass at harvest (Py and Lossois, 1962; Gaillard, 1969; Tan and Wee, 1973; Malézieux, 1988; Malézieux and Sébillotte, 1990b) anything that retards vegetative growth in one season relative to another will either delay harvest or reduce yield.

Seasonal variations in productivity can also be due to the seasonal distribution of rainfall, seasonal and day-to-day variations in forcing success (Wee and Ng, 1968; Wassman, 1991), though with current technology efficacy of forcing should be at worst a minor issue, and in pest and disease pressure. While it is difficult to demonstrate an effect of irradiance on productivity, in Côte d'Ivoire where temperature is relatively stable and irradiance fluctuates significantly throughout the year, potential plant weight at forcing was correlated with irradiance intercepted during vegetative growth (Malézieux, 1988). In these unirrigated conditions, yield was depressed when drought was prolonged.

Drought and water excess

Because pineapple has a great capacity to survive drought (Sideris and Krauss, 1928) it has been grown successfully for the processing industry in a Hawaii environment with less than 600 mm y^{-1} of rain (Noffsinger, 1961). Where the soil is well drained, pineapple also has been cultivated in relatively wet areas in Costa Rica and Guadeloupe where rain can exceed 3500 mm y^{-1} (Py *et al.*, 1968; D.P. Bartholomew, 2012, personal observation). Hence, depending on the location and season, the climatic water balance over some or all of the crop cycle can be positive or negative. In Wahiawa, Hawaii, pineapple was grown for many years without irrigation where class A pan evaporation averaged 1850 mm y^{-1} while rainfall averaged about 1000 mm y^{-1} (Noffsinger, 1961; Ekern, 1965). Now that pineapple is only grown for fresh fruit, fields are irrigated to keep the plantings on a schedule designed to meet market demand in the relatively small Hawaii markets.

Despite the capacity to survive drought, growth and yield are reduced in all areas where water stress occurs (Foote, 1955; Black, 1962; Ekern, 1964; Py, 1965; Huang and Lee, 1969; Malézieux, 1988). Hence, irrigation or polyethylene mulch, or both, are regularly used where water is available and the cost of the mulch can be justified. Though polyethylene mulch has several benefits, its impact on water economy is the main reasons it is used in Côte d'Ivoire (Combres, 1983) and Ghana (D.P. Bartholomew, June 2006, personal observation). Several studies report no effect of irrigation on pineapple growth even where drought occurs (Rao *et al.*, 1974; Tay, 1974; Senanayake, 1978; Htun, 1986). Despite such results, the benefits of irrigation during prolonged drought (Medcalf, 1950; Py, 1965; Huang and Lee, 1969; Combres, 1979, 1983; Kuruvilla *et al.*, 1988; Malézieux, 1992) are sufficient to cause it to be a widely adopted practice. In Guinea, growth of unirrigated plants was delayed by up to 3.5 months at the end of a 5-month dry season (Py, 1965). Despite good evidence of the benefit of irrigation, no quantitative studies were found that would provide the data needed to predict growth or yield reduction in response to water stress.

In regions of high rainfall, water excess can also limit growth as well as increase susceptibility to disease. Water logging impedes gas exchange, resulting in elevated CO_2 and depleted O_2 levels in the soil, which leads to decreased growth by inducing root anoxia. In the Mekong Delta, for instance, beds may become submerged (Le Van Thuong, 1991), and pineapple plants died after 15 days of submersion of the bed beneath 10–15 cm of water. Where rainfall is high, as in the humid tropics, soils used for pineapple cultivation are generally lower in clay and higher in sand to assure good drainage. Such soils have low water-holding capacity and are susceptible to erosion (Ducreux *et al.*, 1980; Ciesiolka *et al.*, 1993; El-Swaify *et al.*, 1993; Ciesiolka, 1994). Ridging practices, widely used in South and West Africa and some areas of Australia and trenching used in Costa Rica (Fig. 6.19), also reduce water-logging problems.

Plant growth and fruit weight are decreased by water logging that occurs in small field depressions in Côte d'Ivoire. While few quantitative data on the effects of water excess on growth under field conditions were found, until relatively recently when effective fungicides became available to control *Phytophthora* spp., loss of plants to root and plant rots was a greater concern than reduced growth due to water logging (Rohrbach and Apt, 1986) (see Chapter 12).

Fig. 6.19. Land preparation and trenching to facilitate drainage for pineapple in the area of Guapiles, Costa Rica where annual rainfall averages more than 4000 mm. Photo D.P. Bartholomew.

Miscellaneous factors

Wind is a minor concern for pineapple growers, though prolonged winds have been reported to reduce plant size by 25% (Nightingale, 1942). It is not known if the reduction was due to the physical effect of wind, to an altered water balance, or to reduced temperature within the canopy. Though wind has little effect on ET when the plant cover is complete, hot, dry winds such as the 'Norte' in Mexico or the 'Harmattan' in Côte d'Ivoire can lead to leaf tip drying (Py *et al.*, 1987). The influence of wind on thermal exchanges in the canopy is significant (Py *et al.*, 1987) and moderate winds can reduce sun injury to fruits and crowns. Strong winds cause leaves to rub against each other and the physical damage provides points of entry for fungi, but neither the physical nor fungal damage are considered economically significant in most instances (Py *et al.*, 1987). Strong trade winds have become a recent problem in Hawaii because the long peduncle of 'CO-2' has resulted in considerable lodging that greatly increases the incidence of sunburn of fruits. In coastal areas, wind-borne salt spray can cause burns, leading to blackish spots near the tips of the leaves (Sideris, 1955). Exceptionally strong winds caused by hurricanes (typhoons) can sand blast and severely damage all parts of the plant resulting in NI (G. Sanewski, Queensland, Australia, personal observation).

Hail is rare in most areas where pineapple is grown but occurs in southern Queensland (Swete Kelly and Bartholomew, 1993) and southern Africa (Metal Box, 1956). Hail can seriously damage the rigid and brittle pineapple leaf, significantly reducing leaf area and inducing flowering. After a severe hailstorm in Swaziland, most of the leaves wilted and died, leaving only a small growing point and leading to a loss of one year's growth (Metal Box, 1956).

Few data on the effects of salinity or salt on vegetative growth of pineapple were found, but pineapple is grown along windward sea coasts in Hawaii with minimal adverse effects, even though the north-easterly trade winds are strong and prolonged.

References

Aragón, C., Pascual, P., González, J., Escalona, M., Carvalho, L. and Amancio, S. (2013) The physiology of ex-vitro pineapple (*Ananas comosus* L. Merr. var. 'MD-2') as CAM or C3 is regulated by the environmental conditions: proteomic and transcriptomic profiles. *Plant Cell Reports* 32, 1807–1818.

Aromose, O.S.A. (1989) Light intensity response in pineapple (*Ananas comosus* (L.) Merr.) seedlings. *Phyton Buenos Aires* 49, 161–165.

Aubert, B. (1971) Effets de la radiation globale sur la synthese d'acides organiques et la regulation stomatique des plantes succulentes exemple d'ananas comosus). *Oecologia Plantarum* 6, 26–34.

Aubert, B. (1973a) *Ecologie de l'ananas.* Réunion Annuelle IRFA, document interne No. 105.

Aubert, B. (1973b) The effect of elevation on the behaviour of pineapple Smooth Cayenne in West Cameroon. *Tropical Ecology* 14, 52–61.

Aubert, B. and Bartholomew, D.P. (1973) Mesures en champ de la temperature de limbes d'ananas par radiometrie infrarouge. *Fruits* 28, 623–629.

Aubert, B., Gaillard, J.P., Lossois, P., Py, C. and Marchal, J. (1973) Influence de l'altutude sur le comportement de l'ananas 'Cayenne Lisse'. Essais realises au pied du mont Cameroun. *Fruits* 28, 203–214.

Azevedo, P.V.D., Souza, C.B.D., Silva, B.B.D. and Silva, V.P.R.D. (2007) Water requirements of pineapple crop grown in a tropical environment, Brazil. *Agricultural Water Management* 88, 201–208.

Bartholomew, D.P. (1982) Environmental control of carbon assimilation and dry matter production by pineapple (*Ananas comosus*). In: I.P. Ting and M. Gibbs (eds), *Crassulacean Acid Metabolism.* Rockville, MD: American Society of Plant Physiologists, pp. 278–294.

Bartholomew, D.P. (2008) Estimating plant weights. *Pineapple News* 15, 2–6. Available at: http://www.ishs-horticulture.org/workinggroups/pineapple/PineNews15.pdf (accessed 6 March 2018).

Bartholomew, D.P. and Kadzimin, S.B. (1977) Pineapple. In: Alvim, P.T. and Kozlowski, T.T. (eds) *Ecophysiology of Tropical Crops.* New York: Academic Press, Inc., pp. 113–156.

Bartholomew, D.P. and Malézieux, E. (1994) Pineapple. In: Schaffer, B. and Anderson, P. (eds) *Handbook of Environmental Physiology of Fruit Crops,* Vol. II. Boca Raton, FL: CRC Press, Inc., pp. 243–291.

Bartholomew, D.P. and Paull, R.E. (1986) Pineapple fruit set and development. In: Monselise, S.P. (ed.) *Handbook of Fruit Set and Development.* Boca Raton, FL: CRC Press, pp. 371–388.

Biscoe, P.V. and Gallagher, J.N. (1977) Weather, dry matter and yield. In: Landsberg, J.J. and Cutting, C.V. (eds) *Environmental Effects on Crop Physiology.* New York: Academic Press, pp. 75–101.

Black, R.F. (1962) Pineapple growth and nutrition over a plant crop cycle in Southeastern Queensland I. Root development and general growth features. *Queensland Journal of Agricultural Science* 19, 435–451.

Black, C.C., Carnal, N.W. and Kenyon, H.W. (1982) Compartmentation and the regulation of CAM. In: Ting, I.P. and Gibbs, M. (eds) *Crassulacean Acid Metabolism.* Rockville, MD: American Society of Plant Physiologists, pp 51–91.

Borland, A.G. and Griffiths, H. (1989) The regulation of citric acid accumulation and carbon recycling during CAM in *Ananas comosus. Journal of Experimental Botany* 40, 53–60.

Borland, A.G., Barrera Zambrano, V.A., Ceusters, J. and Shorrock, K. (2011) The photosynthetic plasticity of crassulacean acid metabolism: an evolutionary innovation for sustainable productivity in a changing world. *New Phytologist* 619–633.

Bouffin, J. (1991) L'ananas en Afrique de Sud. *Fruits* 46, 35–45.

Carnal, N.W. and Black, C.C. (1989) Soluble sugars as the carbohydrate reserve for CAM in pineapple leaves. Implications for the role of pyrophosphate 6-phosphofructokinase in glycolysis. *Plant Physiology* 90, 91–100.

Chapman, K.R., Glennie, J.D. and Paxton, B. (1983) Effect of five watering frequencies on growth and yield of various plant parts of container grown Queensland Cayenne pineapples. *Queensland Journal of Agricultural and Animal Sciences* 40, 75–81.

Chater, C., Peng, K., Movahedi, M., Dunn, J.A., Walker, H.J., Liang, Y.-K., McLachlan, D.H., Casson, S., Isner, J.C., Wilson, I. *et al.* (2015) Elevated CO_2-induced responses in stomata require ABA and ABA signaling. *Current Biology* 25, 2709–2716.

Ciesiolka, C. (1994) Soil conservation for sustainable pineapple farming. In: *Pineapple Industry Farm Committee Pineapple Field Day Book.* Brisbane: Growcom, Brisbane, Australia, pp. 22–29.

Ciesiolka, C.A.A., Drever, I. and Judd, T. (1993) Conservation in the pineapple industry. In: *Pineapple Industry Farm Committee Pineapple Field Day Book.* Brisbane: Growcom, pp. 9–19.

Climate-Data.org (2012) Climate Data for Cities Worldwide. Available at: http://en.climate-data.org (accessed 9 August 2017).

Combres, J.C. (1979) *Economie d'eau ou irrigation intensive.* Réunion annuelle, document Interne No. 57.

Combres, J.C. (1983) *Bilan énergétique et hydrique de l'ananas. Utilisation optimale des potentialités climatiques.* Document interne MENRS/IRFA, Montpellier, France.

Combres, J.C. and Perrier, A. (1976) *Bilan énergétique et hydrique de l'ananas.* Réunion annuelle IRFA, document Interne No. 66.

Combres, J.C. and Perrier, A. (1977) *Bilan énergétique d'une culture d'ananas.* Conférence technique OMM/FAO, Ibadan, 12–16 July 1977.

Commonwealth of Australia: Bureau of Meteorology (2018) Climate Data Online. Available at: www.bom.gov.au/climate/data (accessed 9 August 2017).

Connelly, P.R. (1969) The relative response of the pineapple plant, *Ananas comosus* (L.) Merr, under varying nitrogen rates and carriers under different levels of light intensity. Master's thesis, University of Hawaii, Honolulu, HI.

Connelly, P.R. (1972) The effects of thermoperiod on the carbon dioxide uptake and compensation point of the pineapple plant, *Ananas comosus (L.) Merr.* PhD dissertation, University of Hawaii, Honolulu, HI.

Côte, F.X. (1988) Photosynthese et photorespiration d'une plante a metabolisme crassulaceen: Ananas comosus (L.) Merr. Etude des echanges gazeux. Doctoral thesis, University of Toulouse, Toulouse, France.

Côte, F.X., Andre, M. and Folliot, M. (1989) CO_2 and O_2 exchanges in the CAM plant *Ananas comosus* (L.) Merr. Determination of total and malate-decarboxylation-dependent CO_2 – assimilation rates; study of light O_2 uptake. *Plant Physiology* 89, 61–68.

Côte, F.X., Folliot, M. and Andre, M. (1993) Photosynthetic Crassulacean acid metabolism in pineapple: Diel rhythm of CO_2 fixation, water use, and effect of water stress. *Acta Horticulturae* 334, 113–129.

Dodson, P.G.C. (1968) Effects of spacing, nitrogen and hormone treatment on pineapple in Swaziland. *Experimental Agriculture* 4, 103–115.

Dorey, E., Fournier, P., Lechaudel, M. and Tixier, P. (2015) Validity of the pineapple crop model SIMPIÑA across the climatic gradient in Réunion Island. *European Journal of Agronomy*, 62, 1–12.

Dorey, E., Fournier, P., Léchaudel, M. and Tixier, P. (2016) Modeling sugar content of pineapple under agro-climatic conditions on Reunion Island. *European Journal of Agronomy* 73, 64–72.

Drennan, P.M. and Nobel, P. (2000) Responses of CAM species to increasing atmospheric CO_2 concentrations. *Plant Cell and Environment*, 23, 767–781.

Ducreux, A., Godefroy, J., Lacoeuilhe, J.J. and Marchal, J. (1980) Some problems of working the soil in pineapple plantations in Martinique. *Fruits* 35, 595–604.

Ekern, P.C. (1964) *The Evapotranspiration of Pineapple in Hawaii. VI. Disposition of Net Radiation by Pineapple Plants.* Research Report 104. Private Document, Pineapple Research Institute of Hawaii, Honolulu, HI.

Ekern, P.C. (1965) Evapotranspiration of pineapple in Hawaii. *Plant Physiology* 40, 736–739.

Ekern, P.C. (1967) Soil moisture and soil temperature changes with the use of black vapor-barrier mulch and their influence on pineapple (*Ananas comosus* (L.) Merr.). *Soil Science Society of America Proceedings* 31, 270–275.

Ekern, P.C. (1968) Phyllotaxy of pineapple plants and fruit. *Botanical Gazette* 129, 92–94.

El-Swaify, S.A., Zhang, J., Palis, R., Rose, C.W. and A., C.C. (1993) Erosion problems and conservation needs of pineapple culture. *Acta Horticulturae* 334, 227–239.

Fleisch, H. (1988) Modeling pineapple plant growth and development. PhD dissertation, University of Hawaii, Honolulu, HI.

Foote, M. (1955) *Rate and Frequency of Application of Irrigation Water.* Report of Ayr Regional Experimental Station, Ayr, Queensland, Australia.

Friend, D.J.C. and Lydon, J. (1979) Effects of daylength on flowering, growth, and CAM (Crassulacean acid metabolism) of pineapple (*Ananas comosus* (L.) Merrill). *Botanical Gazette* 140, 280–283.

Gaillard, J.P. (1969) Influence of planting date and weight of slips on the growth of pineapples in Cameroon. *Fruits* 24, 75–87.

Gaillard, J.P. (1970) Recherche d'un equilibre K/N dans la production de l'ananas frais au Cameroun. *Fruits* 25, 11–24.

Galan Sauco, V., Cabrera Cabrera, J. and Rodriguez Pastor, C. (1988) El cultivo de la pina tropical (*Ananas comosus* L. Merr) en Canarias. *Fruits* 43, 35–42.

George, A.P., Nissen, R.J., Chapman, K.R. and Kerslake, R.G. (1984) Measurement of plant water status in pineapple (*Ananas comosus* (L.) Merr. cv. Queensland Cayenne (Clone 13). *Queensland Journal of Agricultural and Animal Sciences* 41, 57–64.

Giacomelli, E.J, and Py, C. (1982) Pesquisas abacaxicolas no brasil. *Ruggiero Dezembro*, 338–340.

Glennie, J.D. (1981) Induction with Ethrel. In: *Pineapple Industry Farm Committee Pineapple Field Day Book.* Brisbane: Growcom, Brisbane, Australia, pp. 34–36.

Htun, R.A. (1986) Vegetative growth, yield, and fruit quality of pineapple as influenced by moisture stress and potassium. PhD dissertation, University of Hawaii at Manoa, Honolulu, HI.

Huang, C.C. and Lee, L.C. (1969) Effect of irrigation on pineapples. *Taiwan Agriculture Quarterly* 5, 50–58.

Ikan, F. (1990) Caracterisation du system racinaire de l'ananas au champ. Memoire de DAAI, Université catholique de Louvain-La-Neuve, Belgium.

Iwaoka, N., Nightingale, G.T. and Clarke, H.E. (1935) Effects of aeration on pineapple roots. *Pineapple Quarterly* 5, 229–233.

Joshi, M.C., Boyer, J.S. and Kramer, P.J. (1965) Growth, carbon dioxide exchange, transpiration, and transpiration ratio of pineapple. *Botanical Gazette* 126, 174–179.

Kadzimin, S.B. (1975) Effects of water stress on pineapple (*Ananas comosus* (L.) Merr). Master's thesis, University of Hawaii, Honolulu, HI.

Kaplan, A. and Björkman, O. (1980) Ratio of CO_2 uptake to O_2 evolution during photosynthesis in higher plants. *Zeitschrift Plantzenphysiology* 96 185–188.

Kenyon, W.H., Severson, R.F. and Black, C.C. (1985) Maintenance carbon cycle in Crassulacean acid metabolism plant leaves. *Plant Physiology* 77, 183–189.

Klug, M. and Ting, I.P. (1978) *Crassulacean Acid Metabolism. Analysis of an Ecological Adaptation.* Ecological Studies, Vol. 30. Berlin: Springer-Verlag.

Kluge, M. (1969) Verändliche markierungmuster bei $^{14}CO_2$-fütterung von *Bryophyllum tubiflorum* zu verschiedenen zeitpunkten der hell/dunkelperiod I. Die $^{14}CO_2$ fixierung unter belichtung. *Planta* 88, 113–129.

Kluge, M., Fisher, A. and Buchanan-Bollig, I.C. (1982) Metabolic control of CAM. In: Ting, I.P. and Gibbs, M. (eds) *Crassulacean Acid Metabolism.* Rockville, MD: American Society of Plant Physiologists, pp. 278–294.

Krauss, B.H. (1930). The transpiration of pineapple plants *(Ananas sativus).* Master's thesis, University of Hawaii, Honolulu, HI.

Krauss, B.H. (1949) Anatomy of the vegetative organs of the pineapple, *Ananas comosus* (L.) Merr. II. The leaf. *Botanical Gazette* 110, 333–404.

Krauss, B.H. (1959) Soil moisture and root distribution in pineapple fields. I. Before and after rains and irrigations during a drought period. *PRI News* 7, 122–133. Private document, Pineapple Research Institute of Hawaii, Honolulu, HI.

Kuruvilla, V., Pillai, G.R., Mathew, J., Santhakumari, G., Gopi, C.S. and Varughese, K. (1988) Effect of irrigation and mulching on the growth and yield of pineapple (*Ananas comosus* L.). *Agricultural Research Journal of Kerala* 26, 85–91.

Lacoeuilhe, J.J. (1974) Pineapple planting density in the Ivory Coast for the export of fresh fruit. *Fruits* 29, 717–720.

Lacoeuilhe, J.J. (1976) Croissance de l'ananas en fonction du type de reje et de la fumure. Bilans de matière fraîche et séche et en éléments minéraux. Réunion annuelle IRFA, document Interne no. 11.

Lacoeuilhe, J.J. (1978) La fumure N-K de l'ananas en Côte d'Ivoire. *Fruits* 33, 341–348.

Lacoeuilhe, J.J. and Sarah, J.L. (1990) L'ananas au Zululand, Reunion Anneulle IRFA, document Interne, Montpellier

Le Van Thuong (1991) Scheme for Pineapple Acreage Zoning. Internal report, Thu Duc University, Ho Chi Minh City, Vietnam.

Lin, Z.F., Peng, C.L., Lin, G.Z. and Li, S.S. (1994) Stable carbon isotope ratio and activities of PEP carboxylase and PEP carboxykinase in pineapple leaves. *Acta Botanica Sinica* 36, 534–538.

Louis, L. and Nightingale, G.T. (1937) The growth responses of planting material of different nitrogen status and carbohydrate content. *Pineapple Quarterly* 7, 1–6.

Maerere, A.P. (1997) Axillary-bud development as it determines suckering in 'Queen Victoria' and 'Smooth Cayenne' pineapple. *Acta Horticulturae* 425, 309–320.

Malézieux, E. (1988) *Croissance et elaboration du rendement de l'ananas (Ananas comosus L. Merr).* Doctoral thesis, INA-PG, Paris.

Malézieux, E. (1991) Recherche de relations entre la biomasse aerienne, la surface foliaire et l'interception du rayonnement solaire chez l'ananas. *Fruits* 46, 523–532.

Malézieux, E. (1992) Effet d'une irrigation localisée et d'une couverture de polyéthyléene sur la croissance et le rendement de l'ananas en Basse Côte d'Ivoire. Compte rendu d'essai, document Interne IRFA.

Malézieux, E. and Sébillotte, M. (1990a) Relations entre les processus d'accumulation de la matiére séche et le rendement chez l'ananas (*Ananas comosus* L. Merr.). Note de réunion annuelle IRFA, document Interne.

Malézieux, E. and Sébillotte, M. (1990b) Relations entre les processus d'accumulation de la matiére séche et le rendement chez l'ananas (*Ananas comosus* L. Merr.). II. Le role de la biomasse présente a l'induction florale. Note de réunion annuelle IRFA, document Interne.

Malezieux, E., Zhang, J., Sinclair, E. and Bartholomew, D.P. (1994) Predicting pineapple harvest date in different environments, using a computer simulation model. *Agronomy Journal* 86, 609–617.

Medcalf, J.C. (1950) *Hawaiian Pineapple Co.'s Irrigation Studies on Lanai. Abstract of Monthly Seminar Summary.* Private document, Pineapple Research Institute of Hawaii, Honolulu, HI, pp. 16–28.

Metal Box (1956) The growing of pineapples, *The Pineapple Industry of South Africa, I. Technical.* The Metal Box Company of South Africa, Ltd, Research Division, Cape Town, South Africa.

Min, X.-J. (1995) Physiological effects of environmental factors and growth regulators on floral initiation and development of pineapple, *Ananas comosus* (L.) Merr. PhD thesis, University of Hawaii at Manoa, Honolulu, HI.

Min, X.-J. and Bartholomew, D.P. (2005) Effects of flooding and drought on ethylene metabolism, titratable acidity and fruiting of pineapple. *Acta Horticulturae* 666, 135–148.

Monteith, J.L. (1977) Climate. In: Alvim, P. de T., and Kozlowski, T.T. (eds) *Ecophysiology of Tropical Crops.* New York: Academic Press, pp. 1–27.

Moreau, B. and Moreuil, C. (1976) L'ananas dans la region de Tamatave (Côte est de Madagascar). Contribution a la connaissance de sa vegetation en condition naturelle et dirigee. *Fruits* 31, 21–30.

Neales, T.F., Patterson, A.A. and Hartney, V.J. (1968) Physiological adaptation to drought in the carbon assimilation and water loss of xerophytes. *Nature* 219, 469–472.

Neales, T.F., Sale, P.J.M. and Meyer, C.P. (1980) Carbon dioxide assimilation by pineapple plants, *Ananas comosus* (L.) Merr. II. Effects of variation of the day/night temperature regime. *Australian Journal of Plant Physiology* 7, 375–385.

Neild, R.E. and Boshell, F. (1976) An agroclimatic procedure and survey of the pineapple production potential of Colombia. *Agricultural Meteorology* 17, 81–92.

N'Guessan, A. (1985) Analyse de l'evapotranspiration reelle de *Ananas comosus* (L.) Merr. en Basse Côte d'Ivoire. Docteur-ingenieur, ENSA-Rennes, Rennes, France.

N'Guessan, A. (1986) Analyse de l'évaporation réelle de l'ananas en Basse Côte d'Ivoire. Réunion annuelle IRFA. Document Interne No. 19. IRFA, Montpellier, France.

Nievola, C.C., Kraus, J.E., Freschi, L., Souza, B.M. and Mercier, H. (2005) Temperature determines the occurrence of CAM or C3 photosynthesis in pineapple plantlets grown in vitro. *In Vitro Cellular & Developmental Biology – Plant* 41, 832.

Nightingale, G.T. (1942) Nitrate and carbohydrate reserves in relation to nitrogen nutrition of pineapple. *Botanical Gazette* 103, 409–456.

Nightingale, G.T. (1949) *The Use and Value of the Plant Log in Pineapple Production.* PRI Seminar Summary, January. Private document, Pineapple Research Institute of Hawaii, Honolulu, HI.

Nobel, P.S. (2010) *Desert Wisdom/Agaves and Cacti: CO_2, Water, Climate Change.* Bloomington, IN: iUniverse.

Noffsinger, T.L. (1961) *Monthly Probabilities of Different Rainfall Levels.* Land Study Bureau Technical Paper No. 4, University of Hawaii, Honolulu, HI.

Norman, J.C. (1977) The effect of plant density on the growth, development and yield of pineapple (*Ananas comosus* (L.) Merr. cv. Smooth Cayenne) in Ghana. *Acta Horticulturae* 53, 349–354.

Nose, A., Miyazato, K. and Murayama, S. (1981) Studies on matter production in pineapple plant. II. Effects of soil moisture on the gas exchange of pineapple plants. *Japan Journal of Crop Science* 50, 525–535.

Nose, A., Matake, S., Miyazato, K. and Murayama, S. (1985) Studies on matter production in pineapple plants. III. Effects of nitrogen nutrition on the gas exchange of shoots. *Japan Journal of Crop Science* 54, 195.

Nose, A., Heima, K., Miyazato, K. and Murayama, S. (1986) Effects of day-length on CAM type CO_2 and water vapour exchange of pineapple plants. *Photosynthetica* 20, 20–28.

Ogura, H., Nakayama, S. and Kushima, T. (1968) Studies on the growth of pineapples under relatively low temperature conditions in a vinyl house without heating. *Bulletin Faculty of Agriculture, Kagoshima* 18, 13–16.

Osmond, C.B. (1978) Crassluacean acid metabolism: a curiosity in context. *Annual Review of Plant Physiology* 29, 379–414.

Priestley, D.H.B. and Taylor, R.J. (1972) On the assessment of surface heat flux and evaporation using large scale parameters. *Monthly Weather Review* 100, 81.

Prigge, M. and Gutiérrez-Soto, M.V. (2014) Pineapple photosynthesis and leaf sap pH as a surrogate of CAM performance in the field. A research advance. *Pineapple News* 21, 18–23. Available at: http://www.ishs-horticulture.org/workinggroups/pineapple/PineNews21.pdf (accessed 6 March 2018).

Py, C. (1959) Etude sur la croissance de l'ananas en Guinee. *Fruits* 14, 3–24.

Py, C. (1965) Approches pour combler le deficit en eau, principal facteur limitant de la culture de l'ananas en Guinee. *Fruits* 20, 315–329.

Py, C., and Lossois, P. (1962) Prevision de recolte en culture d'ananas. Etudes de correlations. Deuxieme partie. *Fruits* 17, 75–87.

Py, C., Tisseau, M.A., Oury, B. and Ahmada, F. (1957) *La culture de l'ananas en Guinee – Manuel du planteur.* Paris: Institute français de recherches fruitieres d'outre-mer (IFAC).

Py, C., Lossois, P. and Karamkam, M. (1968) Contribution to the study of the growth cycle of pineapples. *Fruits* 23, 403–413.

Py, C., Lacoeuilhe, J.J., and Teisson, C. (1987) *The Pineapple. Cultivation and Uses.* Paris: Editions G.-P. Maisonneuve.

Rafaillac, J.P., de Ricaud, J. and Vernéde, A. (1978) *Compacité du sol et enracinement de l'ananas en début de cycle.* Document Interne ORSTOM.

Rainha, N., Medeiros, V.P., Câmara, M., Faustino, H., Leite, J.P. *et al.* (2016) Plasticity of crassulacean acid metabolism at subtropical latitudes: a pineapple case study. *Physiologia Plantarum* 156, 29–39.

Rao, G.G., Sharman, C.B., Chadha, K.L. and Shikhamany, S.D. (1974) Effect of varying soil moisture regimes and nitrogen levels on plant growth, yield and quality of Kew pineapple (*Ananas comosus* (L) Merr.). *Indian Journal of Horticulture* 31, 306–312.

Ravoof, A.A. (1973). Effects of root temperatures and nitrogen carriers on nutrient uptake, growth, and composition of pineapple plants, *Ananas comosus* (L.) Merr. PhD dissertation, University of Hawaii, Honolulu, HI.

Rohrbach, K.G. and Apt, W.J. (1986) Nematode and disease problems of pineapple. *Plant Disease* 70, 81–87.

Romero, C.A., Lopez, G.C. and Gurdian, G.R. (1973) Mineral nutrition of field grown pineapples. Production and fruit quality. *Proceedings Tropical Region American Society for Horticulture Science* 17, 203–219.

Russell, G., Jarvis, P.G. and Monteith, J.L. (1989) Absorption of radiation by canopies and stand growth. In Russell, G., Marshall, B. and Jarvis, P.G. (eds) *Plant Canopies: Their Growth, Form and Function.* Cambridge: Cambridge University Press, pp. 21–39.

Sale, P.J.M. and Neales, T.F. (1980) Carbon dioxide assimilation by pineapple plants, *Ananas comosus* (L.) Merr. I. Effects of daily irradiance. *Australian Journal of Plant Phsiology* 7, 363–373.

Sanford, W.G. (1962) Pineapple crop log-concept and development. *Better Crops Plant Food* 46, 32–43.

Scott, C. (1992) The effect of planting density on spring plant crop/autumn first ratoon crop cycles; Test number 718. In: *Pineapple Industry Farm Committee Pineapple Field Day Book.* Brisbane: Growcom, Brisbane, Australia, pp. 37–45.

Senanayake, Y.D.A. (1978) The effect of plant population and irrigation on growth and yield of pineapple (*Ananas comosus* (L.) Merr.) cv. Kew. *Journal National Agriculture Society of Ceylon* 15, 69–80.

Shiroma, M. (1971) The transpiration of pineapple in Okinawa. *Sciences Bulletin, College of Agriculture, University of the Ryukyus*.

Shiroma, M. (1972) Air temperature in Okinawa and the vegetative growth of pineapple. *Sciences Bulletin College of Agriculture, University of Ryukyus* 19, 363–377.

Shiroma, M. (1973) Evapotranspiration of pineapple in Okinawa measured by heat balance method. *Sciences Bulletin, College of Agriculture, University of the Ryukyus* 20, 169–190.

Shiroma, M. (1977) Effects of solar radiation on pineapple culture in Okinawa. *Nogyo Kisho Journal of Agricultural Meteorology* 33, 1–6.

Shiroma, M., Miyazato, K., Murayama, S. and Nose, A. (1977) Studies on matter production in pineapple plants. I. Effects of light intensity in light period on the CO_2, carbon dioxide, exchange and CO_2 balance of pineapple plants. *Japanese Journal of Crop Science* 46, 580–587.

Sideris, C.P. (1955) Effects of sea water sprays on pineapple plants. *Phytopathology* 45, 590–594.

Sideris, C.P., and Krauss, B.H. (1928) Water relations of pineapple plants. *Soil Science* 26, 305–314.

Sideris, C.P., and Krauss, B.H. (1934) Preliminary studies on the growth and water absorption behavior of pineapple roots. *Pineapple Quarterly* 4, 42–47.

Sideris, C.P. and Krauss, B.H. (1936) The classification and nomenclature of groups of pineapple leaves, sections of leaves and sections of stems based on morphological and anatomical differences. *Pineapple Quarterly* 6, 135–147.

Sideris, C.P. and Krauss, B.H. (1937) The growth of pineapple plants in complete water cultures with either ammonia or nitrate salts. *Growth* 1, 204–210.

Sideris, C.P. and Krauss, B.H. (1955) Transpiration and translocation phenomena in pineapples. *American Journal of Botany* 42, 707–709.

Sideris, C.P., Krauss, B.H. and Young, H.Y. (1936) Effects of light intensity on the time of fruit harvest, yields and quality. *Pineapple Quarterly* 6, 56–66.

Sideris, C.P., Young, H.Y. and Chun, H.H.Q. (1948) Diurnal changes and growth rates as associated with ascorbic acid, titratable acidity, carbohydrate, and nitrogen fractions in the leaves of *Ananas comosus* (L.) Merr. *Plant Physiology* 23, 38–69.

Sinclair, E. (1992) Seasonal growth rates in pineapple. In: *Pineapple Field Day Book*. Pineapple Industry Farm Committee Beerwah, Queensland, Australia, pp. 9–21.

Sison, C.M.A. (2017) Trimming 'MD-2' mother plants for the production of planting materials. *Pineapple News* 24, 30–32.

Srivastava, S.S. and Singh, J.R. (1971) Climatic requirements for the growth and fruiting of pineapple. *Allahabad Farmer* 45, 283–286.

Suwandi, T., Dewi, K. and Cahyono, P. (2016) Pineapple harvest index and fruit quality improvement by application of gibberellin and cytokinin. *Fruits* (Paris), 71, 209–214.

Swete Kelly, D. and Bartholomew, D.P. (1993) Other disorders. In: Broadley, R.H., Wassman, R.C. and Sinclair, E. (eds) *Pineapple Pests and Disorders*. Brisbane: Queensland Department of Primary Industries, pp. 43–52.

Tan, K.M. and Wee, Y.C. (1973) Influence of size of pineapple slips on plant growth, fruit weight and quality in graded and mixed plantings. *Tropical Agriculture* 50, 139–142.

Tay, T.H. (1974) Effect of water on growth and nutrient uptake of pineapple. *MARDI Research Bulletin* 2, 31–49.

Thorne, M.D. (1953) Report on irrigation and trash mulch experiments. *Pineapple News* 1, 64–67. Private document, Pineapple Research Institute of Hawaii, Honolulu, HI.

Tisseau, R. (1971) Experimental pineapple culture. I. The hydroponic system at Anguededou. *Fruits* 26, 279–285.

Tkatchenko, B. (1947) Ecologie de l'ananas dans le Sud Indochinois. *Fruits* 2, 206–212.

US Climate Data (2018) Climate Honolulu – Hawaii. Available at: www.usclimatedata.com/climate/honolulu/hawaii/united-states/ushi0026 (accessed 9 August 2017).

Van de Poel, B., de Proft, M.P. and Ceusters, J. (2009) Determination of pineapple (*Ananas comosus*, 'MD-2' hybrid cultivar) plant maturity, the efficiency of flowering induction agents and the use of activated carbon. *Scientia Horticulturae* 120, 58–63.

Vuillaume, C. (1986) Quelques données sur la culture de l'ananas en Australie. Réunion annuelle IRFA, document Interne No. 48.

Wassman, R.C. (1986) Sequential planting studies. Golden Circle Cannery Test 715 (internal document).

Wassman, R.C. (1990) Effects of seasonal temperature variations on pineapple scheduling for canning in Queensland. *Acta Horticulturae* 275, 131–138.

Wassman, R.C. (1991) The high cost of inadequate fruit induction: Summer forcing for spring harvest. In: *Pineapple Field Day Book*. Pineapple Industry Farm Committee, Beerwah, Queensland, Australia, pp. 37–38.

Wee, Y.C. (1969) Planting density trials with *Ananas comosus* (L.) Merr var. Singapore Spanish. *Malay Agricultural Journal* 47, 164–174.

Wee, Y.C. and Ng, J.C. (1968) Some observations on the effect of month of planting on the Singapore Spanish variety of pineapple. *Malay Agricultural Journal* 46, 469–475.

World Weather Online (2018) Johor Bahru Monthly Climate Averages. Available at: www.worldweatheronline.com/john-bahru-weather-averages/johor/my.aspx (accessed 9 August 2017).

World Weather Online (2018) Pattaya Monthly Climate Averages. Available at: www.wordweatheronline.com/pattaya-weather-averages/chon-buri/th.aspx (accessed 9 August 2017).

Yoder, R.C. (1969) Effects of thermoperiod on the stomatal opening and transpiration of pineapple (*Ananas comosus* (L.) Merr.). Master's thesis, University of Hawaii, Honolulu, HI.

Zhang, J. (1992) Computer simulation of pineapple growth, development, and yield. PhD dissertation, University of Hawaii, Honolulu, HI.

Zhang, J. and Bartholomew, D.P. (1997) Effect of plant population density on growth and dry matter partitioning of pineapple. *Acta Horticulturae* 425, 363–376.

Zhang, J., Malezieux, E. and Bartholomew, D.P. (1997) ALOHA-Pineapple v.2.1: a computer model to predict the growth, development and yield of pineapple. *Acta Horticulturae* 425, 287–296.

Zhu, J. (1996) Physiological responses of pineapple (*Ananas comosus* (L.) Merr.) to CO_2 enrichment, temperatures and water deficit. PhD dissertation, University of Hawaii at Manoa, Honolulu, HI.

Zhu, J., Bartholomew, D.P. and Goldstein, G. (1997a) Effects of temperature, CO_2, and water stress on leaf gas exchange and biomass accumulation of pineapple. *Acta Horticulturae* 425, 297–308.

Zhu, J., Bartholomew, D.P. and Goldstein, G. (1997b) Effect of elevated carbon dioxide on the growth and physiological responses of pineapple, a species with Crassulacean acid metabolism. *Journal American Society for Horticultural Science* 122, 233–237.

Zhu, J., Goldstein, G. and Bartholomew, D. (1999) Gas exchange and carbon isotope composition of *Ananas comosus* in response to elevated CO_2 and temperature. *Plant Cell & Environment* 22, 999–1007.

Zhu, J., Bartholomew, D.P. and Goldstein, G. (2005) Photosynthetic gas exchange and water relations during drought in 'smooth cayenne' pineapple (*Ananas comosus* (L.) Merr.) grown under ambient and elevated CO_2 and three day/night temperatures. *Acta Horticulturae* 666, 161–173.

7 Cultural System*

Jhonny Vásquez Jiménez,[1] Garth M. Sanewski,[2] Domingo Haroldo Reinhardt[3] and Duane P. Bartholomew[4]

[1]Consultant, Alajuela, Costa Rica; [2]Queensland Department of Agriculture and Fisheries, Australia; [3]Embrapa Mandioca e Fruticultura, Cruz das Almas, Brazil; [4]University of Hawaii at Manoa, Honolulu, USA

Land Selection

General requirements

Temperature

The selection of lands for the growing pineapples will mostly be directed by the temperature profile of the areas being considered. Pineapple growth is almost non-existent below 7°C and above 40°C, so lands where such temperatures occur for long periods of time are unsuitable for commercial production of the crop. Night temperatures below 7°C are tolerated for short periods but may promote flower induction as will temperatures around 17–20°C for extended periods. Freezing temperatures must be avoided.

Growth of leaves reaches a maximum at about 32°C and root growth reaches a maximum at 29°C (Sanford, 1962). The optimum mean temperature for growth is believed to be nearer 25°C with an approximately 10°C diurnal temperature range (Neild and Boshell, 1976). This temperature range will assure that growth rates are balanced with a sufficiently high rate of net assimilation to assure that adequate carbohydrate reserves are maintained by the plant, especially near the time of flower induction (see Chapter 6).

Topography

Protection of the soil resource from erosion is an important consideration in all agriculture, and for that reason minimally sloping lands should be selected for pineapple culture. Slopes of no more than a few per cent minimize soil losses and facilitate machinery access while lands with steeper slopes require expensive drainage channels, contouring, and other protective measures to sustain the soil resource and allow all weather access. Soil loss, along with the nutrients and pesticides contained, can be a major problem in pineapple if attention is not paid to plantation design.

In areas that are exclusively managed with hand labour, pineapples can be grown in fields that are extremely steep (Fig. 7.1), but where mechanization is a requirement for land preparation, maintenance, or harvesting, slopes are limited by the requirements of the equipment. Slopes of nearly 40% have been farmed on small holdings with medium-sized machinery in Australia in past years though erosion losses from such fields can be very high (El-Swaify *et al.*, 1993; Ciesiolka *et al.*, 1995).

Most pineapple-growing areas are laid out with consideration of the impact of rainfall on erosion and drainage. In areas that may experience heavy rainfall or have low infiltration rates,

* A revision of the chapter by Anthony Hepton, formerly of Dole Foods, USA. The authors acknowledge the advice provided by Jorge W. Gonzales, Senior Consultant, Pineapples and Bananas, Dole Asia Holdings.

Fig. 7.1. Steep land in Yunnan, China. Photo A. Hepton.

Fig. 7.2. Oats used as a living mulch. Photo Z. Nicholls.

or both, drainage channels should be constructed at intervals that will accommodate the surface runoff and allow this runoff to move off the field with a minimum loss of soil. The objective is to reduce the water's velocity by minimizing slopes and length of run. The details of constructing such drainage channels varies with the characteristics of the soil and the slope of the land (El-Swaify *et al.*, 1982; Hudson, 1995). Drainage and the removal of water are critical to the successful growing of pineapple, as the root system is intolerant of poorly aerated soils. Areas to be avoided are those that accumulate standing water or that have internal barriers to soil moisture movement, such as plough pans or compacted or impervious soil layers. Where drainage is poor, subsoiling, ripping, or installation of internal drainage may be necessary. Ditching and ridging are used to provide adequate soil drainage where high water tables exist or where infiltration rates may be low. Sedimentation ponds should be used to capture potential off-site soil loss and the associated nutrients and pesticide compounds. Bio-reactor trenches can also be used on perimeters of the farm to intercept subsurface horizontal water movement that potentially carries nitrogen off-site. These are narrow trenches dug to just below the impervious barrier and filled with sawdust.

Soils that have high water tables (<1.0 m) should be avoided.

In regions where water is not limiting and soil erosion is likely, living mulches can be used in the inter-rows to slow water movement and reduce erosion. In Australia, nematode-resistant forage sorghums have been used in summer and oats in winter (Fig. 7.2). The living mulch is sprayed with a grass herbicide that does not affect the pineapple before heading and the dried stalks, which eventually decompose, left standing (Fig. 7.3).

Soil types

Pineapples have been grown on a wide variety of soils, from organic peat soils as in Malaysia, volcanic ash soils in Hawaii, many Caribbean islands, and parts of the Philippines, to the very sandy soils found in parts of Australia and South Africa. In between, are a variety of weathered and secondary deposit soils, some with forest top soils, and others converted from farm and pasture lands.

Soils that are ideal for growth have a high organic matter content with excellent internal drainage and a high soil air content to provide optimum amounts of water, nutrients and

Fig. 7.3. Oats used as a living mulch after killing with a grass herbicide and decomposition. Photo Z. Nicholls.

oxygen to plant roots. Soils should have a neutral to acidic pH, although pineapples will grow in slightly alkaline soils if calcium levels are not too high and soil moisture does not favour the growth of water moulds such as *Phytophthora* and *Pythium* species.

Sunlight

Pineapple is most productive in areas with extensive sunlight. An ideal climate will have temperatures below 32°C and cloud-free days; however, most climates will have periods of seasonal cloudiness. High irradiance in combination with high air temperature leads to sunburn, both on the leaves that are directly exposed and to fruit that may lodge at angles that increase fruit surface exposure. There is no day-length requirement for growth or flowering although a short day-length favours flowering. The day-length of pineapple growing areas near the equator vary less than an hour from winter to summer, while near the tropics of Cancer and Capricorn, the variation is somewhat more than two hours. The variation in irradiance and temperature across the range of elevations and latitudes where pineapple is grown has a significant impact on carbohydrate accumulation and subsequent fruit size and yield.

Water and water quality

Pineapple is uniquely adapted to grow in areas with low rainfall. However, where applied water (rainfall + irrigation) is very low, growth of the crop cycle will be lengthened or average fruit weight will be reduced, or both. The very minimum water requirement for growth is about 50 mm (2″) of water per month but in hot, dry environments around 75 mm (3″) is needed. For optimum growth and productivity, 30–70% of the Class Pan A Evapotranspiration is needed during peak growth periods. Water is most efficiently utilized when applied to the root zone at intervals that will keep it adequately supplied with moisture. No more than 50 mm should be applied in a single application. Irrigation efficiency is increased if beds are covered with an impervious mulch and amounts applied should be reduced.

Irrigation water should not be sourced from drainage from fields infected with disease such as *Phytophthora* sp. or *Dickyea* sp.

Pineapple is generally considered to have a medium tolerance to salinity (McCall, 1980) although this will depend on the soil type with tolerance decreasing as soil permeability decreases. Water with a conductivity of 1.28 dS/m (2000 ppm) reduced yield in pineapple. While exhibiting some tolerance to sodium, pineapple is sensitive to chloride. In pot experiments by Liu and Chen (2014), fruit size and fruit firmness decreased with increasing salinity from 0.25 to 0.75%. Fruit aroma was decreased by the highest salt stress. A decrease in calcium and potassium was associated with increasing salinity (increasing sodium).

Alvarez *et al.* (1995) found that sodium chloride at 14 meq l^{-1} and above adversely affected fruit yield and quality. Irrigation water with up to 20 meq l^{-1} $NaHCO_3$ did not affect fruit quality.

Soil surveys

A soil survey should be conducted if an area of more than approximately 8 hectares (20 acres) is being evaluated for pineapple. A survey will provide important information regarding the topography, soil profile, and subsurface variations in soil structure, soil chemistry and drainage. This information will be vital in determining the need for contouring and other erosion control measures, surface drainage, subsoiling, depth of cultivation, need for bed formation, nutrient requirements, and other cultural practices that can be influenced by soil conditions.

The survey reports should include any history of past usage, including cropping and soil management. The physical description of the soil by profile should be supplemented with information on physical and chemical analyses to allow appropriate agricultural practices to be developed.

Maps should be prepared that identify boundaries of the growing areas, areas highly susceptible to erosion, waterways and other drainage channels, roads and pathways, particularly if these are used by the public. The approximate boundaries of each soil type should be identified according to soil classification. Residential and other environmentally sensitive areas should be identified to facilitate compliance with local regulations. All of this information should overlay a topographical map of the area to facilitate development of field layout and drainage systems. If resources permit, the data should be entered into a graphical information system that will facilitate planning and management of the farm or plantation (Jung-Rothenhaeusler *et al.*, 2016). Digital farm maps developed in this way can be used not only for contour mapping and drainage design but can be over-layered with soil analysis, leaf analysis, yield and quality data.

Soil sampling

A composite sample that adequately represents the soil in the field must be taken so that the results can be used to develop soil management information for the crop.

Where soil surveys show that more than one distinct soil type exists in an area to be farmed, each soil type should be sampled separately if the areas are large enough to be fertilized separately. Areas that are not typical of the general area being sampled should be avoided, or if resources permit, sampled separately. Problem areas should be sampled separately, and additional samples may be required from the subsoil in these areas. Topsoil and subsoil samples should be analysed separately and compared with results from the more typical areas.

In new areas, soil samples should be taken from 0–30 and 30–60 cm depths, and in active farms at 0–15 and 15–30 cm depths.

Soil analyses

Most soil analyses can be performed by service providers specializing in this field. Use of these services is often a convenient and professional way to collect this information. More information on soil analyses is contained in Chapter 8.

Soil texture

Soil A and B horizons should be evaluated for soil texture. Aspects that can be documented include:

- coarse fragment analysis
- particle size by hydrometer and hand-sieving
- texture by feel
- moisture content
- water-holding capacity (water content at field capacity and at 1.5 megapascal pressure)
- bulk density.

Soil pH

Soil pH is measured on a saturated paste. This aspect of soil analysis is important for pineapple culture as serious problems can be associated with alkaline soils. The most serious is the proliferation of the pathogens *Phytophthora* sp. and *Pythium* sp. Soils of alkaline pH may also limit the availability of important nutrient elements such as iron.

Cation exchange capacity (CEC)

CEC and base saturation provide measures of the soils' ability to hold nutrients and the quantity retained in a form available for uptake by the plant, and can have an impact on the frequency of fertilizer applications and may also indicate whether the fertilizer should be applied to the soil or as a foliar spray.

Organic matter

The organic matter content can be measured by ashing the dried soil and the volume of organic matter estimated from the loss of weight. As mentioned earlier, pineapple plants respond favourably to soils with a high percentage of organic matter. This is not to be confused with crop residue or other organic material that has not undergone complete decomposition by composting or other means that result in a stable C/N ratio in the soil.

Land and Soil Preparation

Land clearing, field layout and bed design

If the site has not been cropped previously, the first operation will be to remove natural vegetation. After the fields have been surveyed to establish the slope and the need for and frequency and position of drainage channels, these channels should be installed. If drainage channels discharge into adjacent waterways, the need for permanent drop structures should be evaluated. New geo-informatics technology such as drones and satellites can be used to effectively map fields and superimpose the most appropriate layout (Jung-Rothenhaeusler *et al.*, 2016).

Where rock removal is necessary, rocks larger than about 30 cm (1″) in diameter should be removed after ploughing or subsoiling and after final land preparation. Rock removal from stony fields can be expensive and specialized equipment has been developed to mechanize the process. These rocks may be used for the construction of drop structures.

In areas where field operations are machine-assisted, planting areas may be laid out in blocks separated by roads. The dimensions of the blocks are usually designed to accommodate the equipment. Where boom sprayer equipment is to be used, block size is twice as wide as the spray boom is long. Blocks can often be in the order of 24 beds wide.

Once the basic tillage operations have been performed, raised planting beds may be formed if there are known soil limitations or economic advantages (Fig. 7.4). In some areas, pineapple plant growth is enhanced by planting on raised beds due to the increase in the volume of topsoil available to the root system, enhanced aeration, and superior drainage. Raised beds may or may not be covered with plastic mulch. The dimensions of raised beds will depend on the planting density, equipment available and need for manual operations in the field. Where spraying and harvesting operations are performed using boom equipment, the fields will be designed accordingly. Bed centres can vary depending on the equipment available but are usually 1.22–1.52 m. In some cases where capture of sparse rainfall is important, slightly depressed beds direct limited rainfall or overhead irrigation to the planting line. Despite the advantages of raised beds, they are not used where the cost of preparation exceeds the economic benefit. Some high infiltration soils allow the plants to be grown without raised beds even in high rainfall regions. Soil infiltration might be improved by surface organic matter.

Fig. 7.4. Formed beds in long rows ready for planting. Photo S. Newett.

Farm layout

Pineapple farms by their nature need to be highly organized with closely monitored programmes that can be benchmarked with other farms. It is helpful in this endeavour to follow some organizational concepts generally used within the industry.

The 'Gross area' includes all areas of the farm, even areas that cannot be used. The 'Useful area' includes all the areas available, including the space occupied by roads, drainage and installations. The 'Net area' is the area of the farm occupied by pineapple plants. The 'Net area' is divided into 'Fields'. 'Fields' are minimum units of management by their geographical location. 'Increments' are minimum units of management grouped by age or date of establishment or planting. The smallest unit of land managed is called a 'Block'. This is defined as a width twice the length of the spray boom. The number of beds that will form a 'Block' will be equivalent to twice the length of boom length, minus the design overlap, divided by the bed width.

On the outer edges or close to main roads and secondary and primary drains, farms will have half 'Blocks', which also serve as buffers to run off in the main drains and outside the farm.

Treatment of previous crop residues

Knockdown and incorporation

Because of their leaf morphology and succulence, pineapple plants are slow to desiccate, and shoots produced from living plant stems incorporated into the soil can be a serious weed in fields. Thus, it is common to chop standing plants by discing, power flail (slasher) or rotary hoe to hasten their desiccation and decomposition (Fig. 7.5). In the peat soils of Malaysia, plants are killed with the contact herbicide, paraquat to hasten their desiccation. The imazamox molecule has also been shown to desiccate pineapple plants, and while a longer time is required than required by paraquat, it is an alternative for countries that cannot legally use paraquat (Vásquez Jiménez, 2016). Once desiccated, plant residues may be burned or incorporated into the soil if there is sufficient moisture and time for decomposition.

Pineapple plant residue may also be harvested for animal feed or for by-products, such as fuel, fibre or extracts such as bromelain. Removing the crop residue may help to shorten the crop inter-cycle for timely production schedules. However, repeated removal of pineapple residue depletes the soil of essential nutrients and organic matter. As an example, as much as 60% of the whole crops potassium is retained in crop residues. Soils with poor structure, low native organic matter, and low CEC benefit most from residue incorporation and are most adversely affected by residue removal. The benefits of residue incorporation can accrue over several crop cycles. Well-aggregated soils with a stable organic matter fraction and a comprehensive supply of nutrients may show little benefit from residue incorporation in the short term. The removal or burning of plant residues after harvest is not a recommended practice. In addition to the above, residues protect the soil surface from erosion and are critical for the success of no- or minimum-tillage agriculture.

Fig. 7.5. Rotary hoe used to destroy and incorporate the previous crop. Main photo G. Sanewski, inset S. Newett.

Conventional tillage

The chief objective of tillage is to achieve excellent soil tilth which ensures good contact with the planting material to facilitate rapid and sustained root development. Tillage should achieve a permeable soil profile that is free of rocks and large clods and a homogeneous distribution of decomposed residue, amendments and fertilizers. Where fumigation for nematode control is used, fine tilth must be achieved for effective distribution of the fumigant. However before using fumigants, the need should be clearly established as their use can lead to microbiological imbalances.

Compaction due to residue management, tillage operations, and in-field traffic may easily become counter-productive. Compaction needs to be avoided or minimized by selecting the right equipment according to soil type, internal soil structure, soil moisture, organic matter content and the time available to prepare the land.

Subsurface rippers are used to break up impervious or compacted internal layers to improve drainage without changing layering of horizons in the soil profile. Deep ploughing may be used where mixing or redistribution of nutrients or soil layers results in an improved planting and growing environment. Harrows, discs and power-spaders are used to turn over and or break up clods to provide a suitable tilth for planting. Rollers, cultipackers, and leveling boards may be used to finish the tillage operations. These operations will be followed by bed-forming, pre-plant fertilizer application and fumigation with mulch laying where these operations are appropriate. These final steps can be accomplished with one integrated piece of equipment.

Minimum tillage

There are no published examples of minimum tillage being used commercially in pineapple. Minimum and no-tillage combined with a short period of cover cropping were compared experimentally in a tropical region in southern Japan where the normal practice is not to form planting beds (Sugahara *et al.*, 2001). A bed system makes minimum tillage more difficult to implement, particularly in high rainfall tropical regions where soils might not drain freely and bed height needs to be maintained. It is likely, however, that there would be many biological advantages, including improved infiltration (Sugahara *et al.*, 2001). Although it is not suited to every situation, minimum tillage is seen as an attractive cropping system where it is feasible, because it can considerably reduce the cost of land preparation, conserve organic matter from the previous crop, reduce moisture loss during land preparation, preserve and even enhance the balance of microflora and -fauna in the soil profile, and reduce soil erosion. However, there may be valid reasons for not using minimum tillage. For example, tillage might facilitate the control of symphylids and an excellent tilth is essential for good distribution of soil fumigants used to control nematodes. In some soils, tillage may be necessary to provide a tilth that will allow planting. The removal of old plants, planting new material and minimizing compaction are likely to be further obstacles unless systems can be developed to better accommodate the previous crop.

In those soils where tillage is not a prerequisite to subsequent operations, the previous crop residues may be physically or chemically killed and allowed to form an organic mulch through which the pineapple plants are planted. If minimum-tillage practices are used, special consideration should be given to plant nutrition. Fertilization should start early and be directed to the planting lines. Pre-plant incorporation of plant nutrients may not be possible.

In other plantation crops such as sugar cane and ginger where minimum tillage has been tried it generally results in lower yield (Garside *et al.*, 2009; Smith *et al.*, 2011). Minimum tillage might initially be more suited to small-scale, low-technology pineapple production, particularly where permanent beds do not need to be maintained, but with further development could offer many benefits to larger farming operations, especially in the area of soil conservation.

Soil amendments and fertilizers

Soil amendments such as lime, dolomite or gypsum and organic matter influence plant growth indirectly by improving the physical or chemical condition of the soil, although amendments generally also provide plant nutrients. Composted animal manures provide organic matter and improve soil structure while supplying plant nutrients. The incorporation of organic amendments such as sawdust, sugarcane trash, and poultry manure have also been used in pineapple. A relatively large quantity of organic matter (>10 t/ha) has been demonstrated to suppress nematodes in sandy soil (Stirling, 1999) but can lead to high nitrate in processed fruit if nitrogen levels are not managed well.

Lime and dolomite adjust soil pH as well as supplying calcium, and in the case of dolomite, also magnesium. Gypsum will supply calcium without affecting the soil pH. In many cases a pH of 4.5–5.0 is preferred as it assists with the control of *Phytophthora* and is suitable for pineapple growth. Soil amendments can be broadcast immediately after destruction of the old plants so they can be incorporated by tillage operations. Preplant fertilizers can be banded just below or adjacent to the plant line during bed formation to allow early interception by the developing roots.

Mulches and fumigants

Vapour barrier mulches can have several beneficial effects including the retention of soil fumigants and soil moisture and control of weeds. Early mulches were made of heavy craft paper impregnated with materials to slow decomposition. Thin, plastic mulches (Fig. 7.6) that contain inhibitors to slow degradation by ultra

Fig. 7.6. Conventional plastic mulch in Hawaii with planting material ready for hand planting. Photo D. Bartholomew.

violet light replaced these early paper mulches. These mulches can be marked with guides to allow precision planting. However, there is now concern over the contamination of the environment with plastics and biodegradable mulches are preferred. There are several types of these biodegradable mulches, some being truly biodegradable in the soil and others which contain additives that only cause the plastic to fragment and hence are not truly biodegradable. Truly biodegradable films are usually composed of a mixture of polyester and starch. Biodegradable mulches can however photodegrade and are not suited to every soil type. Different formulations would need to be tested for a given site.

To assure proper plant densities, sample lots of mulch should be evaluated after laying in the field to determine the amount of stretch that occurs during application. In some cases plastic or biocomposite films can increase the incidence of *Phytophthora* root disease. Mulch films can also contribute to a higher inter-row concentration of some foliar compounds due to runoff from the film to the inter-row.

Planting material

General requirements

A wide variety of planting material can be used. The more common types are: crowns from the top of the fruit; slips from the peduncle directly below the fruit; hapas, which are borne in the stem–peduncle transition zone; suckers, which originate on the upper part of the stem; and ground suckers, which form on the lower part of the stem with roots that grow into the ground while they are still attached.

Other planting material may be generated by sectioning parts of the plant or by destroying the apical growing point of vegetative plants thus stimulating growth of lateral buds. In addition, plants in the process of flower initiation may be treated with such chemicals as chlorflurenol, which stimulates the production of slips instead of fruit.

Comparisons of various types of planting materials show that early growth rate and plant size at some future time is influenced primarily by the amount of starch reserve in the planting piece, the amount of leaf material present, and the freshness of the piece of material. These variables have rarely been well controlled in field trials comparing various types of planting materials, consequently there remains some controversy over the relative performance of types of planting material. Generally, large propagules with good starch reserves will produce the fastest growth. Some practical aspects of the various types of material are given below.

Regardless of type of planting material, different types should not be mixed even if of the same size (weight). Planting material should be handled carefully to avoid breaking or crushing leaves. Planting material should not be placed in contact with wet soil.

Crowns

Crowns are one of the least expensive forms of planting material. They provide good growth and rooting but are not abundant on farms that produce fresh fruit and sell the fruit with the crown intact. In some cases there might be a sufficient number of crowns coming from fruit rejected to plant a small area. Crowns can be more susceptible to sunburn because the leaves have grown in a more horizontal plane. Crowns can also be more susceptible to burning with solid fertilizers, since their architecture causes salts to more easily reach the white basal tissue.

Crowns can be easily gathered with fruit being harvested for processing although this will rarely provide sufficient crowns to replant the same area as not every planted piece produces a marketable fruit. Crowns can be inexpensive and perform well if planted reasonably fresh but sufficiently cured and protected against heart and butt rot with fungicides. Equipment is available that allows for the treatment of crowns with pesticides at the time of harvest. In some systems, crowns are removed at harvest and transported with the fruit to the processing factory where they are separated from the fruit by forced air.

Crowns are unsuitable as planting material if the growing point has been gouged or if the developing fruit has been treated with a growth regulator to increase fruit weight.

Crown size can vary at different times of the year in subtropical regions due to the large seasonal effects on the fruit size to crown size ratio. In Australia, fruit harvested in winter and spring will be large with a small crown.

Slips and hapas

These shoots are similar in origin in that they develop on the peduncle from the development of lateral buds during fruit initiation. However, they are quite different in shape because slips grow outward and then upward around the base of the fruit, resulting in a sharp curvature at the base of their stem. Hapas emerge well below the fruit and have the same 'arrow' shape as suckers initiated lower on the stem. Slips and hapas usually are left on the plant for a few weeks after fruit harvest and a separate harvesting operation is needed to collect them, so these propagules are more costly than crowns. Not all pineapple varieties develop slips.

Slips are a good source of planting material, but large knobs should be removed before planting, as they may rot in the soil and lead to rot of the stem piece under wet conditions. The performance of slips in terms of months from planting to harvest is intermediate between crowns and suckers, probably because they are intermediate in size between the smaller crowns and larger suckers.

The predominant use of slips as planting material requires a good system of roguing to specifically ensure the collar of slips mutation which produces many slips, does not proliferate.

The production of slips is closely related to the cultivar and environment. In the cultivar MD-2, the number of slips per plant under normal conditions rarely exceeds three. There may be no or only a limited number of slips in very hot environments. Also, slips of the cultivar MD-2 have some susceptibility to the ethephon used to degreen fruit. While this is usually not a major problem, a 50-day waiting period after fruit harvest will allow identification of those that have been induced. On smaller farms, degreening individual fruits can reduce or eliminate the problem.

Slips can be small at the time of fruit harvest and produce only limited growth after fruit harvest with some losses due to detachment. It is best not to use slips if the stem curvature forms an angle close to 90° with the axis of the leaves, as it makes planting difficult.

Hapas are intermediate in size and, like slips, are only present at fruit harvest. They also are prone to induction during the degreening of fruit. Hapas of 'MD-2' are usually planted with suckers.

Ground suckers

Ground suckers have only a very small point of attachment to the mother plant, just below ground. This, coupled with the fact they develop in the shade, could cause the shoot to have low reserves. The lack of reserves can result in poor establishment and if planted with healthy suckers, will seldom keep pace. Ground suckers should never be mixed with plant suckers and usually it is best not to use them. Ground suckers are very likely to carry diseases such as *Pythium* and *Phytophthora*.

Plants of 'MD-2' tend to produce more ground suckers in very warm environments compared to cool environments. Plant sucker number is also lower in very warm environments so there is a tendency to mix the two types of planting material, resulting in increased variability in plant size at the time of forcing.

Suckers

Suckers are grown on the plant for a few to several weeks after fruit harvest, and harvesting is fairly labour-intensive because the suckers must be cut from the plant. Sucker architecture is similar to hapas but they can be much larger and taller. In most tropical areas, suckers are the primary source of planting material because the main cultivars MD-2 and Smooth Cayenne produce few or no slips in tropical environments and many farmers gouge the crowns to increase fruit weight. In such regions, some fields are devoted to sucker production rather than fruit. Suckers can be gathered, stored, and shipped without much loss of vigour, and because they can be large (up to 1.5 kg), they can reach forcing weight in a shorter time than slips or crowns. If large sucker size is a problem, some control over size can be achieved by adjusting planting density, nutrition and the postharvest growth period. Usually one or more suckers are obtained per plant.

Suckers usually have some leaf trimmed off to allow easier planting.

A particular problem with suckers is that they may have undergone floral differentiation before they were harvested, or the shock of harvesting and storing may have induced differentiation. In the case of degreening of fruit, a 50-day waiting period is advised.

Rapid propagation methods

Sectioning

Sectioning is usually used to rapidly multiply a cultivar or to get disease free planting material. In Brazil, sectioning has been used to produce plantlets without contamination by *Fusarium guttiforme* (Cunha and Reinhardt, 2004). Any vegetative shoot of the pineapple plant can be sectioned for propagation if the pieces have two or more axillary buds. For large stems, the leaves are cut off or completely removed (Matos *et al.*, 2009), the stems are mostly quartered lengthwise, and each piece is then cut into sections 3–10 cm in length.

The sections are cured by drying, treated with fungicide and planted in a well-prepared nursery. The soil in the nursery should be fertile at the time sections are planted or the new shoots can be fertilized by foliar feeding with a dilute nutrient solution.

Crowns may be similarly micro-sectioned, producing four or more sections per crown. When such small sections are produced, they are best planted in sterile media and grown in the greenhouse. Small crown sections should be started in a well-drained medium such as coarse sand and must be kept moist. When using a medium of low fertility, foliar feeding should commence soon after the new shoots emerge.

All material produced from sections must be grown to an adequate size for field planting and hardened if they have been grown under shade or greenhouse conditions.

Gouging

Gouging is the mechanical removal of the shoot apex of growing plants, usually with a tool specifically designed for the purpose (Heenkenda, 1993). The young leaves are pulled out prior to gouging. Gouging destroys the plants apical meristem and its dominance, thus releasing axillary buds from their dormancy. These buds form shoots (suckers) that when large enough are harvested and planted in a nursery or field.

Plants as small as 100 g can be gouged but this must be done in a greenhouse with fine tools. When gouging is done in the field, plants should be larger so the gouging can be done by workers from a standing position. The gouge should be shallow, preferably no more than about 1 cm below the apical bud. Treatment of the gouged plants to avoid rot is often necessary. Shoots produced after gouging are allowed to grow to the desired size and their mass at the time of removal depends on the conditions under which they are to be planted. Very small plants, less than 50 g or so, normally would be planted in a greenhouse under shade. Plants weighing 50–100 g could be planted in a propagation bed in the field, while plants weighing 200 g or more are suitable for use in establishing commercial plantings. A well-maintained mother-plant nursery can be productive for 8–12 months, after which it can either be rejuvenated or destroyed and replanted. Rejuvenation involves allowing at least one low-attached shoot to grow until it is of gouging weight. When expanding clones or cultivars, great care must be exercised to keep the area free of pests and diseases as the material is the foundation stock for the farm.

The number of shoots obtained depends on plant vigour, cultivar, environmental conditions and cultural practices: 3–10 suckers weighing 100–400 g have been obtained in different studies and environmental conditions and with different cultivars such as 'Smooth Cayenne' (IRFA, 1984), 'Mauritius' and 'Kew' (Heenkenda, 1993), 'MD-2' (Rebolledo-Martinez *et al.*, 2005) and 'Imperial' (Reinhardt and Nascimento, 2017). More recent studies on gouging of 200–300 g 'Smooth Cayenne' crowns, followed by spraying with gibberellic acid and benzylaminopurine, produced about five 145 g shoots per crown within one year (Coelho *et al.*, 2007). Gouged crowns, inoculated with mycorrhizae fungi, produced 23, 14 and 11 shoots per crown, at least 10 cm long, respectively for the cultivars Pérola, Smooth Cayenne and Jupi (a clone of Pérola) (Santos *et al.*, 2011).

Chlorflurenol propagules

Propagules may be developed by chemical stimulation of the plant during differentiation, so these pieces of planting material resemble crown or slips. Plants are forced following normal farm practice and then treated with a solution of up to 400 ppm chlorflurenol, a plant growth regulator from the morphactin group, in 3000 l ha^{-1}

within a week after forcing (Py *et al.*, 1987). The time interval might be shorter in warm environments where plant development is quickest. Propagules develop in place of the flowers and fruitlets, so the fruit becomes unusable after this treatment. With proper timing and chlorflurenol concentration, up to ten 200 g propagules can be produced on each plant (Glennie, 1981). Propagules should be carefully graded by size before planting, and plants from different production areas should not be mixed if field uniformity is to be maintained. Propagules are an important type of planting material in fresh fruit operations where crowns are sold with the fruit and slip or sucker harvesting is too expensive or inadequate to fulfil planting material requirements. They can also be scheduled to be ready at strategic planting times when crowns are not necessarily available.

Tissue culture

Propagation by tissue culture is discussed in detail in Chapter 5.

Treatments before planting

All planting material is subjected to a variety of treatments prior to planting, including curing, bundling, transportation from the field or growing area, grading, storing, dipping, transport to the planting field and spreading.

Planting material size and uniformity is particularly important. Within any category of planting material, large pieces will usually grow to forcing weight sooner than smaller pieces generated in the same area. For this reason, grading of planting material by size is critical to provide uniform plants at forcing and efficiencies at the time of harvest.

Crowns, slips, hapas, suckers and propagules are very tolerant of storage, and may be stored for months between harvest and planting. However, stored plants continue to grow, stem diameter decreases and storage reserves are consumed, so prolonged storage slows initial growth and increases variability (Py *et al.*, 1987). Therefore, fresh planting material usually grows faster than stored material. However, very fresh material should be treated with a fungicide to avoid black rot caused by *Chalara paradoxa*, especially in humid seasons.

Storage of planting material should be under conditions that allow for good air movement to avoid rot. Storage in piles or other conditions that block exposure to light, should be kept to a minimum as stem etiolation will occur if the material is deprived of light for more than 2 weeks. In Australia, planting material is stored either on plants in the outside rows of fields from where they were harvested if storage is intended for several weeks or in wooden slatted bins if for short periods.

Where planting is done manually, the planting material should be distributed in the field to facilitate the work of planters. Oversupply should be kept to a minimum as additional handling can damage the planting material and it will not perform as well as material only spread once.

Planting

Planting density

Many factors are involved in the determination of planting density, and as a result, densities may vary from as low as 25,000 plants ha^{-1} (10,100 plants a^{-1}) in 'Pérola' fields of Tocantin State in Brazil (Cunha, 1999) to as high as 120,000 plants ha^{-1} (48,600 plants a^{-1}) in 'Queen' fields in Réunion Island (Fournier *et al.*, 1995). Both the cultivar used and the environmental conditions influence plant growth and the competition between plants for available resources. Marketing is an important consideration for the grower because plant growth rates and plant size at the time of forcing influence fruit weight and fruit size distribution. Densities are selected to manage these relationships. Generally, higher densities produce a greater plant mass and yield but smaller fruit. Higher densities can also result in a slightly longer time to maturity, a greater spread of fruit maturities, less sunburn, a greater incidence of blackheart in some climates (Christensen, 1995), less natural initiation of flowering (NI), and fewer suckers and slips per plant but more per hectare (Scott, 1992; Santana *et al.*, 2001). In most studies, fruit quality has not been changed by increasing densities, but in some cases there was a tendency towards an increase in total acids content and, to a less extent, a decrease of soluble solids content, resulting in a smaller ratio, and a lower juice content (Chadha *et al.*, 1974; Mustaffa, 1988; Souza *et al.*, 2009).

Sunlight intensity and duration determine the amount of energy available for photosynthesis. Areas with long periods of cloud-free days will support greater carbohydrate assimilation than will cloudy areas. Seasonal variations will be more accentuated in those growing areas farther away from the equator. Where seasonal changes are large, density changes on a seasonal basis may need to be considered for optimum production (Wassman, 1990). That is to say, in subtropical regions away from the equator, different densities might be necessary depending on the time of year of planting. The effect of higher planting density on plant growth can be expected to be greater further away from the equator where there is a lower sun angle and reduced light penetration through the canopy. In subtropical regions in Australia where temperature is a major consideration, highest planting densities are used where planting is done in spring and, as there is a longer period favourable to plant growth before the following winter, plant growth is therefore rapid and fruit size is too large for efficient processing if not planted at a higher density.

Experiments conducted on the effect of density on average fruit weight and yield have shown quite predictable results. Average fruit weight decreases linearly with increasing density, but the effect is variety- and site-specific (Fig. 7.7). For 'Smooth Cayenne' in Hawaii, fruit weight decreases by 2.4% per 2470 increase in plants per hectare over a wide range of densities. Total yield per unit land area generally increases curvilineal with increasing density (Fig. 7.8), though some studies show that yields decline at very high plant populations. Santana *et al.*. (2001) observed a decline by 102 g in fruit mass and an increase of productivity of 8.3 t per hectare for each increase by 10,000 plants per hectare for cv. Smooth Cayenne in a subhumid tropical climate of coastal tableland in North Bahia, Brazil. Other studies on 'Smooth Cayenne' in Hawaii have shown a 45.4 g reduction in fruit weight for each 400 plants per ha increase above 7000 plants per ha (Sanford, 1964). Each 5 cm increase in spacing between

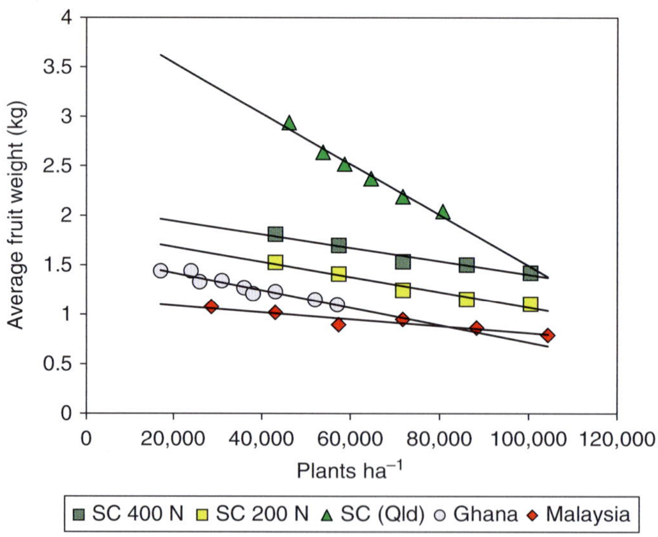

Fig. 7.7. Effect of plant population density on average pineapple fruit weight of 'Queen' grown in Ghana (Norman, 1978), 'Singapore Spanish' grown in Malaysia (Wee, 1969) and 'Smooth Cayenne' grown in Queensland, Australia (Scott, 1992) and Swaziland (Dodson, 1968). Regressions are 'Queen', $y=9\times10^{-6}x+1.59$, $R^2=0.93$, $n=9$; 'Singapore Spanish', $y=-3\times10^{-6}x+1.16$, $R^2=0.835$, $n=6$; Queensland 'Smooth Cayenne', $y=-3\times10^{-5}x+4.05$, $R^2=0.98$; Swaziland 'Smooth Cayenne', 400N, $y=-6.8\times10^{-6}x+2.08$, $R^2=0.95$; 200N, $y-7.6\times10^{-6}x+1.84$, $R^2=0.96$, $n=5$; where y is estimated fruit weight in kg and x is plants ha^{-1}.

20 and 40 cm, increases the fruit weight about 5%, and reduces the number of slips on the plant. The nearer that the forcing is done to the natural differentiation, the larger the plant and the larger the fruit produced. Smaller fruit thus can be obtained by forcing earlier. The optimum density for a given farm or region and variety must be determined by the available technology, environmental resources and market requirements. Manual access to fields requires lower density or leaf trimming.

The optimum planting density will give adequate plant mass at time of induction. Ratios of fruit mass at harvest to plant mass at forcing (harvest index) have been measured in various regions where pineapple is grown. The harvest index can be as low as 0.4 in areas with warm nights and fast growth rates, and can exceed 1.0 where irradiance is high and night temperatures are cool (and hence respiration low). Where growing medium and nutrition were controlled in a variety of climactic conditions (Hepton *et al.*, 1993), stem dry matter content as a percentage of total plant dry matter and harvest index decreased as average night temperature increased (Fig. 7.9). Thus, stem dry matter content has a strong influence on the harvest index of ‘Smooth Cayenne’ (Fig. 7.10) and ‘MD-2’.

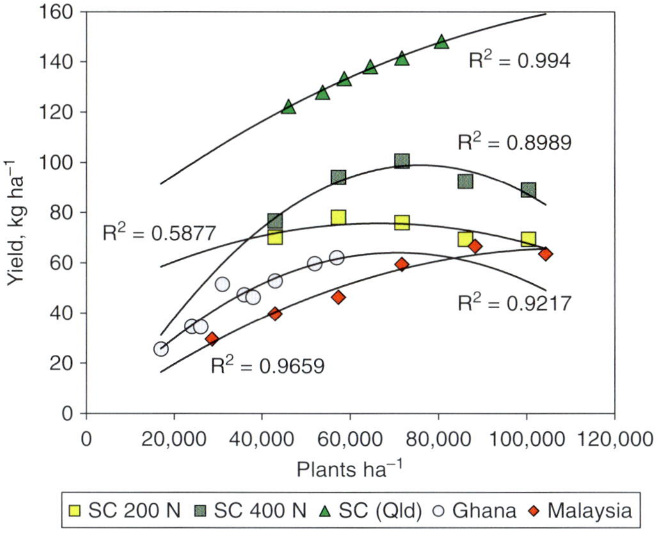

Fig. 7.8. Effects of plant population density on pineapple fruit yield as shown by the cited authors or calculated from their average fruit weight data. Data are for ‘Queen’ grown in Ghana (Norman, 1978), ‘Singapore Spanish’ grown in Malaysia (Wee, 1969) and ‘Smooth Cayenne’ grown in Queensland, Australia (Scott, 1992) and Swaziland (Dodson, 1968).

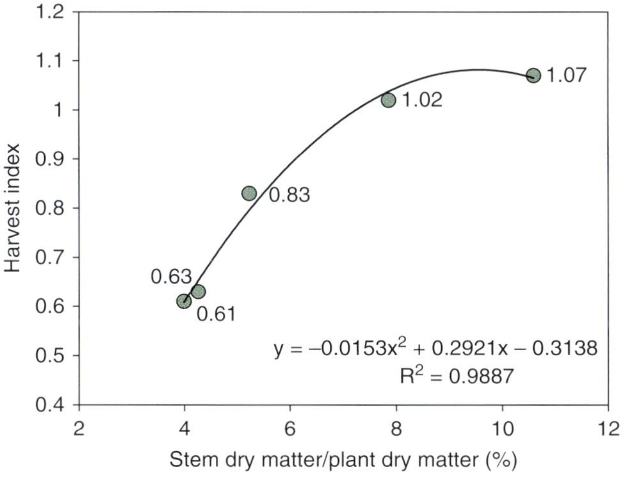

Fig. 7.9. Effect of average night temperature on stem dry matter content (DM) as a percentage of total dry matter and on fruit harvest index calculated from plant fresh weight at forcing and fruit weight at harvest. Redrawn from Hepton *et al.*, 1993.

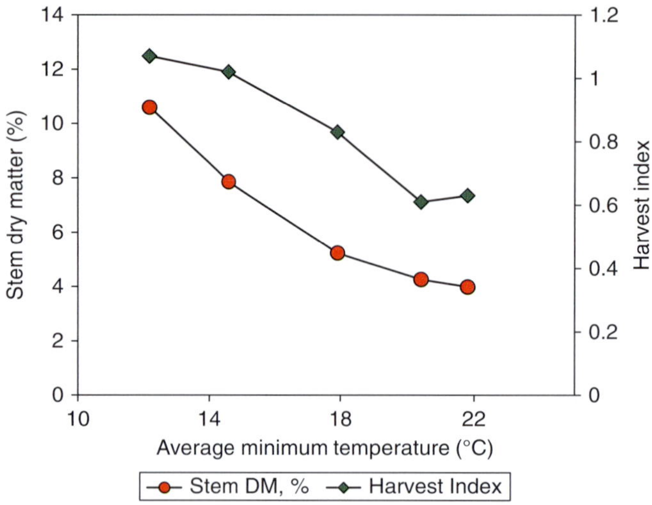

Fig. 7.10. Effect of stem dry matter content on harvest index calculated from plant fresh weight at forcing and fruit weight. Redrawn from Hepton *et al.*, 1993.

These factors should be taken into consideration when selecting a planting density.

The density used generally is determined by the cultivar or clone being grown, the intensity of cultural practices used, and the planned use of the fruit. Planting densities of less than about 50,000 plants ha^{-1} are common in market gardens or where cultivars with long, spiny leaves, e.g. 'Española Roja' and 'Pérola', are grown (Py *et al.*, 1987; Pádua *et al.*, 2013). Under low rainfall and soil fertility conditions of family agriculture and small volume producers in Brazil, 'Smooth Cayenne' pineapples are cultivated at densities of about 40,000 plants ha^{-1} (Reinhardt, 2010), but in most other countries densities for 'Smooth Cayenne' are typically between 50,000 and 75,000 plants ha^{-1} with lower densities being used where fruit are destined for the processor or to be ratooned, while higher densities are more common where fruit is to be sold fresh or where no ratoon crop is planned. Jantee *et al.* (2014) suggested planting densities of 46,900 plants ha^{-1} for a Thai 'Queen' clone and 75,000 plants ha^{-1} for a Thai 'Smooth Cayenne' clone. Pádua *et al.* (2013) recommended 35,700 and 51,200 plants ha^{-1} respectively for cv. 'Pérola' fresh fruit production under rain-fed and irrigated conditions. Leon and Kellon (2012) used a grower survey to establish that planting densities above 70,000 plants ha^{-1} produced the highest yields with acceptable fruit size in 'MD-2' in Costa Rica.

Plant arrangement

Plants can be arranged in the field in single rows or in beds of two, three or four rows. The spacing between plants in the row and between rows is determined by the desired plant population density, type of planting material, planting methods, use of plastic mulch, use of raised beds, equipment available, methods of plant fertilization, and the methods of harvesting. One of the more usual arrangements involves two-row beds with an adequately wide walk space between the beds to allow for field activities, particularly harvesting although in high-density fields workers sometimes walk on top of beds. In this scenario, the distance between the plants within a line, which should not be less than about 20 cm (about 8″), is less than the distance between the lines in the bed.

Row direction has a greater impact on plant growth at higher latitudes and away from the equator and in those regions rows are best run in a north:south orientation where slope permits. In these regions the effect of row orientation is greater on slopes facing away from the sun. In the northern hemisphere these will be northern slopes and in the southern hemisphere will be southern slopes.

Beds can be configured to have long or short rows (Fig. 7.4). Long rows are more efficient for machinery but inter-rows must carry a greater volume of water. When using long rows, it is necessary to have shallow cross drains every 30 m or so to slow down the velocity of running water and avoid soil erosion. These cross drains end in soil catchment ponds (siltation or sediment ponds), one per suitable area depending on rainfall.

Planting depth

An important aspect of planting is obtaining good contact between the planting material and the soil. Where labour is not expensive and risk of *Fusarium* infection is low, basal leaves and knobs on slips can be removed by hand to expose adventitious roots, facilitating good contact with the soil. Good soil tilth and plant setting also contribute to an early and uniform plant start. Placing planting material sufficiently deep into the soil without getting soil into the plant centre is one of the best ways of assuring a good plant start.

Crowns, propagules and plantlets are most sensitive to deep planting, but even those can have up to half of their height buried at planting without adverse effect. Protection of these planting materials against heart rot is necessary in soils with poor drainage, and where soil pH is above 5.5 and thus conducive to the growth of water moulds.

Slips, hapas and suckers may be planted to depths of 10–15 cm (4–6″) without adverse effect. Also, no benefits will be obtained from the removal of basal leaves if the soil has a fine tilth and adequate provisions are made to move the soil, for example with irrigation, into the lower leaf axils.

Mechanical planting

Hand planting is still common where labour is inexpensive or farms small (Fig. 7.11). Driven by the high cost of labour in some pineapple-growing areas, a variety of planters have been developed to plant pineapples (Figs 7.12 and 7.13). These machines usually work best in fields that do not require mulch although some can be used through mulch film. There are several versions of these machines depending on their method of plant spacing and the level of automation. Some two-row machines space planting holes using reciprocating knives, while two people seated behind the machine each take a plant from an overhead hopper and place it in

Fig. 7.11. Hand planting of 'Pérola' in Brazil. Photo D. Reinhardt.

Fig. 7.12. Mechanical planting in Malaysia. Photo D. Reinhardt.

Fig. 7.13. Planter with cam-timed reciprocating knives driven by ground-contact wheels to achieve appropriate plant spacing. Small press wheels are on either side of the two planting lines. Photo G. Sanewski.

the holes. Other machines involve placing the planting material in a holder that will deposit it into a prepared hole or furrow. Planting material may be set by compression wheels that follow immediately behind or by rolling plants after planting. Some planting machines are self-propelled. The quality of planting is usually as good as hand planting, but machines that automate the planting process more can damage the growing point of the planting material. Large suckers are more suited to hand planting.

Attempts to plant through mulch using machines have been marginally successful. Fumigation is usually done at the time the mulch film is laid, making the machine environment undesirable for additional workers without suitable protective clothing.

Setting

To obtain good root initiation, uniform levels of moisture and aeration must be developed in the rooting zone. Too much moisture without aeration leads to rot, while too little moisture leads to uneven root initiation and variability in start of growth.

Good tilth is a prerequisite for good rooting, as it allows fine soil particles to be set around the base of the planting material. Soil can be moved by hand, with the planting tool, by foot (often the heel), by rolling, or with irrigation. Good soil tilth is achieved before planting by adequate soil preparation when the soil is at the correct soil moisture to avoid the formation of clods.

Setting with water has a dual effect. Water moves the soil particles around the base of the planting material, but in addition, provides water for subsequent plant growth. In arid growing areas, this setting irrigation must be supplemented by timely additional rounds of irrigation to sustain the root growth.

Volumes of water used in setting irrigation depend on the method of application, the soil moisture levels at the time of application, the use of mulch material, and soil conditions. A usual quantity would be about 5 cm (2″) by direct overhead application, or somewhat less, if by drip irrigation. Drip irrigation does not move soil particles to any degree.

Planting material without a substantial quantity of dried leaves around the base will be easier to set well.

Water Requirements and Irrigation

Pineapple has a much lower transpiration rate than most other crops and hence a lower requirement for frequent irrigation. The crop coefficient (K value) for pineapple has been estimated at 0.36 (Hanafi *et al.*, 2010a).

In addition to the plants need, irrigation is also useful for establishing newly planted plants and overhead irrigation can be useful for cooling fruiting plants and minimizing sunburn.

The pineapple plant is well adapted to drought and capable of a productive potential in dry soil. Pineapple roots survive drought due to suberization of the root epidermis, shrinkage of the root cortex, cavitation in the xylem vessels and the regulation of trans-membrane water flow. While these strategies enhance survival, they limit growth. However, plants that are grown under ample moisture in well-aerated soils should have well-distributed roots with numerous white root tips.

While much of the world's pineapple is grown without irrigation, long periods of inadequate rainfall seriously compromise both plant growth and fruit development. In dry areas, irrigation is a necessity if predictable yields are to be obtained throughout the year. Uniform and continuous plant growth allows plant size targets and hence flower induction and harvesting targets to be met. Irrigation has been shown to increase yield, increase SSC, and reduce acidity (Souza *et al.*, 2002; Souza *et al.*, 2012). Irrigation has also been shown to at least partially negate the deleterious effects of mealybug wilt-associated virus on yield in the first ratoon crop (Sether and Hu, 2001).

Drip irrigation beneath film mulch can also be used to apply some pesticides (Schneider *et al.*, 1995) although, for safety reasons, this is generally not recommended.

Irrigation was shown to decrease shell firmness in ‘Smooth Cayenne’ (Souza *et al.*, 2012).

Types of irrigation

Types of irrigation include drip irrigation, mini-sprinklers on poly-pipe, aluminium pipes with mini-sprinklers, single jet sprinkler gun on flexible hosing and travelling irrigators such as centre-pivot style (Fig. 7.14). The choice will depend on resources available, slope of land, size of plantation, level of management and water quality.

Overhead irrigation

In Hawaii, overhead booms were found to be the most effective way to saturate dry soil and initiate root growth after planting by using about 2.5–5 cm of water (1–2″) of water. An application of approximately 12 mm (0.5″) immediately after fumigation can be useful for sealing the chemical in the soil. The same overhead booms could be used to irrigate growing plants. The booms should only apply water at a rate that could be absorbed by the soil with each pass. On the island of Lanai, Hawaii pineapple has been planted through mulch in depressed beds, which has helped direct the water to the roots. Other forms of overhead water application have been used with good effect but the irrigation efficiency (water applied/ water available to the crop) is generally not as high as it is with the overhead boom. In Australia, self-propelled, centre-pivot irrigation has been used to irrigate pineapple where volumes of 2–3 megalitres (Ml)/ha/year have been applied. Overhead irrigation should be avoided during flowering to avoid encouraging fruit disease.

Drip irrigation

While the high initial investment associated with drip irrigation may appear prohibitive for pineapple culture, in areas where water is limited or expensive and the need for irrigation may extend for many months, the investment might be worthwhile. Drip irrigation allows precise placement of water in the root zone in volumes that match the crop's need. Drip irrigation is especially efficient during the first few months after planting when a consistent but low-volume supply of water is needed for the developing root system.

Fig. 7.14. Centre-pivot irrigation in Australia. Photo S. Newett.

Drip irrigation does however have limitations and risks. Drip irrigation cannot be used for setting plants after planting. Also, in areas with low or no rainfall for several weeks at a time, drip lines can be damaged by rodents in search of water. This can be a major problem in pineapple because of the potentially dry environments where it is grown and the long duration of the crop. Blockages and breaks can be difficult to detect. Drip irrigation requires a high level of management.

Drip irrigation systems should include a filtration system if particulate matter that can clog emitters is present, chlorination to prevent microbial growth in the drip tubes, a system of in-field distribution pipes that bring water at the correct volumes and pressures to various parts of the fields, and the emitter tubes, which are usually buried just below the soil surface between two lines of plants.

The emitter tubes are usually custom-made for pineapple to allow for a slow release of relatively low volumes of water. If the soil has good hydraulic conductivity, the slow release results in wider spreading of the water in the root zone depending on the soil type. Flow rates will depend on orifice spacing, orifice diameter, and line pressure and these variables will depend on the soil type and cost and availability of water. The length of run for emitter tubes will depend on the slope of the terrain, but should be limited to lengths that will result in output variations along the tubes of no more than 10%. As with overhead irrigation, scheduling includes a consideration of rainfall, with the younger plants taking precedence over the more established ones.

Drip irrigation may be used to deliver water soluble fertilizers, nematicides and fungicides to the root zone. However drip irrigation does not guarantee the desired effects from these materials as their dispersal (especially in sandy soils) and reaction in the soil are largely uncontrolled and may not be as well understood when compared with the standard methods of application.

Oxygenation (12% air/water by volume) of drip irrigation water was shown to produce an increase in fruit yield, a slight reduction in *Phytophthora* incidence and a large increase in soil respiration (Chen *et al.*, 2010; Dhungel *et al.*, 2011).

Volumes to apply and timing

Various published papers collectively suggest different cultivars might differ in the degree to which they express CAM indicating potential cultivar differences in water use efficiency. Likewise there are various estimates of annual water use. It appears pineapple generally requires a minimum monthly rainfall of at least 50 mm (Carr, 2012) but preferably 75 mm depending on the environment. The use of plastic or organic mulch will lessen the need for irrigation. Hot, dry climates will increase the need. Drip irrigation relative to pan evapotranspiration ranged from 0.4 to 0.9 (Ojeda *et al.*, 2012; Patra *et al.*, 2015). Souza *et al.* (2012) reports 0.5 of evapotranspiration provides the greatest efficiency of water use in 'Smooth Cayenne'. Estimates of volumes applied over the life of a crop include 587 mm (Hanafi *et al.*, 2010b) and 596 mm (Souza *et al.*, 2002). Drip irrigation was used in Australia to supply 1.09–1.24 Ml/ha during critical times while 19.63 Ml/ha was supplied as rainfall (Dhungel *et al.*, 2011).

Soil and leaf moisture readings can be used to develop guidelines for irrigation scheduling. Once the guidelines are established, rainfall can be considered in the scheduling. Because all areas cannot be irrigated simultaneously on a large plantation, irrigation is usually scheduled to begin before it is actually needed in the most sensitive areas. These are usually young plant-crop fields which have the highest evapotranspiration. Irrigation is completed after the ideal time in the older ratoon fields, where the crop canopy minimizes soil drying and associated moisture loss.

Most evapotranspiration in un-mulched pineapple fields occurs in the first few weeks and is comprised mostly of loss from the soil (Hanafi *et al.*, 2010a). Daily evaporation can be used in conjunction with available soil moisture to estimate the interval between applications. Information about the type of soil and its maximum water-holding capacity is useful.

It is important to avoid water deficit in the periods 6–10 weeks prior to flower induction to minimize natural induction (NI) and, in an extreme case, to ensure plants are receptive to artificial induction, whereas a good water supply in the period after flowering until 6 weeks prior to harvest is important to ensure good fruit size.

The pineapple leaf is a major water storage organ for the pineapple plant. The cells of the upper mesophyll layer can occupy almost 50% of the leaf volume when full of water. Under conditions of ample water this mesophyll layer is supplied with water to expand to its full thickness. Under conditions of limited water supply, the water in the mesophyll layer is withdrawn by the plant. The subsequent shrinkage of the leaf thickness can be used as an indicator of water deficiency. In practice, representative D-leaves are pulled, nested together in a group of 6–10 leaves and the total thickness is measured and compared with the thickness of fully turgid leaves. A shortcoming of this method is that it does not take into account the viability of the root system and its ability to extract moisture from the soil. Roots infested with nematodes or symphylids, damaged by grubs, or diseased with fungal pathogens may result in leaves showing moisture deficit while soil moisture is sufficient for a healthy plant.

Crop Scheduling

The profitability and viability of large pineapple farms is dependent on strategic scheduling of crops to optimize yield, make efficient use of labour and land and satisfy marketing arrangements. This requires careful planning, appropriate choice of planting material, and monitoring of fields to allow forward projection of harvesting and marketing requirements. Expected induction dates and corresponding plant size can be used to predict harvest dates and expected yield. This can then allow marketing strategies to be devised. Scheduling on large farms requires well-designed spreadsheet or database systems and a reliable system of monitoring. This information forms a record for each field which, when sufficient data is accumulated, can then be used to reveal individual field problems, or, when used on a whole farm basis, provide measures of farm efficiency. This type of software can guide growers on the use of best practices. It is also useful for meeting the traceability needs of retailers and consumers.

Efficient management of an established farm should result in the harvest of approximately the same number of boxes of fruit each year. To achieve this, it is necessary to control the length of each stage of the crop cycle. Environmental factors such as temperature can however have a large impact on the length of some stages of the growing cycle. Among the controllable factors are the date of planting of each block and field, and the date of the commencement of spraying. For small farms using low control levels, the final date of planting of a block is the starting date for future field operations such as spraying. For large farms using low control levels, blocks would be grouped into a larger unit (field) because it is not operationally efficient to make spray applications on the basis of a single block. The starting date for all field operations such as spraying would therefore be the day blocks were grouped to form a field. In low-control farms there is no pressure to do things efficiently and the date of planting is the day planting was completed.

For farms that use high control levels, the planting date is defined as the weighted average date (weighted according to the number of plants planted on each occasion) the blocks in the field were planted. Once the planting of a block is commenced, there must be strict control over the date of completion to maintain the cultivation cycle.

Some farms use software that automatically calculates the weighted age of the block at the completion of planting. This type of software can guide growers on the use of best practices and help improve efficiency and is also useful for meeting the traceability needs of retailers and consumers.

On a high control level farm, a delay at any stage represents a direct loss equivalent to the delayed area in the current cycle. With few exceptions, these delays cannot be compensated for operationally in the future because the budget is based on the annual harvest goal. Any reduction in planted or harvested area results in a negative cash flow equivalent to the unharvested area for that year. It is extremely important therefore to know the approximate duration of each stage (Table 7.1) in the agroecosystem where the farm is located and use this information for scheduling.

The duration of the stages 'planting to forcing' and 'forcing to harvest' and to a lesser extent, 'sucker production to knockdown' (Table 7.1) depend on temperature and other

Table 7.1. Costa Rican industry standard for managing the various stages of pineapple fruit production. Each block could be used for the production of a plant or ratoon crop, but not both.

Stage	Target duration (days)	Maximum duration (days)	Interval between stages
Planting plant crop (start to finish planting)	7	15	Planting to forcing is 240 days
Forcing plant crop (start to finish forcing)	2	2	Forcing to harvest is 150 days
Harvest plant crop (start to finish harvesting)	7	15	
Sucker production (start of grouping until consolidated group)	7	15	Plant block to knockdown is 240 days max. for block. Weighted average is usually 90–120 days
Forcing ratoon crop (finish forcing)	2	2	Establishment of ratoon crop to forcing is 210 days
Harvest ratoon crop (start to finish harvesting)	7	15	Forcing to harvest is 150 days
Knockdown (start knockdown to following soil activity)	25	30	Total duration soil preparation is 120 days

environmental factors. However, the duration of the other stages depends 100% on timely decisions and keeping good records. To keep to schedule, it is helpful to define a production budget in terms of area and times. When the time from planting to forcing and forcing to harvest is known, a production plan can be devised (Table 7.2). The weekly planting budget, and hence forcing budget, depends on the area of the farm and seasonal constraints and hence have not been stipulated in this example.

From Table 7.2, what is planted in week 49 of 2016 will be forced in week 31 of 2017 (planting to forcing is 34 weeks in this environment) and harvested in week 1 of year 2018 (forcing to harvest is 22 weeks in this environment). Also note that the area forced may be different from the area planted in that week. This is normal, due to different growth rates and/or different weights of the planting material, but the forced area will always be equal to the harvested area. At the end of week 52 of 2018, the entire area harvested should be equal to the area planted and forced. A similar table should be developed for the establishment of ratoon crops.

The objective is to obtain an annual production of boxes stable year by year that is compatible with the cultivation practices and climate.

When commencing a new pineapple farm, the faster the area is planted (assuming planting material is obtained externally), the greater will be the volume of pineapple boxes obtained in that first year. In subsequent years, the cycle length and the production of planting material will effectively reduce the annual harvested area compared to the first year, and cultivation processes will be slower due to the need to eliminate the previous crop.

Even though less complex, crop scheduling is also very worthwhile for smaller production areas. Growers will obtain higher economic returns, a better use of labour and a more even market supply if they are able to plan strategic combinations of planting dates, manage planting material sizes and target flower-forcing dates which take into account environmental factors that may favour NI.

Fertilization and Pest, Disease and Weed Control

Relevant topics are covered in Chapters 8, 9 and 12.

Table 7.2. Production plan for a pineapple farm using 'MD-2' and a high level of technology for fresh fruit production on a year-round basis (Costa Rica).

Planting year	Planting week	Planting budget (ha)	Forcing year	Forcing week	Forcing budget (ha)	Harvest year	Harvest week
2016	49	A	2017	31	A +/–	2018	1
2016	50	B	2017	32	B +/–	2018	2
2016	51	C	2017	33	C +/–	2018	3
2016	52	D	2017	34	D +/–	2018	4
2017	1	E	2017	35	E +/–	2018	5
2017	2	F	2017	36	F +/–	2018	6
2017	3	G	2017	37	G +/–	2018	7
2017	17	H	2017	51	H +/–	2018	21
2017	18	I	2017	52	I +/–	2018	22
2017	19	J	2018	1	J+/–	2018	23
2017	20	K	2018	2	K +/–	2018	24
2017	21	L	2018	3	L +/–	2018	25
2017	22	M	2018	4	M +/–	2018	26
2017	47	N	2018	29	N +/–	2018	51
2017	48	O	2018	30	O +/–	2018	52

Depending on the size of the farm and degree of mechanization different systems are used. Large, highly mechanized farms apply most fertilizers and agrochemicals using boom sprayers (Fig. 7.15). Small farms use either backpack sprayers or a variety of motorized aids (Fig. 7.16).

Other Cultural Practices

Slip removal

Some varieties and some clones of 'Smooth Cayenne' have a large number of slips. Though useful as planting material, if left on the plant through fruit development, they can severely reduce fruit yield. Slip numbers tend to be, first, a function of variety or clone, and second, of environmental and forcing conditions. Plants forced with naphthalene acetic acid (SNA) tend to have fewer slips than those forced with ethylene. This difference is attributed to the ability of ethylene to promote floral differentiation of lateral buds as well as the apical meristem. Each lateral bud, once differentiated, becomes a miniature fruit with a crown, i.e. a slip. However, in Brazil, plants forced with ethephon tend to give fewer slips than those forced with calcium carbide (Reinhardt and Cunha, 1982). To minimize the adverse effects of slips on fruit yield, they should be removed as soon as it can be accomplished without damaging the peduncle of the developing fruit. Slips thinning in the week after closure of the last flowers increased the weight of remaining slips, crown and fruit of 'Pérola' pineapples by up to 16% (Lima *et al.*, 2001, 2002).

The effect of slip removal might however be influenced by genotype. Hotegni *et al.* (2015) found slip removal did not improve fruit size, weight, quality or uniformity in the cultivar Sugarloaf (presumed 'Queen' or 'Pérola').

Sucker removal and trimming

The development of suckers appears to be a function of variety and plant vigour. Some varieties, e.g. 'Queen', produce numerous suckers

Fig. 7.15. Applying nutrients by boom sprayer. Photo S. Newett.

Fig. 7.16. Small tractor sprayer in Brazil. Photo D. Reinhardt.

that arise from leaf axils at or above ground level. High numbers of suckers makes ratooning very difficult because ideally the plant should have one large sucker to replace the mother plant, thereby retaining the original planting density.

'Smooth Cayenne' sucker numbers are influenced by planting density, plant weight at the time of forcing, and nutritional status. Environmental factors that result in high carbohydrate accumulation in the stem of the plant-crop plant, will increase the number of sucker buds that develop. Plants located along border rows which receive more sunlight, usually produce suckers earlier and in larger numbers.

Large plants in a population will tend to develop one or more suckers early during fruit development, while sucker development on smaller plants may be significantly delayed. To provide uniformity in the ratoon crop, large suckers may be removed, a standard practice in some fields in Thailand. Additional suckers will develop from the axillary buds that are already present and grow to produce the ratoon crop. Sucker trimming improves the uniformity of ratoon crops, but it will increase the time required to reach the pre-determined average sucker weight at forcing. Trimming of suckers can also increase first ratoon fruit weight but reduce yield per area per unit of time (Sanford, 1964). Consequently, the decision to trim suckers will depend on the purpose for which the crop is being grown, the projected harvest date, and in some areas, the potential for using the trimmed suckers as a source of planting material.

Because sucker trimming is labour-intensive, it is generally practised in growing areas where the cost of removal is offset by the increased value of the larger and more uniform ratoon fruits and the value of suckers as planting material. The value of an earlier harvest from fields planted with suckers must also be taken into consideration.

Crown gouging

When fruit develop under conditions that favour vegetative growth, there is often a significant growth spurt to the crown in the final weeks of fruit development. If the apical growing point of the crown is removed or damaged before this growth spurt occurs, usually 10–12 weeks before harvest, an increase in fruit weight of 6–10% can be obtained.

Crown gouging is not an option in areas where crowns are a primary source of planting material. Also, crowns should not be gouged in fields where high fruit nitrates could be a problem. The continued growth of the crown assures movement of nitrates through the fruit to the expanding crown leaves. If this process is stopped by gouging, nitrates will accumulate in the upper portion of the fruit. As noted previously, high fruit nitrate levels can cause corrosion in metal cans used for processed fruit.

Crown removal

As with crown gouging, crowns may be removed to promote fruit expansion (Sanford, 1964). In multicrowned fruit, removal of all but one crown with a sharp knife will improve fresh fruit marketability. When a crown or crowns are to be removed, the practice should occur earlier than for crown gouging and crowns should be sufficiently well developed to allow for easy removal without damage to the fruit. Crown removal is labour-intensive and may not be practical in all areas, but when it is performed by practised workers, the scar tissue is barely visible when the remaining crown is fully developed. If the crown has grown large enough to shade the shoulders of the fruit, it may be better to retain the crown to prevent fruit sunburn. This problem is serious in areas that experience intense sunlight.

Leaf pruning (Trimming)

In some growing areas, leaves are pruned to facilitate entry into the field for various activities, particularly harvesting. All evaluations of leaf pruning have shown that yields are reduced proportionate to the amount of leaf removed, and the earlier the removal takes place, the greater the impact (Fig. 7.17). Leaf removal also has an adverse effect on the growth of both slips and suckers. The reason for this dramatic impact is twofold: first and most obvious is the reduction in the supply of metabolites from the leaf

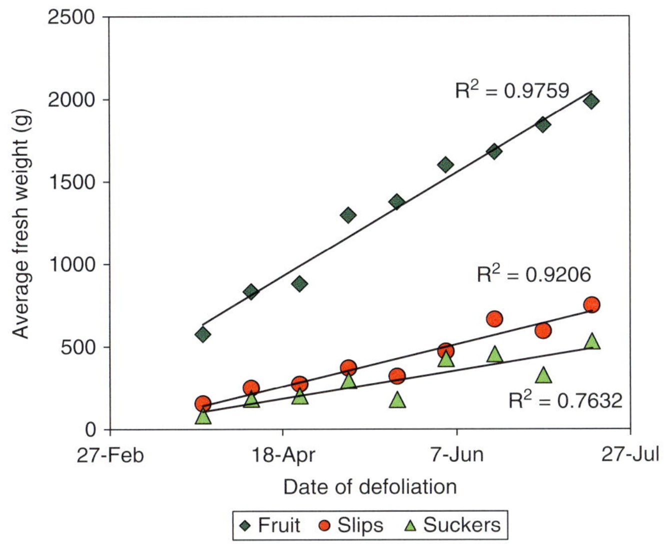

Fig. 7.17. Effect of date of defoliation on average fresh weight of fruit ($y=635.4+12.57x$, $R^2=0.98$), slips ($y=141.17+5.11x$, $R^2=0.92$), and suckers ($y=106.5+3.42x$, $R^2=0.76$) of 'Smooth Cayenne' pineapple in Hawaii; n = 9 for all. From A. Hepton, unpublished results.

canopy; and second is the loss of dry matter (non-structural carbohydrates) already accumulated in the leaves, which would be available for fruit development (King, 1935).

In 'Pérola' pineapple fields of a semi-arid region in Northeast Brazil, the production of at least one ratoon crop was a common practice in the past. After plant crop harvest, plant leaves were pruned to stimulate sucker emission and the suckers arising lower on the stem were selected to form a ratoon crop with lower risk of losses due to fruit lodging. This was, however, shown not to be a viable practice, as it did not favour yield and fruit quality (Alves *et al.*, 1998).

Management of sucker production for planting material

The production of planting material is as important as the production of fruit. It is important to have a system that ensures uniform pieces in terms of vigour, weight, size, type and sanitation are always available and that the propagation area is the minimum area required so the greatest amount of net area is dedicated to the production of fruit. Because pineapple production is highly scheduled, it is necessary to calculate the quantity of suckers that can be harvested per plant per month, depending on the environment. Examples are presented below, based on a 300 ha farm in a cool environment (Table 7.3a) and a 300 ha farm in a warm environment (Table 7.4a).

Tables 7.3a and 7.4a demonstrate that the time for each operation varies with the environment. The environment determines not only the possibility of producing a ratoon crop, but also the productivity. This is because the capacity of the plant to store assimilate in the stem during fruit development, and hence be available for rapid sucker growth, is highly dependent on the environment. Storage of assimilate in the stem only occurs after the requirement of the fruit, the strongest sink, is met. In the warmer environment, sucker development is greatly delayed, making growing a ratoon crop difficult if not impossible. This is because respiration is higher in a warm environment and plant growth rapid, leaving less assimilate available for storage or sucker growth.

Developing a sucker production programme

For logistical reasons, it is necessary to correctly calculate the area of suckers to be harvested weekly. For the farm shown in the cool environment (Table 7.3b), the total area of suckers that must be collected to meet the monthly planting target is 26.7 ha, which equates to 6.68 ha per week. This means that plants sufficient to produce enough suckers to plant 6.68 ha should be established every week. However, if the cycle of

Table 7.3a. The cycle of 'MD-2' pineapple cultivation for a 300-ha farm in a cool environment with mean temperatures around 26°C and a day/night differential of approximately 5°C.

Stage	Time in months	Area (ha)	% of 300 ha
Soil preparation	4	35.24	12%
Planting	0.5	4.41	1%
Plant crop development to induction	8	70.49	23%
Plant crop fruit development	5	44.05	15%
Plant crop harvest	0.25	2.20	1%
Ratoon crop plant development	8	70.49	23%
Ratoon cop fruit development	5	44.05	15%
Ratoon crop harvest	0.25	2.20	1%
Sucker production	3.0	26.86	9%
Total	34.0	300.00	100%

Table 7.4a. The cycle of 'MD-2' pineapple cultivation for a 300-ha farm in a warm environment with mean temperatures above 40°C and a day/night differential of approximately 5°C.

Stage	Time in months	Area (ha)	% of 300 ha
Soil preparation	4	44.08	15%
Planting	0.5	5.51	2%
Plant crop plant development to induction	11	121.22	40%
Plant crop fruit development	4.5	49.59	17%
Plant crop harvest	0.25	2.76	1%
Ratoon crop plant development to induction	0	0.00	0%
Ratoon crop fruit development	0	0.00	0%
Ratoon crop harvest	0	0.00	0%
Sucker production	7.0	76.84	26%
Total	27.2	300.00	100%

Table 7.3b. Farm schedules for 'MD-2' in a cool environment as per Table 7.3a.

Area planted each month (equals area forced, and harvested areas) (ha)	8.81
Total annual area from first harvest (ha)	105.73
Total area from ratoon harvest (ha)	79.03
Monthly establishment of ratoon crops (ha)	6.59
Sucker production (suckers/plant/month)	0.33
Total sucker propagation area required (ha):	26.70
Total annual harvesting area (first and ratoon) (ha)	184.76

Table 7.4b. Farm schedules for 'MD-2' in a warm environment as per Table 7.4a.

Area planted each month (equals area forced, and harvested) (ha)	11.02
Total annual area from plant crop harvests (ha)	132.24
Total area from ratoon harvesting (ha)	0.00
Monthly establishment of ratoon crops (ha)	0.00
Seedling productivity (suckers/plant/month)	0.18
Total sucker propagation area required (ha):	61.22
Total annual harvesting area (first and ratoon) (ha)	132.24

sucker harvesting for the same productivity (suckers/plant/month) is not every month, but every 5 or 6 weeks, more propagation blocks of a smaller unit area would be required but the total area of sucker production blocks will remain at 26.7 ha. The area can be adjusted appropriately if more or fewer than 0.33 suckers are obtained per month per plant.

Due to operational or logistical issues, the suckers can sometimes be left longer than intended. The most important thing is to have only the minimum area necessary devoted to sucker production.

Good sucker propagation practices

Use only pest and disease-free fields that have good drainage. The plants used to establish these fields should be used within the first week after harvesting to encourage rapid root growth.

Fertilization and phytosanitary management

The same general principles of crop nutrition as discussed in Chapter 8 should be followed for sucker propagation blocks. Sucker production is enhanced by high rates of nitrogen and high water volume (2500–3000 l ha^{-1}).

If de-suckering is done every 6 weeks, a programme of three fertilizer applications by spray boom can be done every 10 days and an application of a fungicide at day 40 so that at the time of removal at day 42, the sucker will be protected during the planting and early establishment phase. The application of insecticides should be done based on pest monitoring.

Trimming (Leaf pruning)

Trimming the mother plants is done to increase exposure and stimulate the buds that will give rise to the suckers and to facilitate the passage of workers. Leaves should be trimmed approximately 30 cm above the stem apex, though sometimes plants are trimmed lower to stimulate sucker emergence (Fig. 7.18), but this shortens the useful life of the sucker block and can favour the incidence of diseases. Trimming of the actual suckers is sometimes used to even up growth.

General maintenance

Maintenance consists of eliminating any unproductive shoots present including slips that will not be used and all ground suckers that will compete for space and light with the developing suckers.

Sucker harvesting and handling

The suckers should be harvested as soon as they have reached the desired weight. The production of additional suckers will be delayed if all the suckers large enough are not removed. It is also important to avoid collecting suckers that do not meet the minimum weight for planting as the uniformity of the subsequent planting will be affected. Once harvested, the suckers can be further sorted according to weight. For practical reasons, it is best to manage the smallest number of possible weight ranges. It is recommended to use weight ranges of a maximum 100 g. For example, a farm that uses three weight ranges could be planting suckers of 400–500 g, 501–600 g and 601–700 g.

Soon after harvest the suckers should be treated, preferably by immersion, with the agrochemicals relevant against the pests and diseases likely to be encountered. The cured, treated and sorted suckers should then be distributed throughout the field to be planted usually within the following 24 h rather than left in piles. The treated suckers should not be left stacked for more than 24 h. The plants should be placed with the lesion wound upwards to facilitate drying and healing. To avoid excessive dehydration and the associated loss of weight and turgor useful for root initiation, the suckers should not be left in the above condition for more than 72 h. An alternative approach to immersion in pesticide is to apply the agrochemicals by boom sprayer while the suckers are distributed throughout the field but not yet planted.

Example of data for sucker production

To keep up-to-date information on sucker productivity, information as shown in Table 7.5 should be collected for each minimum handling unit (block).

Fig. 7.18. Leaf trimming to stimulate sucker production in Costa Rican 'MD-2'. Photo S. Newett.

Pre-harvest Fruit Treatments

Post-force and pre-harvest nutrition

Generally, yields from plants that have adequate nutrition during vegetative growth do not respond to additional nutrition during fruit development. However, applications of nitrogen during the later stages of fruit development may increase vegetative growth of developing suckers. When done early, such as 45–90 days after forcing, the number of suckers can increase (Souza and Reinhardt, 2004). This can be important when good sucker growth is required.

Table 7.5. Example of farm sucker propagation data for 'MD-2'.

Block	Area (ha)	Number of suckers
B-7	0.85	16,022
B-4	0.93	13,903
B-10	0.68	15,028
B-14	1.03	12,051
B-11	0.57	11,485
B-21	0.38	6,669
B-19	0.18	3,276
B-15	0.97	18,915
B-18	1.05	12,967
Total	6.64	110,316
Time between harvests (months)		1.41
Planting density		65,000
Suckers/plant/month		0.18

When other nutrients such as boron, iron or calcium are deficient, their addition may increase fruit yield or quality, or both.

Protection against sunburn

In areas with high irradiance during the later stages of development, sunburnt fruit can be a serious problem. Temperatures on the fruit surface can exceed 55°C (130°F), leading to cell death, tissue collapse and dehydration. In some cases, the tissues may be attacked by insects or disease organisms.

To minimize or avoid this problem, a number of solutions have been devised. The techniques all seek to reduce direct radiation of the fruit surface. Popular methods include tying of the longest leaves above the fruit, covering susceptible areas with rice straw or other convenient dry vegetation, the use of shade cloth, or the use of newspaper or other paper caps positioned to cover the shoulders of the fruit (Figs 7.19a and 7.19b). When shade cloth is used, it could have an adverse effect on fruit yield if it covers the entire plant. Reflective sprays containing calcium carbonate and a sticker or UV-blocking compounds have also been used to protect the fruit. A number of commercial mixtures that are either invisible or that can be washed off the fruit after harvest are available.

Where sunburn is a problem, early gouging of crowns should be avoided to retain the benefit of this source of natural shade. The more serious problems of sunburning occur in ratoon fruit during the later stages of development due to lodging which exposes the sides of fruit horizontally to the sun and because lodged fruit are below the level of the canopy and so cannot benefit from convective cooling by winds. The problem of sunburn is accentuated on border rows, where the fruit may be larger, and there are no adjacent plants to help support the fruit. Some relief to the situation can be obtained by planting the outside line at a higher density relative to the rest of the field so as to reduce the fruit weight and provide more mutual support.

Low-volume overhead irrigation is sometimes used to cool plants during periods of high irradiance close to fruit maturity.

Ripening acceleration (degreening)

Ethephon is effective in both accelerating the ripening process and concentrating the harvest

Fig. 7.19. (a) Sun protection using paper hats. (b) Leaf tying. Photo D. Bartholomew.

peak, allowing complete harvest in fewer passes. In some countries, such as Brazil, ethylene application, under the form of ethephon, at the fruit maturation stage is no longer accepted by plant protection legislation. See Chapters 11 and 13 for more information on degreening.

Irrigation during fruit development

As noted previously, pineapple plants require approx. 75 mm (3″) of water per month for optimum growth. This requirement continues during fruit development, and drastic changes in moisture status should be avoided. Drip irrigation can be continued until just prior to harvest. Overhead irrigation should generally be avoided after the onset of the open petal stage of flowering. This will minimize fruit diseases, while having little adverse effect on yield. Low-volume sprinklers have, however, been used in Australia during fruit development to minimize fruit sunburn and russeting.

Fruit enlargement

The chemicals gibberellic acid (GA3) (Li *et al.*, 2011), alpha-naphthalene acetic acid (ANA) and 2-(3-chlorophenoxy) propionic acid (Fruitone CPA) (Rebolledo-Martinez *et al.*, 2002; Othman *et al.*, 2006; Suwandi *et al.*, 2016) have been used to enlarge fruit. The use of chemicals to improve fruit size is usually restricted to small-fruited cultivars such as 'Queen'. It can have undesirable effects on fruit quality and is not used in all countries. See Chapters 11 and 13 for more information on fruit enlargement.

Harvesting

Determining the state of maturity

The state of maturity at which fruit are harvested will depend on the cultivar or clone, the purpose for which the fruit are being harvested, and environmental and growing conditions. Most fruit is harvested according to shell colour although degreening is increasingly being used. The recommended shell colour for harvest is determined by evaluating the relationship between the external appearance and internal characteristics. Often, fruit are sampled at a number of shell colours, and flesh-soluble solids content (SSC%) is measured to determine the degree of flesh maturity at each colour. Minimum SSC levels are required in many markets for fresh fruit. The Revised Codex Standard for Pineapples (CODEX STAN 182-1993 (Rev. 1-1999), FAO/WHO Codex Alimentarius Standards, Vol. 5B, Codex Standards for Tropical Fresh Fruits and Vegetables, Secretariat of the Joint FAO/WHO Food Standards Programme, Food and Agriculture Organization of the United Nations, Viale delle Terme di Caracalla, 00100 Rome, Italy) has set a minimum of 12% SSC for fresh pineapples in international trade.

For fruit destined for both cannery and fresh market, fruit translucence may also be an important factor in determining the shell colour at harvest.

Hand harvesting

Most of the world's pineapples are harvested by hand (Fig. 7.20). For processing, fruit are picked and placed into bags or baskets for removal from the field. The fruit may be accumulated at roadways for transfer into trucks, or loaded directly for transportation to either the processor or a central location, such as a broker, where fruit can be accumulated before transfer to the processor. Harvested fruit should not be left exposed to direct sunlight for more than an hour or so, as

Fig. 7.20. Hand harvesting in Brazil. Photo D. Reinhardt.

the sides and lower parts of the fruit are susceptible to sunburn.

Fruit destined for the fresh market are usually harvested with the crowns, and, in some countries, are cut from the plant with a short length of peduncle remaining attached to the fruit. For local markets, the peduncle length may be 5–10 cm long, but for international trade, peduncle length is limited to 2 cm under the Codex standard referred to above. On some farms and plantations, fresh fruit are packed directly into boxes in the field. Cut surfaces on these fruit may be treated with fungicides to reduce black rot and waxed to help retain fruit quality.

Machine-assisted harvesting

A wide diversity of machines have been developed to assist in the harvesting of pineapples. Truck-mounted systems are mounted on the bin that will receive the fruit, and are moved by the truck carrying the bin. Others are tractor-drawn or self-propelled and move at the same speed as the bin mounted either on a truck or trailer pulled by a tractor. Basically, all harvesters are a series of conveyor belts that assist by moving the fruit from the pickers to the transportation bin (Figs 7.21 and 7.22).

Cannery fruit may have the crowns removed, and the crowns may be left in the field or conveyed in the transportation bin to cushion the fruit against bruising. Later, the crowns may be separated from the fruit at the cannery, and used as planting material. Some machines have been developed for fresh fruit harvesting and in-field packing. These machines are large, heavy, and are designed to be operated in relatively flat areas under dry conditions.

In Australia fresh fruit are sold without crowns. The crowns are removed by an automated circular saw attached at the receiving end of a harvesting boom that has the rubberized belts substituted with belts of interlocking fruit cradles to accurately position the fruit for crown removal (Fig. 7.21).

Transportation of Harvested Fruit

Movement of pineapple from the field to its ultimate destination requires an understanding of the physiology of the fruit after harvest. If the travel time is short, as in the case where the fields are near the processor, the key aspects are to move the volume of fruit efficiently, while minimizing damage. Damage, which reduces slice recovery, can occur from bruising during loading, transportation, unloading, and conveying in the cannery system.

To understand the sources of bruising, extensive studies have been conducted, evaluating the depth of loading in bins, the types of suspension on trucks, road conditions, and the movement of fruit that results in impacts with stationery surfaces or with other fruit. As a result of some of these studies, some fruit are now transported with crowns, while other fruit

Fig. 7.21. A harvesting boom with de-topper as used in Australia. Inset shows fruit in cradle. Photo G. Sanewski.

Fig. 7.22. Conventional fruit harvesting boom without a de-topper. Photo S. Newett

are carefully stacked end to end in layers to minimize damage.

Fruit being transported to fresh markets fall into two major categories. Some are destined for local market, while others are shipped significant distances by air, sea or long surface hauls. If the transportation, including air transportation, takes only one or two days, the fruits may not require refrigeration. However, fruit quality is usually improved if the fruit is picked ripe and held under refrigeration. Fruit that will be transported for three or more days should be refrigerated at temperatures between 7.2°C and 8.0°C (45 and 46.4°F).

Ratooning

The practice of harvesting additional crops from an original planting is known as ratooning. Ratoon crops are important in pineapple crop management where they reduce costs when they can be harvested more economically than replanting the same area. They also provide a range of fruit sizes that complement the plant crop in the cannery and provide crowns, slips and suckers for planting material. Generally, they provide fruit at a lower cost than a plant crop, and in a relatively shorter time.

However ratoon crops have lower productivity and a greater risk of quality issues, especially when considering more demanding export standards. The primary determinants of ratooning capacity will be the health and yields of the preceding plant crop, environment, the soil and the resulting quantity of plant reserves. In some cases the environment might allow adequate accumulation of assimilates for sucker production but the low productivity of the ratoon crop (fewer boxes per hectare) and increased fruit quality risks might not justify the production of a ratoon crop. In such situations it is better to direct all efforts to producing the plant crop in the shortest time possible.

In some growing areas, replanting may be done infrequently if at all, and all fruit are harvested from ground suckers that grow from the previous crop. This practice was once common on the peat soils in Malaysia, which cannot support heavy tillage equipment. On these soils, replanting was limited to replacement of missing plants. A disadvantage of this system is the loss of uniformity and synchronization of the crop in successive years. Suckers were forced individually when they reached an adequate size, and fruit were harvested through most of the year in all fields. An alternative was to desiccate the foliage after harvest with a herbicide, then burn and replant the field with suckers.

The more conventional first and second ratoon crops, grown in areas where sucker growth is fair to good, allows most of the suckers to be forced together. When practices such as sucker pruning have been used, there is an improvement in crop uniformity. Ratoon crops are fertilized, irrigated, forced, ripened and harvested in much the same way as the plant crop, though the amount of fertilizer applied usually is reduced. Practices such as degreening with ethephon may actually be easier if the ratoon fruit being treated is the last before knockdown, and the following suckers will not be used as planting material for another crop.

Successful ratoon production is dependent on a healthy root system. This requirement presents a challenge in areas infested with root-knot or reniform nematodes. Other potential root problems, such as white grubs, symphylids, and root rotting fungi can limit ratoon production. In Brazil, 'Pérola' and 'Smooth Cayenne' ratoon crops have become rare as they represent increasing risks for plant and fruit losses due to further dissemination of fusariosis disease.

Farm Efficiency

Gross margins and individual variable costs can also be used in combination with measures of productivity to assess efficiency. Examples are fertilizer costs/tonne/ha, gross margin/tonne/year, tonnes/man/year, etc. Production systems can also be assessed by many other indicators including the breakdown of fresh market box counts for individual fields, processor reject reports for individual fields etc. At a more basic level, overall farm production efficiency might be determined by the number of exported boxes of optimum size/ha/year.

References

Alvarez, C.E., Carracedo, A.E., Iglesias, E. and Bravo, J.J. (1995) Pineapple yield and quality on a banana soil of the Canary Islands irrigated with acid and saline water. *Tropical Agriculture* 72, 220–224.

Alves, A. A., Reinhardt, D.H., Alcântara, J.P., Souza, L.F.S. and Caldas, R.C. (1998) Manejo e avaliação da soca do abacaxi ´Pérola` nas condições do semi-árido de Itaberaba, Bahia. *Revista Brasileira de Fruticultura* 20(3), 323–331.

Carr, M.K.V. (2012) The water relations and irrigation requirements of pineapple (*Ananas comosus* var. *comosus*): a review. *Experimental Agriculture* 48, 488–501.

Chadha, K.L., Melanta, K.R. and Shikhamany, S.D. (1974) High density planting increases pineapple yields. *Indian Agriculture* 18, 3–5.

Chen, X., Dhungel, J., Bhattarai, S., Torabi, M. and Midmore, D.J. (2010) Impact of oxygation on soil respiration and crop physiological characteristics in pineapple. *Paiguan Jixie Gongcheng Xuebao/Journal of Drainage and Irrigation Machinery Engineering* 28, 543–547.

Christensen, D. (1995) Quality, efficiency and yield. *Pineapple Field Day Book.* Queensland Fruit and Vegetable Growers, Growcorn, Brisbane, Australia, pp. 15–19.

Ciesiolka, C.A.A., Coughlan, K.J., Rose, C.W. and Smith, G.D. (1995) Erosion and hydrology of steeplands under commercial pineapple production. *Soil Technology* 8, 243–258.

Coelho, R.I., Carvalho, A.J.C. de, Lopes, J.C., Teixeira, S.L. and Marinho, C.S. (2007) Coroa do abacaxi 'Smooth Cayenne' na produção de mudas do tipo rebentão. *Ciência e Agrotecnologia* 31(6), 1867–1871.

Cunha, G.A.P. da (1999) Implantação da cultura. In: Cunha, G.A.P. da, Cabral, J.R.S., Souza, L.F. da S. (eds) *O abacaxizeiro: cultivo, agroindústria e economia*. Brasília, Brazil: Embrapa Comunicação para Transferência de Tecnologia, pp. 139–167.

Cunha, G.A.P. da and Reinhardt, D.H. (2004) Propagação do abacaxizeiro. Coleção Plantar, 11. EMBRAPA. Brasilia, Brazil.

Dhungel, J.K., Midmore, D.J., Walsh, K.B., Bhattarai, S.P., Subedi, P.P. and Xinming, C. (2011) Oxygation enhanced pineapple yield and quality. *Acta Horticulturae* 889, 551–556.

Dodson, P.G.C. (1968) Effects of spacing, nitrogen and hormone treatment on pineapple in Swaziland. *Experimental Agriculture* 4, 103–115.

El-Swaify, S.A., Dangler, E.W. and Armstrong, C.L. (1982) *Soil Erosion by Water in the Tropics.* Research Extension Series No. 024. Honolulu, HI: Hawaii Institute of Tropical Agriculture and Human Resources.

El-Swaify, S.A., Zhang, J., Palis, R., Rose, C.W. and Ciesiolka, C.A.A. (1993) Erosion problems and conservation needs of pineapple culture. *Acta Horticulturae* 334, 227–239.

Fournier, P., Bouffin, J. and Vanniere, H. (1995) Cultivation of pineapple Queen Victoria at a very high density in Reunion: intra-plot variability. *2nd Symposium International Ananas, Abstracts.* Trois-Ilets, Martinique.

Garside, A.L., Poggio, M.J., Park, G., Salter, B. and Perna, J. (2009) Long-term Ingham and Mackay farming system experiments: comparisons between permanent non-tilled beds and re-formed beds. 31st Annual Australian Society of Sugar Cane Technologists Conference, pp. 13–26.

Glennie, J.D. (1981) Pineapple slip production using the morphactin multiprop applied after flower induction with different chemicals. *Australian Journal Experimental Agriculture and Animal Husbandry* 21, 121–128.

Hanafi, M.M., Shahidullah, S.M., Niazuddin, M., Abd Aziz, Z. and Mohammud, C.H. (2010a) Crop water requirement at different growing stages of pineapple in BRIS soil. *Journal of Food Agriculture & Environment* 8, 914–918.

Hanafi, M.M., Shahidullah, S.M., Niazuddin, M., Aziz, Z.A. and Mohammud, C.H. (2010b) Potential use of sea water for pineapple production in BRIS soil. *International Journal of Agriculture and Biology* 12, 396–400.

Heenkenda, H.M.S. (1993) Effect of plant size on sucker promotion in 'Mauritius' pineapple by mechanical decapitation. *Acta Horticulturae* 334, 331–336.

Hepton, A., Ingamells, L., Macion, E., Gonzales, J. and Sampongse, D. (1993) Pineapple plant and fruit growth and development in fertilized native soil and artificial root medium. *Acta Horticulturae* 334, 131–139.

Hotegni, V.N.F., Lommen, W.J.M., Agbossou, E.K. and Struik, P.C. (2015) Influence of weight and type of planting material on fruit quality and its heterogeneity in pineapple *Ananas comosus* (L) Merrill. *Frontiers in Plant Science* 5(798), 1–16.

Hudson, N. (1995) *Soil Conservation*. Ames, IA: Iowa State University Press.
IRFA (1984) *La Culture de l'ananas d'exportation en Côte d'Ivoire : manuel du planteur*. IRFA, Montpellier, France, pp. 65–66.
Jantee, C., Hama, J., Sangudom, T. and Saradhuldhat, P. (2014) Effects of plant density, fertilization and crown removal on fresh pineapple production for exporting. *Acta Horticulturae* 1024, 205–210.
Jung-Rothenhaeusler, F., Traut, K., Gauggel, C., Smith, A.B. and Umali, R. (2016) Pineapple plantation information system – application of geo-informatics and unmanned aerial sensing technology for efficient land use management. *Acta Horticulturae* 1111, 255–265.
King, N. (1935) Some physical aspects of pineapple plant development. *Pineapple Quarterly* 5, 48–59.
Leon, R.G. and Kellon, D. (2012) Characterization of 'MD-2' pineapple planting density and fertilization using a grower survey. *Horttechnology* 22, 644–650.
Li, Y.-H., Wu, Y.-J., Wu, B., Zou, M.-H., Zhang, Z. and Sun, G.-M. (2011) Exogenous gibberellic acid increases the fruit weight of 'Comte de Paris' pineapple by enlarging flesh cells without negative effects on fruit quality. *Acta Physiologiae Plantarum* 33, 1715–1722.
Lima, V.P. de, Reinhardt, D.H. and Costa, J.A. (2001) Desbaste de mudas tipo filhote do abacaxi cv. Pérola – 1. Produção e qualidade do fruto. *Revista Brasileira de Fruticultura* 23(3), 634–638.
Lima, V.P. de; Reinhardt, D.H. and Costa, J.A. (2002) Efeito do desbaste de mudas do tipo filhote na cultura do abacaxi 'Pérola' sob irrigação. Anais. XVII Congresso Brasileiro de Fruticultura, Embrapa, Belém, Brazil, pp. 1–4.
Liu, C. and Chen, J. (2014) Effects of salt stress on growth, ion concentration and quality of pineapple. *Communications in Soil Science and Plant Analysis* 45, 14.
Matos, A.P. de, Reinhardt, D.H., Sanches, N.F., Souza, L.F. da S., Teixeira, F.A. *et al.* (2009) *Produção de mudas sadias de abacaxi*. Circular Técnica, 89. Cruz das Almas, Brazil: Embrapa Mandioca e Fruticultura.
McCall, W.W. (1980) *The Salt Tolerance of Plants*. General Home Garden Series. No. 21. Hawaii Cooperative Extension Service. University of Hawaii, Honolulu, Hawaii.
Mustaffa, M.M. (1988) Influence of plant population and nitrogen on fruit yield and quality and leaf nutrient content of 'Kew' pineapple. *Fruits* 43 (7–8), 455–458.
Neild, R.E. and Boshell, F. (1976) An agroclimatic procedure and survey of the pineapple production potential of Colombia. *Agricultural Meteorology* 17, 81–92.
Norman, J.C. (1978) Responses of 'Sugarloaf' pineapple, *Ananas comosus* (L.) Merr., to plant population densitites. *Gartenbauwissenschaft* 43(5), 237–240.
Ojeda, M., Pire, R., Pérez De Camacaro, M. and Mogollón, N. (2012) Effects of irrigation on growth, flowering and fruit quality of pineapple 'Red Spanish'. *Acta Horticulturae* 928, 171–178.
Othman, Z., Madom, M.S., Osman, A.. and Mohamed, S. (2006) The post-harvest sensory and characteristics of cold-stored chemical pineapples pre-treated with different Fruitone concentrations. *Journal of the Science of Food and Agriculture* 86, 2067–2075.
Pádua, T.R.P., Reinhardt, D.H. and Cunha, G.A.P. (2013). Plantio. In: Sanches, N.F. and Matos, A.P. de (eds) *Abacaxi: o produtor pergunta, a Embrapa responde*, 2nd rev. edn. Brasília, Brazil: Embrapa, pp. 51–66.
Patra, S.K., Pramanik, S. and Saha, S. (2015) Techno-economic feasibility of micro-irrigation in pineapple under the Gangetic alluvial plain of West Bengal. *Indian Journal of Horticulture* 72, 329–333.
Py, C., Lacoeuilhe, J.J. and Teisson, C. (1987) *The Pineapple. Cultivation and Uses*. Paris: Editions G.-P. Maisonneuve.
Rebolledo-Martinez, A., Uriza-Avila, D., Rebolledo-Martinez, L., Becerril-Román, A.E. and Ruiz-Posadas, L. del M. (2002) Fruitone CPA para retardar la maduración en piña *Ananas comosus* (L.) Merr., cv. Cayena lisa, Cosechada en primavera. *Revista Brasileira de Fruticultura* 24 (2), 354–358.
Rebolledo-Martinez, L., Uriza-Avila, D., Rebolledo-Martinez, A. and Zagada, G. (2005) Suckers production of MD-2 hybrid pineapple by three methods: gouging, leaf pruning and a growth regulator. *Acta Horticulturae* 666, 277–285.
Reinhardt, D.H. (2010) Pineapple crop management in Brazil in response to market demands. *Acta Horticulturae* 864, 143–155.
Reinhardt, D.H. and Cunha, G.A.P. da (1982) Efeitos do ethephon combinado com uréia na indução floral do abacaxizeiro. In: *Congresso da sociedade americana de ciências hortícolas, região tropical*, 29. Campinas, Brazil, vol. 25, pp. 29–34.
Reinhardt, D.H. and Nascimento, L.S. do (2017) Produção de mudas de abacaxi cv. Imperial por supressão do meristema apical, em condições de campo. *Anais Congresso Brasileiro de Fruticultura*, 25. Porto Seguro, Brazil.

Sanford, W.G. (1962) Pineapple crop log-concept and development. *Better Crops Plant Food* 46, 32–43.

Sanford, W.G. (1964) Influence of the removal of one or more parts of the pineapple plant on the growth and yield of the remaining parts. *Research Report* 110, 35.

Santana, L.L. de A., Reinhardt, D.H., Cunha, G.A.P. da and Caldas, R.C. (2001) Altas densidades de plantio na cultura do abacaxi cv. Smooth Cayenne, sob condições de sequeiro. *Revista Brasileira de Fruticultura* 23, 353–358.

Santos, P.C., Freitas, S.J., Freitas, M.S.M., Sousa, L.B. and Carvalho, A.J.C. (2011) Produção de mudas do tipo rebentão, utilizando coroas de três cultivares de abacaxi inoculadas com fungos micorrízicos. *Revista Brasileira de Fruticultura* 33(3), 954–961.

Schneider, R.C., Green, R.E., Wolt, J.D., Loh, R.K.H., Schmitt, D.P. and Sipes, B.S. (1995) 1, 3-dichloropropene distribution in soil when applied by drip irrigation of injection in pineapple culture. *Pesticide Science* 43, 97–105.

Scott, C. (1992) The effect of plant density on spring plant crop/autumn first ratoon crop cycles; Test number 718. In: *Pineapple Field Day Book.* Pineapple Industry Farm Committee, Beerwah, Queensland, pp. 37–45.

Sether, D.M. and Hu, J.S. (2001) The impact of pineapple mealybug wilt-associated virus-1 and reduced irrigation on pineapple yield. *Australasian Plant Pathology* 30, 31–36.

Smith, M.K., Smith, J.P. and Stirling, G.R. (2011) Integration of minimum tillage, crop rotation and organic amendments into a ginger farming system: impacts on yield and soilborne diseases. *Soil and Tillage Research* 114, 108–116.

Souza, L.F.D., de Almeida, O.A., Caldas, R.C. and Reinhardt, D.H. (2002) Effect of soil moisture and fertilization on 'Pérola' pineapple in Coastal Tableland areas of Brazil. *Tropical Agriculture* 79, 83–87.

Souza, L.F. da S. and Reinhardt, D.H. (2004) *A adubação do abacaxizeiro após a indução floral.* Comunicado Técnico, 103. Cruz das Almas, Brazil: Embrapa Mandioca e Fruticultura.

Souza, O.P. de, Teodoro, R.E.F., Melo, B. de and Torres, J.L.R. (2009) Qualidade do fruto e produtividade do abacaxizeiro em diferentes densidades de plantio e lâminas de irrigação. *Pesquisa Agropecaria Brasileira* 44(5), 471–477.

Souza, O.P. de, Zanini, J.R., Torres, J.L.R., Barreto, A.C. and Souza, E.L.C. (2012) Pineapple production and physical quality under different irrigation frequencies and water depths. *Irriga* 17, 534–546.

Stirling, G.R. (1999) Increasing the adoption of sustainable, integrated management strategies for soilborne diseases of high-value annual crops. *Australasian Plant Pathology* 28, 72–79.

Sugahara, K., Ohwaki, Y. and Banzai, K. (2001) Erosion control in pineapple fields on the island of Ishigaki. *Jarq-Japan Agricultural Research Quarterly* 35, 91–96.

Suwandi, T., Dewi, K. and Cahyono, P. (2016) Pineapple harvest index and fruit quality improvement by application of gibberellin and cytokinin. *Fruits* 71(4), 209–214.

Vásquez Jiménez, J. (2016) Burndown of pineapple plants in Costa Rica. *Pineapple News* 23, 26–30. Available at: http://www.ishs-horticulture.org/workinggroups/pineapple/PineNews23.pdf (accessed 6 April 2018).

Wassman, R.C. (1990) Effects of seasonal temperature variations on pineapple scheduling for canning in Queensland. *Acta Horticulturae* 275, 131–138.

Wee, Y.C. (1969) Planting desnity trials with *Ananas comosus* (L.) Merr., var. Singapore Spanish. *Malay Agricultural Journal* 47, 164–174.

8 Plant Nutrition*

Jhonny Vásquez-Jiménez[1] and Duane P. Bartholomew[2]
[1]Consultant, Alajuela, Costa Rica; [2]University of Hawaii at Manoa, Honolulu, USA

Nutrient Management Philosophy

In order to design a good nutrient management strategy, it is essential to understand soil chemical and physical characteristics and climatic conditions, as well as how water availability and pests and diseases impact plant growth and appearance.

While there are many reports on the effects of individual nutrients, only one study was found on nutrient management practices used by pineapple growers. That study showed that best management practices are not used by the small to medium-sized farms included in the study (Leon and Kellon, 2012). The survey of practices used by farmers in the study, all growing 'MD-2', showed that there was no clear relationship between fertilizers applied and planting density or yield and the farmers did not use soil or tissue analysis to guide nutrient management practices. The lack of good management would reduce yields where too little fertilizer was used and potentially degrade environmental quality where excessive amounts of fertilizers were applied.

Best nutrient management practices begin after the soil has been tilled to eliminate physical limitations for root development and soil water is adequate to sustain active root growth. The pineapple root system plays an important role in a plant's nutrition and growth and roots do not grow well into dry or infertile soil, but grow longer and deeper when nutrients, especially nitrogen, phosphorus and oxygen, are available.

Soil characteristics associated with soil mineralogy are important factors in root development. While pineapples are cultivated in a great diversity of soils, the best soil type for pineapple depends on the climate, the available resources and the area under cultivation. When the resources for growing pineapples do not limit land purchases, fertilizer, irrigation equipment and technical assistance, one should opt for soil physics rather than soil fertility. Rarely are the most fertile soils available for growing pineapple and even if they were, they could lack porosity, the physical property essential for the aeration and drainage required by the pineapple root system. Very sandy or rocky soils, generally considered poor agricultural soils, are the soils most likely to have the physical properties required by the pineapple root system. Remarkably, it is possible to produce as large a pineapple crop on infertile soils as could be produced on fertile soils if the grower has a good knowledge of the soil's chemical and physical properties and of the nutrient requirements of the pineapple plants to be grown in those soils.

The nutritional status of the pineapple plant has a large influence on plant growth, yield and, mostly indirectly, on fruit quality. For pineapple, soil analysis and plant indicators, which include visual deficiency symptoms and optimum nutrient levels in reference tissues, are used to manage fertilization of the pineapple crop. The alternative to these methods is to conduct calibrated fertilizer trials in each area

* A revision of the chapter by Eric Malézieux and Duane P. Bartholomew. The authors acknowledge the advice provided by Jorge W. Gonzales, Senior Consultant Pineapples and Bananas, Dole Asia Holdings.

where the crop is grown. Calibration trials are more common where pineapple is grown on small farms and where access to technology is limited (Souza, 1999). To sustain growth and obtain good yields, it is important to provide adequate supplies of all nutrients in proper balance based on the principles of best management practices. Such practices ensure that excess nutrients are not applied and that excess amounts of one nutrient do not induce deficiencies of others or, in the case of N and P, lead to environmental degradation.

The plant reference tissue for pineapple is the D-leaf, which if growth is uninterrupted is, from the soil, the tallest leaf on the plant. Sideris and Krauss (1936) categorized leaves as: 'A', does not grow after planting; 'B', elongates after planting with the new growth indicated by greater leaf width; 'C', develops after planting, is difficult to pull from the plant and has a leaf base wider than the green tissue above it; 'D', 1–3 leaves that are nearly mature, have immature non-chlorphyllous basal tissue, are easily pulled from the plant and width of the basal tissue is approximately the same as the green tissue above it; 'E' and 'F', progressively younger whorls of leaves (Fig. 8.1). The D-leaf on an actively growing plant can be quickly identified by bunching the tallest leaves on the plant together and selecting the leaf that stands above the others. The D-leaf is used for tissue analysis because it reflects the plant's current access to nutrients with acceptable accuracy.

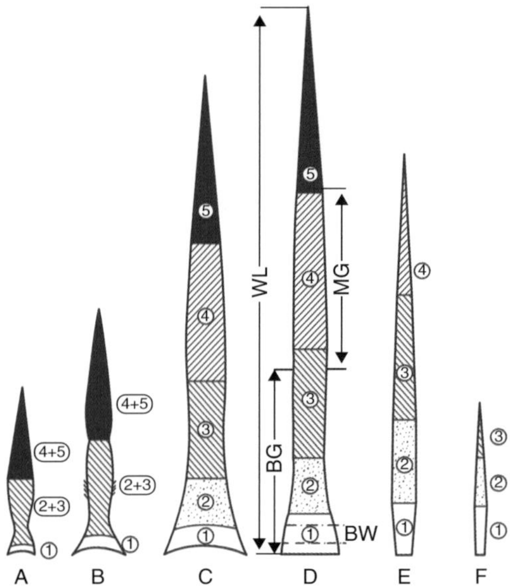

Fig. 8.1. Pineapple leaf categories (A–F) and within-leaf sections 1, basal non-chlorophyllous; 2, transitional subchlorophyllous; 3, lower cholorophyllous; 4, intermediate chlorophyllous; 5, terminal chlorophyllous. The D-leaf sections used in tissue analysis are the middle one-third of the basal white tissue (BW); lower one-third of the basal green tissue (BG); middle one-third of the green tissue (MG); the whole D-leaf (WL). Modified from Sideris and Krauss, 1936.

Soils and Soil Analysis

Soil minerals and texture

Among the soils where pineapple grows well with proper fertilization are those where kaolinite, which has a 1:1 layered lattice structure, is the primary clay mineral (see Table 8.1). The average particle size of kaolinite is larger than that of other clays so they are more porous and water drains away rapidly. However, such soils have a very low cation exchange capacity (CEC) so they represent a soil selected for its good physical properties.

Soils of oxides and hydroxides of Fe and Al are highly weathered with low base saturation and low CEC, but generally have good to excellent physical properties and low plasticity, aeration and drainage characteristics so water infiltration is rapid and the soils are easy to till and have good water storage capacity.

Soils containing 2:1 clays are less well suited for pineapple because they compact easily and have poor drainage, so are prone to becoming waterlogged. They also shrink during dry periods, which causes roots to break leaving them vulnerable to pathogens. Even with loam or sandy textural classes of soils with 2:1 clays, the land can be difficult to prepare and on sloping terrain after rain slippery roads can delay spraying or result in poor coverage.

Soils with blocky or prismatic structure and with clay contents of 30–50%, even with low-to-medium fertility, offer better conditions for pineapple because their greater porosity results in good aeration and drainage. These soils also retain enough water to make irrigation more

Table 8.1. Main characteristics of the soils that pineapple is grown in and their suitability for the production of pineapple. From J. V. Jimenez, based on extensive field experience.

Soil order	Main clay mineral	Physical properties for pineapple	Common structure	CEC (cmol/kg)	Surface area (m/g)	Percentage clay and suitability for pineapple			
						10–20	20–30	30–40	>50
Ultisols, Alfisols, Oxisols	Kaolinite, 1:1, (non-expanding)	Excellent	Blocky, Prismatic, Columnar	3–15	5–20		Medium	High	Medium
Ultisols, Alfisols, Oxisols	Oxides, etc. of Fe & Al, mainly amorphous	Excellent	Blocky, Prismatic, Columnar	10–25	–		Medium	High	Medium
Andisols	Allophane, amorphous	Intermediate	Single grain	10–50	–	Low	Medium	Medium	Low
Vertisols, Alfisols, Inceptisols	Illite, 2:1, non-expanding	Intermediate	Single grain, Granular,	10–40	50–200	Low	Medium	Medium	Low
Vertisols, Alfisols, Inceptisol, Entisols	Vermiculite, 2:1, expanding	Poor	Single grain, Granular, Platy	100–150	500–700	Medium	Low	Very low	None
Vertisols, Alfisols, Inceptisols	Montmorillonite, 2:1, expanding	Very poor	Massive	80–150	700–800	Medium	Low	Very low	None

manageable, and store enough water to sustain the crop without irrigation during a drought with only a slight compromise in productivity. Additional advantages of these soils are that their structure and clay content make them less susceptible to erosion. Their main disadvantage is that land preparation operations must be suspended while the soil is wet to avoid destroying the soil structure.

When the same class of soils has clay contents of less than 15% and high sand and silt contents, for example, alluvial soils, plant growth can be very vigorous so long as the soil moisture is optimum. However, the limitations of these soils include very low water storage capacity and the granular or crumbly structure makes them susceptible to erosion by heavy rains or poorly managed sprinkler irrigation. Where dry periods subject plants to water stress severe enough to begin collapsing the leaf water storage tissue, plant growth will slow or cease unless irrigation can be provided. An advantage of these soils, however, is that water percolates rapidly so delays in field operations are relatively short.

In general, when choosing land for the cultivation of pineapple it is best to avoid light-textured soils with a crumbly or laminar structure, such as loams, silts and silt loams, because they can become compacted after the crop has been planted. Soil compaction decreases root extension and downward penetration, with the result that plants wilt and show other stress responses, such as the development of spines in 'MD-2' and 'Smooth Cayenne', due to the lack of a functioning root system rather than a lack of water. In general, these soils are not recommended for growing of pineapple, but if used, plastic mulch will help to reduce compaction.

Soil pH and liming

Soil acidity is the most important chemical characteristic when managing a soil for pineapple production. All pineapple cultivars seem equally well adapted to a pH range of 4.5–5.5 and maintaining a low soil pH ensures an optimum supply of micronutrients (Py *et al.*, 1987). Managing soil pH requires a good knowledge of soil chemistry, soil buffering capacity, and the acidity that needs to be neutralized to increase soil pH by tenths of a unit. The acid-neutralizing capacity of carbonate rock (limestone) is measured by its calcium carbonate equivalent (CCE with calcite at 100%), and the CCE informs a buyer about the product quality.

Generally, the soil pH should be maintained above 4.5. In soils where *Phytophthora* spp. root or heart rots are a problem, the pH should be kept below 5.5.

The solubility of Al and Mn are strongly influenced by soil pH. While most pineapple cultivars are tolerant of high levels of soluble Al, there are cultivar differences. 'Queen', relative to 'Smooth Cayenne', grew better in acid soils, had less aluminum in leaves and, in that study, did not benefit from liming (Guong *et al.*, 1997). However, 'Smooth Cayenne' was less sensitive to soluble Al than was 'Tainung 17' (Lin, 2010) and Mota *et al.* (2016) reported that 'Fantástico' was less affected by Al concentration than 'Vitória'.

While pineapple is generally considered tolerant of high Mn, large cultivar differences exist (Py *et al.*, 1987; Lin, 2010). In all cultivars, high concentrations of Mn interfere with the utilization of iron by the plant, which results in iron chlorosis. Py *et al.* (1987) reported that a D-leaf Fe:Mn ratio below 0.4 resulted in visible iron deficiency in 'Smooth Cayenne' pineapple due to very high soil soluble Mn levels. Vásquez-Jiménez (2010) recommended that the Fe:Mn ratio in the lower one-third of the D-leaf BG tissue (Fig. 8.1) be no lower than 0.2 for 'MD-2', which may indicate that it has greater tolerance to soluble Mn than does 'Smooth Cayenne'. An Fe:Mn ratio less than 0.2 indicates soil Mn is high enough to require liming to reduce the soluble Mn level or foliar Fe sprays to increase tissue Fe levels. A soil test should be used to determine the amount of lime to be applied to keep the pH within the recommended range and to maintain cation balances.

Possible calcium (Ca) sources include gypsum, which will not change the pH (Silva *et al.*, 2006), and calcium carbonate, calcium silicate, calcium sulfate, and dolomite, all of which will increase the soil pH. The best option could be a mixture of two of these.

Liming requirements

Soil sampling should be done at least 6 months before planting to provide time for receipt of the analysis results and application of any soil amendments (Du Plessis *et al.*, 1989). Where the soil characteristics are known, for example soils with constant charge, and Ca level falls below 50 mg kg^{-1}, the lime requirement can be estimated from soil test results (Swete Kelly, 1993; neutralized ammonium acetate method).

However, some soils in areas where pineapples are grown have developed under intensive weathering and can have low soil pH but do not respond to liming in the same way as do temperate soils. Fassbender and Bornemisza (1987) found that such soils had both fixed and variable charger constituents that depended on the solution pH and ionic strength. As a result, they could have the same pH levels, but they could have different degrees of cation saturation and, therefore, their properties relating to acidity were also different. This caused Fassbender and Bornemisza (1987) to recommend liming using the approach of Kamprath (1970) that consisted of liming in proportion to the sum of fixed and variable charges present in the soil.

Sufficient lime should be applied to supply Ca for plant growth and to counteract the acidifying effect of fertilizers applied over the life of the crop. Determining the lime requirement of a soil requires an understanding of soil chemistry, including pH, CEC, how the levels of Ca, Mg and K and their ratios (Ca:Mg, Ca:K, Mg:K and Ca + Mg:K) affect plant availability and the saturation and concentration of Mn. The calculation of the lime requirement should include consideration of the typical 30 cm rooting depth for pineapple and soil bulk density. Measurement of soil pH is independent of soil type so recommendations about an ideal pH range can be provided for all soil types.

If the soil test shows that the soil pH is below 4.0 or bases are badly imbalanced, or both, as much as 10 megagrams (Mg) ha^{-1} of lime or dolomite could be required to raise the pH or restore cation balances. While it may be possible to apply such large quantities of lime (J. Vásquez-Jiménez, 2009, unpublished data), for practical, technical and budgetary reasons, 1–2 Mg ha^{-1} will usually meet the needs of the current crop. While not sufficient to rebalance soil bases, a smaller amount can be part of a plan to gradually achieve a balance over multiple cycles of production. A technical reason for not applying very large quantities of the lime is that in some cases it can reduce plant growth. Ortiz (2008) reported a reduction in 'MD-2' plant weights when more than 2 Mg ha^{-1} of lime were added to a soil with an initial pH of less than 4.5. If possible, lime should be broadcast and ploughed in 3 months before planting (Du Plessis *et al.*, 1989).

An alternative to broadcasting lime over soils low in Ca and/or Mg, is to lime only the root zone. If beds are used to facilitate drainage, the amendment can be applied only to the top of the bed. While reducing the amount of product required, the chemical balance of the soil would not be uniform throughout the cultivated area.

Soil sampling and chemical analysis

Soil chemical analysis should be done every 2 years when there is no ratoon crop (Py *et al.*, 1987), otherwise at the end of the crop cycle. It is essential that the prepared field contains a balanced supply of nutrients distributed throughout the rooting zone to support good root development. Soil nutrient content varies with soil depth, as well as across fields, so samples should be taken to the rooting depth of the crop at representative sites across the field. Precision agriculture practices are designed to deal with soil spatial variability and their use is widespread in temperate-zone agriculture. Geoinformatics can be applied to many aspects of pineapple culture (Balasundram *et al.*, 2008a,b; Das and Sudhakar, 2014; Jung-Rothenhaeusler *et al.*, 2016), but are especially well suited to dealing with the spatial variability of pests and diseases as well as soil chemical and physical characteristics. Soil sampling on a grid will produce data that can used to generate field nutrient maps and can be integrated into a graphical information system. If samples are not collected on a grid, the sampling procedure must produce samples that are representative of the soil characteristics in the field areas sampled.

The soil chemical analysis provided by the laboratory should estimate the natural fertility of the sampled soil. This estimate, which ideally has been developed through calibration using 'indicator plants', uses extractive solutions that simulate the absorption of the elements by the roots under the particular conditions of each soil. The most suitable soil extractants for analysis can therefore vary with soil type. Consequently, there is no universal optimum soil level for each macronutrient. Further information on analytical methods can be found on the internet (Sikora and Moore, 2014).

Soil analyses always include the levels of P, K, Ca and Mg, and pH, but may also include B, Zn, Fe, Mn and Cu (Wang *et al.*, 2004) as well as soluble Al and salinity. Where nitrate in fruits is a concern, incubated soil N may also be analysed (Swete Kelly, 1993; Keeney and Bremner method).

The information provided by a soil analysis is used to guide the development of a fertilization programme. If the laboratory is familiar with the nutrient requirements of the crop it may provide growers with soil fertilizer recommendations. If the quantities of soluble Al or Mn, or both, are high, information on the quantity of amendments to manage these problems should be provided. Lacking such information from the laboratory, growers will need to develop their own fertilizer plan.

Plant tissue analysis and tissue levels

While soil analysis estimates the available nutrients in the soil, foliar analysis estimates the total amount of each nutrient in the whole D-leaf (WL, French method) or some fraction of it. Other tissues sampled include the middle one-third of the basal white tissue (BW, Hawaiian method), the lower one-third of the D-leaf (BG, Costa Rican method) (Fig. 8.1). The middle green (MG, Fig. 8.1) was also used in Hawaii for leaf chlorophyll determinations. Foliar analyses are performed to diagnose a possible nutrient deficiency or imbalance and to provide historical records that will be used for long-term monitoring. The cost of tissue sampling and analysis can only be justified if growth is not limited by other factors.

Tissue sampling and analysis

No publication was found that described a universally applicable 'best practice' leaf tissue sampling method suitable for all types of planting material (crowns, slips, suckers) in all environments where pineapple is grown. Lacking that, the following approximates an industry 'best practice' method of sampling D-leaves that could apply across propagule type and environment. As with soil sampling, tissue sampling can be done on a grid (this is best) or in a stationary plot that is representative of the field where it is located. The stationary plot must be large enough to be suitable for the collection of all current and future soil samples as well as samples to assess plant growth and D-leaf nutrient contents. It also must exclude field border beds as well as beds in the centre of a block where the block width is twice that of the spray boom. The plot could be split to allow for sampling across the field.

Around 20–30 D-leaves should be collected in a zigzag pattern within the stationary plot. The first tissue sample should be taken after approximately one-third of the spray fertilizations are made or after about one-third of the vegetative growth period has passed. A second sample should be taken when approximately two-thirds of fertilizer sprays have been made or a comparable portion of the vegetative growth period has passed. The third and last sample should be taken after the pre-forcing foliar spray programme has been completed or 4–6 wks before forcing to allow time for sample collection, laboratory analysis and to make any needed spray fertilizations. If cost or other factors limit the number of analyses to one, it should be the one taken before forcing. For ratoon crops, a first sample should be taken after about half the spray fertilizations are completed or half the vegetative growth period has passed and the last should be as for the plant crop.

Comprehensive monitoring of plant development and use of tissue analysis or nutrient deficiency symptoms, or both, to guide fertilization of pineapple began in the 1940s (Sanford, 1962). These methods work well for pineapple because the long vegetative growth period, 6–12 months or more, provides ample time to correct deficiencies (Sanford, 1962; Py *et al.*, 1987; Newett and Rigden, 2015).

Diagnosis and Recommendation Integrated System (DRIS) norms have also been developed for pineapple (Angeles *et al.*, 1990; Teixiera *et al.*, 2009; Sema *et al.*, 2010). However, the long-term nature of the pineapple crop, which provides ample time to manage nutrition problems, makes it unlikely DRIS provides any benefit to growers already using soil and tissue analyses to guide amendment and fertilization practices. In both systems, if soil and tissue analyses are to provide 'best practice' results, soil and D-leaf analyses must be collected on schedule and recommendations must be followed.

Leaf tissue nutrient concentrations

Leaf tissue norms have been developed for N, P, K, Ca and Mg and sometimes also for B, Cu and Zn (Table 8.2). Guides to the optimum leaf nutrient levels that can be used to diagnose the sufficiency of nutrients have been produced for at least five cultivars (Table 8.2).

When doing foliar analysis for pineapple it is necessary to take into account the following aspects:

- optimum levels could be cultivar specific and possibly even clone specific so data may be required for each cultivar;
- optimum foliar nutrient levels can only be compared if they are expressed on a similar dry or fresh mass basis;
- the D-leaf must be used; but
- all D-leaf tissue types can be used after calibration.

The limited data (Table 8.2), indicate that the variation within cultivars is like that between cultivars and levels in the different tissues are also similar. No direct comparisons of tissue levels in different cultivars were found so there is no way of knowing if any differences between cultivars represent significant differences. When tissue levels for a cultivar or clone are not available, growers often use information developed for another cultivar, as is the case where practices for 'Smooth Cayenne' were used for 'Queen' in Australia (C. Scott, 2017, Queensland, personal communication) and for 'MD-2' in the Philippines and Kenya (A. Lucas, Mindanao, Philippines; C. Olende, 2017, Kenya; personal communications).

Distribution of nutrients within the plant

The nutrient content of any tissue depends upon its physiological age. However, leaves are the focus of most of the discussion that follows because 80–90% of a vegetative pineapple plant fresh or dry mass is leaves (Py *et al.*, 1987).

Table 8.2. Published leaf nutrient levels, some reported as norms or optimum levels, for 'Pérola', 'Queen' (Q), 'Smooth Cayenne' (SC), 'MD-2' and 'Imperial' (Imp C, complete and Imp def, deficient). Macronutrients (N to S) are in g kg^{-1} and micronutrients (B to Mo) are in mg kg^{-1}. Values below are on a dry mass basis and if originally fresh mass, conversion assumed 12% dry matter.

Cultivar	Tissue[a]	N	P	K	Ca	Mg	S	B	Cu	Fe	Mn	Zn	Mo	Ref[b]
Pérola	BW	20–22	2.1–2.3	25–27	3–4	4–5	2–3	30–40	9–12	100–200	50–200	10–15		8
Q 6 month	WL	21	1.5	24	2–5	2.5	-	15–25	>5	200–300	50–300	10–40		1
SC veg	BW	-	1.2	42–48	1.8	3								2
SC force	BW	-	2.4	36	1.8	3								2
SC def.	BW	-	1.1	24	0.5	1.8								2
SC	WL	15–17	1.0	22–30	8–12	3		30	8	100–200	50–200	10		3
SC	BW	12–30	2.0–3.5	25–50	1.3–3.5	2.5–3.5				133–240	<100	40–200		4
SC	MG	8.2–14.9	0.9–1.0	9.6–27.2	3.1–5.2	2.3–4.2		7.3–9.2		65.9–83.9	324–454	6.2–7.3		5
'MD-2'	BG	15–18	2.0	27–30	2.5–3.0	2.5–3.0	1.3–1.5	35–40	10–15	100–200	75–100	30–35	0.8	6
Imp C	WL	14.8	1.37	20	4.4	2.3	1.45	18.4						7
Imp def	WL	9.7	0.67	11.6	0.9	0.7	0.45	5.8						7

[a] The D-leaf tissues are middle 1/3 of the basal white tissue (BW), the whole leaf (WL), middle 1/3 (MG) or the lower 1/3 of the leaf (BG). Micronutrients are analysed in the middle 1/3 of the green tissue D-leaf. To convert SI units to per cent of element divide g kg^{-1} by 10.

[b] References: 1, Du Plessis, 1985; 2, Swete Kelly, 1993; 3, Dalldorf and Langenegger, 1978; 4, C. Olende, personal communication (Kenya); 5, Spironello *et al.*, 2004; values are ranges for NPK study; 6, Vásquez-Jiménez, personal communication (Costa Rica); 7, Mota Ramos *et al.*, 2011; 8, Malavolta *et al.*, 1997, as cited by Siebeneichler *et al.*, 2002.

Young, actively growing leaves younger than the D-leaf generally have higher nutrient levels than older leaves and the nutrient content of any given leaf increases from the base (youngest tissue) to the tip (older tissue) (Sideris and Young, 1946). Leaf dry matter contents of the basal portions of 'C' leaves are higher than in younger leaves and the dry matter content of all leaves increases from the base to the tip while that of the stem decreases from the base to the young and still growing tissue at the top.

The total amount of nutrients in the plant increases with age but the concentration in any given tissue can increase, decrease, or remain unchanged depending on the nutrient and the tissue. Assuming no other limitations, nutrient concentration in the plant tends to reflect the supply available, especially where the supply is limiting, whether from foliar sprays or the soil. When fertilizer applications are made, D-leaf nutrient concentration may temporarily increase in response to those applications even though the long-term trend of nutrient concentration in the D-leaf basal tissue is downward with increasing time after planting. Seasonal changes in tissue nutrient concentrations occur (Sanford, 1962), but they may result from changes in the plant growth rate rather than from changes in nutrient availability. Much of the available data on plant nutrient composition over time were collected before pineapple was planted throughout the year so it is uncertain if changes with season are caused by weather-induced changes in growth rate or by plant age.

Plant Macro- and Micronutrients

Visual deficiency symptoms were developed by growing plants in inert media or in a solution culture that included all essential elements except the one being investigated. Deficiency symptoms can also be identified on plants growing in field soils that are severely deficient in a specific nutrient. Visual deficiency symptoms of each nutrient have been obtained for 'Smooth Cayenne' (Sideris and Young, 1951, 1956; Cibes and Samuels, 1961; Tisseau and Tisseau, 1963), although 'Singapore Spanish' (Kanapathy, 1959) and 'Red Spanish' (Cibes and Samuels, 1958) were also studied. These visual symptoms and the leaf concentrations of the nutrients associated with them can be used to diagnose nutritional problems in the field. However, multiple deficiencies may occur in the field that can make diagnosis difficult without confirming plant tissue and soil analyses. Also, not all nutrient deficiency symptoms are diagnostic.

Nutrition and fruit quality

In the context of this chapter, fruit quality is primarily determined by soluble solids content (SSC), titratable acidity (TA), the SSC:TA ratio and volatile flavours, which are largely determined by solar irradiance and temperature (Py *et al.*, 1987; Dorey *et al.*, 2016a,b). As long as plant nutrient levels are near optimum at the time of forcing, the grower has little opportunity to manage quality beyond harvest maturity and postharvest handling.

Plant nutrient characteristics, typical deficiency levels and symptoms

Nitrogen (N)

Nitrogen is required by pineapple in large amounts because of the very large amount of fresh biomass, ~175 Mg ha^{-1} (70,000 plants × ±2.5 kg at forcing) that accumulates during the vegetative growth. It is essential to provide adequate supplies of N to rapidly growing plants to maintain high rates of growth and produce good yields. Both leaf size and number may decrease when N is deficient. Fruit and crown mass are consequently reduced and slips may be abnormally absent. No consistent direct effect of N on fruit quality was found. However, foliar sprays applied after forcing may reduce fruit acidity (Py *et al.*, 1987) and result in fruit nitrate levels that make the fruit unsuitable for canning (Scott, 1993, 1994). No information was found that correlated pre- or post-forcing leaf N with fruit nitrate levels.

Plant indices for N include leaf colour, nitrate-nitrogen in leaf basal white tissue, total N in green tissue, and leaf chlorophyll content. Leaf colour is a diagnostic index for N (Sanford, 1962). When N is deficient, leaves are yellow-green (No. 1 leaf colour in Fig. 8.2). Nitrogen is

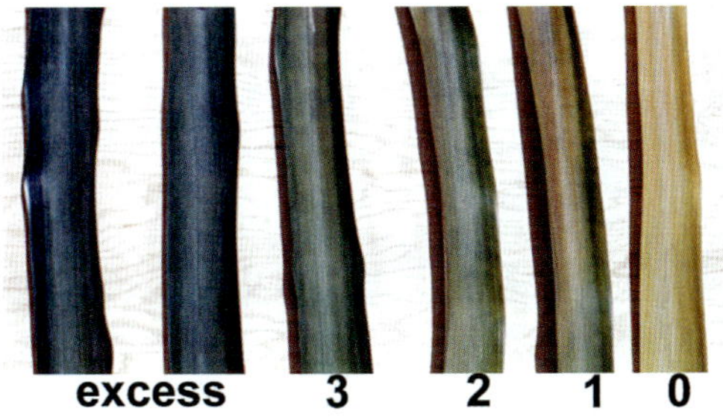

Fig. 8.2. 'Smooth Cayenne' pineapple leaf colours associated with excessive (left) to severely deficient (right) N supply. The black-green colour (left) could also result from growth in shade. Leaf colour 1 corresponds to the no. 1 colour referred to by Sanford, 1962. Photo W.G. Sanford. Reproduced with permission of Claire Sanford.

a mobile element so a well exposed N-deficient plant will have yellow older leaves because the N will have been translocated to the younger leaves. However, N deficiency is seen on young leaves of pineapple plants in the field. Older leaves remain green despite the removal of N from them because of mutual shading of lower and older leaves by adjacent plants. So N deficiency is seen on younger leaves when N is insufficient.

The N in foliar sprays should be applied in amounts and at frequencies approximately related to the increase in plant mass over time and, in fact is the most common method for managing the N supply because scheduling of equipment on large plantations is such a critical component of their operation. The frequency and quantity of N applied will determine the average leaf canopy greenness. With longer intervals between applications and smaller quantities of N per application, leaves will become greener after the spray and then will gradually become more yellow as the plants grow. Research in Hawaii showed that mild N deficiencies during vegetative growth had little impact on plant weight at forcing or on average fruit weight as long as the plant canopy leaf colour was fully green (15% No. 1 colour for the canopy = 85% No. 2 and 3 colour) when plants were forced (Sanford, 1962).

Leaf greenness is a measure of leaf chlorophyll content so deficiencies of Fe and Mg and the rare over-application of herbicides can reduce chlorophyll levels and interfere with the use of leaf colour as an indicator of N sufficiency. The elements N, Fe and Mg are directly or indirectly involved in leaf chlorophyll synthesis. Therefore the successful use of leaf colour as an index is dependent on eliminating, minimizing or recognizing all factors other than N that also can influence leaf colour.

Young (1961) proposed that measures of both chlorophyll and total leaf N would provide a more quantitative substitute for the leaf colour index. For plants that were 6–11 months old, as leaf N in the D-leaf MG tissue increased from approximately 14 to 30 g kg^{-1}, leaf chlorophyll content in the same tissue increased linearly from <200 to 400 mg kg^{-1} (Table 8.3). However, Young (1961) reported that as plants grew older the chlorophyll increased by as much as 10% while leaf N remained approximately constant. The author stated that differentiation took place at 14 months, which would normally be during the winter when solar radiation would be about half that received in summer. It seems more likely that the higher chlorophyll levels in the leaves of older plants at a given N level was due

Table 8.3. Relationship between chlorophyll (mg kg^{-1} fresh mass) in the middle 1/3 of 'D'-leaf green tissue, total nitrogen (g kg^{-1}, dry basis) in the same tissue at two plant age ranges (Young, 1961). Regression for chlorophyll × N for 6–11-month-old plants, $N = 7.6384^{e0.0032x}$, $R^2 = 0.91$ and for 12–15-month-old plants, $N = 5.8381^{e0.0038x}$, $R^2 = 0.97$. Data reproduced with permission of D. Nellis, of the Pineapple Growers Association of Hawaii.

Chlorophyll	Leaf N (6–11 mth)	Leaf N (12–15 mth)
400	31	28
380	27	24
360	20	25
340	23	20
320	21	19
300	20	18
280	19	17
260	18	16
240	17	15
220	16	14
200	15	13
180	13	11

to decreased solar radiation rather than to increased plant age. It is well known that at a given N level shaded leaves are greener and have higher chlorophyll levels than do leaves growing in full sun. Such results indicate that remote sensing of leaf chlorophyll could be diagnostic for N sufficiency.

Much has been published on the use of chlorophyll as an N sufficiency index and Cammarano *et al.* (2014) successfully estimated leaf canopy N in wheat using remotely sensed chlorophyll content. In 2015, Jung-Rothenhaeusler *et al.* (2016) reported that unmanned remote sensing technology was used to manage soil erosion and to maximize fertilization efficiency on a pineapple plantation in the Philippines. Since leaf greenness is determined by leaf chlorophyll content, the well-documented relationship between leaf chlorophyll and leaf N would place N on the list of efficiency improvements.

In terms of leaf N levels, dry mass basis, a 'Smooth Cayenne' plant having 12 g kg^{-1} N in the D-leaf MG would be quite deficient, while values at forcing (12–15 months in Table 8.3) should be about 22–23 g kg^{-1}. Py *et al.* (1987) stated that a WL N level less than 10 g kg^{-1} would limit growth, while all N greater than that was assumed to be available to support new growth.

In the case of 'MD-2', a D-leaf BG N content, g kg^{-1}, are deficient, <15; low, 15–17; optimal, 17.5–18.5; and excessive due to over fertilization, >20. Corresponding amounts for MG tissue are, in g kg^{-1}, <17, 17–19, 20–23 and >25. Boom spray overlap in beds can result in very high leaf N levels.

Variations in N concentrations in successive D-leaves are correlated with the growth of each D-leaf. The quantity of N available for growth can thus be determined in relation to the N content in the leaf. Leaf analyses made in a large number of fertilization trials showed that the critical mass of N required in the D-leaf at flower induction to obtain a fruit of 1.8 kg (without crown) was 100 mg (Marchal, 1975, as cited by Py *et al.*, 1987).

Nitrogen in soil is generally not measured, but pre-plant soil N was measured in Australia to help keep nitrate levels in fruit to be canned below the critical level for processed fruit (Newett and Rigden, 2015). The optimum level of N in soil based on water extraction after a 2-week incubation was ≥120 mg kg^{-1} nitrate (Swete Kelly, 1993). Pre-plant N recommendations were based on pre-plant levels of soil nitrate (Table 8.4). Total plant-crop requirements for N for pineapple range from 250 to 700 kg ha^{-1} (i.e. 4–10 g $plant^{-1}$) depending on the soil and ecology of the site, plant population density, expected fruit mass, and other environmental or management factors. Calibration experiments, leaf colour indices, and tissue indices can all be used to guide growers to the correct amount of N required for optimum growth.

Nitrogen moves in the soil to plant roots mainly by mass flow, though, as root systems grow, root interception could contribute to N uptake. Pineapple has a very low transpiration rate so water flux to the roots is low. No studies of N movement to roots were found but it is clear that mass flow is an important transport mechanism when pineapples are fertilized by soil or basal leaf applications. Soil applications of soluble forms of N, a practice more common on small, non-mechanized farms, require water to transport them to the root system. Transport of ammoniacal forms of N needs to be relatively rapid to avoid N losses due to volatilization.

While mature pineapple leaves tolerate high concentrations of fertilizer salts well, soil application of excessive amounts can injure or kill the roots by desiccation. Spray applications of fertilizers can cause plant damage if the

Table 8.4. Pre-plant nitrogen and potassium recommendations for 'Smooth Cayenne' pineapple in Australia with varying amounts of residual soil nitrate nitrogen and potassium (Swete Kelly, 1993). Used with permission of the Department of Agriculture and Fisheries, Queensland. Australia.

Pre-plant N (mg kg^{-1})	N to apply (kg ha^{-1})	Pre-plant K (mg kg^{-1})	K to apply (kg ha^{-1})
125	0	150	0
100	20	120	75
75	45	100	125
50	70	80	175
25	95	60	225
0	100	40	275
		20	325
		0	375

concentration and volumes are both high. Most fertilizer sprays are applied in high volume, usually 2000–2500 l ha^{-1}, but the macronutrient and micronutrient concentrations do not exceed 5% and 1% respectively. The concentration of macronutrients can be as high as 20% in low volume sprays because the volume, usually 350–500 l ha^{-1}, is too low to flow to the bases of the very young leaves. However, if low-volume sprays dry on the leaf surface, uptake would be delayed until the leaf surface is wetted by dew or rain. In environments with hot and dry periods during the year, it may be necessary to alter the N source or reduce its concentration, or both, in spray mixes to decrease the danger of fertilizer burn.

Nitrogen sources can be quite versatile. Ammonium sulfate can be used to lower soil pH while calcium or potassium nitrate can be used to adjust soil cation balances. However, adding a large quantity of an ammoniacal fertilizer to the soil will release large amounts of ammonia, a cation, which could interfere with the absorption of other cations. If all N is applied to the soil, anion and cation retention could be managed by using ammoniacal fertilizers (a cation source) in the wet season and nitrate fertilizers (an anion source), which are more susceptible to leaching, during dry periods. In reality, most N is applied in foliar sprays, which guarantees delivery to the site of utilization. The benefit of that is that there is minimal risk to the environment. Reinhart (2000) found that soil N posed a threat to ground water while large amounts of N applied as foliar sprays did not. Also, N applied in fertilizer sprays has little or no effect on soil chemistry.

Phosphorus (P)

The visual symptoms of P deficiency are similar to those seen on plants suffering from drought or root injury due to nematode damage, insect feeding or mealybug wilt. In cultivars that contain anthocyanins in their leaves, which includes 'Smooth Cayenne' and 'CO-2' ('73–50'), but not 'MD-2', young leaves appear dark green while older leaves are purple-red in colour with yellow margins (Fig. 8.3). With severe deficiency, the distal section of older leaves die (leaf-tip dieback). Plants with a phosphorus content of 1.08 g kg^{-1}, in the basal white tissue of the D-leaves will have the described symptoms. However, critical values are difficult to define because leaf P tends to decrease when growth is rapid.

Fig. 8.3. Phosphorus deficiency in 'CO-2' on left and healthy plant on right. Photo G. Sanewski.

Mycorrhizal associations exist in pineapple roots (Mourichon, 1981). However, mycorrhizal fungi apparently do not contribute significantly to the P nutrition of pineapple, except when soil P is very low, <0.02 mg l^{-1} of soil solution (Aziz *et al.*, 1990). In a study, which did not include data on P in soil or leaves, average fruit mass and yield per 12 m^2 plot were significantly greater than the control where plants were inoculated with *Glomus mosseae* and *Glomus manihotis* or to either mycorrhizae species alone (Thamsurakul *et al.*, 2000). If mycorrhizae did facilitate P uptake, it would only postpone the time when P fertilization would be required. Soil P stimulates root growth and the soil environment can have a large impact on P uptake. For example, the Entisol in Table 8.5 has a high P content but leaf P can be low if plants have a poor root system or the soil becomes very dry.

Table 8.5. Effect of soil type, pH and extraction method on plant-available phosphorus in the lower 1/3 of 'MD-2' pineapple D-leaves (BG). Values are means of 12 determinations (J. Vásquez-Jiménez, 2017, unpublished data).

Soil	Soil pH (KCl)	P (mg kg^{-1}) (resins)	P (g kg^{-1}) (KCl- mod Olsen)	P_2O_5. (kg ha^{-1})	Leaf P (g kg^{-1})
Entisol	3.87	104.52	31.50	150	1.6
Inceptisol	4.18	17.67	5.67	150	1.9
Ultisol	3.77	15.22	5.83	150	1.2

The range of P in 'Smooth Cayenne' and 'Queen' cultivars is 1.0–1.6 g kg^{-1} except for 'Smooth Cayenne' plants grown in a cool, high elevation environment where it ranged from 2.0–3.5 g kg^{-1} (Table 8.2). For 'MD-2', a P content of less than 2.0 g kg^{-1} (BG) of the D-leaf is considered deficient while anything greater than 2.0 is considered adequate (J. Vásquez-Jiménez, 2017, unpublished data). In 'MD-2', foliar P contents greater than 2.0 are usually associated with soils with high availability of P. If available P in the soil is low, spray applications of P will bring leaf levels above 2.0 g kg^{-1} (BG).

Phosphorus moves in the soil mainly by diffusion on a concentration gradient from the soil to the root zone. There must be adequate soil moisture, both for nutrient diffusion and root uptake. Placement of P fertilizer in the root zone will increase plant access to that nutrient. Soil type also affects P sorption and availability over time and volcanic and highly weathered soils such as oxisols and ultisols have high P sorption capacities. In general, such soils don't represent a problem for pineapple because it has a low P requirement and growth is relatively slow.

Growers should have their soils tested to determine P levels and to estimate the amount to be applied. Because of the variable chemistry of the variety of soils where pineapple is grown, it is important to choose a phosphate fertilizer appropriate to the soil. Phosphorus fixation can occur within 35 days after application and if fixation is a concern, banding of the fertilizer below and to one side of the plant line will provide an adequate supply to the crop (Newett and Rigden, 2015). If the pH of the soil is less than 4 and liming isn't an option, diammonium phosphate would be the preferred fertilizer. However, if the pH is 4.5–5.5, mono-ammonium phosphate or triple superphosphate can be considered, with triple superphosphate used where Ca is limiting. In soils where S may be required, simple superphosphate could be an option if the cost is justified.

Phosphorus applied as a foliar spray is more effective than plant dry basal applications, but if applied with $FeSO_4$ can result in severe iron deficiency (Sanford, 1959). Foliar sprays of P are used by pineapple growers (J. Gonzales, 2017) and presumably would be of most benefit to a ratoon crop.

Phosphorus is one of the most expensive fertilizers and where cost represents a problem, rock phosphate, which was used extensively in Hawaii, represents a low-cost slow-release source of P. Because of its slow-release characteristics, the fertilizer is best suited to acid soils since they solubilize the P most rapidly.

Potassium (K)

During the crop cycle, the pineapple requirement for K is greater than for any other nutrient. In some countries, there is an assumption that K is the element of quality, a claim that caused some to apply excessive amounts of the nutrient. As with P, the primary index for exchangeable K, the form most available to plants, is the level in the soil measured by soil analysis. The high requirement of pineapple for K has made it relatively easy to reproduce deficiency symptoms. Below-optimum levels of K are associated with shorter leaves that are narrower in relation to their length, growth is reduced, pale yellow spots develop on the leaves and leaf tips die back (Py *et al.*, 1987; Swete Kelly, 1993). During the early stages of K deficiency, leaves are dark-green and narrow (Fig. 8.4), but, if the deficiency is prolonged, leaves eventually become yellow.

Visual K deficiency symptoms are evident when, on average, there is less than 24 mg kg^{-1} of K in leaf tissue on a dry-mass basis (Table 8.2). However, there may be small differences among cultivars. The critical leaf K level at flower induction for 'Smooth Cayenne' was reported to be 36 mg kg^{-1} (BW) (Swete Kelly, 1993) or about 30 mg kg^{-1} (WL) (Dalldorf, 1975). Swete Kelly (1993) recommended a leaf K for plants 3–5 months old of 3500–4000 ppm (BW fresh basis; 29–33 g kg^{-1}, dry basis).

To meet the high plant requirement for K, large quantities of K fertilizer are applied to the soil, or as a foliar spray, or both. The optimum soil K level at planting is 150 mg kg^{-1} (neutral ammonium acetate extraction) and deficiency symptoms are observed when the soil level is below 60 mg kg^{-1} (Swete Kelly, 1993). Soil analysis should be done to guide soil application of K because the recommended amounts of pre-plant K vary with amounts in the soil (Table 8.4) and the soil chemistry. Potassium is well retained by most soils, but soils with a low CEC cannot retain the quantities of K required by a pineapple crop.

Fig. 8.4. Symptoms of potassium deficiency (very narrow, dark green leaves) for 'Smooth Cayenne' (left) and 'MD-2' (right). Photos W.G. Sanford (left) and J. Vásquez-Jiménez (right).

Soil losses can be reduced by making split applications to the soil (Junqueira *et al.*, 2011) or by applying part of the K as a foliar spray.

A large pre-planting application of a K fertilizer can bring productive benefits, but it also could result in nutrient imbalances that interfere with the absorption of Mg, Ca or Zn, or of more than one of them. If the soil analysis indicates a potential imbalance the nutrient balance can be maintained by applying up to half the required K to the soil and the remainder by foliar sprays. On most farms, N will be applied as a spray in any case and adding K to the spray solution will pose no problems for field operations or for plant nutrition.

Potassium, which reaches roots primarily by diffusion, can be supplied to the pineapple crop and to the soil as KCl, K_2SO_4 or KNO_3 with KCl being the least expensive and most used source. The sources would not significantly affect soil pH and the benefits from the added N or sulfur in the other forms do not justify their cost. Results regarding possible reduction in growth caused by KCl are not consistent. Junqueira *et al.* (2011), found that pineapple plant mass decreased linearly as leaf Cl increased even though no chlorine damage symptoms were found on the leaves. Fertilization with KCl also decreased biomass accumulation as compared to K_2SO_4. Lacoeuilhe (1979) also reported an improvement in fruit weight, as well as a higher SSC and lower acidity with K_2SO_4 than with KCl. On the other hand, Tapchoi (1990) found no significant effect of KCl, K_2SO_4 and KNO_3 on plant weight, leaf weight, stump weight, leaf numbers, leaf width or leaf length. However, plants from the KNO_3-treated plot tended to have greater fresh weight than the other treatments and had the highest yield. Fruit from plants in a KCl+KCl treatment had the highest titratable acids (0.41%) while fruits from plants in the K_2SO_4 treatment had only 0.34% acid. There was no significant treatment effect on crown weight, fruit diameter, fruit length, core diameter or soluble solids content (SSC).

Calcium (Ca)

Pineapple has a very low requirement for Ca and the level in soils for optimum growth should be >100 mg kg^{-1} (neutral ammonium acetate extraction), a level approximately one-tenth that normally recommended for most crops. Calcium is particularly important in the initial stages of crop establishment because cell division and differentiation are dependent on an adequate supply. Calcium is also important after forcing because it also is a period of rapid cell division and growth. Calcium may improve cell structure and reduce fruit translucence. The fruit translucency index (TI) of a low-acid hybrid decreased as the amount of Ca applied to the soil was increased and 64% of the decrease in TI was accounted for by applied Ca (Silva *et al.*, 2006). The TI also was negatively and significantly correlated with extractable soil Ca, Ca in BW and MG tissues, and Ca in the fruit. Higher levels of fruit Ca are also associated with reduced incidence of storage disorders (Wilson Wijeratnam *et al.*, 1996; Selvarajah *et al.*, 1998; Herath *et al.*, 2003).

Deficiency symptoms are observed when soil Ca is less than 25 mg kg^{-1}. Deficiencies can occur in highly weathered soils low in basic cations and where the soil pH has been lowered by long-term use of acidifying fertilizers, such as ammonium sulfate. Calcium moves in the soil by mass flow and moves apoplastically in the root to the plant xylem and then is transported to the stem and leaves in the transpiration stream. Once Ca is incorporated into the plant tissue, it is immobilized so deficiency symptoms appear first on younger tissues. Because Ca moves in soil primarily by mass flow it is a difficult nutrient to manage in agroecosystems where periodic dry periods occur and irrigation water is not available. When water stress slows growth and nutrient uptake, mobile nutrients can move from older to younger tissue to support growth to the extent that water supply permits while no such movement is possible for Ca.

Visual symptoms of Ca deficiency on vegetative and reproductive plants were developed in sand culture (Sanford, 1962) and observed in the field where an early symptom of deficiency was a grimy grey-green leaf appearance rather than the normal clean yellow-green or green colour (Fig. 8.5), (W.G. Sanford, 1992, Honolulu, HI, personal communication). Initially growth of Ca-deficient pineapple plants may not be stunted, but growth depression becomes evident as the deficiency becomes more severe. Symptoms of Ca deficiency, as with those of boron, are likely to be seen first on the fruit because the demand for both Ca and B in the growing point is highest during reproductive development. Severe Ca deficiency interrupts cell division in the apical meristem and other cells tear apart because of weak cell walls. As a result, new leaves may appear to have been cut off or are tipless, can have serrations or scalloping along the margins, and leaves may develop callus, be abnormally thick and have streaks of corky tissue running parallel to their length. With extreme deficiency, the growing point may die, resulting in the growth of side-shoots, which initially may be symptom-free. Plants deficient in Ca may fail to produce an inflorescence and the leaves of such plants become progressively shorter and roots are thicker than normal, making plants exceptionally difficult to pull from the ground. Fruits may be abnormal in size and shape (Fig. 8.5).

Symptoms were observed on 'Smooth Cayenne' plants in Hawaii at a D-leaf (BW) Ca concentration of 0.24 g kg^{-1} while Swete Kelly (1993) reported the critical concentration (BW) at flower induction for 'Smooth Cayenne' was 1.8 g kg^{-1} and deficiency symptoms developed when Ca was <0.48 g kg^{-1}. The Ca requirement

Fig. 8.5. Symptoms of calcium deficiency of a 'Smooth Cayenne' vegetative plant (a) and fruits (b) from a Ca-deficient (left) and Ca-adequate (right) plant. Photo A by Pineapple Research Institute with permission of D. Nellis, of the Pineapple Growers Association of Hawaii; Photo B by W.G. Sanford and reproduced with permission of Claire Sanford.

for 'MD-2' appears to be higher than for 'Smooth Cayenne' and the concentration in the leaf (BG) depends on the availability of Ca in the soil. Where soil Ca is very available, tissue levels can exceed 10 g kg^{-1}; however, in the typical soil where 'MD-2' is grown (utlisols and oxisols with udic moisture regimes) it is unusual for leaf Ca (BG) to exceed 4 g kg^{-1} and levels of less than 2.4 g kg^{-1} are common. For fresh fruit of 'MD-2', (J. Vásquez-Jiménez, 2017, personal observation) recommended >2.5 g kg^{-1} (BG, dry basis). In a study where soil Ca (0–15 cm) ranged from 169 to 2038 mg kg^{-1} near the time of forcing, Ca in the BW tissue was more than twice that found in the MG tissue. However the correlation between the two tissues was 0.89 and was significant at the $P < 0.01$ level (Silva *et al.*, 2006). In the whole D-leaf, leaf Ca should be 1.0 g kg^{-1} of dry matter at the time of induction (Py *et al.*, 1987).

If best soil management practices are followed, growers should almost never encounter plants with signs of Ca deficiency. However, if such symptoms are observed, or, though unlikely, if the last tissue analysis prior to forcing shows Ca to be dangerously low, a high-volume spray application of calcium nitrate can be applied. The high-volume spray will carry calcium nitrate to auxiliary roots that surround the stem as well as to roots near the soil surface, which should help meet plant Ca requirements. Soil application is another option, however, with typical plant populations, access to the plant base-soil surface area is almost impossible and without rain, calcium nitrate will not reach the root system.

Magnesium (Mg)

Magnesium is a component of the chlorophyll molecule, and a deficiency will reduce chlorophyll concentration, photosynthesis and growth. Magnesium is mobile in the plant and the predominant visual symptom of Mg deficiency is bright yellow older leaves (Fig. 8.6), particularly leaves or parts of leaves that are exposed to sunlight. However, they retain their green colour if shaded, so the presence of green bands where older leaves are shaded by younger leaves is a characteristic symptom of this deficiency (Py *et al.*, 1987; Swete Kelly, 1993; Newett and Rigden, 2015). Sanford (W.G. Sanford, 1992, Honolulu, HI, personal communication) stated that the symptoms of Mg deficiency are most pronounced just prior to floral differentiation and all leaves may be yellow during fruit development if the problem is not corrected. Probably because of the deficient plant's reduced capacity to assimilate CO_2, the stems of Mg deficient plants are short and have a small diameter, plant root systems tend to be weak so plants are easily pulled from the soil and fruits are low in acidity, sugar content and aroma (Py *et al.*, 1987).

At planting time, the optimum level of Mg in soil is 50 mg kg^{-1} (neutral ammonium acetate extraction) and Mg deficiency occurs at levels below 10 mg kg^{-1} (Swete Kelly, 1993). Tissue concentrations in the 'Smooth Cayenne' D-leaf are 1.8 g kg^{-1} when Mg is deficient while the optimum concentration is ≥3.0 g kg^{-1}. Concentrations for other cultivars are similar (Table 8.2).

Fig. 8.6. An 'MD-2' pineapple plant with symptoms of magnesium deficiency. See green bands in photo on right where the leaf has been shaded. Photos J. Vásquez-Jiménez (left) and S. Newett (right).

Magnesium reaches plant roots by mass flow, by root interception, or both, but is also readily absorbed by leaves. If foliar analysis shows that Mg levels in pineapple leaves are below optimum, the disorder can be corrected by soil applications of granular kieserite or foliar sprays of magnesium sulfate. Dolomite also contains Mg, but is not recommended as a way to correct a magnesium deficiency. Plants that are marginally deficient in Mg, if growing rapidly while exposed to high irradiance (>22 Mj m^{-2} d^{-1}) may show the symptoms of yellowing described above. In such conditions, high doses of magnesium sulfate (>40 kg ha^{-1}) as spray fertilizations every 15–30 days will reduce the symptoms. Adding 50–90 kg ha^{-1} of a kaolin reflectant to the spray will protect leaves against high irradiance.

Some herbicides will cause leaf yellowing and the effect can be counteracted by applying 40 kg ha^{-1} of magnesium sulfate as a spray the day following application of the herbicide, either alone or in a mixture with 20 kg ha^{-1} of molasses.

Sulfur (S)

Sulfur deficiency is rare and leaf tissue levels typically are at or above the normal levels indicated in Table 8.2. Deficient plants have bright lemon yellow leaves that are broader than normal. Sulfur is classified as an immobile nutrient in plants; however, both young and old leaves of S-deficient vegetative pineapple plants are yellow. If the deficiency persists, later-formed leaves become narrow, plants are stunted and fruit size is reduced (Py *et al.*, 1987). Leaf S (WL) for 'Imperial' pineapple was normal at 1.45 g kg^{-1} and deficient at 0.45 g kg^{-1} (Mota *et al.*, 2011).

The S requirement for pineapple is low relative to other macro-elements and applications are not common because it is usually the accompanying ion when K, Mg and most micronutrients except B are applied as spray fertilizers. Sulfur moves in soils by mass flow and is taken up by roots as the SO_4^{-2} ion. As with other nutrients that depend on water for their movement, when soils dry out irrigation is required to facilitate plant uptake.

When ammonium sulfate was a common source of N for pineapples, plants tolerated the high levels of sulfate well. Pineapples are one of the few crops that can be profitable when grown in acid sulfate soils, though probably because they tolerate soluble Al well (Cho *et al.*, 2002; Guong *et al.*, 1997).

Iron (Fe) and manganese (Mn)

Iron is an immobile nutrient, so deficiency symptoms always appear first on younger leaves. The initial symptoms include interveinal chlorosis (yellow-green or yellow) while leaf veins, which run parallel to the length of the leaf, remain green (Py *et al.*, 1987) but leaves also become yellow with green mottling (Newett and Rigden, 2015). If the deficiency becomes more severe, the entire surface of the leaves may be pale yellow (Fig. 8.7) or creamy white, leaves are soft and leathery, rather than rigid, and considerable red pigmentation can be seen at the terminal ends of cultivars that have anthocyanin pigmented leaves. Fruits on plants with severe Fe deficiency will be small, hard and reddish in colour, have cracks between the fruitlets and crowns will be light yellow or creamy white in colour.

Visual symptoms are used to diagnose the disorder in Queensland, Australia because 'soil and leaf tissue analyses can be misleading' (Newett and Rigden, 2015). Also, visual symptoms of Fe deficiency will be seen before any measureable effect on yield is found and spraying iron sulfate will quickly cure the disorder (Swete Kelly, 1993).

Where soluble Mn is high, common in soils having high Mn and low pH, the soluble Fe:Mn

Fig. 8.7. Symptoms of iron deficiency in 'CO-2' slips. Inset shows mild symptoms in 'MD-2'. Main photo D.P. Bartholomew, inset G. Sanewski.

ratio in the soil and in the D-leaf (WL) is more important than the absolute amounts of either element (Hopkins *et al.*, 1944; Py *et al.*, 1987). According to Py *et al.* (1987), Fe deficiency occurs and visual symptoms are observed where Fe in the D-leaf (WL) is 60–475 mg kg^{-1} and the Fe:Mn ratio is less than 0.4 for 'Smooth Cayenne' while a ratio of 0.2 (BG) is recommended for 'MD-2' (Vásquez-Jiménez, 2010). Fe deficiency has also been observed in Queensland, Australia when cold, wet soils prevent the uptake of Fe or where roots have been damaged by pests (Newett and Rigden, 2015).

Iron sulfate sprays, often applied fortnightly on plants grown in soils high in soluble Mn, are used to manage Fe deficiency. To be effective, the Fe in iron sulfate sprays must be in reduced form and dry storage conditions are required to prevent oxidation.

Observations about an innate inability of pineapple to absorb Fe (Collins 1949), have transcended the borders of several countries producing pineapples and nowadays Fe applications are included in fertilization programmes 'by default'. As a result, it is assumed, not always correctly, that pineapple plants have problems absorbing Fe even in conditions of high availability. Vásquez-Jiménez (2010) hypothesized that regular iron sulfate sprays, a practice brought to Costa Rica by Hawaiian pineapple companies, might not be required in all of the diverse soils in Costa Rica. The hypothesis was tested using a 'total fractionation of soil elements' method to estimate the amount of Fe in three soil orders in Costa Rica. With the 'total fractionation' method, the forms of the elements in their different states were determined. In contrast, the principle that governs conventional analyses, for example the modified Olsen and Mehlich 3 methods, is that they provide an estimation of soil bioavailability of the element, in this case Fe, to plants. The inconsistency between Fe availability estimated by conventional methods of soil analysis, which showed levels to be higher in the entisol than in the inceptisol, and leaf tissue analysis results, where levels were much higher in plants grown in the inceptisol than in the entisol, is striking (Table 8.6). The inconsistency is explained by the greater amount of assimilable Fe (the sum of Fe in soluble, exchangeable and organic matter fractions) in the inceptisol than in the entosol. Where only assimilable forms are considered, there is good correspondence between assimilable Fe in the soils, in mg kg^{-1}, 2024 in the entisol and 5333 in the inceptisol, and Fe in leaf tissue. In this research, three common soil extractants did not measure all forms of plant available Fe and such inconsistent results could lead one to apply Fe sprays based on inaccurate information. The methodology used in this research can also be used for all metallic elements.

Table 8.6. Iron contents of two Costa Rican soil orders measured by three conventional methods and by measurement of soil Fe fractions and uptake by pineapple leaves. All units are mg kg^{-1} (N = 12 for each group). Data from Vásquez-Jiménez, 2010.

Form	Entisol	Inceptisol
Total fractionation of iron in the soils		
Soluble	0.74±0.16	0.09±0.09
Exchangeable	0.28±0.18	1.57±0.19
Associated with organic matter	2,023±250	5,331±1,570
Associated with amorphous oxides	601.3±373	3,646±1,492
Associated crystalline oxides	14,937±1,868	29,351±2,575
Remaining	8,833±3,320	13,817±4,672
Total	26,396±2,677	52,146±4,032
Conventional methodologies results		
Olsen method	342±114	176±40
Mehlich 3 method	419.0±29	105.1±24
Bioavailable (DTPA method)	450.4 ±131	67.0±28
Leaf analysis results		
Fe (BG)	48 mg kg^{-1}	109 mg kg^{-1}

Manganese deficiency is rare, but does occur in soils high in Ca and with a high pH. Py *et al.* (1987) reported that Mn deficiency symptoms are not specific. Affected leaves are marbled with pale green areas, mainly where vessels are located. Mn deficiency occurs and visual symptoms are observed if Mn in the D-leaf (WL) is between 29 and 78 mg kg^{-1} and the Fe:Mn ratio is greater than 10.5 (Py *et al.*, 1987). Despite high levels of soluble Mn, 1000 mg kg^{-1} or higher in many tropical soils (Krauss and Young, 1967; Py *et al.*, 1987), symptoms of Mn toxicity have not been observed in pineapple. Pineapple plants growing in acid soils tolerate high levels of both soluble Mn and Al where other plants either will not grow or show symptoms of toxicity. In acid, high-Mn soils, soluble Mn interferes with Fe absorption, or utilization, or both (Sideris, 1950). As noted above, the Fe:Mn ratio is more important than the absolute amount of either element. When high Mn contents in the soil interfere with Fe uptake, it is good practice to amend the soil with a Ca source (see 'Soil pH and liming' section on p. 177 for more details).

Zinc (Zn)

Zinc deficiency occurs in soils with a pH of 6.0 or higher, with low organic-matter content or where lime or phosphorus were not well incorporated or were applied in excessive amounts (Ca- or P-induced Zn deficiency). When the deficiency is severe, the centre cluster of leaves is mildly-to-sharply curved (Fig. 8.8). When the deficiency develops in older plants, the surface of older leaves develop yellowish-brown pinhead-sized dashes (Fig. 8.9) that coalesce to form elevated, blister-like spots (Newett and Rigden, 2015). The centre leaves may have rips or serrations on their margins. In less severe cases, the blisters occur only on the older leaves and the centre leaves are only slightly curved. Occasionally, the curved leaves will be seen without blisters. At times, Zn deficient plants, like Ca-deficient plants, have been observed to remain continuously vegetative. Zn deficiency is distinguished from Ca deficiency by the curved central leaves and the presence of blisters on the leaf surfaces.

Leaf tissue does not provide diagnostic information about Zn deficiency in pineapple because no relationship was found between Zn visual deficiency symptoms and leaf tissue Zn levels (Py *et al.*, 1987). Zn in stem tissue was diagnostic, however, it required sacrificing the sampled plants. A level of less than 33 mg kg^{-1} in stem apical tissue is considered deficient and would be associated with typical Zn-deficiency symptoms. For leaf tissue, Zn deficiency symptoms have not been observed when values for 'MD-2' (BG) were higher than 25–30 mg kg^{-1} (J. Vásquez-Jiménez, 2017, personal observation) while symptoms for 'Smooth Cayenne' would not be expected if the level (WL) was greater than 8 mg kg^{-1} (Py *et al.*, 1987). Newett and Rigden (2015) recommend adding 7 kg ha^{-1} of zinc to soils with less than 3 mg kg^{-1} of the element. Zn deficiency also is easily correctable with sprays of zinc sulfate. In Costa Rica it is common to find Zn values below 30 mg kg^{-1} so two or more sprays may be required to meet crop requirements.

Fig. 8.8. Symptoms of severe zinc deficiency on 'Smooth Cayenne' (a) and moderate deficiency on 'MD-2' (b) pineapple. Photo A by Pineapple Research Institute of Hawaii with permission of D. Nellis of the Pineapple Growers Association of Hawaii; Photo B by J. Vasquez.

Fig. 8.9. Early symptoms of zinc deficiency. Photo S. Newett.

Boron (B)

Boron deficiency in leaves and fruit of several pineapple cultivars has been induced in sand and solution culture. Boron deficiency symptoms, especially cracking followed by corking at fruitlet margins (Fig. 8.10) are characteristic of its role in forming and strengthening cell walls and symptoms are found first on pineapple fruits. The symptoms are similar across cultivars (Smith *et al.*, 1962; Tay, 1974; Siebeneichler *et al.*, 2008; Mota Ramos *et al.*, 2009). Other symptoms have been described and attributed to pathogens (Py *et al.*, 1987) but in Hawaii corking between fruitlets is seldom found since B was added to fertilizer sprays.

Boron deficiency on vegetative leaves includes death of the tips of the youngest leaves, sometimes with serrations on the margins, while with severe deficiency the growing point will die. In such plants the central leaves will be stiffer and shorter than normal and lateral buds will eventually develop. It has been proposed that there is a relationship between B and Ca because the symptom of deficiency of both elements is disruption of cell development. Hernandez-Medina and Lugo-Lopez (1958) studied the relationship between the two elements in 'Red Spanish' pineapple in sand culture fertilized with solutions containing complete nutrients apart from Ca, which was supplied at 5, 50, 100, 250 and 500 mg l^{-1}, and B, which was supplied at 0, 0.001, 0.25, 1.0 and 5.0 mg l^{-1}. In the absence of B, plants receiving 5 mg l^{-1} of Ca produced more than those receiving 500 mg l^{-1} of Ca and a concentration of 50 mg l^{-1} Ca was optimal. Plants fertilized with 5 mg l^{-1} B and 250 or 500 mg l^{-1} Ca produced the highest fruit yield, indicating that there was a metabolic disturbance with no B and high Ca.

Boron transport in the soil to the plant is by mass flow, however, the B requirement of pineapple is so low that leaf sprays of 3–5 kg·ha^{-1} are sufficient to meet plant requirements. When forcing pineapples with ethephon, the addition of 0.5% borax to the spray will prevent B deficiency and improve forcing in Australia (Sinclair, 1994). Sanford *et al.* (1961) stated that very low amounts of B, Mo and Cu were required to correct deficiencies so the cost of establishing critical indices could not be justified. Correcting a possible deficiency of B is easily done by spray fertilization, however, extreme care should be exercised because there is a narrow margin between sufficiency and toxicity.

Copper (Cu)

The leaves of plants deficient in Cu are lighter green than those of normal plants and are narrow and the 'U' shape in cross-section is more

Fig. 8.10. Symptoms of inter-fruitlet corking typical of boron deficiency. Photos G. Taniguchi.

distinct than for normal leaves (Newett and Rigden, 2015). Tips of leaves curve downward instead of being erect. The deficiency is moderately common in the heavily leached sandy soils of southern Queensland, Australia (Swete Kelly, 1993), and has also been observed on pineapples grown in peat soils in Malaysia. The optimum range in the D-leaf tissue (WL) is 5–15 mg kg^{-1} (Table 2) and a whole-leaf level below 5 mg kg^{-1} is considered deficient (Py *et al.*, 1987). The optimum soil level (DPTA method) for Queensland is ≥2 mg kg^{-1} (Swete Kelly, 1993).

Copper deficiency can be corrected by a foliar spray of 0.5% neutralized copper sulfate or a pre-planting soil spray application of 7.5 kg ha^{-1} Cu as copper sulfate (Newett and Rigden, 2015). Newett and Rigden (2015) do not recommend applying foliar sprays of copper salts because of their phytotoxicity. In Costa Rica, applications of copper sulfate that have been reformulated into liquid products for foliar application have caused severe phytotoxicity to 'MD-2' plants at doses higher than 1.0 l in 2000 l ha^{-1} of water (Vásquez-Jiménez, 2017, personal observation). To avoid major damage to fields of 'MD-2' plants, a test application of the Cu spray should be made on 20–30 plants, 20–30 days prior to whole-field application. The absence of necrotic lesions on D, E and F leaves after 30 days indicates the spray can be safely applied. If a multi-element solution that includes Cu is to be sprayed on the field, the above-mentioned test should be performed with the final spray formulation.

Molybdenum (Mo)

There are no known reports of visual symptoms of Mo deficiency on pineapple and also no indication of a pineapple growth response to Mo. Molybdenum is essential for proper functioning of the nitrate reductase enzyme, and it was reported that a foliar spray of 11.7 mg $plant^{-1}$ of Mo at 2 and 4 months after flowering increased Mo in D-leaf tissue (BW) at harvest and reduced the nitrate level in pineapple fruits without any effect on SSC or fruit acidity (Chongpraditnun *et al.*, 2000). Chairidchai (2000) applied 25 ml of a solution containing from 0 to 5 mg of Mo to the tops of 'Smooth Cayenne' plants at the red bud stage and reported that nitrate in crowns of 'Smooth Cayenne' fruits increased linearly with the Mo content in D-leaf tissue (MG), which ranged from 0.7 to 1.4 µg g^{-1}, while pineapple juice nitrate content decreased quadratically with increasing Mo in D-leaf tissue. Research in Australia failed to demonstrate any change in fruit juice nitrate levels as a result of spraying plants with Mo (Scott, 2000).

No studies with Mo were found for 'MD-2', nor was a critical level for the element reported for any cultivar. Steiner and Zoz (2015), citing work of others, reported that for potatoes, Mo levels less than 0.6 mg kg^{-1} reduced growth while levels sustaining normal growth were 0.6–10 mg kg^{-1}. They reported that when 0–104 g ha^{-1} of Mo was sprayed on sunflower plants at the eight-leaf stage, Mo in leaf tissue at flowering was linearly related to the amount applied over the range 26–104 g ha^{-1} and leaf N concentration was linearly related to Mo applied. The results of 107 foliar analyses (BG) of 'MD-2', done for multiple growers located in different agroecosystems in Costa Rica (J. Vásquez-Jiménez, 2017, unpublished data), showed that Mo ranged from 0.11 to 1.74 mg kg^{-1} (BG) while leaf N ranged from 8 to 23.2 g kg^{-1}; however, the correlation between Mo and N in leaf tissue was only 0.014.

Silicon (Si)

Silicon (SiO_2) is a major component of most soils and most crop plants contain some silica. Large increases in sugarcane yields were obtained after application of calcium silicate to highly weathered mineral soils low in Si and organic soils (Fox *et al.*, 1967; Anderson *et al.*, 1986). Other crops also benefit from Si amendments (Heckman, 2013).

Silica bodies are found in the epidermis of pineapple leaves and the size of silica bodies was reduced when plants were grown in a nutrient solution lacking silica (Krauss, 1949). Fewer than ten papers were found that contained information about Si in pineapple, mostly mentioning its presence in plant ash, 4.02% of total ash in Jamaica and 9.4% in Hawaii (Bowrey, 1896; Stewart *et al.*, 1927). Melo *et al.* (2016) reported that Si was not among the abiotic resistance inducers that reduced the incidence of fusariosis on pineapple.

Thiagalingam *et al.* (1977) reported a 6% increase in vegetative yield of 'Smooth Cayenne' plants grown in pots in a Humic Ferruginous Latosol amended with the equivalent of up to 8.8 Mg ha^{-1} of calcium silicate. However, there was no consistent effect of Si on the dry weights of plants grown in either soil, possibly because growth, only about 25% of that for field-grown plants over the same time period, was restricted by the small pot environment and twice-daily watering. Silica in leaves increased with increasing Si applied in both soils while leaf Ca increased and Mn decreased with increasing Si in Paaloa soil (Table 8.7). Thiagalingam (1971) concluded that in pineapple tissue (BW), a low Si level would be 1.0–1.6 g kg^{-1} and an adequate level would be 3.0 or higher.

It is unlikely that pineapple would benefit from applications of Si when grown on soils with adequate amounts of soluble Si. No data on Si in soils in which pineapples were grown were found and such data are needed to help pineapple growers determine the value, if any, of Si application to pineapples. Also, the critical tissue levels of Si in pineapple reported by Thiagalingam (1971) require confirmation. Calcium silicate is an excellent, though probably expensive, liming material and potassium silicate also provides that essential mineral element. Therefore, the potential benefits of these added elements on soil chemistry and plant nutrition cannot be ignored and mixtures that contain Si as well as other plant nutrients cannot be used to test for the specific benefits of Si unless controls also contain equivalent amounts of any elements in a mixture.

Deficiency problems

Deficiencies of N, P, K, Ca and Mg are likely to occur anywhere that pineapple is grown if quantities removed by the crop and lost by leaching are not replaced through fertilization. Deficiencies of S, Fe, Zn, B, Cu, Mn, Mo and Cl are likely to be localized in specific areas where pineapples are grown. The use of inorganic fertilizers containing S make it unlikely that S would become deficient. Since the main function of Mo in higher plants is related to nitrate reduction, Mo is unlikely to become deficient as long as urea and other ammoniacal forms of N continue to be applied to leaves. Boron and Cu deficiency already are found in a few countries and these two nutrients may eventually become limiting in other pineapple growing regions. Iron deficiency is common in many pineapple growing regions while Zn deficiency seems somewhat less widespread.

As long as fertilizers are readily available and inexpensive relative to the value of the crop, and the technology is available to apply these fertilizers after the deficiencies appear, deficiencies of most nutrients should not limit the productivity of pineapple where the crop is grown for commercial purposes because most deficiencies are readily correctable by foliar spray.

Table 8.7. Effect of soil type and silica applied as calcium silicate slag (Mg ha^{-1}) on plant dry mass and mineral composition (g kg^{-1} for Si, P, K, Ca and Mg and mg kg^{-1} for Fe Mn and Al) of leaves of 'Smooth Cayenne' pineapple grown in 3.78 l pots in a greenhouse for 420 days. Data from Thiagalingam (1971) are by means of two replications.

Soil type		Kapaa				Paaloa			
Si (Mg ha^{-1})		0	2.2	5.6	8.8	0	2.2	5.6	8.8
Plant wt.		149	116	139	158	152	174	160	190
	Nutrient								
Leaves	Si	0.5	0.8	1.4	2.6	0.6	1.2	1.9	4.2
	P	1.2	1.3	1.1	0.9	1.2	1.2	1.2	0.9
	K	7.9	11.4	12.1	11.9	23.3	20.6	24	19.8
	Ca	8	7.6	7.7	8.4	3	3.8	5.4	8
	Mg	1.6	1.8	1.6	1.4	1.5	1.4	1.5	1.7
	Fe	150	141	82	79	255	315	252	265
	Mn	87	124	126	99	204	179	165	90
	Al	85	104	78	75	143	162	175	176

Meeting Plant Requirements

Crop residue

The residue that remains at the end of the pineapple crop cycle varies with the culture system, but has been estimated to range from 40 to 60 Mg ha^{-1}, dry-mass basis (Py *et al.*, 1987; Pena Arderi and Dominguez Martin, 1988). Pineapple residue is approximately 75% leaves and 25% stems, with the root system representing a relatively insignificant amount. Most of the dry residue consists of organic compounds, such as sugars, starch, cellulose and proteins. The residue contains relatively large quantities of plant nutrients and Ingamells (1981) reported whole-plant amounts of macronutrients and Si, in g kg^{-1}, were N, 10.3; P, 1.9; K, 19.4; Ca, 3.1; Mg, 1.9; S, 0.7; and Si, 3.1. Micronutrients in the same study were, in mg kg^{-1}, Fe, 180; Mn, 270; Zn, 30; and Cu, 10. As long as pineapple residue is not used as feed for livestock, nutrient removal is limited to what is carried away in pineapple fruits and their crowns.

For many years, at least some growers burned the residue to reduce the fallow period. The current best practice is to incorporate the residue to return the nutrients and organic matter to the soil. Bacterial cultures can speed decomposition before incorporation into the soil which benefits livestock health by reducing sites where stable flies can reproduce (Alpizar *et al.*, 2013). Herbicides have also been used to kill pineapple plants (Lee, 1973; Vásquez-Jiménez, 2016) to speed desiccation and also to make sure that stems don't become a weed in the next crop.

Crop residue nutrient content probably is not of great interest to growers because managing soil nutrients based on what is returned to the soil in the residue doesn't eliminate the need for soil and tissue analyses. The rate of crop nutrient return to the soil pool is mainly based on residue carbon:nitrogen ratio, particle size and soil temperature and moisture content, which makes the rate of return somewhat unpredictable. The process took approximately 26 weeks in Hawaii (Ingamells, 1981) and 8–12 weeks in Costa Rica. Early in the decomposition process, much of the N and probably some of the S and P would be utilized by decomposer microorganisms so could not be relied on as sources of those nutrients for the pineapple crop. On the other hand, available K, which is unbound in plant tissue, would be released to the soil relatively early in the decomposition process while Mg and the tightly bound Ca would be released later.

Ingamells (1981) grew pineapple in four soil series with and without pineapple residue and reported that there was no significant effect of residue on the concentration of N in the plants. However, after 16 weeks, plants grown in soil with residue had higher concentrations of P, K, Ca, Mg and Si in leaves than did plants grown in soil without residue. Since incorporation of residue into the soil also increased the water-holding capacity of the soil, the increase in nutrient elements in leaves may be due to the combination of higher levels of nutrients and increased availability of soil moisture.

The benefits from residue incorporation into the soil are difficult to evaluate, especially for a long-term crop like pineapple. Certainly there is ample time during the crop cycle for complete decomposition of the residue. As with any incorporated organic matter, pineapple residue can improve soil physical, chemical and biological characteristics, but there probably is not an obvious effect on the following crop.

Nutrient absorption and growth

Pineapple plant N and K requirements are low until about 4 months after planting (Lacoeuilhe, 1978; Ingamells, 1981), after which requirements increase with growth until flower induction. Lacoeuilhe (1978) supplied individual plants with 8 g of N and 20 g of K prior to forcing. When plants were forced 10 months after planting, plant dry mass was 375 g and contained 5 g N and 11 g K. Plant N content remained constant from forcing until harvest while K absorption continued to increase from forcing to harvest, at which time 13 g were accumulated in a dry mass of 800 g. If growth is reduced by drought, low temperature or root anoxia, the plant's nutrient requirement is also reduced (Py *et al.*, 1987).

Total uptake of a particular nutrient does not necessarily indicate a plant's requirements for that nutrient. Pineapple plants may take up more, the same as or less than the amounts required for optimum growth, depending on the factors indicated above and the specific nutrient in question. In a field high in available N and K, total uptake over a 30-month period was highest for K, followed by N and P. The efficiency of pineapple in extracting K from the soil is high and, if readily available, more K is accumulated than is required for optimum growth, commonly known as luxury consumption. The uptake of N also shows luxury consumption, generally being proportional to the amount of N fertilizer applied during vegetative growth (Scott, 1993). Luxury consumption of Ca (Godefroy *et al.*, 1971), Mg, S, B, Cl, Cu and Mn also occurs in pineapple when these elements are readily available. An overdose of B and Zn in leaf sprays applied to 'MD-2' can lead to luxury consumption of those elements with levels exceeding 100 mg kg^{-1} (BG) (J. Vásquez-Jiménez, personal observation).

Conversely, P uptake is not proportional to the supply available, but reflects the plant's requirement and this is also true for Fe as long as soluble Mn levels don't interfere with Fe uptake or utilization.

The elements Si, Na and Al also are accumulated when levels are high in the soil. It appears that pineapple is not an accumulator of heavy metals because fruits of 'Giant Kew' (syn. 'Smooth Cayenne') grown on sand tailings in Malaysia 'had negligible contamination of all heavy metals except the insignificant presence of Hg' (Ang *et al.*, 2011). Cadmium was among those elements not accumulated but fertilizer contaminated with Cd resulted in unwitting Cd contamination of canned pineapples produced in South Africa that exceeded permissible levels (Hill *et al.*, 2012).

Relatively high concentrations of Na and Cl are found in soils in close proximity to the ocean, where seawater sprays are prevalent (Sideris and Young, 1954; Sideris, 1955). Sea-water sprays can cause leaf-tip dieback, which results in alternating dark and light brown bands running across the leaves (Sideris, 1955; Py *et al.*, 1987) and Marinho *et al.* (1998) reported that leaf-tip necrosis occurred at salinity levels >4 dS m^{-1}. Pot studies of sodium chloride tolerance of pineapple showed growth was reduced at salt concentrations >2 dS m^{-1} (Wambiji and El-Swaify, 1974, 1976).

Important Nutrients Issues

Environmental quality

The major environmental quality issues are soil contamination and nutrient losses to ground water or in runoff from agriculture fields. The nutrients of greatest concern are N and P and when growers strive for maximum production, the large amounts of fertilizer applied can result in nutrient losses, especially nitrate, in drainage water. Reinhart (2000) found that nitrate leaching in pineapple fields below a depth of 30 cm was greatest where pre-planting applications of more than 300 kg ha^{-1} of inorganic N fertilizers were made to a soil with high residual N. Losses were negligible where manure was used and foliar fertilization with over 700 kg N ha^{-1} did not increase leaching losses (Reinhart, 2000). Phosphorus losses are of little concern because pineapples have a low P requirement and many of the soils pineapples are grown in have a high capacity to fix P. As a result, soil P levels generally are low and losses through leaching are likely to be very small.

Soil erosion from pineapple fields is a major environmental concern. Fernández *et al.* (2012) found yearly soil loss due to erosion of 37 Mg ha^{-1}, of which 9% was lost during the preparation phase and 91% during the first 6 months after planting. The highest soil loss was 26.8 Mg ha^{-1} from roads and 6.6 Mg ha^{-1} in the drainage system. However, soil erosion has caught the attention of management and companies are researching methods to better manage soils in pineapple fields (López *et al.*, 2012; Jung-Rothenhaeusler *et al.*, 2016; Young, 2017). Jung-Rothenhaeusler *et al.* (2016) reported success in using aerial surveys and GIS technology to bring soil erosion to an acceptable level.

Chemical losses through runoff and soil loss may also be significant when pineapple is cultivated on ridges or slopes without soil cover. Losses from volcanic soils in Martinique (kg ha^{-1} $year^{-1}$) were 63 N, 44 K and 56 Ca (Khamsouk, 2001). As long as most N is applied as foliar sprays, losses will be small (Reinhart, 2000). While K losses may be significant, K is not a serious environmental contaminant. However, disproportionate applications can generate serious cation imbalances that from a more rigorous perspective could also be called pollution.

Leon and Kellon (2012), surveyed growers of 'MD-2' pineapple and found that applications of NPK elements varied greatly among producers. Average and high amounts (kg ha^{-1}) were for N, 632 and 950; for P, 129 and 630; and for K, 400 and 750. The results illustrate opportunities for improvement of fertilizer efficiency and a lower environmental impact as fertilization practices are better aligned with crop requirements.

The best methods for making sure that nutrient losses from pineapple culture are kept to a minimum are to develop and follow plans for the control of soil erosion and only apply the amounts of nutrients required by the crop as indicated by soil and foliar analysis.

References

Alpízar M.D., Fallas G.M., Aguilar B.F., Arrieta C.R. and Fallas G.R. (2013) Bacthon® accelerates post-harvest decomposition of pineapple residue and decreases populations of *Stomoxys calcitrans* in Costa Rica. *Pineapple News* 20, 42–45. Available at: http://www.ishs-horticulture.org/workinggroups/pineapple/PineNews20.pdf (accessed 7 March 2018).

Anderson, D.L., Jones, D.B. and Snyder, G.H. (1986) Response of a rice-sugarcane rotation to calcium silicate slag on everglades histosols. *Agronomy Journal* 79, 531–535.

Ang, L.H., Tang, L.K., Ho, W.M., Fui, H.T. and Ramli, M. (2011) Bio-accumulation of mercury, lead, arsenic and cadmium by pineapple grown as an agroforestry crop for ex-tin mines in Peninsular Malaysia. *Acta Horticulturae* 902, 313–318.

Angeles, D.E., Sumner, M.E. and Barbour, N.W. (1990) Preliminary nitrogen, phosphorus, and potassium DRIS norms for pineapple. *HortScience* 25, 652–655.

Aziz, T., Yuen, J.E. and Habte, M. (1990) Response of pineapple to mycorrhizal inoculation and fosetyl-Al treatment. *Communications in Soil Science and Plant Analysis* 21, 19–20.

Balasundram, S.K., Husni, M.H.A. and Ahmed, O.H. (2008a) Application of geo-informatic tools to quantify spatial variability of selected soil chemical properties from a cultivated tropical peat. *Journal of Agronomy* 7, 82–87.

Balasundram, S.K., Husni, M.H.A. and Ahmed, O.H. (2008b) Spatial variability of pineapple yields on a tropical peat. *Crop Management* 7 (1). DOI:10.1094/CM-2008-0418-01-RS.

Bowrey, J.J. (1896) Pine apple: analysis and manure. *Bulletin of the Botanical Department Jamaica, New Series* 3, 236.

Cammarano, D., Fitzgerald, G.J., Casa, R. and Basso, B. (2014) Assessing the robustness of vegetation indices to estimate wheat N in Mediterranean environments. *Remote Sensing* 6, 2827–2844.

Chairidchai, P. (2000) The relationships between nitrate and molybdenum contents in pineapple grown on an inceptisol soil. *Acta Horticulturae* 529, 211–216.

Cho, K.M., Ranamukhaarachchi, S.L. and Zoebisch, M.A. (2002) Cropping systems on acid sulphate soils in the central plains of Thailand: constraints and remedies. *17th World Congress of Soil Science*, Bangkok, Thailand, 14–20 August 2002. Bangkok: Soil and Fertilizer Society of Thailand.

Chongpraditnun, P., Luksanawimol, P., Limsmuthchaiporn, P. and Wasunun, S. (2000) Effect of fertilizers on the content of nitrate in pineapple fruit. *Acta Horticulturae* 529, 217–220.

Cibes, H.R. and Samuels, G. (1958) Mineral-deficiency symptoms displayed by Red Spanish pineapple plants grown under controlled conditions. *Agricultural Experiment Station Technical Paper* 25. San Juan, Puerto Rico: University of Puerto Rico.

Cibes, H.R. and Samuels, G. (1961) Mineral-deficiency symptoms displayed by Smooth Cayenne pineapple plants grown under controlled conditions. *Agricultural Experiment Station Technical Paper* 31. 5. San Juan, Puerto Rico: University of Puerto Rico.

Collins, J. (1949) History, taxonomy and culture of the pineapple. *Economic Botany* 3, 335–359.

Dalldorf, D. (1975) Potassium fertilisation of Smooth Cayenne pineapples. Information Bulletin, Citrus and Subtropical *Fruit Research Institute* 32, 2.

Dalldorf, D.B. and Langenegger, W. (1978) Macroelement fertilization of Smooth Cayenne pineapples. *Farming in South Africa, Pineapple Series* E. 2. Pretoria, South Africa: Department of Agricultural Technical Services.

Das, P.T. and Sudhakar, S. (2014) Land suitability analysis for orange & pineapple: a multi criteria decision making approach using geo spatial technology. *Journal of Geographic Information System* 6, 40–44.

Dorey, E., Fournier, P., Léchaudel, M. and Tixier, P. (2016a) Modeling sugar content of pineapple under agro-climatic conditions on Reunion Island. *European Journal of Agronomy* 73, 64–72.

Dorey, E., Fournier, P., Léchaudel, M. and Tixier, P. (2016b) A statistical model to predict titratable acidity of pineapple during fruit developing period responding to climatic variables. *Scientia Horticulturae* 210, 19–24.

Du Plessis, S. (1985) Pynappels. norme vir blaarontedings.. *Citrus and Subtropical Fruit Research Institute Information. Bulletin* 153, 5–6.

Du Plessis, S.F, Koen, T.J. and Dalldorf, D.B. (1989) Fertilising pineapples. *Farming in South Africa, Pineapple Series* E. 2. Pretoria, South Africa: Department of Agricultural Technical Services.

Fassbender, H.W. and Bornemisza, E. (1987) *Química de suelos con énfasis en suelos de América*. San José, Costa Rica: Instituto Interamericano de Cooperación para la Agricultura.

Fernández, M., Gadea-Rivas, A., Boniche, J.Y. and Furcal Berigüete, P. (2012) Estimación de la erosión en una plantación de piña en la zona norte de Costa Rica. Abstracts, VII Congreso Nacional de Suelos. *Pineapple News* 20, 47. Available at: http://www.ishs-horticulture.org/workinggroups/pineapple/PineNews20.pdf (accessed 7 March 2018).

Fox, R.L., Silva, J.A., Younge, O.R., Plucknett, D.L. and Sherman, G.D. (1967) Soil and plant silicon and silicate response by sugar cane. *Soil Science Society of America Journal* 31, 775–779.

Godefroy, J., Poignant, A., and Marchal, J. (1971) First results from a pineapple liming trial on a lower Ivory Coast soil. *Fruits* 26, 103–113.

Guong, V.T., Tinh, T.K., Trang, T.T.T. and Moi, L.T. (1997) Effect of phosphorus, lime and potassium fertilization on aluminum uptake and pineapple yield in acid sulphate soils in the Mekong Delta, Vietnam. *Acta Horticulturae* 425, 403–409.

Heckman, J. (2013) Silicon: a beneficial substance. *Better Crops* 97(4), 14–16.

Herath, H.M.I., Bandara, D.C. and Banda, D.M.G.A. (2003) Effect of pre-harvest calcium fertilizer application on the control of internal browning development during the cold storage of pineapple 'Mauritius' (*Ananas comosus* (L.) Merr.). *Journal of Horticultural Science and Biotechnology* 78, 762–767.

Hernandez-Medina, E. and Lugo-Lopez, M.A. (1958) Effects of the calcium-boron relationship on growth and production of the pineapple plant. *Journal of Agriculture of the University of Puerto Rico* 42, 207–223.

Hill, G., Fraser, G. and Baiyegunhi, L. (2012) The impact of contaminated fertilizer on pineapple growers in the Eastern Cape, South Africa. *African Journal of Agricultural Research* 7, 3898–3905.

Hopkins, E.F., Pagan, V. and Silva, F.J.R. (1944) Iron and manganese in relation to plant growth and its importance in Puerto Rico. *Journal of Agriculture of the University of Puerto Rico* 28, 43–101.

Ingamells, J.L. (1981) The effects of pineapple residue (trash) on mineralization and early growth of pineapple. PhD dissertation, University of Hawaii at Manoa, Honolulu, HI.

Jung-Rothenhaeusler, F., Traut, K., Gauggel, C., Smith, A.B. and Umali, R. (2016) Pineapple plantation information system – application of geo-informatics and unmanned aerial sensing technology for efficient land use management. *Acta Horticulturae* 1111, 255–265.

Junqueira, L.A., Quaggio, J.A., Cantarella, H. and Mellis, E.V. (2011). Potassium fertilization for pineapple: effects on soil chemical properties and plant nutrition. *Revista Brasileira de Fruticultura* 33, 627–636.

Kamprath, E.J. (1970) Exchangeable aluminum as a criterion for liming leached mineral soils. *Soil Science Society of America Proceedings* 34, 252–254.

Kanapathy, K. (1959) Visual symptoms of major nutrient deficiencies of the Singapore Spanish pineapple. *Malaysian Agriculture Journal* 42, 157–160.

Khamsouk, B. (2001). Impact des systèmes de culture bananiers sur l'érosion, le bilan hydrique et les pertes en nutriments sur un sol volcanique en Martinique. PhD thesis, ENSA Montpellier, France.

Krauss, B.H. (1949) Anatomy of the vegetative organs of the pineapple, *Ananas comosus* (L.) Merr. II. The leaf. *Botanical Gazette* 110, 333–404.

Krauss, B. and Young, H.Y. (1967) Studies on maganese, iron, and nitrogen as related to the growth of pineapple. Part II. Soil- and leaf-manganese and plant response. *PRI News* 15, 91–96.

Lacoeuilhe, J.J. (1978) La fumure N-K de l'ananas en Côte d'Ivoire. *Fruits* 33, 341–348.

Lacoeuilhe, J.J. (1979) N-K manuring of the pineapple in the Ivory Coast. *Potash Review* 27(89), 3.

Lee, S.A. (1973) Ratoon clearing of pineapple with paraquat. *Proceedings of the 4th Asian-Pacific Weed Science Society Conference* Rotorua, pp. 400–406.

Leon, R. and Kellon, D. (2012). Characterization of 'MD-2' pineapple planting density and fertilization using a grower survey. *HorTechnology* 22(5), 644–650.

Lin, Y. (2010) Effects of aluminum on root growth and absorption of nutrients by two pineapple cultivars [*Ananas comosus* (L.) Merr.]. *African Journal of Biotechnology* 9, 4034–4041.

López, A., Matarritá, G., y and Hidalgo, E. (2012) Estimación de la erosión de suelos en plantaciones de piña MD-2 bajo diferentes prácticas de manejo, en la Región Pacífico Sur de Costa Rica. Abstracts, VII Congreso Nacional de Suelos. *Pineapple News* 20, 47–48. http://www.ishs-horticulture.org/workinggroups/pineapple/PineNews20.pdf

Marinho, F.J.L., Fernandes, P.D. and Gheyi, H.R. (1998) Desenvolvimento inicial do abacaxizeiro, cv. Smooth Cayenne, sob diferentes condicoes de salinidade da agua. *Revista Brasileira de Engenharia Agricola e Ambiental* 2, 1–5.

Melo, L.G.D.L., Candido E Silva, E.K., Campos Neto, J.R.M., Lins, S.R.D.O., Rodrigues, A.a.C. and Oliveira, S.M.a.D. (2016) Abiotic resistance inducers for control of pineapple fusariosis. *Pesquisa Agropecuária Brasileira* 51, 1703–1709.

Mota, M.F.C., Pegoraro, R.F., Batista, P.S.C., Pinto, V.D.O., Maia, V.M. and Silva, D.F.D. (2016) Macronutrients accumulation and growth of pineapple cultivars submitted to aluminum stress. *Revista Brasileira de Engenharia Agrícola e Ambiental* 20, 978–983.

Mota Ramos, M.J., Monnerat, P.H., Carvalho, A.J.C.D., Pinto, J.L.D.A. and Silva, J.a.D. (2009) Visual symptoms of macronutrients and boron deficiency in 'Imperial' pineapple. *Revista Brasileira de Fruticultura* 31, 252–256.

Mota Ramos, M.J., Monnerat, P.H., Pinho, L.G.D.R. and Silva, J.a.D. (2011) Macronutrients and boron deficiency in 'Imperial' pineapple: mineral composition. *Revista Brasileira de Fruticultura* 33, 261–271.

Mourichon, X. (1981) Mise en évidence d'une association endomychorhizogéne chez l'ananas en Côte-d'Ivoire. *Fruits* 36, 745–749.

Newett, S. and Rigden, P. (2015) *The Pineapple Problem Solver Field Guide.* Queensland, Australia: State of Queensland.

Ortiz, E.F. (2008) Evaluación del efecto de la cal dolomita sobre algunas características químicas del suelo y la absorción de nutrientes en el cultivo de piña (*Ananas comosus*) (L) Merr. híbrido MD-2 en Finca Ganadera La Flor S.A, en Río Cuarto, Grecia, Costa Rica. Associate degree thesis, ITCR, San Carlos, Costa Rica.

Pena Arderi, H. and Dominguez Martin, Q. (1988) Materia organica y nutrientes que se incorporan al suelo con la demolicion de la pina, *Ananas comosus* (L.) Merr, cultivar Espanola Roja. *Centro Agricola* 15, 84–87.

Py, C., Lacoeuilhe, J.J. and Teisson, C. (1987) *The Pineapple: Cultivation and Uses.*Paris: Editions G.-P Maisonneuve.

Reinhart, A. (2000) Assessment of and potential to reduce nitrate leaching in the Pearl Harbor watershed. PhD dissertation, University of Hawaii, Honolulu, HI.

Sanford, W.G. (1959) Review of research on nutrient foliar sprays, 1916–1959. *Pineapple Research Insitute of Hawaii Research Report* 60, 43.

Sanford, W.G. (1962) Pineapple crop log-concept and development. *Better Crops with Plant Food* 46, 32–43.

Sanford, W.G., Smith, D.H., Krauss, B.H. and Young, H.Y. (1961) Plant nutrition and soil fertility aspects of pineapple culture. *Pineapple Research Institute of Hawaii Research Report* 84, pp. 1–129.

Scott, C. (1993) The effects of: 1. Artificial flower inductants; 2. Nitrogen rates on juice nitrates and yield in pineapples. In: *Pineapple Field Day Notes*, Queensland Fruit and Vegetable Growers, Beerwah, Queensland, Australia, pp. 21–29.

Scott, C. (1994) Update of nitrate trials. In: *Pineapple Field Day Notes.* Queensland Fruit and Vegetable Growers, Beerwah, Queensland, Australia, pp. 9–21.

Scott, C. (2000) The effect of molybdenum applications on the juice nitrate concentration of pineapples. In: *Pineapple Field Day Notes.* Queensland Fruit and Vegetable Growers, Beerwah, Queensland, Australia, pp. 23–27.

Selvarajah, S., Herath, H.M.W., Bandara, D.C. and Banda, D.M.G.A. (1998) Effect of pre-harvest calcium treatment on post-harvest quality of pineapple. *Tropical Agricultural Research* 10, 214–224.

Sema, A., Maiti, C.S., Singh, A.K. and Bendangsengla, A. (2010) DRIS nutrient norms for pineapple on Alfisols of India. *Journal of Plant Nutrition* 33, 1384–1399.

Sideris, C.P. (1950) Manganese interference in the absorption and translocation of radioactive iron in *Ananas comosus* (L.) Merr. *Plant Physiology* 25, 307–321.

Sideris, C.P. (1955) Effects of sea water sprays on pineapple plants. *Phytopathology* 45, 590–594.
Sideris, C.P. and Krauss, B.H. (1936) The classification and nomenclature of groups of pineapple leaves, sections of leaves and sections of stems based on morphological and anatomical differences. *Pineapple Quarterly* 6, 135–147.
Sideris, C.P. and Young, H.Y. (1946) Effects of nitrogen on growth and ash constituents of *Ananas comosus* (L.) Merr. *Plant Physiology* 21, 247–270.
Sideris, C.P. and Young, H.Y. (1951) Growth of *Ananas comosus* (L.) Merr., at different levels of mineral nutrition under greenhouse and field conditions. II. Chemical composition of the tissues at different growth intervals. *Plant Physiology* 26, 456–474.
Sideris, C.P. and Young, H.Y. (1954) Effects of chlorides on the metabolism of pineapple plants. *American Journal of Botany* 41, 847–854.
Sideris, C.P. and Young, H.Y. (1956) Pineapple chlorosis in relation to iron and nitrogen. *Plant Physiology* 31, 211–222.
Siebeneichler, S.C., Monnerat, P.H., Carvalho, A.J.C.D. and Silva, J.a.D. (2002) Composicao mineral da folha em abacaxizeiro: efeito da parte da folha analisada. *Revista Brasileira de Fruticultura* 24, 194–198.
Siebeneichler, S.C., Monnerat, P.H. and Da Silva, J.A. (2008) Boron deficiency in pineapple 'Pérola'. *Acta Amazonica* 38, 651–656.
Sikora, F.J. and Moore, K.P. (2014) Soil test methods from the Southeastern United States. *Southern Cooperative Series Bulletin* No. 419. Available at: http://aesl.ces.uga.edu/sera6/PUB/MethodsManualFinalSERA6.pdf (accessed 25 July 2017).
Silva, J.A., Hamasaki, R., Paull, R., Ogoshi, R., Bartholomew, D.P. *et al.* (2006) Lime, gypsum, and basaltic dust effects on the calcium nutrition and fruit quality of pineapple. *Acta Horticulturae* 702, 123–131.
Sinclair, E. (1994) Borax or Solubor for Ethrel induction? In: *Pineapple Field Day Notes*. Queensland Fruit and Vegetable Growers, Beerwah, Queensland, Australia, pp. 64–66.
Smith, D.H., Pfenninger, K. and Krauss, B. (1962) Boron deficiency symptoms of pineapple – a preliminary report. *PRI News* 10, 95–97.
Souza, L.F.d.S. (1999) Correção de acidez e adubação. In: d.C.G.A. Pinto, J.R. Santos Cabral and L.F.d.S. Souza (eds) *O abacaxizeiro: Cultivo, agroindústria e economia.* Brasília, Brazil: EMBRAPA-SCT, pp. 169–202.
Spironello, A., Quaggio, J.A., Teixeira, L.a.J., Furlani, P.R. and Sigrist, J.M.M. (2004) Pineapple yield and fruit quality effected by NPK fertilization in a tropical soil. *Revista Brasileira de Fruticultura* 26, 155–159.
Steiner, F. and Zoz, T. (2015) Foliar application of molybdenum improves nitrogen uptake and yield of sunflower. *African Journal of Agricultural Research* 10, 23–28.
Stewart, G.R., Thomas, E.C. and Horner, J.M. (1927) The composition of the pineapple plant at various stages of growth. *PRI News* 31, 368–387.
Swete Kelly, D.E. (1993) Nutritional disorders. In: Broadley, R.H., Wassman, R.C., and Sinclair, E.R. (eds) *Pineapple Pests and Disorders.* Brisbane: Department of Primary Industries, pp. 33–42.
Tapchoi, K. (1990) Effects of potassium forms on growth, yield and fruit quality of pineapple. Thesis, Kasetsart University, Bangkok, Thailand (abstract only; original not seen).
Tay, T.H. (1974) Boron deficiency in pineapple. *MARDI Research Bulletin* 2, 46–48.
Teixeira, L.a.J., Quaggio, J.A. and Zambrosi, F.C.B. (2009) Preliminary DRIS norms for 'Smooth Cayenne' pineapple and derivation of critical levels of leaf nutrient concentrations. *Acta Horticulturae* 822, 131–138.
Thamsurakul, S., Nopamornbodi, O., Charoensook, S. and Roenrungroeng, S. (2000) Increasing pineapple yield by using VA mycorrhizal fungi. *Acta Horticulturae* 529, 199–202.
Thiagalingam, K. (1971) Effects of calcium silicate on yield and nutrient uptake and mechanism of silicon transport in plant. PhD dissertation. University of Hawaii, Honolulu, HI.
Thiagalingam, K., Silva, J.A., and Fox, R.L. (1977) Effect of calcium silicate on yield and nutrient uptake in plants grown on a humic ferruginous latosol. *Conference on Chemistry and Fertility of Tropical Soils.* Kuala Lumpur, Malaysia: Malaysian Society of Soil Science, pp. 149–155.
Tisseau, M.A. and Tisseau, R. (1963) The setting up of an experiment on the flowering and mineral nutrition of pineapples. *Fruits* 18, 33–36.
Vásquez-Jiménez, J. (2010) Evaluación de la Necesidad de Hierro del Cultivo de Piña *Ananas comosus* (L) Merr., var MD-2 en Tres Órdenes de Suelo del Norte y Caribe Norte de Costa Rica. MSc thesis, Costa Rica University, San José, Costa Rica.

Vásquez-Jiménez, J. (2016) Burndown of pineapple plants in Costa Rica. *Pineapple News* 23, 26–30. Available at: http://www.ishs-horticulture.org/workinggroups/pineapple/PineNews23.pdf (accessed 8 March 2018).

Wambiji, H. and El-Swaify, S.A. (1974) Effects of soil salinity on pineapple. I. Growth parameters. *Departmental Paper* 22. Honolulu, HI: Hawaii Agricultural Experiment Station, College of Tropical Agriculture, University of Hawaii.

Wambiji, H. and El-Swaify, S.A. (1976) Effects of soil salinity on pineapple. II. Chemical composition. *Department Paper* No. 25. Honolulu, HI: Hawaii Agricultural Experiment Station, College of Tropical Agriculture, University of Hawaii.

Wang, J.J., Harrell, D., Henderson, R.E. and Bell, P.F. (2004) Comparison of soil-test extractants for phosphorus, potassium, calcium, magnesium, sodium, zinc, copper, manganese, and iron in Louisiana Soils. *Communications in Soil Science and Plant Analysis* 35, 145–160.

Wilson Wijeratnam, R.S., Abeyesekere, M., Hewajulige, I.G.N. and Suganthini, R. (1996) Studies on the black heart disorder, electrolyte leakage and endogenous calcium content in Kew and Mauritius variety pineapples grown in Sri Lanka. In: Vijaysegaran, S., Pauziah, M., Mohamed, M.S. and Ahamad Tarmizi, S. (eds) *Proceedings of the International Conference on Tropical Fruits*, Vol. I. Kuala Lumpur, Malaysia: Malaysian Agricultural Research and Development Institute, pp. 423–430.

Young, H.Y. (1961) The iron-manganese-nitrogen problem. In Sanford, W.G., Smith, D.H., Krauss, B.H. and Young, H.Y. (1961) Plant nutrition and soil fertility aspects of pineapple culture. *Pineapple Research Institute of Hawaii Research Report* 84, pp. 45–69.

Young, R.A. (2017) Pineapple production on Dole farms in Latin America. *Acta Horticulturae* 1111, 227–229.

9 Organic Production

Alain Soler,[1] Domingo Haroldo Reinhardt,[2] Aristoteles Pires de Matos[2] and Tullio Raphael Pereira de Pádua[2]

[1]CIRAD, UPR GECO, Réunion, France, and GECO, University of Montpellier, France; [2]Embrapa Mandioca e Fruticultura (Embrapa Cassava and Fruits), Cruz das Almas, Brazil

Agroecology applies principles of ecology to agrosystems to develop practices relevant for organic farming. Agroecology and organic farming, both take into account conservation of natural resources, adapted soil fertility management and conservation of agrobiodiversity in sustainable cultural systems. More specifically, organic farming means no use of synthetic chemicals, antibiotics nor genetically engineered organisms. Agroecology and organic farming also take into account the social situation for the rural populations. These often suffer of social exclusion due to poverty and they are also the first concerned by chemical pollution in conventional production systems (Altieri, 1989). These populations, particularly the small-scale farmers, are severely impacted by the cost of inputs and organic production presents a good opportunity for them. In some countries, the small-scale farmers have contributed to a rapid increase in the market for organic pineapple (Gibbon and Bolwig, 2007; Jha, 2007; Brenes *et al.*, 2011) even though large-scale farms, such as in Costa Rica, are now also producing organic pineapple. In most producing countries, including developing countries, farmers are concerned about environmental degradation and food quality (Hong, 1994; Fairweather, 1999; Kleemann and Abdulai, 2012; Kleemann *et al.*, 2014).

While there is a considerable body of published literature on organic farming, there is little dedicated to pineapple. A recent publication of note, *La Piña*, reviews organic production of pineapple (Garita Coto, 2014). Also, Embrapa has recently made available recommendations for organic production of pineapple in Brazil based upon studies in the Chapada Diamantina region, a tropical highland of the State of Bahia (Pereira de Pádua *et al.*, 2016b).

Specifications and Certifications for Organic Production

While most countries producing organic pineapple have associations for organic farmers, the systems of certification might only be recent. However, farmers must comply with their regulations of the countries to which they export and hence often rely on them for procedures.

In the USA, organic farming is certified by USDA and a National Organic Program (NOP) validates the products and practices. The products also need to be registered by the Environmental Protection Agency (EPA). These organizations rely on a review of products by the Organic Material Review Institute (OMRI), an international non-profit organization that determines which input products can be used.

In Europe, Communal Regulations have been established but with some specificities for each European country. In France, under the Ministry of Agriculture, ITAB (French Institute for Organic Agriculture) is in charge of the organization of research on organic farming. ANSES (French Agency for Food, Environmental and Occupational Health & Safety) establishes the list of authorized chemical inputs for organic farming.

In Japan, organic farming began as a social movement, *sanshôteiki* or *Teiki*. Teiki means 'contract between farmers and consumers'. These associations were local networks with no official standards. Today, there is a Japan Organic Agriculture Association (JOAA) under the Ministry of Agriculture, Forests and Fisheries to guarantee the quality of all organic products.

In Africa, the African Organic Network (AfrONet) is coordinating organic agriculture related activities throughout the continent. The International Society of Organic Agriculture Research (ISOFAR), Swiss Aids, supports these activities. The International Centre for Research in Organic Food Systems (ICROFS), originally a Danish research centre on organic farming in Europe, also supports organic farming in developing countries particularly Africa.

In Brazil, the Ministry of Agriculture, Food Supply and Livestock has established standards and given support to organic production and research carried out by Embrapa (Brazilian Corporation for Agricultural Research) and other organizations (MAPA, 2014).

An International Federation of Organic Agriculture Movements (IFOAM) federates the local organizations for organic farming. It acts as an international network for local 'Organic federations', providing information on standards and specifications, and helping in homogenization of standards throughout the world. It also helps with the transition from conventional to organic farming. According to IFOAM, in its General Assembly (2008), the definition of Organic Agriculture is:

> Organic Agriculture is a production system that sustains the health of soils, ecosystems and people. It relies on ecological processes, biodiversity and cycles adapted to local conditions, rather than the use of inputs with adverse effects. Organic Agriculture combines tradition, innovation and science to benefit the shared environment and promote fair relationships and a good quality of life for all involved.'
>
> This definition points out the four main principles of the organic agriculture:
>
> **Health:** Healthy soil, plants, animals and humans mean a healthy planet;
>
> **Ecology:** Emulating and sustaining natural systems;
>
> **Fairness:** Equity, respect and justice for all living things;
>
> **Care:** For the generations to come.
>
> (IFOAM, n.d.)

Organic Cultural Systems

One of the main challenges in organic farming is to find alternatives to pesticides. Servicing plants (SP), in rotation or association with pineapple, are a good way to restore a natural balance to pest populations. Some of these SP help with integrated pest management but might also be natural sources of new biopesticides. In addition, biocontrol technologies have been developed to replace some conventional pesticides and improve growth. These include biopesticides, biostimulators of growth, and bacteria and mycorrhizal fungi which stimulate systemic resistances.

Land selection and transition

While soil requirements for conventional and organic productions are similar, organic farming relies heavily on soil microflora for natural regulations. Large microbial populations are necessary to mineralize organic matter and manage soil-borne pathogens. A transitional period is needed to allow beneficial microbial populations to increase as a farm moves from conventional to organic farming. The transition may be as long as 3 years before new organic farms can be certified. The transition should decrease chemical residues and increase the soil organic matter content and availability of nutrients. However, soil fertility parameters such as K and P will need sufficient time to be reflected in soil analyses (Devadas and Kuriakose, 2005). Transitional practices include cover crops, and the application of organic and mineral amendments, organic fertilizers as composted animal manures, liquid humic acids, and other composts (Murray and Manicom, 2009; Pereira de Pádua *et al.*, 2016a; Soler and Dorey, 2017).

Land and soil preparation

The destruction of old pineapple plants is a critical issue, as the stems bear one viable bud at the base of each leaf (Soler and Dole, 2006). Coarse shredding of the stems might allow the development of undesired plantlets which can maintain a relatively high level of inoculum of soil-borne pests including nematodes, symphylids and mealybugs. To avoid this, the old plants should be finely shredded using machines such as wood chippers or modified rotary slashers. An alternative is to burn or remove the stems, but this should not be a usual practice as high amounts of organic matter and N are lost. Where there is a very low infestation, the conventional techniques for dealing with the old crop are satisfactory. Another alternative is to use the pineapple field residues for production of silage for cattle feed.

Servicing Plants (SP)

In organic systems, nutrition and management of pest and diseases partially relies on SP. SP help to restore agrosystems that control pest and disease as well as producing organic matter.

SP are selected for certain characteristics depending on the specific problems encountered on the pineapple farm (Jannoyer *et al.*, 2011; Damour *et al.*, 2014; Soler *et al.*, 2016). Pedoclimatic conditions such as photoperiod and temperature susceptibility of the SP, and the type of soil will determine the expression of desired traits. For example, *Crotalaria* spp. reach their flowering stage a few weeks after sowing during winter instead of 6 months in longer day-length periods. A shorter growth period means less biomass, less value as a cover crop and insufficient time to reduce the nematodes or symphylids.

Combining different SP species with different traits might help restore more efficient ecosystemic functions. Relationships between ecosystemic services and SP traits are proposed in Table 9.1 (Soler *et al.*, 2016).

Legumes such as *Crotalaria juncea* have been successfully tested as cover crops for weed management, to control the nematode *R. reniformis*, and to increase the organic matter content of the soil (Pinon and Champanhet, 1988; Ko and Schmitt, 1993; Wang *et al.*, 2002; Wang *et al.*, 2003; Murray and Manicom, 2009; Pereira de Pádua *et al.*, 2016a). They also contribute to the quantitative and dynamic bioavailability of nutrients through their redistribution of soil nutrients from deeper to upper layers.

Table 9.1. Expected ecosystemic services provided by the agrosystems and related traits of cover crops integrated in rotation or association with pineapple.

Expected ecosystemic services	Related cover crop traits
Clearing the soil of nematodes and symphylids before pineapple is planted	Host unfavourable to pathogens, allelopathy, biofumigation after incorporation
Cleansing the soil to the depth at which symphylids can be found	The root system should be capable of exploring the soil profile beyond the depth usually explored by the pineapple root system (30–40 cm)
Enhancing the suppressive effect of soil on pests (balance between microflora and microfauna communities, increased mycorrhization potential)	Plant capable of mycorrhization, satisfactory biomass production, quality of organic matter
Controlling weeds (many plants host soil-borne parasites)	Cover crop, allelopathy
Improving the soil structure and organic matter content	Appropriate root system, satisfactory biomass production
Improving the quantitative and dynamic bioavailability of nutrients	Satisfactory biomass production, quality of organic matter, ability to fix nitrogen, appropriate root system
Control of aerial parasites of pineapple or associated cash crops	Attraction of auxiliaries, repulsive effect on aerial parasites

SP help to restore natural balances between beneficial and pathogenic microbial populations which are believed to help establish stronger below- and above-ground interactions (rhizospheric microbiome–plants) and thus give better growth and yields (Orrell and Bennett, 2013; Lakshmanan *et al.*, 2014).

Servicing plants for pest management

Many plants have been tested as cover crops for weed control and as unfavourable hosts or repellents of pineapple pests. So far, *Crotalaria* spp. have been one of the most studied SP with pineapple and are known to have a direct impact on pineapple nematodes (Wang *et al.*, 2002; Wang *et al.*, 2003; Wang *et al.*, 2011) and symphylids (Soler *et al.*, 2016).

Soil-borne parasites

In different situations, the same nematode can have a different behaviour towards pineapple. *Pratylenchus brachyurus* is one the main pathogenic nematodes in West Africa but in the French Caribbean it does not impact pineapple although present in many plants. Similarly, the interactions between pathogenic nematodes and the SP may also be different with the situation. Thus, the choice of SP to control nematodes should not rely on the literature only. The selection should be done after inoculations of monospecific populations of the local pests under controlled conditions (Fig. 9.1) and/or in field tests in experimental rotation systems (Fig. 9.2).

Figure 9.2 shows that a rotation between pineapple and *Crotalaria retusa* reduced the populations of *R. reniformis* to lower level than under the natural weed fallow. This, together with data in Table 9.2, suggests that a large number of the weed species were good hosts for *R. reniformis*.

Root nodulation is an important characteristic of *Crotalaria* sp. This process is necessary for the production of the toxic alkaloid pyrrolizidines (PAs) (Irmer *et al.*, 2015). The appropriate *Rhizobium* species must be used for inoculation. Although the nodulation occurs early after germination, the length of the cycle

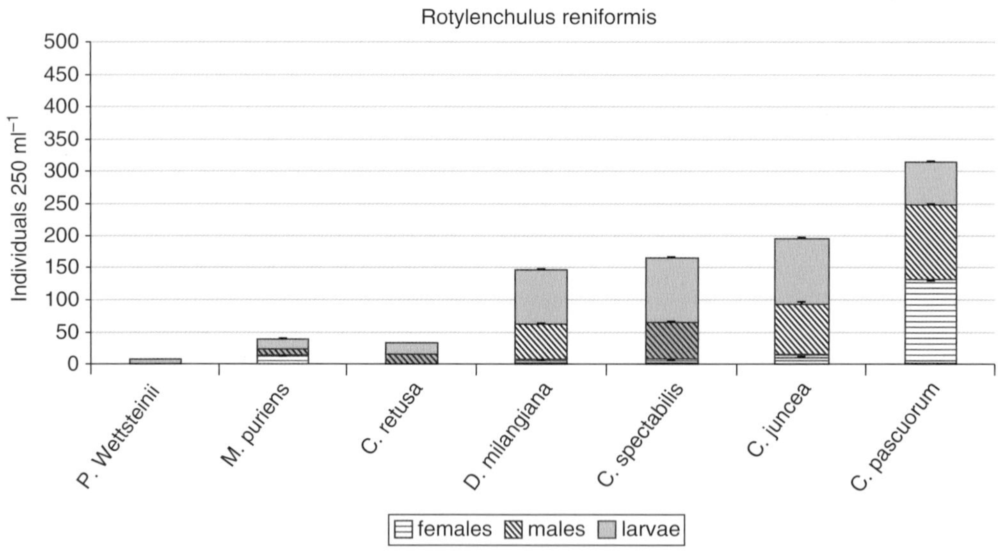

Fig. 9.1. Host status for *R. reniformis* of a selection of rotation crops under controlled conditions (Marie-Alphonsine *et al.*, 2017). In culture chamber, the SP were directly sown in sterile soil (250 ml/pot; 10 replicates), then they were inoculated with 4400 individuals of a monospecific reared population. SP were grown for 45 days before extracting and counting the nematodes by centrifugation to determine in each sample, male, female and larvae according to Coolen and d'Herde (1972). Personal communication, P.A. Marie-Alphonsine.

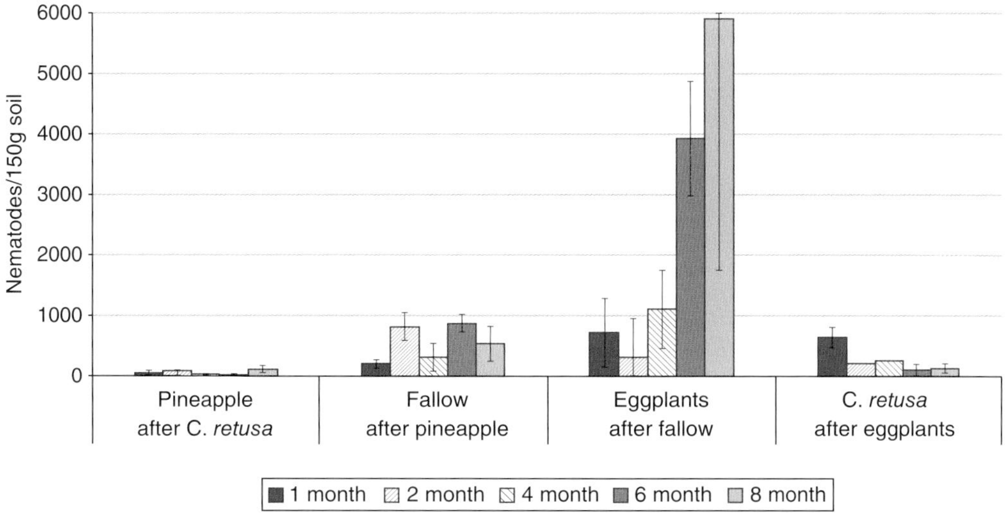

Fig. 9.2. *R. reniformis* populations monitored under the different crops in a rotation in an experimental field. Monitoring during 8 months, and then the plots left 4 months before destruction and planting a rotation crop. The fields in two replicates were monitored during 3 years/rotations. The area was not favourable for symphylids. Reproduced with permission of the International Society of Horticultural Science from Soler, A., Marie-Alphonsine, P.A., Corbion, C., Fernandes, P., Portal Gonzalez, N., Gonzalez, R., Repellin, A., Declerck, S. and Quénéhervé, P. (2016). Bioprotection of pineapple in ecological cropping systems. *Acta Horticiculturae* 1111, 159–168. DOI: 10.17660/ActaHortic.2016.1111.24.

Table 9.2. Weeds hosts of *Rotylenchulus reniformis* growing in natural fallow after pineapple in Martinique. Reproduced with permission of the International Society of Horticultural Science from Soler, A., Marie-Alphonsine, P.A., Corbion, C., Fernandes, P., Portal Gonzalez, N., Gonzalez, R., Repellin, A., Declerck, S. and Quénéhervé, P. (2016). Bioprotection of pineapple in ecological cropping systems. *Acta Horticulturae* 1111, 159–168. DOI: 10.17660/ActaHortic.2016.1111.24. https://doi.org/10.17660/ActaHortic.2016.1111.24

Number of crops	Weeds under short fallow	Weeds under long fallow
Known as hosts	10	5
Potential hosts	7	5
Non-hosts	2	2
Not determined	5	10
Total	*24*	*22*

may have an impact on the production of the toxic PAs.

Crotalaria spp. that have been tested in pineapple systems show different traits offering potentially different ways to use them. *C. juncea* is known for its high-speed germination and growth, overtaking many weeds. It produces more dry matter than *C. spectabilis*. It is an unfavourable host for nematodes and symphylids. *C. retusa* grows in many places over the world as a ruderal species. This *Crotalaria* sp. has the highest content of PAs (12,200 ppm dwt) and is very toxic to soil-borne pathogens, suggesting good potential for rotation systems (Thoden and Boppré, 2010). However, *C. retusa* is also toxic for cattle and other farm animals. Other *Crotalaria* spp. are good forages. *C. spectabilis* with large leaves is a good cover crop capable of shading out *Cyperus rotundus* and *C. esculentus*. This species has a relatively slow growth (germination and beginning of the cycle) and is less toxic (PAs content ~600 ppm dwt) than *C. juncea* or *C. retusa*.

Crotalaria spp. possess extra-floral nectaries (Guimaraes *et al.*, 2006) and good quality pollen (Venzon *et al.*, 2006) making them potential candidates for flower strips that support conservation biological control principles (Fig. 9.3).

Crotalaria spp. have their own parasites that need management. In Africa and Indian ocean

Fig. 9.3. Rotation/association with *C. spectabilis* as flower strips for conservation biological control, between pineapple and Solanaceae. From A. Soler, CIRAD.

areas, they are strongly attacked by Miridae. *Moissonia importunitas* appears on very small plantlets after germination in Réunion (Fig. 9.4) (Ratnadass *et al.*, 2018). In Mozambique, another species of Miridae, *Eurystylus oldi* Poppius, is present on the flowers and cloves of *C. juncea* (Maldes and Ratnadass, 1998). Some of the Miridae species are actually predators of the phytophagous ones and could help minimize leaf symptoms. In French Caribbean, the caterpillar of *Utetheisa ornathrix* is the main parasite on *Crotalaria* spp. (Fig. 9.4). Btk insecticide may control this caterpillar if applied just when the population starts increasing. Both parasites reduce strongly the growth of the *Crotalaria* spp., leading to a lower biomass production and a shorter time for field sanitization.

Sometimes other pests that might be found on *Crotalaria* spp. include other Miridae bugs (*Lygus* spp.), Bryoba mites, and Pyrales, but also fungi such as *Microsphaera diffusa*, *Colletotrichum curvatum*, *Fusarium* spp, and *Ceratocystis* spp.

Other SP species for management of soil-borne pathogens

Several plants non-host of *R. reniformis* have been tested for nematode management in pineapple including Rhodes grass (*Chloris guayana*), Pangola grass (*Digitaria decumbens*), *Tagetes patula* (Caswell *et al.*, 1991), *Mucuna* spp. and *Prueraria* spp.

SP selected for management of soil-borne parasites on pineapple ideally should be non-hosts for both nematode and symphylid. Tests using *R. reniformis* (Fig. 9.5) and symphylids (Fig. 9.6) were made in Martinique for *Crotalaria juncea, Crotalaria spectabilis, Tithonia diversifolia, Raphanus sativus, Mucuna deeringiana, Tagetes erecta* and *Chloris guayana* (Marie-Alphonsine and Soler, 2013). *M. deeringiana* and *Bracharia decumbens*, known as non-hosts of *R. reniformis*, were not suitable for symphylid management. On the contrary, *Tagetes erecta* was a non-host for symphylids but a good host for *R. reniformis*. The plants that did not allow the multiplication of symphylids (small number of larvae) exhibited a differentiated toxicity towards adults (Fig. 9.6).

Asteraceae and Boraginaceae

These SP do not produce a high biomass quick enough to be used as cover crops but they are known to attract auxiliaries and repel many insects. *Asteraceae* are potential candidates as flower strips. These plants also produce multiple

Fig. 9.4. (a) *Utetheisa ornathrix* on *C. spectabilis*; (b) *Moissonia importunitas* and symptoms on *C. juncea* . Photos: (a) A. Soler, CIRAD; (b) A. Ratnadass, CIRAD.

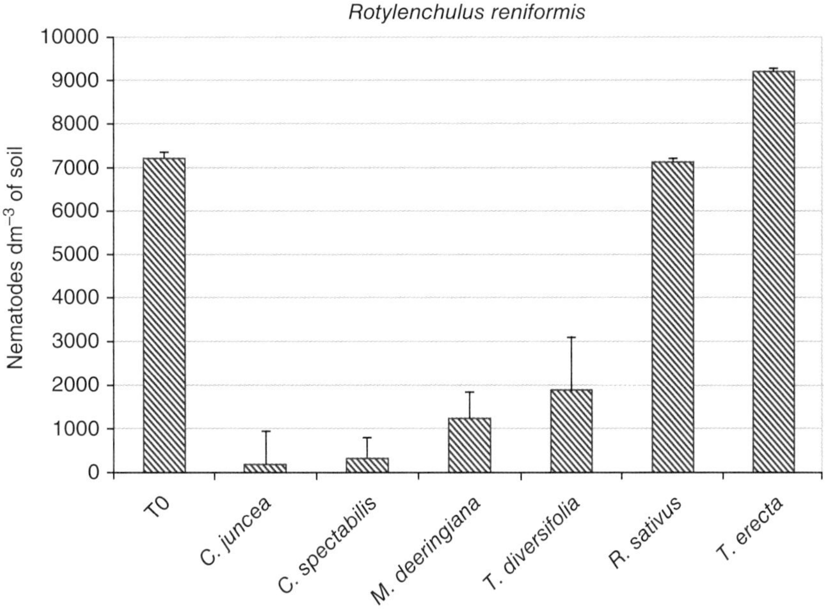

Fig. 9.5. Evolution of *R. reniformis* inoculated populations on different servicing plants (SP). In greenhouse, *Vigna unguiculata* was grown for 2 months to multiply the nematodes in 25 l pots of soil from an experimental pineapple field infested with *R. reniformis*. Finally, the selected SP were sown into the pots (10 replicates per SP) then grown for 45 days before taking a representative sample of 250 ml of soil to extract and count the populations by elutriation. Personal communication, P.A. Marie-Alphonsine.

types of PAs and some of them could be tested as SPs for pineapple, at least for strip barriers limiting the extension of parasite populations (Thoden and Boppré, 2010). There is not a general repellency of the PAs in these species towards nematodes and not all the PAs produced are toxic for all nematode species.

A special mention must be made of composts made from SPs as organic amendments, because they may suppress soil-borne parasites but also enhance the microorganism populations that are direct antagonists (Akhtar and Malik, 2000).

Mealybugs, ants, other insects and mites

There are more than 150 compounds identified as phenols and sesquiterpene-lactones in Mexican marigold (*Tithonia diversifolia*) that can act as repellents. A fermented water extract of the green part of the plant will repel many different insects including termites (Adoyo *et al.*, 1997). For pineapple, the water extract may be incorporated into the soil through drip irrigation system or sprayed at the base of the plant.

Tansy (*Tanecetum vulgare*) has not been tested on pineapple but is used as an ant repellent in organic vegetable farms in la Réunion or as flower strips.

For the best pest management and nutritional or cover crop effect, it is possible to mix several SPs.

Servicing Plants for Weed Control and Soil Erosion

Cover crops should be fast to germinate and grow and possess large leaves and/or high biomass to limit weed growth and reduce soil erosion. Inhibition of weed seed germination by an allelopathic mechanism might also be possible. Some species suitable as cover crops for pineapple that have shown allelopathic effect include *Crotalaria* spp. (Skinner *et al.*, 2012), *Mucuna* spp. and pigeon pea (*Cajanus cajan*) (Hepperly *et al.*, 1992), pearl millet and sorghum (Weston *et al.*, 2013).

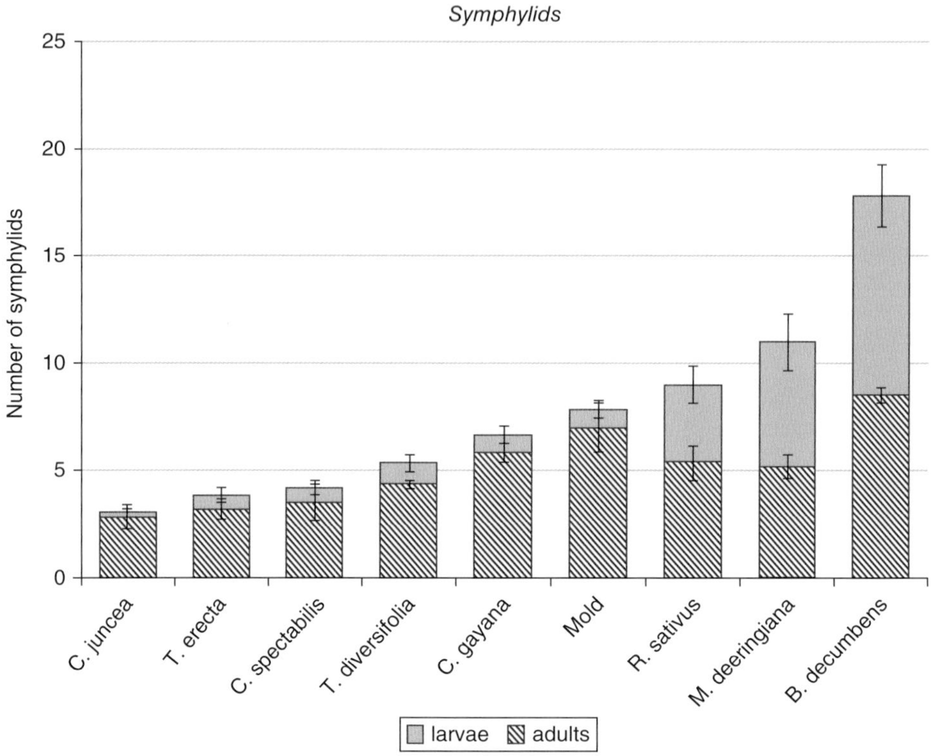

Fig. 9.6. Evolution of populations of symphylids inoculated on different servicing plants (SP). In a controlled culture chamber plants were grown 3 months in 250 ml pots, then ten individuals were inoculated and symphylids were counted 35 days after. Personal communication, P.A. Marie-Alphonsine.

Field trials in Brazil showed that millet (*Pennisetum americanum*) controlled weeds but did not affect the pineapple plants. Nutrients were detected in higher levels in these pineapple plants compared with those grown under a conventional production system (Matos and Reinhardt, 2009).

To date the evaluation of cover crops in pineapple cultural systems has focused on weed control but the role of allelopathy has not been clearly assessed. Where allelopathy is sought it is important to also assess the possible negative effects of root exudates on the following pineapple crop.

Erosion can be a major problem in pineapple production including organic farming systems. Trimming the cover crop and allowing the mulch to protect the soil, besides the control of weeds, protects the soil against erosion, improves its physical and chemical properties, prevents nutrient leaching, retains and recycles nutrients, harbours natural enemies contributing to integrated management of pests and diseases, and conserves soil moisture. The spatial arrangement of different plots in a field might help reduce soil erosion by creating discontinuity in the field (see 'Spatial Organization of the Plots' section on p. 214).

Servicing plants and nutrition (nitrogen)

SPs when used as green manures can be a good source of organic matter and nitrogen (N). The carbon-to-nitrogen ratio (C/N) is an indicator of the degree of decomposition of organic matter in the soil (Giroux and Audesse, 2004). Similarly, the C/N ratio of plants may help in evaluating their impact on soil C/N and the speed of their mineralization once incorporated into the soil.

C. juncea has a C/N ratio of about 16.7, as is the case for many other *Fabaceae*. Due to their N-fixing capabilities they contain more N (3–5%) relative to C (50%), than *Poaceae* spp., which contain about 0.8% N vs 49% C (Van Soest *et al.*, 1991). Nevertheless, there is a large variability of the C/N in plants and organic matter in soils and the speed of the mineralization does not depend only on this ratio as other nutrients and bacterial populations in the soil also play a role.

The *Crotalaria* spp., *C. juncea* and *C. spectabilis*, and mixed sowing of these two, or *Crotalaria* sp. with pearl millet (*Pennisetum glaucum*) are still in test. Data collected in la Réunion showed that there is a large range of biomasses produced by different *Crotalaria* sp. A lower biomass production was observed (6–10 tons dwt ha^{-1}) with *C. juncea* as a green manure in la Réunion compared with Martinique (12–16 tons dwt ha^{-1}). This corresponds to averages of ~240 kg N.ha^{-1} vs ~420 kg N.ha^{-1} incorporated into the soil. In Brazil, biomass reached 15–21 tons dwt ha^{-1}, corresponding to an average of ~540 kg N ha^{-1}.

Pearl millet (*Pennisetum glaucum*) can produce ~10 tons dwt ha^{-1} and ~80 kg N ha^{-1}. Sorghum (*Sorghum bicolor*) can produce ~35 tons dwt ha^{-1} and ~280 kg N ha^{-1} (Pereira de Pádua *et al.*, 2016b). Sorghum is however a host for *Pratylenchus* spp., so it is unsuitable where *P. brachyurus* is a major pest of pineapple as in West Africa.

Velvet bean (*Mucuna aterrima*) produces ~5–12 tons dwt ha^{-1} and ~150–360 kg N ha^{-1} (Pereira de Pádua *et al.*, 2016b). Other *Mucuna* species (*M. pruriens* and *M. deeringiana*) are also good cover crops. They increase soil organic matter content and are non-hosts of *R. reniformis*, although less efficient against symphylids (Soler *et al.*, 2016). Jack Bean (*Canavalia ensiformis*) produces ~20–35 tons DM ha^{-1} and, at up to 4% N, ~1000 kg N ha^{-1}. These two *Fabaceae* cover the soil well while producing high biomasses (Pinon and Champanhet, 1988; Pereira de Pádua *et al.*, 2016b).

The four SPs, *Mucuna aterrima*, *Pennisetum glaucum*, *Canavalia ensiformis* and *Sorghum bicolor* have been proposed as a 25% mix each in Brazil to increase organic matter and N before planting pineapple (Pereira de Pádua *et al.*, 2016a). Comparative data for these four species is shown in Table 9.3.

Mexican marigold (*Tithonia diversifolia*) can provide high levels of nutrients averaging 3.5% N, 0.37% P and 4.1% K, with 5 tons dwt ha^{-1} (Jama *et al.*, 2000). The green biomass can be harvested twice a year leading to an increase of soil microbial biomass. In that study, a significant increase occurred in saprophytic nematode populations associated with organic matter degradation. Mexican marigold has been used as a rotation crop with bananas in the French Caribbean several years ago but the practice has been abandoned because considerable labour was required to prepare and plant it.

Table 9.3. Nutrients in the phytomass of the aerial part of different green manure crops compared with spontaneous vegetation. Lençóis, Bahia, 2013. Adapted from Xavier *et al.*, 2016.

Plant cover	Biomass (t ha^{-1} dry mass)	N (kg ha^{-1})	P (kg ha^{-1})	K (kg ha^{-1})	Ca (kg ha^{-1})	Mg (kg ha^{-1})	S (kg ha^{-1})
Millet	6.5	116.13^a	12.35^a	80.11^a	13.52^e	19.29^b	15.75^b
Sorghum	6.0	78.33^a	9.14^a	56.18^b	7.72^e	15.61^b	10.93^b
Crotalaria	3.5	101.07^a	4.86^b	41.36^b	93.74^b	15.05^b	9.43^b
Jack bean	5.0	101.66^a	6.75^b	45.99^b	143.78^a	27.32^a	27.91^a
Velvet bean	5.3	105.45^a	6.03^b	50.75^b	70.77^c	17.69^b	14.14^b
50% Velvet bean + 50% Sorghum	5.5	149.23^a	10.46^a	51.61^b	39.07^d	17.20^b	12.21^b
*NSV	4.0	96.98^a	6.45^b	50.33^b	40.67^d	22.31^a	11.69^b

*NSV (native spontaneous vegetation). Average values followed by the same letter do not differ significantly among them by the Scott-Knott test at 5%.

Biopesticides

'Biopesticide' is an inclusive term encompassing microorganisms, natural pesticides from plants and microorganisms, pheromones, plant defence stimulators such as bacteria, mycorrhizae or natural chemicals and beneficial macroorganisms such as nematodes or insects.

Bioherbicides

Essential oil of green mint (*Mentha spicata*) contains a terpene (carvol) which is used as a seed germination moderator or inhibitor in French organic farms.

Pelargonic acid (from *Pelargonium* spp.) is a contact herbicide that belongs to the fatty acid C7-C20 family. This chemical alters cell membrane permeability producing cell content leakage and rapid tissue death. This natural herbicide has been tested in pineapple and banana fields. The product is authorized for conventional farming in Europe (Annex IV of the CE regulation n°396/2005) but it is not yet authorized under French regulations for organic farming.

Bioinsecticides

Pyrethrin from flowers of the *Asteraceae*, *Tanacetum cinerariifolium* and *Chrysanthemum cinerariifolium*, is a bioinsecticide against aphids and caterpillars that can be used to control pests of SPs, including *Crotalaria* spp.

Azadirachtine (as neem extract, neem oil or as a chemical formulation) is a repellent for many insects that is authorized in Europe for organic farming.

Bacillus thuringiensis kurstaki (*Btk*) exhibits a certain specificity between the strain of Bt and the *Lepidopter* sp. Upon ingestion of Btk by young larvae, the caterpillar's digestive system triggers the Btk bacterium to release an endotoxin which paralyses the caterpillar's digestive tract. Bt has been found also effective against white grubs (pineapple weevils) or pineapple fruit borer. The management of pineapple fruit borer in Brazil relies on monitoring and spraying *Bacillus thuringiensis* from open heart to dry petal stage (Sanches, 1999; Inclán *et al.*, 2008). The pest will be minimized if the pineapple orchard is situated away from rainforest, inflorescences are protected with coverings and those that are attacked are removed and destroyed and old crop residues are removed (Noronha *et al.*, 2009; Noronha *et al.*, 2016).

Rotenone is produced from Malesian *Derris* spp., Peruvian *Lonchocarpus* spp. and *Terphrosia* spp. As with nicotine, it is one of the historical bioinsecticides. It is a powerful inhibitor of cellular respiration. It was formerly used in organic farming but has recently been forbidden under European regulations due to its toxicity to fish and unacceptable operator risk. It is still allowed in USDA-certified organic farming on the 'Allowed with Restriction' list. The National Organic Program is however exploring possible regulation changes limiting or prohibiting the use of rotenone.

Citrus oil extracts (limonene and linalool) are used as repellents for many insects in organic farming.

Beauveria bassiana and *Metarhizium anisopliae* are entomopathogic fungi which are commonly used to control insects including whiteflies, aphids, thrips, grasshoppers and certain types of beetles but there is little information for insects of pineapple apart from white grubs in Cuba (M.L. Cisne, personal communication, 2015) and pineapple fruit borer (Inclán *et al.*, 2008). *Metarhizium* spp. have been successfully tested on pineapple mealybugs *Dysmicoccus brevipes* (Cockerell) (Liang, 2008).

Kairomones (cinnamyl alcohol mixed with eugenol) have been tested in yellow plastic traps to trap white grubs in South Africa (Petty, 2012).

Ferreira *et al.* (2015) obtained a high mortality of pineapple mealybugs using the beneficial nematodes Heterorhabditidae and Steinernematidae. This could form an important part of pineapple mealybug wilt management in organic farming systems.

The mite fungal pathogen *Hirustella thompsoni* has been used to control the pineapple red mite (Joy *et al.*, 2013).

Fungicides

Trichoderma spp. is a soil dwelling genus of fungi prevalent worldwide. Several strains of

Trichoderma have been tested on pineapple to control other pathogenic fungi such as *Phytophthora nicotianae*, *Rhizoctonia solani* and *Fusarium subglutinans* on the cultivar MD-2 (Hernandez Mansilla *et al.*, 2006; Sierra Pena *et al.*, 2007). *Trichoderma* is competitive with other fungi but also releases chitinases that interfere with their cell wall.

Muscodor albus, an endophytic fungus isolated from the wild pineapple *Ananas* var. *microstachys*, is an efficient *in vitro* antagonist of *Phytophthora* spp. and some other pathogens through the production of volatile organic compounds (VOC) (Banguela-Castillo *et al.*, 2015).

Eugenol (an essential oil of *Eugenia caryophyllus*) and limonen (a sesquiterpene in essential oil of sweet orange) are authorized for organic farming under French regulations (respectively AMM 210075 and AMM 210127).

Stimulation of natural defences

The natural defence mechanisms, Induced Systemic Resistances (ISR) and Systemic Acquired Resistances (SAR) are promising tools for pest management in organically farmed pineapple. ISRs act through the jasmonic acid pathway and the physiological process named 'Priming' (immune resistance). This targets necrotrophic organisms. SARs act through the salicylic acid pathway which targets biotrophic microorganisms.

The efficiency of a bioprotection strategy for pineapple that uses systemic resistances (RS) depends on several factors (Soler *et al.*, 2013b). First, the RS efficiency appears dependent on favourable conditions provided by rotations with cover crops (Soler *et al.*, 2016). Second, it is important that only pineapple cultivars capable of developing RS are grown. Finally, the appropriate RS stimulator must be used (Soler *et al.*, 2013a). Some microorganisms are known to be stimulators of RS. These microorganisms are likely to modulate the plants' tolerance of a wide range of pathogens (Zamioudis and Pieterse, 2011).

Trichoderma harzianum and mycorrhizal fungi, and different diazotrophic bacteria endophytes of the pineapple root system can stimulate ISR against *R. reniformis* in the greenhouse in 'MD-2' but not 'Smooth Cayenne' (Soler, unpublished). Stimulation of ISR by methyljasmonate and endophytic bacteria has also been successfully tested on 'MD-2' and 'Victoria' ('Queen') pineapple under rotation with *Crotalaria* sp. (Soler, unpublished).

Stimulations of systemic defences in 'Smooth Cayenne' against *R. reniformis* and *M. javanica* with other chemicals, known as SAR stimulators, such as salicylic acid, and amino butyric acid, acibenzolar-s-methyl and others look promising (Chinnasri *et al.*, 2006). Preliminary results have been obtained with SAR on *Fusarium subglutinans* in pineapple in Brazil (Matos *et al.*, 2006).

Using biologicals to enhance nutrition

Diazotrophic bacteria (Tapia-Hernández *et al.*, 2000; Weber *et al.*, 2003; Weber *et al.*, 2004) occur naturally in pineapple. Inoculation of selected strains of some of these bacteria (*Asaia bogorensis*, strain 219) has been used to produce organic 'Champaka' (syn. 'Smooth Cayenne') pineapple in Brazil (Weber *et al.*, 2010). Diazotrophic bacteria such as *Azotobacter chroococcum* (González *et al.*, 2011; González Rodríguez *et al.*, 2012), might also be useful for reducing the time of acclimatization and enhancing the growth of tissue-cultured plants.

Planting Material

It is very important to use only disease-free planting material as the management of mealybugs and the associated mealybug wilt disease is difficult under the specifications of organic farming. During the long cycle of pineapple, mealybugs and ants have enough time to proliferate. Thus, the next generation of plants may see wilt symptoms expand rapidly into the fields. In countries where pesticides such as pyrethrum are authorized, the technique of disinfection previously described applies.

When no insecticides are available, the use of tissue-cultured plants or the stem-sectioning technique to produce clean planting material is important. This technique is essential for growing 'Pérola' pineapple in Brazil, a cultivar susceptible to fusariosis disease. Control of this disease under organic cultivation is based on the identification and eradication of contaminated plants in the field and production and use of healthy planting material (Padua *et al.*, 2017).

Spatial Organization of the Plots

The spatial arrangements of the different plots bearing pineapple, cover crops and sometimes an additional cash crop might enhance biodiversity on farms. The arrangements will be different for small farms compared with more mechanized large farms.

On small farms, it is relatively easy to organize the rotation of the different crops as small plots. That will ensure a maximum of biodiversity per surface unit. The system is close to conservation biological control, where flowering plants attract and maintain the auxiliaries and beneficial organisms or the repulsive effect against pests (Fig. 9.7). On large-scale farms, the arrangement may be based on long interleaved strips of the different crops. The simplest design would be alternate strips of pineapple and cover crops. Thus, the width of the strips is conditioned by the equipment of the farms. In both cases, the farmer has to synchronize the length of the pineapple cycle with the cover crops cycle plus the additional cash crop.

The larger the plots (as in conventional farming), the less effective the biodiversity. Thus, the cultural system is closer to conventional farming and pest and disease management will rely mainly on biopesticides and less on agroecology principles.

Fig. 9.7. Spatial arrangement on small farm enhancing biodiversity. Photo A. Soler, CIRAD.

Plant Nutrition

Assessment of nutritional requirements

The assessment of nutritional requirements and the application of mineral or organic amendments as described for conventional systems also apply to organic farming. Indeed, many of the fertilizers used in conventional systems of production comply with the requirements of the organic production except for nitrogen (N). Natural sources of potassium (K) and phosphorus (P) are authorized for organic production. Therefore, it is necessary to seek sources of organic N that could supply the quantity of N required for a pineapple crop (350–600 kg ha^{-1}).

The contribution of organic matter to soil fertility and plant nutrition is well known (Pichot, 1975; Py *et al.*, 1987; Pieri, 1989; Jama *et al.*, 2000). Mineral N actually supports the long-term fertility of the soil, which is not the case with many organic fertilizers. Nevertheless, it depends on the stability of the organic fertilizers in the soil. A comparison of the index of stability for different types of organic matter showed that composts are much more stable than green or animal manures (Fig. 9.8; Rothé and Poletti, 2015). A soil with a high organic matter content is usually richer in stable humus and will have a better retention capacity for water and mineral nutrients as K. This soil will support life of micro- and macroorganisms that participate in organic matter mineralization, which is a critical issue in organic farming.

Sources of organic nitrogen and mode of application

The main challenge is to identify approved sources of N that comply with organic farming specifications and are consistently available. The farmer must then determine the appropriate mode of application in order to manage the timing of N availability. The main useful sources of organic nitrogen are mulches used to control weeds and preserve soil humidity but these are

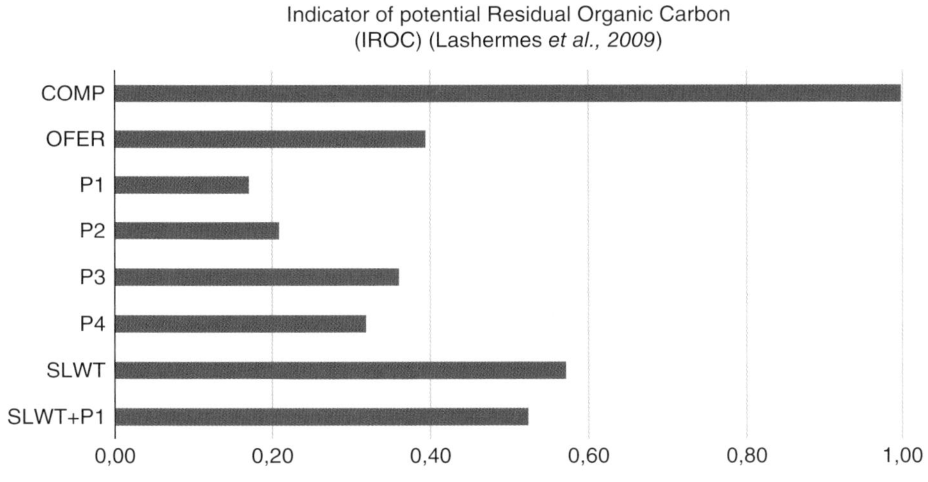

Fig. 9.8. Stability Index of different sources of organic carbon. IROC (Indicator of potential Residual Organic Carbon) measured according to Lashermes *et al.* (2009) for compost (COMP), organic fertilizer (OFER), cover crops (P1-P4), slaughterhouse waste (SLWT), and a mixture of slaughterhouse waste plus cover crop P1 (SLWT+P1). Results normalized against COMP. Reproduced from Rothé and Poletti, 2015, with permission.

difficult to apply on a schedule. Side-dressing at the base of old leaves is not very reliable due to the low solubility of most of the organic fertilizers. Therefore, the most appropriate mode of application is pre-plant either broadcasted or banded with other fertilizers.

Organic N may come from residue of the previous pineapple crop, blood and bone meals, legume cover crops, composts, animal manures, preferably composted, or commercial liquid formulations to be sprayed on leaves, although the last are generally too expensive. The N content of these different sources is variable and calculations for application should rely on specific chemical analyses.

The pineapple residue after harvest may provide a large amount of N as the amount of fresh organic matter may exceed 200 tons fwt. ha^{-1} from varieties like Smooth Cayenne (Py *et al.*, 1987). 'Victoria' ('Queen') in la Réunion produced 20–40 tons dwt ha^{-1}, with a potential N source of 170–320 $kg.ha^{-1}$. In Brazil, a biomass of residue exceeding 100 tons fwt ha^{-1} for 'Pérola' was reported by Alves *et al.* (1998). However, the quick mineralization of pineapple leaves means the requirement of growing plants during the first 9–10 months cannot be met. Part of this organic matter will contribute to restore the soil organic status and to increase the growth of the cover crops used in rotation systems.

Composts, including manures from organically fed cattle (Kiggundu *et al.*, 2014) or chicken (Aiyelaagbe *et al.*, 2012; Fawole *et al.*, 2015) are interesting solutions if well composted to correct their high pH. Self-produced composts have been used in Ghana using banana and pineapple residue from the organic farm itself but mixed with large amounts of aquatic grasses collected from the Volta River (Soler, 2007).

Spent mushroom substrate, rich in nutrients and left after mushroom harvest, has been successfully tested for pineapple in Nigeria (Adedokun and Orluchukwu, 2013).

In Brazil, the production of Bokashi-type organic compounds, enriched with beneficial microorganisms from the cultivation region, allows soil colonization with organisms that contribute to recycle and solubilize soil nutrients. They also improve protection against pathogens by biological equilibrium (Padua *et al.*, 2017).

An example of comparison of these types of materials has been given by Fawole *et al.* (2015) in an African context (Table 9.4). Garita Coto (2014) also compared the composition of different organic fertilizers and amendments in Costa Rica. A guide for organic fertilization specific for

Table 9.4. Some chemical properties of compost material, composts and organic wastes used for N enrichment. From Fawole *et al.*, 2015.

Parameters	Blood meal	Bone meal	Hoof meal	Horn meal	Tithonia meal	Neem meal	Cow dung	Poultry droppings	Sawdust	CDSD compost	PDSD compost
pH	6.2	10.5	6.2	8.2	8.2	6.5	8.0	8.4	8.4	8.4	8.2
Total N (g/kg)	78.8	2.0	58.1	44.2	82.0	98.7	53.9	79.2	1.5	64.0	42.2
Total C (g/kg)	325.6	7.8	405.4	182.6	339.1	407.8	222.9	327.2	334.3	258.0	158.0
Total P (g/kg)	3.2	241.4	8.9	256.3	26.8	11.7	268.2	86.4	2.2	10.0	23.0
Ca (g/kg)	12.2	10.9	12.1	10.0	9.7	9.2	9.5	13.1	10.4	13.0	11.5
M (g/kg)g	6.1	5.8	6.6	6.0	6.2	5.9	6.0	6.7	6.4	6.8	6.2
K (g/kg)	4.4	4.1	5.0	4.1	5.1	4.1	4.4	5.2	4.8	6.1	5.4
Na (g/kg)	1.1	0.8	0.7	0.7	0.5	0.5	0.8	0.8	0.5	1.3	0.4

CDSD = cow dung + sawdust
PDSD = poultry droppings + sawdust

the context of la Réunion has been published (Chabalier *et al.*, 2006), and detailed preliminary recommendations for 'Victoria' ('Queen') pineapple are available on the CIRAD website of a FEDER Project (COSAQ) supported by EU (Rothé and Poletti, 2015). Padua *et al.* (2017) provided detailed information with the nutrient contents of different organic sources available, such as fresh and dry plant crop residues, industrial residues, animal manures, Bokashi composts and natural minerals.

While composts can be a good source of N, it can be difficult to apply these solid, bulky materials. A large Ghanaian farm used water extracts of crude cattle manure that was macerated in large water tanks. The extracts were collected at the base of the tank, and then filtered before application on pineapple through a drip irrigation system under plastic mulch. The remaining solid matter was applied as a banded pre-plant fertilizer.

Although green manure crops are quickly mineralized in the soil compared to composts, they can still be a worthwhile source of N, particularly if they derive from plants containing 3–5% dwt of N. The high biomass should also be composted resulting in a longer release of N than the green manure itself. Rothé *et al.* (personal communication) compared the rate of N mineralization for different SPs, composts and other sources of organic N (Fig. 9.9).

Ramial Chipped Wood (RCW) are chipped branches <7.0 cm in diameter that can be used to cover the soil and control weeds. This fresh material is a source of N but with a high C/N ratio. This technique is becoming popular, including for small private horticulture gardens, but it can induce a proliferation of ants and mealybugs (Soler and Dorey, 2017).

Flower Initiation

Artificial forcing of pineapple is important for managing the harvest. Different countries have different options available. As an example, ethephon formulations are authorized in Kerala (India) for 'Mauritius' (syn. 'Queen') pineapple (Pineapple Research Station, 2014), and Ghana for 'Sugarloaf', possibly 'Pérola' (AGROECO and FAO, 2008), but not permitted in other countries. Gaseous ethylene, commonly used in large scale farms in West Africa (ethylene being injected in water with activated charcoal), is not authorized for organic farming by French authorities. Consequently, a technique based on minimal use of ethylene loaded on dry activated charcoal has been proposed (Soler *et al.*, 2006) but is not yet approved. In other areas of production, small scale farmers are still authorized to use the old technique of calcium carbide in water as in Brazil (MAPA, 2014). Finally, ice block or chilled water induction, aimed at stimulating the production of ethylene by the plant itself, have not given consistent results (González, 2006).

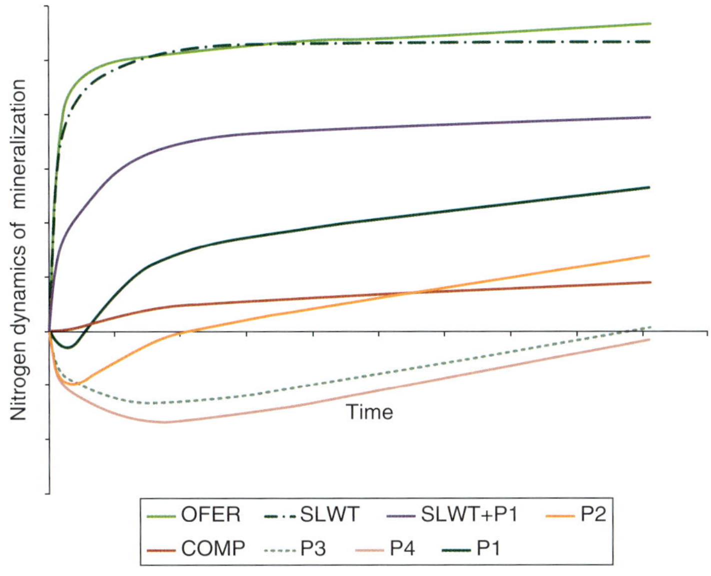

Fig. 9.9. Dynamics of mineralization of different N sources. Nitrogen dynamics measured in controlled conditions in the laboratory according to the French standard AFNOR XP U 44 163. The X axis represents the incubation time (max ~6 months), while Y axis represents the amount of N mineralized from the organic materials on test (%). Under the 0 line = immobilization of N (temporary or long-lasting); above 0: net mineralization of N. Compost (COMP), organic fertilizer (OFER), cover crops (P1–P4), slaughterhouse waste (SLWT), and a mixture of slaughterhouse waste plus cover crop P1 (SLWT+P1). Personal communication M. Rothé.

Pre-harvest Fruit Treatments, Harvesting and Postharvest

Except for synthetic fungicides, most of the practices used in conventional postharvest systems comply with organic specifications. These include brushing for mealybugs, hot water treatment to reduce fungi development such as black rot (*Thielaviopsis* sp.) (Wilson Wijeratnam *et al.*, 2005), and natural wax treatments to prolong shelf-life.

The yeast *Pichia guilliermondii* and a mixture of other yeasts already present on the fruit were shown to give the same level of control of postharvest moulds and fungi as the cold storage or fungicide (Reyes *et al.*, 2004). The different yeasts used appear to use different mechanisms to minimize growth of the pathogens including competition for space and/or nutrients. One of the yeasts, *Pichia guilliermondii*, exhibits a strong attachment to fungal hyphae. Wisnewski *et al.* (1991) showed in apples that this attachment could be linked to a β–1–3 glucanase attack of the cell wall of *Botrytis cinerea*.

Organic production of pineapple is feasible but most easily on a small scale in a mixed cropping situation. Government, consumer and grower interest is, however, increasing and this can only help to encourage further research in this field.

References

Adedokun, O.M. and Orluchukwu, J. (2013) Pineapple: organic production on soil amended with spent mushroom substrate. *Agriculture and Biology Journal of North America* 4, 590–593.

Adoyo, F., Mukala, J.B. and Enloya, M. (1997) Using Tithonia concoctions for termite control in Busia District, Kenya. *ILEIA Newsletter* (December), 24–25.

AGROECO and FAO (2008) Improve your Sugarloaf pineapple production. In: *Illustrated Manual for Organic Farmers in Ghana.* Agro Eco-Louis Institute, Accra. Ghana, p. 22 .

Aiyelaagbe, I.O.O., Oshuniyi, A.A. and Adegoke, J.O. (2012) Response of 'Smooth Cayenne' pineapple to organic fertilizer in south western Nigeria. *Acta Horticulturae* 933, 261–264.

Akhtar, M. and Malik, A. (2000) Roles of organic soil amendments and soil organisms in the biological control of plant-parasitic nematodes: a review. *Bioresource Technology* 74(1), 35–47.

Altieri, M A. (1989) Agroecology: a new research and development paradigm for world agriculture. *Agriculture, Ecosystems and Environment* 27(1), 37–46.

Alves, A.A., Reinhardt, D.H., Alcântara, J.P., Souza, L.F.S. and Caldas, R.C. (1998) Manejo e avaliação da soca do abacaxi 'Pérola' nas condições do semi-árido de Itaberaba, Bahia. *Revista Brasileira de Fruticultura* 20(3), 323–331.

Banguela-Castillo, A., Ramos-González, P.L., Peña-Marey, M., Tanaka, F.A.O., Blassioli-Moraes, M.C. *et al.* (2015) The same host but a different Muscodor: a new *Muscodor albus* isolate from wild pineapple (*Ananas ananassoides*) with potential application in agriculture. *Crop Protection* 78, 284–292.

Brenes, L., Piedra, B., Blanco, R., Salazar, D., Linares, J. and Rodriguez, A. (2011) Developing a small grower based organic pineapple export system [electronic resource]. *Acta Horticulturae* 906, 99–108.

Caswell, E.P., DeFrank, J., Apt, W.J. and Tang, C.S. (1991) Influence of nonhost plants on population decline of *Rotylenchulus reniformis. Journal of Nematology* 23(1), 91–98.

Chabalier, P.F., Kerchove, V.V.D. and Macary, H.S. (2006) *Guide to Organic Fertilization in Reunion.* Réunion: Centre de Coopération Internationale en Recherche Agronomique pour le Développement (CIRAD), Delegation de la Réunion.

Chinnasri, B., Sipes, B.S. and Schmitt, D.P. (2006) Effects of inducers of systemic acquired resistance on reproduction of *Meloidogyne javanica* and *Rotylenchulus reniformis* in pineapple. *Journal of Nematology* 38(3), 319–325.

Coolen, W. and d'Herde, C. (1972) *A Method for the Quantitative Extraction of Nematodes from Plant Tissue.* State Agriculture Research Centre, Belgium, pp. 1 –77.

Damour, G., Dorel, M., Quoc, H.T., Meynard, C. and Risède, J.M. (2014) A trait-based characterization of cover plants to assess their potential to provide a set of ecological services in banana cropping systems. *European Journal of Agronomy* 52, 218–228.

Devadas, V.S. and Kuriakose, K.P. (2005) Standardisation of leaf sampling for nutrient analysis in pineapple var. Mauritius. *Acta Horticulturae* 666, 91–192.

Fairweather, J.R. (1999) Understanding how farmers choose between organic and conventional production: results from New Zealand and policy implications. *Agriculture and Human Values* 16, 51–63.

Fawole, F.O., Ayodele, O.J., and Adeoye, G.O. (2015) Nitrogen contents of composts enriched with organic nitrogen sources. *The 3rd African Organic Conference, 'Achieving Social and Economic Development through Ecological and Organic Agricultural Alternatives'*, Lagos, Nigeria, pp. 263–267.

Ferreira, K.D. de S., Ferreira, T. de F., Souza, R.M. de and Dolinski, C. (2015) Potential of entomopathogenic nematodes (Rhabditida) for control of pink pineapple mealybug adult females, *Dysmicoccus brevipes* (Hemiptera: Pseudococcidae), under laboratory conditions. *Nematoda* 2, e072015.

Garita Coto, R.Á. (2014) *La Piña.* Editorial Tecnologica de Costa Rica, Cartago, Costa Rica.

Gibbon, P. and Bolwig, S. (2007) *The Economics of Certified Organic Farming in Tropical Africa: A Preliminary Assessment.* DIIS Working Paper No. 2007/3. Danish Institute for International Studies, Copenhagen, Denmark.

Giroux, M. and Audesse, P. (2004) Comparaison de deux méthodes de détermination des teneurs en carbone organique, en azote total et du rapport C/N de divers amendements organiques et engrais de ferme. *Agrosol* 15(2): 107–110.

Gonzalez, M.V.T. (2006) Alternativas orgánicas para inducir la floración en piña (*Ananas comosus*) (L.) Merr. híbrido MD-2. Graduate project report, Costa Rica Institute of Technology, San Carlos, Costa Rica.

González, R., Laudat, T., Arzola, M., Méndez, R., Marrero, P. *et al.* (2011) Effect of *Azotobacter chroococcum* on in vitro pineapple plants' growth during acclimatization. *In Vitro Cellular & Developmental Biology – Plant* 47(3), 387–390.

González Rodríguez, R., Iglesias, A., Lorenzo, J. and Dibut, B. (2012) Selection of *Azotobacter chroococcum* strains for its application on *in vitro* pineapple plants acclimatization. *Biotecnología Vegetal* 12(3), 157–164.

Guimaraes, P.R., Raimundo, R.L.G., Bottcher, C., Silva, R.R. and Trigo, J.R. (2006) Extrafloral nectaries as a deterrent mechanism against seed predators in the chemically protected weed *Crotalaria pallida* (Leguminosae). *Austral Ecology* 31(6), 776–782.

Hepperly, P., Aguilar-Erazo, H., Perez, R., Diaz, M. and Reyes, C. (1992) Pigeon pea and velvet bean allelopathy. In: Rizvi, S.J.H. and Rizvi, V. (eds) *Allelopathy: Basic and Applied Aspects.* Dordrecht: Springer, pp. 357–370.

Hernandez Mansilla, A. Sierra, P.A. and Carr Pérez, A. (2006) Evaluacion in vitro del antagonismo de especies de Trichoderma sobre hongos fitopatogenos que afectan las vitroplantas de pina (*Ananas comosus* (L.) Merr.). *Fitosanidad* 10(2), 105–108.

Hong, C.W. (1994) *Organic Farming and the Sustainability of Agriculture in Korea.* Extension Bulletin – ASPAC, Food and Fertilizer Technology Center, Taipei, Taiwan .

IFOAM (n.d.) IFOAM Organics International: cultivating change. Available at: http://www.ifoam.bio/en/Organic-landmarks/principles-Organic-agriculture (accessed 8 March 2018).

Inclán, D.J., Bermúdez, F.J., Alvarado, E., Ellis, M., Williams, R.N. and Acosta, N. (2008) Comparison of biological and conventional insecticide treatments for the management of the pineapple fruit borer, *Strymon megarus* (Lepidoptera: Lycaenidae) in Costa Rica. *Ecological Engineering* 34(4), 328–331.

Irmer, S., Podzun, N., Langel, D., Heidemann, F., Kaltenegger, E. *et al.* (2015) New aspect of plant–rhizobia interaction: alkaloid biosynthesis in Crotalaria depends on nodulation. *Proceedings of the National Academy of Sciences of the United States of America* 112(13), 4164–4169.

Jama, B., Palm, C.A., Buresh, R.J., Niang, A., Gachengo, C. *et al.* (2000) *Tithonia diversifolia* as a green manure for soil fertility improvement in western Kenya: a review. *Agroforestry Systems* 49(2), 201–221.

Jannoyer, M.L., Le Bellec, F., Lavigne, C., Achard, R. and Malézieux, E. (2011) Choosing cover crops to enhance ecological services in orchards: a multiple criteria and systemic approach applied to tropical areas. *Procedia Environmental Sciences* 9, 104–112.

Jha, K. (2007) Cultivation of pineapple for upliftment of rural economy in Nagaland, India. *VI International Pineapple Symposium* 822, 99–106.

Joy, P., Anjana, R. and Soumya, K. (2013) *Pests of Pineapple and their Management.* Pineapple Research Station, Kerala Agricultural University, Muvattupuzha, Kerala. India

Kiggundu, M., Kabi, F., Mette, V., Nalubwama, S. and Odhong, C. (2014) Management and use of dairy cattle feed resources on smallholder certified organic pineapple farms in Central Uganda. *Journal of Agriculture and Environment for International Development* 108(2), 207–225.

Kleemann, L. and Abdulai, A. (2012) *Organic Certification, Agro-ecological Practices and Return on Investment: Farm Level Evidence from Ghana.* Kiel Working Papers No. 1816. Kiel Institute for the World Economy, Kiel, Germany.

Kleemann, L., Abdulai, A. and Buss, M. (2014) Certification and access to export markets: adoption and return on investment of organic-certified pineapple farming in Ghana. *World Development* 64, 79–92.

Ko, M.P. and Schmitt, D.P. (1993) Pineapple inter-cycle cover crops to reduce plant-parasitic nematode populations. *Acta Horticulturae* 334, 373–382.

Lakshmanan, V., Selvaraj, G. and Bais, H. (2014) Functional soil microbiome: belowground solutions to an aboveground problem. *Plant Physiology* 166(22), 689–700.

Lashermes, G., Nicolardot, B., Parnaudeau, V., Thuries, L., Chaussod, R. *et al.* (2009) Indicator of potential residual carbon in soils after exogenous organic matter application. *European Journal of Soil Science* 60(2), 297–310.

Liang, C.C. (2008) Detection of pathogenicity of metarhiziums against *Dysmicoccus Brevipes* (Cockerell) in laboratory. *Journal of Anhui Agricultural Sciences* 29, 12775; 12872.

Maldes, J.M. and Ratnadass, A. (1998) *Crotalaria juncea* (Fabaceae) citée pour la première fois comme hôte alternatif d'Eurystylus oldi important ravageur africain du Sorgho et du Ricin (Hem,Miridae). *Bulletin de la Société entomologique de France* 103(3), 272.

MAPA (2014) *Ministério de Agricultura, Abastecimento e Pecuária do Brasil. DOU 20.06.2014.* Brazil.

Marie-Alphonsine, P.A. and Soler, A. (2013) *Host Status for Symphylids of Different Cover Crops Tested in Pineapple Cultural Systems* (FEDER, EU), CIRAD, Martinique.

Marie-Alphonsine, P.A., Soler, A., Gaude, J.M., Berimey, M.A. and Quénéhervé, P. (2017) Effects of some rotation crops on the populations of *Rotylenchulus reniformis* and Hanseniella sp., two soil borne parasites on pineapple crop in Martinique. *IX International Pineapple Symposium.* Havana, Cuba.

Matos, A.P. de and Reinhardt, D.H. (2009) Pineapple in Brazil: characteristics, research and perspectives. *Acta Horticuturae* 822, 25–36.

Matos, A.P. de, Querino, C.M.B., Cabral, J.R.S. and Caldas, R.C. (2006) Preliminary report on systemic acquired resistance in pineapple plants to control *Fusarium subglutinans*. *Acta Horticulturae* 702, 167–171.

Murray, D.N.A. and Manicom, B. (2009) Development of an organic pineapple cultivation strategy for the Eastern Cape Region of South Africa. *Acta Horticulturae* 822: 107–116.

Noronha, A.C.S., Silva, T. de A., Tavares, G. de S., Kramer, V.R. and Matos, A.P. de (2009) *A Broca do Fruto do Abacaxi Strymon Megarus no Estado do Pará*. [Annals of Congress] Seminário brasileiro de produção integrada de frutas 11; Seminário sobre sistema agropecuário de produção integrada, 3. Embrapa Semiárido, Petrolina, Brazil.

Noronha, A.C.S., Lemos, W.P., Fazolin, M., Sanches, N.F. and Garcia, M.V.B. (2016) Abacaxi. In: Silva, N.M.A. and Zucchi, R.A. (ed.) *Pragas agrícolas e florestais na Amazônia*. Embrapa Semiárido, Petrolina, Brazil, pp. 22 –43.

Orrell, P. and Bennett, A.E. (2013) How can we exploit above–belowground interactions to assist in addressing the challenges of food security? *Frontiers in Plant Science* 4 (October), 432.

Padua, T.R.P., Matos, A.P. de and Cordeiro, Z.J.M. de (2017) Sistema orgânico para a produção de abacaxi para Lençóis, Chapada Diamantina – BA. *Embrapa Mandioca e Fruticultura, Sistema de Produção* 45.

Pádua, T.P., Matos, A.P. Reis, R., Cardoso, V.E.S., Rosa, R., Castro, C.X. F., da Silva, A (2016a) Cultivation of pineapple in organic system – recommendations based on studies in the Semiarid Region of the Chapada Diamantina, Bahia, Brazil. *Pineapple News* 23, 16–20.

Pádua, T.P., Matos, A.P. Reis, R., Cardoso, V.E.S., Rosa, R., Castro, C.X. F., da Silva, A. and Reinhardt, H.D. (2016b) Organic Pineapple Production System for the Region of Lençóis, Chapada Diamantina, Bahia, Brazil. *Pineapple News* 23, 9–16.

Petty, G. (2012) Adult beetle kairomones as attractants for the monitoring and control of *Popillia bipunctata* (Coleoptera: Rutelinae), a Whitegrub pest of Pineapples in South Africa. *Pineapple News* 19, 49.

Pichot, J.P. (1975) Rôle de la matière organique dans la fertilité du sol. *L'Agronomie Tropicale* 2, 170–175.

Pieri, C. (1989) *Fertilité des terres de Savanes: bilan de trente ans de recherche et de développement agricoles au sud du Sahara*. Paris: Ministère de la Cooperation et du Developpement.

Pineapple Research Station (2014) Ad-hoc recommendation for the organic production of the pineapple variety Mauritius. Available at: http://prsvkm.tripod.com/Html/AdhocOrganicMauritius.htm (accessed 23 May 2018).

Pinon, A. and Champanhet, F. (1988) Trial on the introduction of a rotation with soil-improving fodder plants into a pineapple monoculture. *Fruits Paris* 43(5), 275–286.

Py, C., Lacoeuilhe, J.J. and Teisson, C. (1987) *The Pineapple: Cultivation and Uses*. Paris: Editions G.-P. Maisonneuve.

Ratnadass, A., Soler, A., Chabanne, A., Tullus, G., Techer, P. *et al.* (2018) First records of *Moissonia importunitas* as a pest of rattle box (*Crotalaria* spp.) in Réunion Island (Hemiptera, Miridae). *Bulletin de la Société entomologique de France* 123(1), 59–64.

Reyes, M.E.Q., Rohrbach, K.G. and Paull, R.E. (2004) Microbial antagonists control postharvest black rot of pineapple fruit. *Postharvest Biology and Technology* 33(2), 193–203.

Rothé, M. and Poletti, S. (2015) *La Fertilisation de L'ananas Victoria à la Réunion*. CIRAD, Paris, pp. 1 –9.

Sanches, N.F. (1999) Pragas e seu contrôle. In: Cunha, G.A.P. de, Cabral, J.R.S. and Souza, L.F.S. (ed.) *O abacaxizeiro – Cultivo, Agroindústria e economia*. Brasília, Brazil: DF Embrapa Comunicação para Transferência de Tecnologia, pp. 307–341.

Seinhorst, J.W. (1962) Modifications of the elutriation method for extracting nematodes from soil. *Nematologica* 8, 117–128.

Sierra Pena, A., Hernandez Mansilla, A.A., Diaz, J.A. and Carr Perez, A. (2007) Control *in vitro* with *Trichoderma* spp. of phytopathogenus of pineapple vitroplants (*Pineapples comosus* (L). Merr.). *Centro Agricola* 34(4), 19–23.

Skinner, E.M., Díaz-Pérez, J.C., Phatak, S.C., Schomberg, H.H. and Vencill, W. (2012) Allelopathic effects of sunnhemp (*Crotalaria juncea* L.) on germination of vegetables and weeds. *HortScience* 47(1), 138–142.

Soler, A. (2007) *VREL, BioFair Fruit Ltd, Organic Pineapple Production, General report* [Mission survey]. CIRAD, Paris.

Soler, A. and Dole, B. (2006) Pineapple multiplication: practical techniques for small farms. *Pineapple News* 13, 23–27.

Soler, A. and Dorey, E. (2017) ANANABIO: a project to design organic pineapple cropping systems through a participative approach between research and producers. *Pineapple News* 24, 21–29.
Soler, A., Teisson, C., Dole, B. and Marie-Alphonsine, P.A. (2006) Forcing in pineapples: what is new? *Pineapple News* 13, 27–30.
Soler, A., Marie-Alphonsine, P.A., Corbion, C. and Quénéhervé, P. (2013a) Differential response of two pineapple cultivars (*Ananas comosus* (L.) Merr.) to SAR and ISR inducers against the nematode *Rotylenchulus reniformis*. *Crop Protection* 54, 48–54.
Soler, A., Marie-Alphonsine, P.A., Corbion, C., Fernandes, P., Portal Gonzalez, N. *et al.* (2013b) A strategy towards bio-protection of tropical crops: Experiences and perspectives with ISR on pineapple and banana in Martinique. *6th Meeting of IOBC-WPRS Working Group 'Induced Resistance in Plants Against Insects and Diseases': Leaping from Success in the Lab to Success in the Field.* IOBC, Avignon, France.
Soler, A., Marie-Alphonsine, P.A., Corbion, C., Fernandes, P., Portal Gonzalez, N. *et al.* (2016) Bioprotection of pineapple in ecological cropping systems. *Acta Horticulturae* 1111, 159–168.
Tapia-Hernández, A., Bustillos-Cristales, M.R., Jiménez-Salgado, T., Caballero-Mellado, J. and Fuentes-Ramírez, L.E. (2000) Natural endophytic occurrence of *Acetobacter diazotrophicus* in pineapple plants. *Microbial Ecology* 39(1), 49–55.
Thoden, T. and Boppré, M. (2010) Plants producing pyrrolizidine alkaloids: sustainable tools for nematode management? *Nematology* 12(1), 1–24.
Van Soest, P.J., Robertson, J.B. and Lewis, B.A. (1991) Symposium: carbohydrate methodology, metabolism, and nutritional implications in dairy cattle. Methods for diatery fiber, neutral detergent fiber, and nonstarch polysaccharides in relation to animal nutrition. *Journal of Dairy Science* 74(10), 3583–3597.
Venzon, M., Rosado, M.C., Euzébio, D.E., Souza, B. and Schoereder, J.H. (2006) Suitability of leguminous cover crop pollens as food source for the green lacewing *Chrysoperla externa* (Hagen) (Neuroptera: Chrysopidae). *Neotropical Entomology* 35, 371–376.
Wang, K.H., Sipes, B. and Schmitt, D.P. (2002) Crotalaria as a cover crop for nematode management: a review. *Nematropica* 32(1), 35–57.
Wang, K.H., Sipes, B.S. and Schmitt, D.P. (2003) Intercropping cover crops with pineapple for the management of *Rotylenchulus reniformis*. *Journal of Nematology* 35(1), 39–47.
Wang, K.H., Sipes, B.S. and Hooks, C.R.R. (2011) Sunn hemp cover cropping and solarization as alternatives to soil fumigants for pineapple production. *Acta Horticulturae* 902, 221–232.
Weber, O.B., Correia, D., Rocha, M., Alvez, G.C., de Oliveira, E.M. and Sa, E.G. (2003) Response of pineapple plantlets to inoculation with diazotrophic bacteria in greenhouse. *Pesquisa Agropecuaria Brasileira* 38(12), 1419–1426.
Weber, O.B., Terao, D., Rocha, L.S., Correia, D. and Santos, F.J. de S. (2004) Efeito de bactérias diazotróficas na produção de abacaxizeiro 'Cayenne Champac', sob irrigação, em dois níveis de adubação nitrogenada. *Revista Brasileira de Fruticultura* 26(2), 249–253.
Weber, O.B., Lima, R.N., Crisóstomo, L.A., Freitas, J.A.D., Carvalho, A.C.P. and Maia, A.H. (2010) Effect of diazotrophic bacterium inoculation and organic fertilization on yield of Champaka pineapple intercropped with irrigated sapota. *Plant and Soil* 327(1–2), 355–364.
Weston, L.A., Alsaadawi, I.S. and Baerson, S.R. (2013) Sorghum allelopathy – from ecosystem to molecule. *Journal of Chemical Ecology* 39(2), 142–153.
Wilson Wijeratnam, R.S., Hewajulige, I.G.N. and Abeyratne, N. (2005) Postharvest hot water treatment for the control of Thielaviopsis black rot of pineapple. *Postharvest Biology and Technology* 36(3), 323–327.
Wisnewski, M., Biles, C., Droby, S., McLaughlin, R., Wilson, C. and Chalutz, E. (1991) Mode of action of the postharvest biocontrol yeast, *Pichia guilliermondii*. I. Characterization of attachment to *Botrytis cinerea*. *Physiological and Molecular Plant Pathology* 39, 245–258.
Xavier, F.A.S., Dias, F.P.M., Farias, P.S. and Rosa, R.C.C. (2016) Efeito de adubos verdes sobre frações de carbono orgânico do solo na Chapada Diamantina, Embrapa, Brazil. *Boletim de Pesquisa e Desenvolvimento* 76, 35.
Zamioudis, C. and Pieterse, C.M.J. (2011) Modulation of host immunity by beneficial microbes. *Molecular Plant–Microbe Interactions* 25(2), 139–150.

10 Production for Other Uses

Fernanda Vidigal Duarte Souza[1], Everton Hilo de Souza[2], Alfredo Rodrigues Sena Neto[3], José Manoel Marconcini[4] and Sandra Aparecida de Assis[5]

[1]Embrapa Mandioca e Fruticultura, Cruz das Almas, Brazil; [2]Universidade Federal do Recôncavo da Bahia, Cruz das Almas, Brazil; [3]Universidade Federal de Lavras, Minas Gerais, Brazil; [4]Embrapa Instrumentação, São Paulo, Brazil; [5]Universidade Estadual de Feira de Feira de Santana, Bahia, Brazil

The pineapple plant has many uses besides fresh and processed fruit products, where it stands out not only for its flavour, but also for its nutritional properties (Matsuura and Rolim, 2002). These alternate uses include ornamental purposes (Duval *et al.*, 2001; Chan, 2006; Coppens d'Eeckenbrugge and Duval, 2009; Coppens d'Eeckenbrugge *et al.*, 2011; Sanewski, 2009; Souza *et al.*, 2009, 2012a, 2014; Taniguchi *et al.*, 2015; Costa Júnior *et al.*, 2016a), reinforcing fibres for varied products (Mohanty *et al.*, 2000; Zah *et al.*, 2007; Leão *et al.*, 2009; Sena Neto *et al.*, 2013, 2015), fibres for paper (Yusof *et al.*, 2012), enzymes with proteolytic action, secondary metabolites (Marques *et al.*, 2007; Manetti *et al.*, 2009; Yusof *et al.*, 2012) and animal feed (Fagundes and Fagundes, 2010; Santos *et al.*, 2014).

The potential alternative uses of pineapple are mainly associated with crop residues, thus opening the possibility of obtaining two or more products from the same planted area (e.g., production of fruit and fibres; production of cut flowers and fibres; production of fruit and bromelain; extraction of fibres and residues for animal feed). The utilization of crop residues and by-products from processing can result in additional gains for growers, with potential use of virtually the entire plant during the cultivation cycle.

Ornamental

The flower and ornamental plant segment is dynamic with a constant search for novelty. In this context, ornamental pineapple growing is an alternative that can offer many unique, innovative products, including cut flowers for floral arrangements (Fig. 10.1a), potted plants (Fig. 10.1b), foliage plants (Fig. 10.1c), and plants for landscaping, living fences and mini-fruits (Fig. 10.1d) (Souza *et al.*, 2012a, 2014). Pineapple plants have interesting characteristics for the floriculture segment, not only for the beauty and uniqueness of their fruits, but also for the postharvest durability of the flowers, which can last an average of 14 days (Costa Júnior *et al.*, 2016b).

Among the advantages of ornamental pineapple plants is their great plasticity of use. Souza *et al.* (2012a) characterized pineapple germplasm according to use qualities. Accessions were identified for use as cut flowers (peduncle + syncarp), potted plants, and plants for landscaping, live fences, mini-fruits and foliage.

The commercial production of ornamental pineapple plants in Brazil is concentrated in the Northeast, mainly in the states of Ceará and Rio Grande do Norte. Most of the output is exported to the USA and Europe (Sebrae, 2015).

South Africa, Kenya, Ivory Coast, France and Malaysia also cultivate ornamental pineapple but no production data are available (Duval *et al.*, 2001; Chan, 2006; Sanewski 2009; Coppens d'Eeckenbrugge *et al.*, 2011; Hunter, 2012). The commercial cultivation in Brazil only involves three botanical varieties: *A. comosus* var. *erectifolius* (accounting for 75% of sales), *A. comosus* var. *bracteatus* and *A. comosus* var. *microstachys* (Quirino *et al.*, 2009; Carvalho *et al.*, 2014).

The CIRAD Fruit and Horticultural Crops Department (CIRAD-FLHOR) selected 30 ornamental hybrids, with three being considered suitable for the cut flower market ('Luciana', 'Geliana' and 'Carmina'). These three hybrids are attractive for their small fruits, erect stems and good blooming, with floral induction starting 6 months after planting in Martinique (Duval *et al.*, 2001). In Australia, six cultivars were selected from 5000 seedlings. These hybrids were found to present strong and bright colours, ranging from pink to red, and three were considered appropriate for cut flowers and three for landscaping (Sanewski, 2003).

In Brazil, improvement of ornamental pineapple genotypes started in 2003. Careful analysis for this purpose, based on the large genetic variability preserved in the Active Germplasm Bank of the Embrapa Cassava and Fruits research unit (Embrapa Mandioca e Fruticultura), permitted the identification of various accessions with ornamental potential as parents (Souza *et al.*, 2006, 2007, 2009, 2012a, 2014; Taniguchi *et al.*, 2015). As a consequence, controlled crosses were performed, leading to the selection of 16 hybrids for cut flowers, 17 for landscaping, 4 for mini-fruits, 2 for potted plants and 1 for foliage, with 13 being recommended for more than one of these categories (Souza *et al.*, 2014). Recently, two cultivars of this selection, 'BRS Anauê' (Fig. 10.1e) and 'BRS Boyrá' (Fig. 10.1f), were commercially launched for the cut flower, landscaping and mini-fruit segment (Souza *et al.*, 2016) and are being grown in Northeast Brazil.

Hybrids with potential for use as potted plants are also being evaluated. Traits sought include a compact crown, short peduncle, small leaves and a small fruit (Souza *et al.*, 2012a,

Fig. 10.1. Different uses, cultivars and hybrids of ornamental pineapple plants developed by the genetic improvement programme of Embrapa Mandioca e Fruticultura, Cruz das Almas, Bahia, Brazil. (a) Cut flowers for floral arrangements. (b) Potted plant. (c) Foliage for floral arrangements. (d) Mini-fruits used in floral arrangements. (e) BRS Anauê cultivar. (f) BRS Boyrá cultivar. (g) Hybrid for potted growth with recurring flowering. (h) Hybrid with sinuous stem. (i) Hybrid with almost black colour. Photos F. Souza and E. de Souza.

2014; Taniguchi *et al.*, 2015) (see Fig. 10.1g). The search for these traits in pineapple has not been easy due to the plant's size, general shape and morphology. Therefore, the harmonization of a pineapple plant for potted growth, considering all the traits to obtain a product with good marketable quality, has been carried out by adjustments in the production system, which is complex, laborious and slow (Taniguchi *et al.*, 2015).

Good potential exists for incorporating a range of fruit colours in ornamental pineapple. Colours can range from white through pink and red to almost black (Fig. 10.1i) (O.S. Lima *et al.*, unpublished results). In this respect, based on a new characterization process, pineapple plants with sinuous stems (see Fig. 10.1h) and/or 'black' fruits are being selected and can be an interesting novelty in the national and international floriculture market.

In relation to the cultivation and production aspects, ornamental pineapple plants present some differences, mainly in relation to the spacing, plant density, harvest and postharvest handling. The spacing and planting density is one of the most important production factors, since this is directly related to the yield and production cost (Carvalho *et al.*, 2014). Planting for ornamental varieties is denser than for food crops, ranging from 60,000 to 80,000 plants per hectare, depending on the size and growth habit of the cultivar (Carvalho *et al.*, 2014).

In Brazil, the use of soil cover of various types including carnauba palm leaves and branches, chopped grass, chopped coconut palm leaves, sugarcane bagasse, sawdust, leaves and ashes of various plants, and coconut or babaçu seed fibre, are commonly used depending on regional availability (Carvalho *et al.*, 2014). Floral induction is achieved by spraying 30 ml of a solution containing 500 ppm of Ethrel® and 3% urea per plant in the coolest part of the day to the apical bud region.

The complete development of the stem for harvest is variable for each cultivar, but usually 50–70 days after induction. The inflorescences should be cut at the base of the stem immediately after the closing of the last flower to avoid excessive growth of the fruit. The postharvest procedures are crucial for the ornamental segment. This includes careful stem treatment to make the flowers more attractive to consumers. The stems with flowers are cut when they have reached a suitable length, and placed in buckets of water, then sent to a packing house to be treated, stored, packaged and shipped. When arriving at the packing house, each floral stem is carefully inspected, mainly for removal of small dried flowers adhered to the infructescence, damaged leaves and undeveloped buds near the infructescence. Then the stems are washed with a strong stream of water and placed in the shade to dry. The products are separated according to standard parameters before packaging (Souza *et al.*, 2012b; Carvalho *et al.*, 2014). The standard parameters adopted for classification include length of the stem and infructescence. For stem length, the acceptable range is 35–60 cm, generally with classification into three categories: up to 40 cm; 40–50 cm; and 50–60 cm. The ideal infructescence size varies according to the consumer market.

Fibre

The use of pineapple fibre for production of rope, twine and baskets, among other artisanal applications dates to the colonial period in Brazil (Leão *et al.*, 2009). Among the leading countries in commercial exploitation are the Philippines, Thailand and India, where it is used for various purposes, including traditional clothing (Mukherjee and Satyanarayana, 1986).

Malaysia is one of the largest pineapple producers in Asia and the use of the fibre to make fabric dates to the colonial period when the fruit was introduced by the Spanish (Asim *et al.*, 2015). The variety most traditionally used is 'Singapore Spanish', which has long leaves and is widely cultivated, generating an estimated 335,488 metric tons of residues in 2012 (Aditiya *et al.*, 2016). The defibring process is artisanal, slow and laborious, resulting in a product with high value, used particularly to make clothing for festive occasions and ceremonies such as weddings. Because of this artisanal defibring process, large-scale production does not exist, but it is a valued traditional activity and a symbol of the country's cultural identity.

The Philippines also has a tradition of using pineapple plant fibre to make textiles. The thread is soft and shiny, similar to silk, and can be woven into cloth without chemical treatment, resulting in a final product that is light, luxurious and easy to maintain (Fitzgerald, 2017). These exceptional characteristics of pineapple fibre for textile use have won over a growing contingent of upscale consumers in Europe and the USA, boosting cultivation and creating a new business niche composed of specialized firms making fabrics for fine clothes and alternative leather, among other products.

The use of pineapple fibre for papermaking can be considered a relatively new industrial use. This is particularly true of Malaysia, where it is expanding in the country, based on studies to improve the quality of the paper (Nayan *et al.*, 2013; Laftah and Rahaman, 2015). In general, the pulp and paper sector is considered to be one of the main consumers of natural resources requiring large quantities of chemicals and water, causing significant environmental impacts. The demand for paper produced through sustainable processes is increasing, prompting the search for alternatives such as pineapple. Pineapple leaf fibre is a workable alternative to wood pulp and other traditional fibre sources, for the production of paper with good properties (Laftah and Rahaman, 2015).

The possible uses of pineapple fibre go beyond textiles and paper. The search for renewable and sustainable sources of raw materials is a premise for the development of any technology. In particular, the biodegradability and renewability of plant fibres makes them good candidates to replace synthetic fibres as reinforcement in different types of matrixes with a wide variety of industrial applications (Ferreira *et al.*, 2012). With respect to this, pineapple fibre has some potential to replace glass fibre in composite materials. Considering that global demand for fibreglass in 2010 was 4.35 million metric tons (Research in China, 2010), the pineapple leaf fibres produced in Brazil could supply 5.5% of global demand for fibreglass.

Plant fibres have several advantages over the synthetic fibres used for mechanical reinforcement in composites: lower costs, renewable source, lower density and abrasiveness, possibility of using large volumetric fractions in relation to polymer matrix (resulting in high rigidity and superior elastic modulus and specific tensile and flexural strength), wide variety of available fibres throughout the world, and generation of rural income (Saheb and Jog, 1999; Sanadi, 2004; Shanks *et al.*, 2006; Chattopadhyay *et al.*, 2009).

Pineapple leaf fibre has been studied for mechanical reinforcement in composites due to its plentiful availability, generation of rural income and the ecological appeal of the composites produced. Composites or conjugated materials are produced by physical blending or rational combination of two or more materials that differ in chemical composition and/or form, and that are essentially mutually insoluble (Chawla, 1998; Agarwal *et al.*, 2006). The main limitations of plant fibres for mechanical reinforcement are high moisture absorption (reducing the mechanical properties), unsatisfactory interaction with hydrophobic resins, lower degradation temperature than the composite processing temperature, seasonal variability and tendency to form agglomerates during the feed stage of processing (Saheb and Jog, 1999; Sanadi, 2004; Shanks *et al.*, 2006; Chattopadhyay *et al.*, 2009). In comparison with other plant fibres, pineapple fibre has high tensile strength, elastic modulus values and degradation onset temperatures (Mohanty *et al.*, 2000; Marques *et al.*, 2007; Satyanarayana *et al.*, 2007; Leão *et al.*, 2009; Sena Neto *et al.*, 2013, 2015). These three fibre properties are the factors considered essential for mechanical reinforcement in composites.

The curauá plant is a wild pineapple variety (var. *erectifolious*) whose fibre is widely used to make baskets, twine and musical strings, as well as for mechanical reinforcement (Pematec Triangel, n.d.; Satyanarayana *et al.*, 2007; Leão *et al.*, 2009). However, the majority of studies of pineapple fibre have only investigated the main fresh or processing varieties, failing to exploit the genetic variability of the genus *Ananas*, although some research groups have studied wild varieties and hybrids along with commercial varieties (Sena Neto *et al.*, 2013, 2015). According to the results observed, differences exist in the thermal and mechanical properties among pineapple genotypes, providing options to choose a genotype depending on the intended use of the fibre and processing method. Despite this variability, the leaf fibres of all the varieties studied, including *Ananas comosus* var. *comosus*, have presented satisfactory properties for use as mechanical reinforcement in polymer composites (Sena Neto *et al.*, 2013, 2015).

Despite the good potential to use this fibre, there is still a need to establish a supply chain, starting with choice of the most suitable cultivars for the envisioned use, passing through agronomic aspects (planting, management, harvesting and postharvest handling), to industrial aspects, such as extraction and processing of the fibres and final production. In Fig. 10.2 it is possible to see the different steps in the pineapple fibre extraction process, as well as several products ranging from tubes (containers to grow small plants), paper, composites (a mix of pineapple fibres and other resins or fibres such as fibre glass, plastic etc.) to handicraft items made from these products.

According to Aquino (2006), each hectare can produce 4–5 metric tons of pineapple leaves. With these numbers it is possible to estimate the output of pineapple leaf fibre (around 300,000 tons in Brazil in 2014). This fibre yield per hectare of 4–5 tons compares favourably with the national average fibre yield of cotton, of some 1.4 tons per hectare (Guerra *et al.*, 2012).

The research results confirm that pineapple leaf fibres have potential for use in new materials, due to their high cellulose fraction, high cellulose crystallinity index and small microfibril angle. These fibres also have adequate technical quality in terms of mechanical resistance and degradation temperature for processing involving thermal cycles to obtain polymeric composites. Further research is necessary to develop new pineapple cultivars with more resistant fibres, along with establishment of agroindustrial productive chains for greater economic insertion of these fibres to generate jobs and income. Finally, it should be considered that one of the greatest limitation for the adoption of pineapple fibre in the textile industry, despite all quality, is the inadequacy of the spinning and weaving machines, which were developed to use for cotton fibre.

Fig. 10.2. Process and products. (a) Cultivation of *Ananas comosus* var. *erectifolius*. (b) Cut leaves to fibre extraction. (c) Extraction process in *paraibana* machine. (d–e) Drying. (f) Dry and processed fibre. (g) Different products: 1 – rustic fibre; 2 – nanocellulose; 3 – cloth; 4 – plant tubes from different composites; 5 – papers from different composites. (h) handcraft. Photos (a), (b), (d) and (f) A. Sena Neto; (c) J. Marconcini; (e) F. Souza; (g) and (h) E. de Souza.

Bromelain

Bromelain is an enzymatic complex composed mainly of proteases, obtained from the tissues of species of the family *Bromeliaceae*, especially pineapple. Its proteolytic activity makes it suitable for various applications, especially therapeutic, as described by Taussig and Batkin (1988) in a review of the theme, reporting both anecdotal observations and applications confirmed in previous studies.

Bromelain is a mixture of thiol endopeptidases and other components such as phosphatases, glycosidases, peroxidases, cellulases, escarases, glycoproteins, carbohydrates, and many protease inhibitors (Maurer, 2001; Pavan *et al.*, 2012). Stem bromelain (EC 3.4.22.32), ananain (EC 3.4.22.31), and comosain are extracted from pineapple stems, while fruit bromelain (EC 3.4.22.33) is obtained mainly from fruit juice (Rowan *et al.*, 1990; Arshad *et al.*, 2014). Similar proteases are also present in pineapple peel, core, crown, and leaves, with the highest proteolytic activity and protein contents detected in the extract of pineapple crown (Ketnawa *et al.*, 2012; Arshad *et al.*, 2014).

This enzyme has been widely used in the food industry, in processes such as meat tenderizing, production of protein hydrolysates and clarification of beverages, among others. The results of these uses are comparable to those of other exogenous enzymes (Sullivan and Calkins, 2010; Arshad *et al.*, 2014). It is also used in the textile, leather, cosmetic and especially the pharmaceutical industries (Bala *et al.*, 2012). In the bakery industry, it is used to make dough lighter and to enable more uniform fermentation and expansion, as well as to produce hypoallergenic flour to make baked goods for people who are allergic to wheat (Tanabe *et al.*, 1996; Watanabe *et al.*, 2000; Kong *et al.*, 2007; Arshad *et al.*, 2014).

The enzyme's therapeutic use is mainly based on its proteolytic activity. Several studies have demonstrated that its anti-inflammatory activity is similar to that of traditional drugs used for this purpose (Hale *et al.*, 2005; Arshad *et al.*, 2014; Manzoor *et al.*, 2016). Other therapeutic applications are to treat autoimmune diseases and allergies, for reversible inhibition of platelet aggregation, alleviation of bronchitis and healing of surgical wounds. It also has been shown to improve absorption of drugs, particularly antibiotics (Kelly, 1996; Murakawa *et al.*, 2002; Roep *et al.*, 2002; Secor *et al.*, 2009; Bresolin *et al.*, 2013), and it has been used in dentistry for teeth whitening and removal of stains and plaque (Chakravarthy and Acharya, 2012). It also has been employed as a digestive enzyme to treat intestinal disorders, pancreatectomy and exocrine pancreas insufficiency (Rathnavelu *et al.*, 2016).

Another promising use of bromelain is to treat malignancies, such as leukaemia and cancer of the lungs, ovaries, breast and skin (Manzoor *et al.*, 2016). The enzyme has the ability to modify the key pathways that support malignancy and to modulate the immune, inflammatory and haemostatic systems (Chobotova *et al.*, 2010; Pavan *et al.*, 2012). In a recent study, Miranda *et al.* (2016) evaluated the extraction of bromelain from different parts of a wild pineapple variety (stems, crowns, peels, pulp and leaves) to investigate its *in vitro* pharmacological activity on murine melanoma cells (B16F10). The pulp and peel extracts presented higher enzyme activity and the *in vitro* assays demonstrated 100% inhibition of tumour cells. Majid *et al.* (2016) studied the effects of bromelain, from different parts of the 'Gandul' cultivar on the inhibition of cancerous cells, in particular A2780 (breast) and HT29 (colon). They found that pulp juice was most effective with IC_{50} of 280.54 µg/ml after 24 hours and 104.95 µg/ml after 48 hours for A2780 and HT29, respectively. The authors concluded that ingestion of a 250 ml glass of pineapple juice is recommended to maintain health. However, despite testing in various experimental and clinical studies, which have demonstrated its anti-cancer and anti-inflammatory effects, the precise molecular mechanisms are not fully understood.

In veterinary medicine, the enzyme can protect animals from diarrhoea caused by bacterial enterotoxins from *Escherichia coli* and *Vibrio cholerae* (Mynott *et al.*, 1997; Rathnavelu *et al.*, 2016), by acting as an anti-adhesion agent that modifies the receptor attachment sites and influences the intestinal secretion signalling pathways (Mynott *et al.*, 1996; Chandler and Mynott, 1998; Rathnavelu *et al.*, 2016).

Because of these various uses, demand for bromelain is growing, making it necessary to develop better production methods, with fewer steps, able to turn out a purer and less expensive product. One way to do this is to optimize the use of residues in pineapple production regions. The material from the crown, peel and leaf contains lower quantities compared to the stem and fruit. In this respect, stem bromelain (E.C. 3.4.22.32) and fruit bromelain (E.C. 3.4.22.33) are already commercially available (Ketnawa *et al.*, 2011). The proportion of each of the components in the plant can vary according to the soil composition, climate conditions, plant development stage, geographic location, variety and extraction process, causing variations in the pharmacological and other activities (Taussing and Batkin, 1988; Tochi *et al.*, 2008). In light of this, it appears reasonable to use known pineapple varieties whose cultivation is standardized, to minimize the variations in the active components.

Animal Feed

Crop plant residues and the bagasse left after extraction of fibres can also be used for animal feed. The volume of these materials can reach 50 tons of green mass per hectare (Santos *et al.*, 2014). The fruit sold for food represents only 38% of the plant, generating significant agricultural residue (Santos *et al.*, 2014). Therefore, various ways have been proposed to use this material, as hay or silage, with the aim of supplying food when forage is scarce.

Several studies have reported the use of pineapple field residues (Gutierrez *et al.*, 2003; Herrera *et al.*, 2009; Fagundes and Fagundes, 2010; Santos *et al.*, 2014; Zainuddin *et al.*, 2014), and by-products of fruit processing (Lousada Júnior *et al.*, 2006; Silva *et al.*, 2014) for animal feed. The use of the leaves as animal feed can add value to pineapple production.

Plant fibre is considered a functional food, promoting the metabolism of lipids and carbohydrates, improving the action of the gastrointestinal tract, slowing the absorption of nutrients and causing a sensation of satiety (Oliveira and Marchini, 1998).

Stock breeders often face shortages of voluminous feed during certain periods of the year, reducing the productivity of herds and/or increasing the expense to purchase concentrated feed to supplement forage and correct nutritional deficiencies (Silva *et al.*, 2014). The use of alternative sources of feed can generate a significant positive impact on stock breeding.

In Brazil, the use of leaves of various plants, including pineapple, as animal feed has been increasing in recent years particularly in semi-arid regions. The preparation does not require complicated technology, consisting basically of drying and chopping to produce silage (Fig. 10.3).

Furthermore, animals kept in confinement depend heavily on corn silage which can be expensive at certain times of the year. Under these circumstances, pineapple silage obtained from non-transplanted seedlings, crowns and stalks, which would otherwise be discarded, can be used. This silage has high protein and sugar content, and has been shown to have good nutritional quality, besides being voluminous and palatable (Portal DBO, 2016).

Pineapple crop residues can be used by animal breeders irrespective of the size of the operation, and can be supplied to the animals soon after harvest without processing (and can be kept for approximately 5 days without losing quality). For large producers, the stored residues can serve as silage, with the possibility of storage for up to a year after fermentation without losing quality (Canal Rural, 2014).

Efforts are also under way in Malaysia and other Asian countries to improve the properties of pineapple leaf silage. Among these efforts are

Fig. 10.3. Use of pineapple leaves in the production of silage for animal feed. Photos E. de Souza.

studies involving pressing of biomass into pellets to increase the apparent density, to reduce storage and transport costs and facilitate handling by farm machinery. For example, Zainuddin *et al.* (2014) evaluated the effect of moisture on the physical properties of pellets made of pineapple residues.

The multitude of potential uses combined with the ability of the pineapple plant to survive in dry environments make pineapple suitable for many rural communities with limited resources. These qualities and utility also highlight the need to protect germplasm collections of this valuable and unique natural resource.

References

Aditiya, H.B., Chong, W.T., Mahlia, T.M.I., Sebayang, A.H., Berawi, M.A. and Nur, H. (2016) Second generation bioethanol potential from selected Malaysia's biodiversity biomasses: a review. *Waste Management* 47, 46–61.

Agarwal, B.D., Broutman, L.J. and Chandrashekhara, K. (2006) Analysis and performance of fibre composites. Hoboken, NJ: John Wiley & Sons.

Aquino, M.S.D. (2006) Desenvolvimento de uma desfribadeira para obtenção da fibra da folha do abacaxi. MSc thesis, Universidade Federal do Rio Grande do Norte, Natal, Brazil.

Arshad, Z.I.M., Amid, A., Yusof, F., Jaswir, I., Ahmad, K. and Loke, S.P. (2014) Bromelain: an overview of industrial application and purification strategies. *Applied Microbiology and Biotechnology* 98, 7283–7297.

Asim, M., Abdan, K., Jawaid, M., Nasir, M., Dashtizadeh, Z., Ishak, M.R. and Hoque M.E. (2015) A review on pineapple leaves fibre and its composites. *International Journal of Polymer Science*. DOI: 10.1155/2015/950567.

Bala, M., Ismail, N. A., Mel, M., Jami, M. S., Salleh, M. H. and Amid, A. (2012) Bromelain production: current trends and perspective. *Archives des Sciences* 65, 369–399.

Bresolin, I.R.A.P., Bresolin, I.T.L., Silveira, E., Tambourgi, E.B. and Mazzola, P.G. (2013) Isolation and purification of bromelain from waste peel of pineapple for therapeutic application. *Brazilian Archives of Biology and Technology* 56, 971–979.

Canal Rural (2014) Palhada da Lavoura de Abacaxi é boa Fonte de Volumoso. Available at: http://www.canalrural.com.br/noticias/rebanho-gordo/palhada-lavoura-abacaxi-boa-fonte-volumoso-8861 (accessed 12 July 2017).

Carvalho, A.C.P.P., Souza, F.V.D. and Souza, E.H. (2014) *Produção de abacaxizeiro ornamental para flor de corte.* Fortaleza, Brazil: Embrapa Agroindústria Tropical.

Chakravarthy, P.K. and Acharya, S. (2012) Efficacy of extrinsic stain removal by novel dentifrice containing papain and bromelain extracts. *Journal of Young Pharmacists* 4, 245–249.

Chan, Y.K. (2006) Hybridisation and selection in pineapple improvement. The experience in Malaysia. *Acta Horticulturae* 702, 87–92.

Chandler, D.S. and Mynott, T.L. (1998) Bromelain protects piglets from diarrhoea caused by oral challenge with K88 positive enterotoxigenic *Escherichia coli*. *Gut* 43, 196–202.

Chattopadhyay, S.K., Khandal, R., Uppaluri, R. and Ghostal, A.K. (2009) Influence of varying fiber lengths on mechanical, thermal, and morphological properties of MA-g-PP compatibilized and chemically modified short pineapple leaf fiber reinforced polypropylene composites. *Journal of Applied Polymer Science* 113, 3750–3756.

Chawla, K. (1998) *Composite Materials: Science and Engineering*. New York: Springer.

Chobotova, K., Vernallis, A.B. and Abdul Majid, F.A. (2010) Bromelain's activity and potential as an anticancer agent: Current evidence and perspectives. *Cancer Letters* 290, 148–156.

Coppens d'Eeckenbrugge, G. and Duval, M.F. (2009) The domestications of pineapple: context and hypotheses. *Pineapple News* 16, 15–26.

Coppens d'Eeckenbrugge, G., Sanewski, G.M., Smith, M.K., Duval, M.F. and Leal, F. (2011) Ananas. In: Kole, C. (ed.) *Wild Crop Relatives: Genomic and Breeding Resources. Tropical and Subtropical Fruits*. Berlin/Heidelberg: Springer Verlag, pp. 21–41.

Costa Júnior, D.S., Souza, E.H., Costa, M.A.P.C., Pereira, M.E.C. and Souza, F.V.D. (2016a) Clonal evaluation of new ornamental pineapple hybrids to use as cut flowers. *Acta Scientiarum. Agronomy* 38, 475–483.

Costa Júnior, D.S., Souza, E.H., Pereira, M.E.C., Costa, M.A.P.C. and Souza, F.V.D. (2016b) Influence of washing and plastic packaging on post-harvest quality of ornamental pineapple stalks. *Pineapple News* 23, 22–25.

Duval, M.F., Coppens d'Eeckenbrugge, G., Fontaine, A. and Horry, J.P. (2001) Ornamental pineapple: perspective from clonal and hybrid breeding. *Pineapple News* 8, 13–14.

Fagundes, N.S. and Fagundes, N.S. (2010) Restos culturais do abacaxizeiro na alimentação de ruminantes. *Revista Eletrônica Nutritime* 7, 1243–1247.

Ferreira, S.R., Lima, P.R.L., Silva, F.A., Toledo Filho, R.D. (2012) Influência de ciclos molhagem-secagem em fibras de sisal sobre a aderência com matrizes de cimento Portland. *Revista Matéria* 17, 1024–1034.

Fitzgerald, B. (2017) Piña couture: pineapple fiber makes fabric in the philippines. Available at: http://www.lesouk.co/articles/material-inspiration/pina-couture-pineapple-fiber-makes-fabric-in-the-philippines (accessed 8 July 2017).

Guerra, J., Aguiar, G.A.M., Lima Filho, R.R., Torres Jr, A.M. (2012) Índices e resultados econômicos. Agro Analysis. Available at: http://www.agroanalysis.com.br/3/2012/mercado-negocios/algodao-indices-e-resultados-economicos (accessed 20 March 2018).

Gutiérrez, F., Rojas-Bourrillón, A., Dormond, H., Poore, M. and Ching-Jones, R.W. (2003) Características nutricionales y fermentativas de mezclas ensiladas de desechos de piña y avícolas. *Agronomía Costarricense* 27, 79–89.

Hale, L.P., Greer, P.K., Trinh, C.T. and James, C.L. (2005) Proteinase activity and stability of natural bromelain preparations. *International Immunopharmacology* 5, 783–793.

Herrera, M.L., Wingching-Jones, R. and Rojas-Bourrillón, A. (2009) Características fermentativas y nutricionales del ensilaje de rastrojo de piña (*Ananas comosus*). *Agronomía Costarricense* 33, 1–15.

Hunter, NT. (2012) *The Art of Floral Design*, 3rd edn. Stamford, CT: Cengage Learning.

Kelly, G.S. (1996) Bromelain: a literature review and discussion of its therapeutic applications. *Alternative Medicine Review* 1, 243–257.

Ketnawa, S., Chaiwut, P. and Rawdkuen, S. (2011) Aqueous two-phase extraction of bromelain from pineapple peels ('PhuLae' cultv.) and its biochemical properties. *Food Science and Biotechnology* 20, 1219.

Ketnawa, S., Chaiwut, P. and Rawdkuen, S. (2012) Pineapple wastes: a potential source for bromelain extraction. *Food and Bioproducts Processing* 90, 385–391.

Kong, X., Zhou, H. and Qian, H. (2007) Enzymatic hydrolysis of wheat gluten by proteases and properties of the resulting hydrolysates. *Food Chemistry* 102, 759–763.

Laftah, W.A. and Rahaman, W.A.W.A. (2015) Chemical pulping of waste pineapple leaves fiber for kraft paper production. *Journal of Materials Research and Technology* 4, 254–261.

Leão, A.L., Machado, I.S., Souza, S.F. and Soriano, L. (2009) Production of curaua fibers for industrial applications: characterization and micropropagation. *Acta Horticulturae* 822, 227–238.

Lousada Júnior, J.E., Costa, J.M.C., Neiva, J.N.M. and Rodriguez N.M. (2006) Caracterização físico-química de subprodutos obtidos do processamento de frutas tropicais visando seu aproveitamento na alimentação animal. *Revista Ciência Agronômica* 37, 70–76.

Majid, F.A.A., Teh, L.C., Yap, L.F., Poh, S.S., Mashudin, M., Gani, M.A. (2016) Pineapple is a super fruit, but how much can it be consumed? *Acta Horticulturae* 1111, 203–211.

Manetti, L.M., Delaporte, R.H. and Laverde Junior, A. (2009) Metabólitos secundários da família Bromeliaceae. *Química Nova* 32, 1885–1897.

Manzoor, Z., Ali, N., Hamid M., and Haq, I. (2016) Bromelain: methods of extraction, purification and therapeutic applications. *Brazilian Archives of Biology and Technology* 59, 1–16.

Marques, G., Gutiérrez, A. and Del Rio, J.C. (2007) Chemical characterization of lignin and lipophilic fractions from leaf fibers of curaua (*Ananas erectifolius*). *Journal of Agricultural and Food Chemistry* 55, 1327–1336.

Matsuura, F.C.A.U. and Rolim, R.B. (2002) Avaliação da adição de suco de acerola em suco de abacaxi visando a produção de um 'blend' com alto teor de vitamina C. *Revista Brasileira de Fruticultura* 24, 138.

Maurer, H.R. (2001) Bromelain: biochemistry, pharmacology and medical use. *Cellular and Molecular Life Sciences* 58, 1234–1245.

Miranda, Í.K.S.P.B., Miranda, A.F.S., Souza, F.V.D., Vannier-Santos, M.A., Pirovani, C.P., Pepe, I.M.P., Rodowanski, I.J., Ferreira, K.T.S.E., Vaz, L.M.S. and Assis, S.A. (2016) The biochemical characterization, stabilization studies and the antiproliferative effect of bromelain against B16F10 murine melanoma cells. *International Journal of Food Science and Nutrition* 68, 442–454.

Mohanty, A., Tripathy, P.C., Misra, M., Parija, S. and Sahoo, S. (2000) Chemical modification of pineapple leaf fiber: graft copolymerization of acrylonitrile onto defatted pineapple leaf fibers. *Journal of Applied Polymer Science* 77, 3035–3043.

Mukherjee, P. and Satyanarayana, K. (1986) Structure and properties of some vegetable fibres. *Journal of Materials Science* 21, 51–56.

Murakawa, T., Yamagata, H., Tsuruta, H. and Aizono, Y. (2002) Cloning of cold-active alkaline phosphatase gene of a psychrophile, *Shewanella* sp., and expression of the recombinant enzyme. Bioscience, *Biotechnology, and Biochemistry* 66, 754–761.

Mynott, T.L., Guandalini, S., Raimondi, F. and Fasano, A. (1997). Bromelain prevents secretion caused by *Vibrio cholerae* and *Escherichia coli* enterotoxins in rabbit ileum in vitro. *Gastroenterology* 113, 175–184.

Mynott, T.L., Luke, R.K. and Chandler, D.S. (1996) Oral administration of protease inhibits enterotoxigenic *Escherichia coli* receptor activity in piglet small intestine. *Gut* 38, 28–32.

Nayan, N.H.M., Razak, S.I.A., Rahman, W.A.W.A. and Majid, R.A. (2013) Effects of mercerization on the properties of paper produced from Malaysian pineapple leaf fiber. *International Journal of Engineering & Technology* 12, 1–6.

Oliveira, J.E.D. and Marchini, J.S. (1998) *Ciências Nutricionais*. São Paulo, Brazil: Sarvier.

Pavan, R., Jain, S., Shraddha and Kumar A. (2012) Properties and therapeutic application of bromelain: a review. *Biotechnology Research International* 2012, 1–6.

Pematec Triangel (n.d.) Fibra natural. Available at: http://www.pematec.com.br/produtos/3 (accessed 13 March 2018).

Portal DBO (2016) Silagem de abacaxi é alternativa ao milho para confinamento. Available at: http://www.portaldbo.com.br/Revista-DBO/Noticias/Silagem-de-abacaxi-e-alternativa-ao-milho-para-confinamento/16162 (accessed 12 July 2017).

Quirino, Z.B.R., Barboza, S.B.S.C., Viegas, P.R.A. and Ledo, A.S. (2009) *Multiplicação in vitro do abacaxizeiro ornamental, Ananas comosus var. erectifolius, em meio líquido e gelificado*. Aracaju, Brazil: Embrapa Tabuleiros Costeiros.

Rathnavelu, V., Alitheen, N.B., Sohila, S., Kanagesan, S. and Ramesh, R. (2016) Potential role of bromelain in clinical and therapeutic applications. *Biomedical Reports* 5, 283–288.

Research in China (2010) *China Glass Fiber Industry Report, 2010*. Available at: http://www.researchinchina.com/Htmls/Report/2010/6012.html (accessed 22 July 2017).

Roep, B.O., Engel, N.K., Halteren, A.G., Duinkerken, G. and Martin, S. (2002) Modulation of autoimmunity to beta-cell antigens by proteases. *Diabetologia* 45, 686–692.

Rowan, A.D., Buttle, D.J. and Barrett A.J. (1990) The cysteine proteinases of the pineapple plant. *Biochemical Journal* 266, 869–875.

Saheb, D.N. and Jog, J. (1999) Natural fiber polymer composites: a review. *Advances in Polymer Technology* 18, 351–363.

Sanadi, A. (2004) Natural fibers as fillers/reinforcements in thermoplastics. In: Johnson, M. (ed.) *Low Environmental Impact Polymers*. Shrewsbury, UK: Rapra Technology Ltd, pp. 105–139.

Sanewski, G.M. (2003) Ornamental pineapples. *Pineapple News* 10, 5.

Sanewski, G.M. (2009) Breeding *Ananas* for the cut-flower and garden markets. *Acta Horticulturae* 822, 71–78.

Santos, S.C., Fernandes, J.J.R., Carvalho, E.R., Gouvea, V.N., Lima, M.M. and Dias, M.J. (2014) Utilização da silagem de restos culturais do abacaxizeiro em substituição à silagem de cana-de-açúcar na alimentação de ovinos. *Ciência Animal Brasileira* 15, 400–408.

Satyanarayana, K., Guimarães, J. and Wypych, F. (2007) Studies on lignocellulosic fibres of Brazil. Part I: Source, production, morphology, properties and applications. *Composites Part A: Applied Science and Manufacturing* 38, 1694–1709.

Sebrae (2015) *Flores e plantas ornamentais:* série estudos mercadológicos. Brasília, Brazil: Serviço Brasileiro de Apoio as Micro e Pequenas Empresas.

Secor, E.R. Jr, Singh, A., Guernsey, L.A., McNamara, J.T., Zhan, L. *et al.* (2009) Bromelain treatment reduces CD25 expression on activated CD4+ T cells in vitro. *International Immunopharmacology* 9, 340–346.

Sena Neto, A.R., Araujo, M.A.M., Souza, F.V.D., Mattoso, L.H.C. and Marconcini, J.M. (2013) Characterization and comparative evaluation of thermal, structural, chemical, mechanical and morphological properties of six pineapple leaf fiber varieties for use in composites. *Industrial Crops and Products* 43, 529–537.

Sena Neto, A.R., Araujo, M.A.M., Barboza, R.M.P., Fonseca, A.S., Tonoli, G.H.D. *et al.* (2015) Comparative study of 12 pineapple leaf fiber varieties for use as mechanical reinforcement in polymer composites. *Industrial Crops and Products* 64, 68–78.

Shanks, R., Hodzic, A. and Ridderhof, D. (2006) Composites of poly (lactic acid) with flax fibers modified by interstitial polymerization. *Journal of Applied Polymer Science* 99, 2305–2313.

Silva, A.M., Oliveira, R.L., Ribeiro, O.L., Bagaldo, A.R., Bezerra, L.R. *et al.* (2014) Valor nutricional de resíduos da agroindústria para alimentação de ruminantes. *Comunicata Scientiae* 5, 370–379.

Souza, F.V.D., Cabral, J.R.S., Cardoso, J.L. and Benjamin, D.A. (2006) Identification and selection of ornamental pineapple plants. *Acta Horticulturae* 702, 93–99.

Souza, F.V.D., Cabral, J.R.S., Souza, E.H., Santos, O.S.N., Santos-Serejo, J.A. and Ferreira, F. R. (2007) Caracterização morfológica de abacaxizeiros ornamentais. *Magistra* 19, 319–325.

Souza, F.V.D., Cabral, J.R.S., Souza, E.H., Silva, M.J., Santos, O.S.N. and Ferreira, F.R. (2009) Evaluation of f1 hybrids between *Ananas comosus* var. *ananassoides* and *Ananas comosus* var. *erectifolius. Acta Horticulturae* 822, 79–84.

Souza, E.H., Souza, F.V.D., Costa, M.A.P.C., Costa Júnior, D.S., Santos-Serejo, J.A. *et al.* (2012a) Genetic variation of the *Ananas* genus with ornamental potential. *Genetic Resources and Crop Evolution* 59, 1357–1376.

Souza, F.V.D., Carvalho, A.C.P.P. and Souza, E.H. (2012b) Abacaxi Ornamental. In: Paiva, P.D.O. and Almeida, E.F.A. (Org.) *Produção de Flores de Corte.* Lavras, Brazil: UFLA, pp. 19–39.

Souza, E.H., Costa, M.A.P.C., Santos-Serejo, J.A. and Souza, F.V.D. (2014) Selection and use recommendation in hybrids of ornamental pineapple. *Revista Ciência Agronômica* 45, 409–416.

Souza, F.V.D., Souza, E.H., Santos-Serejo, J.A. and Pereira, M.E.C. (2016) BRS Boyrá e BRS Anauê: cultivares de abacaxizeiros ornamentais destinados para flor de corte, paisagismo e minifrutos. In: *XXIV Congresso Brasileiro de Fruticultura.* São Luiz, Brazil: Sociedade Brasileira de Fruticultura, pp. 1–4.

Sullivan, G.A. and Calkins, C.R. (2010) Application of exogenous enzymes to beef muscle of high and low-connective tissue. *Meat Science* 85, 730–734.

Tanabe, S., Arai, S. and Watanabe, M. (1996) Modification of wheat flour with bromelain and baking hypoallergenic bread with added ingredients. *Bioscience, Biotechnology, and Biochemistry* 60, 1269–1272.

Taniguchi, C.A.K., Castro, A.C.R.D., Silva, T.F.D. and Café, F.B.D.S. (2015) Development of pineapple as an ornamental potted plant. *Acta Horticulturae* 1087, 379–384.

Taussig, S.J. and Batkin, S. (1988) Bromelain, the enzyme complex of pineapple (*Ananas comosus*) and its clinical application: an update. *Journal of Ethnopharmacology* 22, 191–203.

Tochi, B.N., Wang, Z., Xu, S.Y. and Zhang, W. (2008) Therapeutic application of pineapple protease (bromelain). *Pakistan Journal of Nutrition* 7, 513–520.

Watanabe, M., Watanabe, J., Sonoyama, K. and Tanabe, S. (2000) Novel method for producing hypoallergenic wheat flour by enzymatic fragmentation of the constituent allergens and its application to food processing. *Bioscience, Biotechnology, and Biochemistry* 64, 2663–2667.

Yusof, Y., Ahmad, M.R., Wahab, M.S., Mustapa, M.S. and Tahar, M.S. (2012) Producing paper using pineapple leaf fiber. *Advanced Materials Research* 383–390, 3382–3386.

Zah, R., Hischier, R., Leão, A.L. and Braun, I. (2007) Curaua fibers in the automobile industry – a sustainability assessment. *Journal of Cleaner Production* 15, 1032–1040.

Zainuddin, M.F., Rosnah, S., Noriznan, M.M. and Dahlan, I. (2014) Effect of moisture content on physical properties of animal feed pellets from pineapple plant waste. *Agriculture and Agricultural Science Procedia* 2, 224–230.

11 Inflorescence and Fruit Development and Yield*

Duane P. Bartholomew[1] and Garth M. Sanewski[2]
[1]University of Hawaii at Manoa, Honolulu, USA; [2]Queensland Department of Agriculture and Fisheries, Nambour, Australia

The reproductive phase of the pineapple begins in response to natural or growth-regulator (PGR) forced induction of reproductive development (natural induction (NI) and forced or forcing used hereafter). Because the inflorescence of pineapple is terminal, when reproductive development begins, formation of new leaves ceases. Expansion of previously initiated leaves continues, but not all expand fully and some, much reduced in size are distributed along the fruit peduncle.

Once induced or forced, inflorescence and fruit development continues without interruption until the fruit matures. However, development can be interrupted by disease or slowed by water or temperature stress. Forcing generally is highly successful because the available PGRs are equally effective if growers have the knowledge required to work with the variation in plant sensitivity to forcing. The effect of temperature on fruit development basically determines the days from forcing to harvest which for 'Smooth Cayenne' can range from 130 in tropical Ghana to nearly 300 is southern South Africa. Pineapple cultivars have always been susceptible to NI but the year-round market demand for fresh pineapple has greatly increased the significance of NI. Consumers expect that pineapples will be available for purchase in all months of the year and growers have expanded to meet the demand. However, year-round production and the introduction of new cultivars highly sensitive to NI has forced growers and researchers to seek controls for this problem. As discussed below the perfect solution has yet to be found and may come from plant breeders or biotechnologists possibly using gene editing or genetic transformation.

Year-round production has also heightened grower awareness of the effect of seasonal changes in environment on fruit quality, especially the SSC:TA ratio (soluble solids content:titratable acidity). The introduction of, and almost overwhelming consumer acceptance of, the low acid 'MD-2' significantly reduced fruit quality problems, rapidly expanded profits for those first to market with the cultivar and nearly collapsed markets for growers of 'Smooth Cayenne.' While acidity in 'MD-2' fruit still varies with season, its initial low acidity results in good acceptability even when acidity reaches its highest level.

There is some evidence that the use of information technology on larger farms and plantations is expanding. If available crop models for pineapple are refined and their utility can be integrated into farm practice, applications relevant to this chapter include improved scheduling and forecasting of fruit maturation.

* A revision of the chapter by Duane P. Bartholomew, Eric Malézieux, Garth M. Sanewski and Eric R. Sinclair.

Inflorescence and Fruit Morphology and Growth

Early development

The meristematic area of the vegetative 'Smooth Cayenne' pineapple stem is about 0.36 mm wide (Kerns *et al.*, 1936) while the stem can be several cm in diameter. At the onset of reproductive development, the size of the apical dome increases to about 0.65 mm and the number of primordia bordering it increases greatly (Kerns *et al.*, 1936; Bartholomew, 1977) (compare Fig. 11.1a and b). The transition from leaf to flower production occurs within days. As a result, a relatively large number of leaf primordia present at induction are later found on a peduncle that was not present at the time they were initiated.

The first structure of the inflorescence to develop is a bract and each flower forms in the axil of a bract (Fig. 11.1c). Shortly after the bracts are initiated, three sepal and petal primordia develop (Fig. 11.1d) and shortly thereafter, six stamen primordia develop. The flower parts all originate at the same level and grow away from their point of origin. As they grow upward, a cavity is formed that eventually closes over and

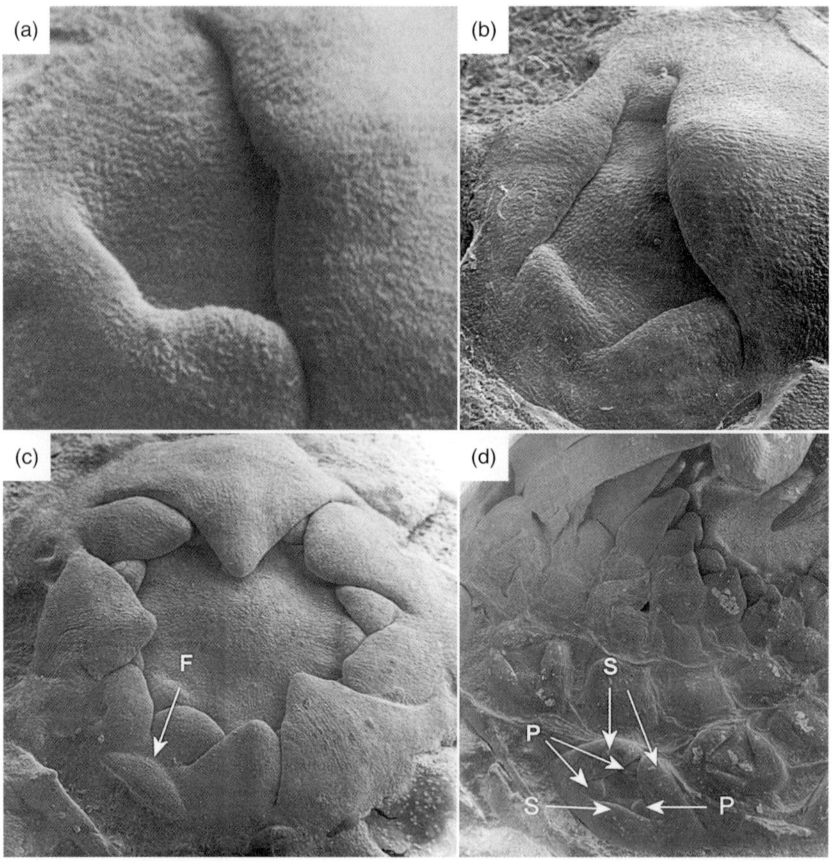

Fig. 11.1. Scanning electron photomicrographs of stem apical development of 'Smooth Cayenne' pineapple from 0 to 22 days after forcing with ethephon. (a) 0 days, vegetative apex bordered by two leaf primordia (about 185×); (b) 8 days, apex showing the apical dome bordered by four primordia (about 133×); (c) 11 days, apex showing numerous primordia and where a bract was removed, the first floret primordia (about 84×); (d) 22 days, apex with floret bracts removed to show three each of sepal (S) and petal (P) primordia (about 40×) From Bartholomew, 1977.

three carpels are formed within it. Once flower initiation ceases there is a gradual transition region where bracts but no flowers are produced (Okimoto, 1948) as the apical dome reverts to its original diameter and leaf primordia that form the crown begin to develop.

The apex of the vegetative plant stem is broad and flat, but as the first flowers begin to develop, the stem apex narrows and the fruit stem (peduncle) begins to elongate (Fig. 11.2).

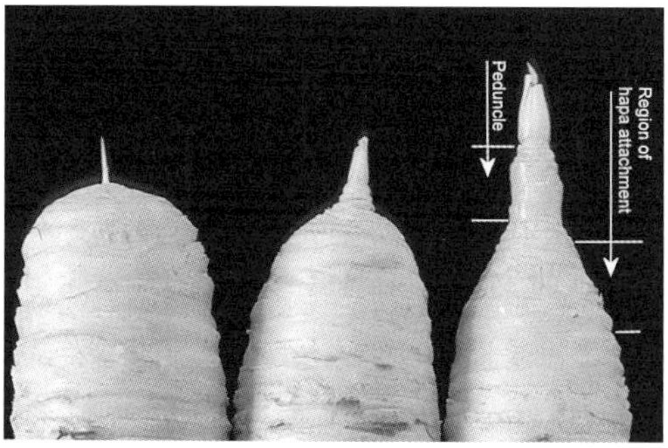

Fig. 11.2. Vegetative (left) and reproductive stems of 'Smooth Cayenne' plants after forcing. Note the progressive elongation of the upper portion of the stem as well as the fruit stem (peduncle). Photos D.P. Bartholomew.

Shoots referred to as hapas form in the stem-peduncle transition zone (Collins, 1960). The elongation of the stem tip observed in a stem that has been split longitudinally through the centre is diagnostic of inflorescence development. Rate of development is temperature dependent and for 'Smooth Cayenne' in Hawaii stem elongation is obvious 10–14 days after forcing. Development of flowers at the base of the young inflorescence continues at the same time that additional flowers are being produced higher on the young inflorescence (Fig. 11.3). The number of flowers initiated on the inflorescence varies considerably with the cultivar, the size and vigour of the plant at induction, plant population density and the quality of forcing. As the flowers progressively develop, the sepals are partially covered by the subtending bract and the petals, while the stamens and style are enclosed by the sepal primordia and are no longer visible without dissection.

Physiologically each flower is independent of adjacent flowers and anthesis (flowering) and maturation of each flower occurs in the same sequence as its initiation, so when individual fruitlets at the base of the fruit are mature the youngest fruitlets at the top are still immature. Flower formation ceases about 30 days after

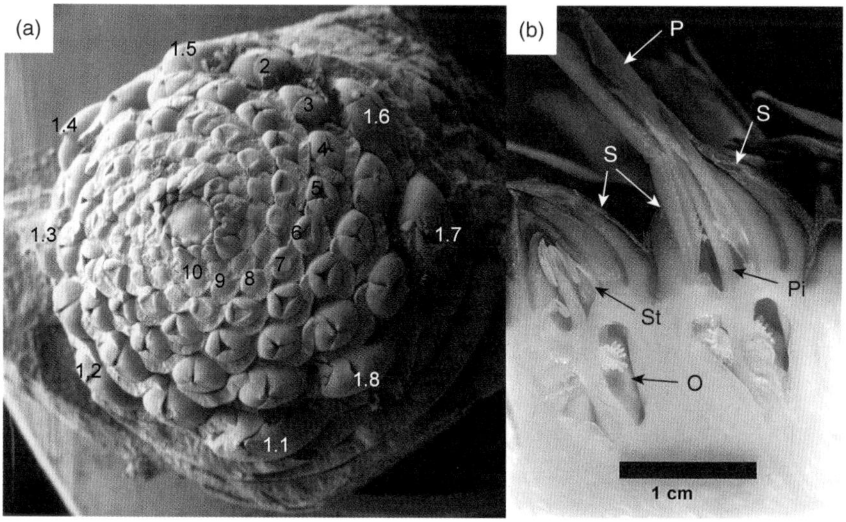

Fig. 11.3. (a) Young developing pineapple inflorescence (20×) 25 days after forcing during the summer in Hawaii. Bracts below each floret were removed to show eight long spirals typical of *Ananas comosos* var. *comosus* with 9–10 identifiable floret primordia in each spiral. (b) Section through 'Smooth Cayenne' floret at anthesis: S, fleshy sepals; P, petals; St, stamen; Pi, pistil; O, locule of ovary with ovules. Photos D.P. Bartholomew.

forcing of a 'Smooth Cayenne' plant, the diameter of the apical dome gradually transitions back to that of a vegetative plant and crown leaf primordia develop. Several weeks later, all flower structures are present and anthesis occurs (Fig. 11.3b)

The rate of inflorescence development is determined almost entirely by temperature (Medcalf, 1949; Fleisch and Bartholomew, 1987). In the cool tropics (Hawaii), flower initiation for 'Smooth Cayenne' ceased about 34 days after forcing in mid-summer (Bartholomew, 1977) and after 37 days in the cool season (December) (Kerns, 1936). In the warmer Côte d'Ivoire environment flower number is fixed at 30 days after forcing (E. Malézieux, 1992, unpublished data). Cell division continues in the fruit until after anthesis after which fruit development is primarily the result of cell enlargement (Okimoto, 1948).

Stages of fruit development defined on the basis of morphology include the beginning and end of flower initiation (all flowers formed), the beginning and end of anthesis, and fruit maturity. Other 'stages' were defined for convenience in estimating progress towards maturation and to establish possible stages susceptible to disease organism entry to the fruit (Rohrbach and Taniguchi, 1984). The earliest stage seen in the centre of the leaf whorl without sacrificing the plant is commonly referred to as 'open heart', or 'red heart', which would be green for cultivars lacking anthocyanins, for example 'MD-2' (Fig. 11.4). By the 'mid-cone' stage the sepals and bracts of the flower are visible and flower numbers can be counted. The stages occur at approximately weekly intervals in summer in Hawaii and anthesis begins approximately 1 week after 'late cone'. The tip of the corolla, blue for 'Smooth Cayenne', can usually be seen by the late-cone stage. Anthesis usually takes from 15 to 25 days, depending on the flower number and the average temperature. The increase in fruit fresh and dry weight from induction to maturity is sigmoid (Sideris and Krauss, 1938) and growth rate is primarily dependent on temperature.

Fruit growth and ripening

The time from induction to maturity is a characteristic of the cultivar, possibly of the clone, and is about 3 weeks longer for 'MD-2' than for 'Smooth Cayenne' in Kenya (C. Olende, Kenya, 2016, personal communication) and 'Smooth Cayenne' and 'Pernambuco' take longer to mature than do 'Spanish' or 'Queen' (Py *et al.*, 1987). However, Chan and Lee (2000) found that earliness among F1 progenies was primarily associated with the time required to reach a plant weight suitable for induction while the time from induction to harvest was similar.

The main biochemical changes associated with maturation of pineapple fruit, such as accumulation of sugars and carotenoids, occurs in the last weeks of development (Gortner *et al.*, 1967). With 'Smooth Cayenne' and 'MD-2', the external signs of approaching maturity are the flattening and increased gloss of the surface of individual fruitlets, which mature progressively from the base to the top of the fruit. As ripening occurs, the fruitlet epidermis typically turns from green to, depending on the cultivar, shades ranging from yellow to red. However, in some locations or at certain times of the year, the shell colour remains green as the fruit matures and green-ripe fruit are produced (Py *et al.*, 1987). Large fruit weighing 2.0 kg or more can have basal fruitlets that are overmature by the time the upper fruitlets ripen so overall fruit quality is near optimum when large fruit are about half-yellow. The fruit does not show a respiratory climacteric during ripening. Aspects of the physiology of fruit development are discussed later in this chapter and Chapter 13.

Factors Affecting Induction and Inflorescence Development

Plant sensitivity to NI and forcing affects the success of forcing and the incidence of NI. Forcing success is one of the critical factors determining average fruit weight and yield per unit land area (yield). Success is measured by the percentage of plants forced, quality of the induction (fruitlets initiated per unit of plant weight) and uniformity of induction, which determines the sharpness of the harvesting peak. Once flower number is established, growth and enlargement of the individual fruitlets determines the final fruit weight.

Fig. 11.4. Qualitative staging of inflorescences from soon after emergence until 'late flowering'. (a) Open heart (2.5 cm). (b) Early cone. (c) Mid-cone. (d) Late cone. (e) Early anthesis. (f) Mid-anthesis. (g) Late anthesis. (h) Dry petal. Photos G. Sanewski.

Natural induction (NI) of flowering

The factors known to affect natural induction (NI) of pineapple, in approximate order of significance, are cultivar sensitivity, plant size, photoperiod, and assorted stresses, which include temperature, nutrients, primarily nitrogen, and water. Pineapple plants generally must attain some minimum plant weight, and possibly age, before NI can occur (Py *et al.*, 1987). However, sensitivity of plants to NI is a moving target determined by weather and plant stresses so in a given year there may be no NI regardless of plant size, while in other years moderately small plants thought not to be susceptible to NI are induced. In a general sense, size is a factor in plant sensitivity and for 'MD-2' plants stunted in growth due to an uncharacterized stress the minimum was about 1.0kg (D. Bartholomew, Hawaii, 2011, personal observation). Without stress the minimum size for 'MD-2' plants grown from crowns in Kenya was 1.4kg (C. Olende, Kenya, 2016, personal communication). No information was found for other cultivars and discovery of a minimum plant size is more by accident than by design because fruit from small plants has little or no market. There are few regions where environmental conditions are constantly optimum for growth so seasonal patterns in NI are evident in most regions. These patterns have long been recognized and were the basis for commercial pineapple production before the discovery of chemical forcing agents.

Once some minimum plant weight is reached, the factors that promote flowering are mainly those that retard vegetative growth. For 'Smooth Cayenne' grown in subtropical regions there is a strong wave of NI in the autumn-winter season; however, NI can occur at other times of the year (Millar-Watt, 1981; Sinclair, 1997, 1998, 1999), so plants that do not flower naturally in the winter may flower in mid-summer.

Definite peaks in NI are also recognized in tropical regions (Aubert, 1977; Wee, 1978; Giacomelli *et al.*, 1984). The main peak usually occurs when either minimum or maximum temperatures are at their lowest. This peak may also be associated with a decline in the hours of solar radiation. A small peak also occurs when maximum temperatures are at their highest (Teisson, 1972). Gowing (1961) indicated that, depending on the kind of planting material used and the time of planting, NI could occur in any month of the year in Hawaii.

Inhibition of natural induction is associated with those factors that promote rapid vegetative growth. For example, vigorous growth stimulated by high nitrogen, absence of water stress and warm night temperatures inhibits flowering (Nightingale, 1942; Conway, 1977).

Since forcing results in very uniform induction, the strength of the forcing 'signal' is assumed to be much stronger than the one associated with NI.

Photoperiod

The earliest work on photoperiod (Van Overbeek, 1946) showed that pineapple responded like a short-day plant. Gowing (1961) later established that 'Smooth Cayenne' was a quantitative, short-day plant so NI could occur in any photoperiod but is accelerated by an 8-hour photoperiod. Friend and Lydon (1979) showed that an 8-hour photoperiod was more inductive for 'Smooth Cayenne' than photoperiods of 10, 12 or 16 hours. Photoperiod is a significant factor in regions above and below the equator and in Hawaii, natural induction of 'MD-2' and 'CO-2' can occur any time after the photoperiod is less than about 11 hours (Bartholomew, 2014). The high correlation between photoperiod and temperature makes it difficult to establish the relative importance of temperature and photoperiod in field studies.

Solar radiation

There is little evidence showing that solar radiation *per se* has any direct role in NI. It seems plausible that NI in subtropical regions results from a combination of shortened photoperiods and cool temperatures. However, NI also occurs in equatorial regions where photoperiod is nearly constant and temperatures generally are higher. In equatorial regions such as Côte d'Ivoire (Lat. 5–10° N), and Malaysia (Lat. 1° 17′ N), the periods of greatest natural flowering coincided with decreases in maximum (Teisson, 1972) or minimum temperature (Wee, 1978). In Malaysia there also was a correlation with a decrease in hours of sunshine (Wee, 1978). Bourke (1976) found no correlation between sunshine hours and NI in 'Queen' pineapples in Papua New Guinea (Lat. 4° 21′ S) once the variance associated with low temperature was removed.

An indirect effect of solar radiation is the observation that NI occurs first at field borders where plants are most exposed. Greater exposure increases radiation interception by individual plants, which would increase plant growth rate relative to less well-exposed plants. Consistent with this observation is the finding of Scott (1992) that in subtropical Queensland the incidence of NI of the ratoon crop increased consistently from 3.5 to almost 12% as plants ha^{-1} decreased from 80,695 to 46,112. No leaf or plant weight data were taken for the ratoon crop so it isn't possible to rule out differences in plant weight as a factor in the study.

Temperature

While 'Smooth Cayenne' pineapple is considered a short-day plant, cool temperatures particularly enhance the flowering response and the same is true for other cultivars. It is worth remembering that the optimum temperatures for growth of pineapple are considered to be close to 30°C day and 20°C night with an optimum mean of about 24°C (Neild and Boshell, 1976; Bartholomew and Malézieux, 1994).

Van Overbeek and Cruzado (1948a) established that a night temperature of approximately 16°C over 30 days induced flowering in 88% of 'Red Spanish' plants during late summer in Puerto Rico; 28% of plants held at a night minimum temperature of 22°C also flowered. So while a night minimum temperature of 16°C was more inductive than 22°C for 'Red Spanish'; the optimum temperature for flowering was not demonstrated.

Gowing (1961), using 'Smooth Cayenne', compared night temperatures of 15, 23 and 26°C and found that a night temperature of 15°C induced flowering when in combination with a short daylength for a period of 30 days. At normal daylengths (*c*.12.5 hrs) 30 days at a low night temperature did not induce flowering. However, exposure to a constant 18°C for 9 weeks was claimed to be more effective. The longer period of exposure was probably the main factor accounting for the improved response and a more recent study by Sanewski *et al.* (1998) supported this observation. Sanewski *et al.* (1998) found that a constant 20°C for 10 weeks induced 100% of 'Smooth Cayenne' plants but a constant 10 or 15°C for up to 12 weeks was not inductive. These studies were conducted when NI was most prevalent in the field and while photoperiod decreased from 12.5 to 10.5 hours. A subsequent study (G. Sanewski, 1999, unpublished data) confirmed that a constant 20°C or a day/night combination of 25/20°C was more inductive than 20/15°C or 25/15°C for plants in those environments for up to 64 days.

Friend and Lydon (1979) found that with a photoperiod of 8 hours and day temperatures of 25 or 30°C, a night temperature of 20°C was more inductive of 'Smooth Cayenne' than were night temperatures of 15, 25 or 30°C. Flowering was induced at night temperatures of 15 and 25°C but not at 30°C. Bourke (1976) concluded that NI of 'Queen' pineapple in Papua New Guinea was closely associated with minimum temperatures of 20–23°C. A night temperature of 20°C in combination with a day temperature of 20–25°C was close to optimal. In Thika, Kenya, at 1.04° S latitude and over 1600 m elevation, the average night temperature ranges from about 23 to 19°C and in that environment NI of 'MD-2' pineapple plants begins 9 months after planting if plants from crowns weigh more than 1.4 kg, those from slips weigh more than 1.55 kg and those from suckers more than 1.9 kg. The problem is even greater during the cool season when the minimum temperature can reach 10°C (C. Olende, Kenya, 2016, personal communication).

Nightingale (1942) observed that sudden brief periods of very low temperature, 5–10°C, could result in NI, and Yow (1959) reported that 5°C for 2–3 days was correlated with NI of 'Smooth Cayenne' in Taiwan. The effect of temperature perturbations can, however, be difficult to interpret. No data were found on the effects of a sudden drop in temperature on pineapple physiology or NI, however, it is possible that the sudden interruption in growth or increased ethylene production as a result of chilling could cause NI. Relative to this, data of Sanewski *et al.* (1998) suggests warming after exposure to cool temperatures has a direct positive effect on NI. In their study they collected and potted mature suckers of 'Smooth Cayenne' from the field in late winter and placed them at 5, 10 or 28°C for

3 days. Exposure of plants to 28°C induced flowering in 75% of plants.

NI will also occur after brief periods of high temperature. In Australia NI has been associated with diurnal maximums of 40°C over 2 days (Sinclair, 1997) and is probably the result of heat-induced ethylene production.

Water and nutrient availability

Water stress and water excess often produce a flowering response in tropical and subtropical species and NI of pineapple also is induced by a seasonal water deficit and also in response to water excess (Py, 1964), due to growth inhibition or enhanced ethylene production, or possibly both. Tay (1974) reported that both water excess and water deficit delayed NI of pineapple grown in peat soil while growth and NI were enhanced by increased watering frequency. Min (1995) also found that a severe water deficit decreased the susceptibility of 'Smooth Cayenne' to ethephon and also did not induce flowering. Mild water deficit was not studied. Water excess increased ethylene production of the basal-white tissue of immature leaves by approximately 100% but did not induce flowering. Water logging, a more severe degree of water excess, had no effect on ethylene production or flowering (Min, 1995).

Much of the inconsistency in the effect of plant water status on NI is probably due to differences in the severity of deficit or excess as well as the transition time from sufficiency to severe deficit or excess. Pot studies impose water deficits so quickly that there isn't time for the physiological adjustment that occurs in the field.

Nutrient stress, primarily nitrogen, also increases plant sensitivity to NI. Nutrient stress reduces the rate of plant growth relatively more than photosynthesis which would support increases in plant dry matter content and stem starch reserves (Nightingale, 1942). When pineapple growers relied on NI to initiate fruit development, and some still do, it was common practice to withhold nitrogen fertilization to increase the NI percentage.

Genetic variation

Natural induction in wild types such as *Ananas comosus* var. *microstachys* and *A. comosus* var. *parguazensis* is usually annual (Leal and Coppens d'Eeckenbrugge, 1996), which might also be the case for most cultivars if they were grown under the same conditions. In an *Ananas* spp. germplasm collection held in subtropical Australia (Lat. 27° S), plants of 'Pérola', 'Perolera', 'Queen' and *A. comosus* var. *microstachys* all flower naturally more frequently than does 'Smooth Cayenne'. Williams (1987) reported that some hybrids flower when they reach a plant weight of about 1.5 kg, regardless of the time of year.

Van Overbeek and Cruzado (1948a) reported that 'Red Spanish' was less responsive to daylength than 'Cabezona', a triploid of 'Spanish' (Samuels, 1969), but more responsive to low temperature. In Malaysia, 'Smooth Cayenne' is more difficult to force than is 'Singapore Spanish' (Dunsmore, 1957), the main canning variety there. Among the commercially important fresh fruit cultivars that have been grown in the same environment, their rank from most to least susceptible to NI is 'MD-2' > 'CO-2' > 'Smooth Cayenne'. 'Queen', 'Pérola' and 'Tainung 17' are more sensitive to NI than is 'Smooth Cayenne'.

Geotropic stimulation

Van Overbeek and Cruzado (1948b) reported that 'Cabezona' was geotropically stimulated to flower when plants were held horizontally for 3 days or more. It was theorized that flowering was induced by the accumulation of endogenous auxin in the lower longitudinal half of the stem apex.

Forced induction of flowering

Late in the 19th century it was discovered that pineapple could be forced with smoke (Collins, 1960) and later research showed that the active ingredient in smoke was ethylene (Rodriquez, 1932). Work in Hawaii (Collins, 1960) showed that acetylene gas also could force pineapple and water-saturated solutions of acetylene or ethylene sprayed over plants could deliver the required quantity of either gas. Green leaf tissue is required for forcing with ethylene (Traub *et al.*, 1940), but its contribution has not been identified. Calcium carbide, which releases acetylene on contact with water, is commonly used in forcing. Calcium carbide must be placed in the centre of each plant, which precludes mechanization.

A variety of synthetic auxins also forced 'Smooth Cayenne' pineapple (Collins, 1960). The auxin used commercially was the sodium salt of naphthaleneacetic acid (NAA) (Collins, 1960). Burg and Burg (1966) showed that auxins applied to pineapple stimulated natural ethylene production and concluded that ethylene forced flower development. NAA was once widely used, however, its shortcomings and the technical and managerial difficulties associated with the use of acetylene and ethylene kept alive the search for better forcing agents. None of hundreds of compounds screened were superior to NAA, acetylene and ethylene (Gowing and Leeper, 1959). Gowing and Leeper (1955) reported good results forcing with β-hydroxyethylhydrazine (BOH) but use ceased when it was identified as a carcinogen. Later, Palmer *et al.* (1967) showed that BOH decomposed to produce ethylene and its greater efficacy at night indicates that ethylene was the active component. Ethephon, which degrades to produce ethylene, was identified in the 1960s and extensive research has since demonstrated its efficacy as a forcing agent (Bartholomew, 2014).

Forced induction sharpens the harvest peak within a field so most of the fruit can be harvested within a fortnight. This period can be shortened by degreening mature-green fruit with acidified ethephon. While pineapple is generally easy to force, plant sensitivity to forcing varies considerably. It is important to understand how plant sensitivity varies because such knowledge can help increase forcing success.

Variation in plant sensitivity to forcing in different months of the year has been documented for 'Singapore Spanish' (Wee and Ng, 1968; Wee and Rao, 1977) and 'Smooth Cayenne' (Py *et al.*, 1987), and it is likely that other varieties also show such variation. Conway (1977) studied the effects of plant size and night temperature on sensitivity of Champaka F153 ('Smooth Cayenne') to forcing with ethephon (Table 11.1). Plants were grown in 3 l pots for 4 months and transferred to 5 l pots at the beginning of the study. The main treatments were plant size-age and exposure to night temperatures of 20, 25 and 30°C for the indicated duration. At the end of the temperature treatment plants were treated by pouring 50 ml of a urea-ethephon solution with 44 mg ethephon a.i. into the leaf whorl. Incandescent lamps were used to maintain a 13-hour photoperiod throughout the study. The 6-month-old plants were not forced. At 8 months, plants forced decreased as night temperature increased in all treatments; however, after 14 days of treatment, the decrease was most marked at 25 and 30°C. At 10 months, at least 5 plants in each treatment were forced

Table 11.1. Effect of plant size-age, night temperature and days of exposure on sensitivity to forcing with ethephon (n = 10 in most cases). Forcing was done in November and every other month thereafter. Nitrogen (N) (g kg^{-1}) dry mass, was measured in the middle one-third of the D-leaf on the day of forcing. (Conway, 1977).

Days	Temp (°C)	6 months	8 months	N	10 months	N	12 months	N
7	20	0	5	17.1	7	10.9	7	11.5
7	25	0	4	17.6	6	11.6	6	10.3
7	30	0	3	17.9	5	11.3	5	9.1
14	20	0	6	15.0	6	10.1	8	11.2
14	25	0	1	14.5	6	11.4	5	9.1
14	30	0	1	16.7	5	11.9	0	13.1
21	20	0	5	15.1	7	10.6	8	15.0
21	25	0	4	15.2	6	9.2	3	12.0
21	30	0	2	13.6	5	12.5	2	15.3
	Ambient	0	8/9		10/10		10/10	
Avg wt, kg		0.47	0.81		1.36		2.16	

and there were no consistent effects of night temperature. The lack of an effect was attributed to low leaf nitrogen caused by a proportioning pump malfunction. However, the pump failure demonstrated that nitrogen stress negated at least some of the effects of warm night temperature on plant sensitivity. At 12 months the marked decrease in plants forced after 14 or 21 days at night temperatures of 25 and 30°C was again evident. Plants held at ambient temperatures remained vegetative until treated with ethephon and after 6 months almost all plants were forced.

Van de Poel *et al.* (2009) determined that 'MD-2' plants grown from suckers averaging 475 g were not forced with ethylene 1 month after planting, 27% were forced after 2 months and 100% were forced at 3 or more months after planting. During the first 3 months, there were no changes in D-leaf length or weight. Research on root development shows that root mass increases faster than plant mass after planting (Py *et al.*, 1987) and stem dry weight and stem dry matter content decreased during the first 52 days after planting (PRI Experiment 1.9.J.1, unpublished data). It is likely that all available resources during the first 2 months were consumed by the plant establishment phase thus rendering plants insensitive to forcing. Attached crowns, slips and suckers of 'MD-2' are susceptible to being forced by ethephon applied to degreen fruit (observations of D. Bartholomew in Hawaii and J. Vásquez Jiménez in Costa Rica), which indicates that for the highly sensitive 'MD-2', the main requirement for forcing is a healthy and actively growing shoot.

In commercial fields, once the plant size has reached the target weight required to produce a marketable fruit, plant sensitivity to forcing shows little variation with increasing weight. Above some upper limit, which has not been well characterized, 'overly large' plants are reported to be more difficult to force than those considered to be of optimum size (Py *et al.*, 1987). The cause of this apparently reduced sensitivity is not known, but it may be related to a rapid growth rate, intense mutual shading, difficulty in delivering an effective dose of growth regulator to susceptible tissues, or to more than one factor. The issue is a minor one because growers normally force plants before sensitivity begins to decline.

Especially for 'Smooth Cayenne', plants of sufficient size are most sensitive to forcing during the time of year when natural induction normally occurs. Outside of this time, the sensitivity of all widely grown cultivars to forcing is believed to be determined primarily by the prevailing temperature (Wee and Rao, 1977; Bartholomew and Malézieux, 1994). In warm tropical regions, sensitivity to forcing, even during the time of natural induction, tends to be low because conditions favour rapid growth and dry matter content remains low. However, if well grown plants can be forced, fruitlet numbers would be high and, if conditions for fruit filling are favourable, large fruit are produced at harvest.

Genotypic variation

As with natural induction, comparative studies show that 'Smooth Cayenne' is less easily forced than are most other pineapple cultivars (Wee and Ng, 1970; Py *et al.*, 1987). Breeding to improve plant and fruit characteristics has resulted in cultivars that have better fresh fruit characteristics and are more sensitive to NI and also are easier to force than 'Smooth Cayenne' under challenging conditions. However, sensitivity to NI and ease of forcing are not always desired characteristics. 'MD-2', now the leading fresh fruit cultivar in the world, is relatively easy to force but also is so sensitive to NI that precocious fruiting is a serious problem (Bartholomew, 2014). As a result private industry breeding efforts have focused on replicating many of the qualities of the hybrid while reducing the sensitivity to NI (Young and Gonzales, 2010; Firoozabady and Young, 2013). No studies of the biochemical or genetic basis for differences in cultivar sensitivity to NI or forcing were found.

Environment

The environmental factors that promote natural induction also increase sensitivity to forcing. Variation in plant sensitivity was seen when plants were forced with acetylene, especially with NAA and ethephon. NAA is not covered here because it is much less effective than ethephon. When plant sensitivity to NI is high, forcing is almost always successful. Forcing success outside of the season where sensitivity to NI is high or in environments where it is low can challenge the best forcing practices, especially for 'Smooth Cayenne'. Where it is possible to plan ahead, moderate nitrogen stress established by withholding nitrogen fertilizer for 4–8 weeks prior to forcing will increase plant sensitivity to forcing.

The effect of moderate water stress on sensitivity to forcing apparently is similar to that of moderate nitrogen stress (Evans, 1959), but no controlled studies were found. However, severe water stress should be avoided because plants are unresponsive under such conditions (D.F. Williams, Fort Collins, CO, 1985, personal communication).

Outside the season of natural induction, and particularly during the warmest summer months, a short-term reduction in sensitivity to forcing has been observed. This sensitivity is associated with the short-term effect of high day temperatures, usually above 32°C. Under such conditions, leaf temperature may be 20°C higher than air temperature (Bartholomew and Malézieux, 1994) and high leaf temperatures reduce leaf carbon assimilation rates (Zhu *et al.*, 1999). This type of reduced sensitivity, which is often seen during January and February in southern Queensland, determines not only whether plants can be forced, but also the 'quality' of the force (Glennie, 1979; Wassman, 1991). 'Quality' of forcing is a subjective term used to characterize the intensity of the plant response to the growth regulator as reflected in the time from forcing to appearance of the inflorescence in the plant heart and the number of fruitlets initiated by plants of comparable size within a given environment. When the quality of the force is poor, fewer fruitlets per fruit will be initiated, fruit emergence is delayed, and fruit weight and yield are markedly reduced (Wassman, 1991). However, if the following day is cooler, sensitivity may be high and forcing is easily accomplished (Wassman, 1991). No data were found regarding what step in the induction process, i.e. fruitlet initiation or fruit development, is delayed on hot days when forcing is poor.

In tropical countries that have high average night temperatures, 'Smooth Cayenne' is difficult to force with ethephon (Py *et al.*, 1987). High night temperatures can reduce the percentage of plants forced (Conway, 1977) and will also reduce fruit size (Min and Bartholomew, 1997). Night temperatures of 25–30°C and higher can reduce natural ethylene production, the intensity of crassulacean acid metabolism (CAM) (Min and Bartholomew, 1997), total carbon assimilation (Zhu, 1996) and fruitlet number (Min and Bartholomew, 1997) of plants forced with ethephon. Weak induction can also result in long peduncles. It seems likely that any environmental factor that reduces net assimilation, either by reducing carbon assimilation or increasing respiration, at the time of forcing will reduce fruitlet number and thus fruit weight at harvest.

Forcing with plant growth regulators

The main growth regulators used in forcing pineapples, acetylene and calcium carbide, ethephon and ethylene, are about equally effective if the best practices are used for each chemical. Forcing efficacy is greatest for all growth regulators if the dose is delivered near or into the leaf whorl. Applications into the leaf whorl are not practical for large-scale growers, but can increase effectiveness and on smaller farms can reduce the cost of materials.

Acetylene or calcium carbide, its common source, is very effective in forcing pineapples. The main benefits of calcium carbide are its low cost and relative ease of use. A main concern is that when placed in the leaf whorl it can raise the leaf whorl solution pH as high as 12, which can increase the incidence of rots caused by *Phytophthora* spp. (Py *et al.*, 1987). If *Phytophthora* spp. are present, another forcing agent should be used.

Small pellets of calcium carbide, about 1 g plant^{-1}, are used. The effectiveness of calcium carbide probably is related to the timing and speed of release of acetylene. If no water is present in the leaf whorl, water may be poured into the whorl to release acetylene.

Calcium carbide, applied as a 1.0% solution with 0.5% activated carbon, significantly increased forcing efficacy (Maruthasalam *et al.*, 2010). If forcing is expected to be difficult, dry calcium carbide or the water solution can be applied in late evening or early morning when it is cool and the stomata are likely to be open. A double application can also increase efficacy.

Ethylene applied late in the evening or at night to facilitate uptake through open stomata (see Chapter 6) is an excellent forcing agent. Specialized equipment is required so that the pressurized gas can be injected directly into the spray boom to apply an ethylene saturated water solution to plants. High-volume sprays are used to ensure adequate delivery of sufficient ethylene into the leaf canopy to force plants. In Queensland ethylene was applied in one application of a saturated solution in 6500 to 9000 l ha^{-1} of water or two applications of 4500 l ha^{-1} 24 hours apart. Py *et al.* (1987) state that 800 g of ethylene is applied in 6000–8000 l ha^{-1} of water. Activated charcoal at 0.5% or 1% bentonite was added to the water to increase absorption of the ethylene in the solution. A plantation in Ecuador growing 'MD-2' applied 2.27 kg ethylene and 20 kg activated carbon in 7000 l ha^{-1} of water (Van de Poel *et al.*, 2009). Special equipment is required to prepare the solution for application (Py *et al.*, 1987; Soler *et al.*, 2006).

Ethephon was initially considered to be less effective than ethylene for forcing pineapple. However, as experience was gained in its use, it can be about as effective as ethylene and is widely used on many large commercial farms in most environments. An important advantage of ethephon is that under ideal conditions for forcing it can be applied with a boom sprayer during the day or night with equal effectiveness. Ethephon is typically applied in a 2–5% (weight:volume) urea solution. Nitrogen sources such as ammonium nitrate and ammonium sulfate were less effective than urea (Yamane and Ito, 1970a). The mechanism by which urea enhances the activity of ethephon is not known. Studies with radio-labelled-ethephon show that urea does not increase absorption (Turnbull *et al.*, 1999).

High solution pH hastens the rate of ethephon breakdown to ethylene, which may enhance forcing efficacy. The addition of 2.9 l of ethephon (39.5% a.i.) to 1794 l of 2.7% urea-water solution raised the pH from 2.90 to 3.20 (Yamane and Ito, 1970a) and Glennie (1979) reported that an ethephon solution with a pH of 3.0 had a half-life of 75,000 hours. Alkalization at 25°C reduced the half-life of an ethephon solution from 761 hours at pH 5 to 12 hours at pH 9 (Cox, 1979). At the summer temperatures of Australian fields, ethylene release from an ethephon solution is rapid at pH 9.2, which allows the use of lower quantities of ethephon. However, when the air temperature is high and field conditions are adverse, forcing is better with a solution pH of only 7 (Sinclair, 1991). These high pHs are achieved by adding calcium or sodium carbonate (Dass *et al.*, 1976) or borax or Solubor® to the solution.

If forcing is expected to be difficult, the ethephon dose could be increased or even doubled. Tests of high amounts have been reported to reduce fruit weight (Dalldorf, 1979; Iglesias, 1979); however, others (Yamane and Ito, 1969; Guyot and Py, 1970) found no reduction in fruit weight, length, or other evidence of phytotoxicity when more than 16 kg ha^{-1} were applied. Based on the accumulated research results, the maximum amount likely to be considered under adverse conditions would be less than 3.0 kg ha^{-1}.

Forcing and cold shock

In an effort to comply with principles of organic production, the use of non-chemical methods of flower induction have been sought. Cold shock with 1–2 applications of ice water to 'Tainung 17' plants increased their sensitivity to NI and 3–4 treatments induced flowering as well as the standard CaC_2 treatment (Lin *et al.*, 2009).

Prevention of natural flowering

When sensitivity to natural induction is high in subtropical regions, inhibiting NI can be a great concern. The pineapple fresh fruit industry literally exploded with the introduction of 'MD-2' in 1996 (Bartholomew, 2009), but may now be close to market saturation (Loeillet *et al.*, 2011). To meet demands for fresh fruit, the industry rapidly shifted to forcing and harvesting fruit on an almost daily basis. Forcing throughout the year exposed large plants of sensitive cultivars, greater than about 1.5 kg for 'MD-2' and 'CO-2' in Hawaii, to inductive conditions of reduced photoperiod and cool temperatures. 'CO-2' has replaced 'MD-2' in Hawaii because of reduced sensitivity. However, without protection, NI in susceptible fields averages 35%. In other areas, precocious fruiting percentages can range from a few percent of the fruit to nearly 100% for 'Tainung 17' in Taiwan (Kuan *et al.*, 2005). Fruit that ripen off schedule increases harvesting costs, variability in fruit sizes, can overtax harvesting and packing operations and result is lost sales when market prices are high because of gaps in supply.

Natural induction is less likely to be a problem in warm, humid tropical environments, however with 'MD-2', some precocious fruiting occurs in almost every environment. As noted previously, one solution to the problem of NI was to breed a less sensitive cultivar (Young and Gonzales, 2010).

Cultural controls

During the time of year when NI normally occurs, cultural practices provide some degree of control. These controls include timing of planting to insure that only relatively small, rapidly growing plants are in the field when NI is most likely to occur. Plants scheduled to be forced a few months after NI normally occurs should be grown with optimum levels of nutrients and water to minimize stresses and promote rapid vegetative growth. The severity of NI in the sensitive cultivars, 'MD-2' in Central America, 'Queen' in South Africa and China, 'Pérola' in Brazil, 'CO-2' in Australia and Hawaii and 'Tainung 17' in Taiwan, led to a search for more effective cultural controls. In Taiwan, 90% shade and extra urea fertilization completely controlled NI of plants less than 11 months old (Lin *et al.*, 2015). When the shading fabric was present from November to early April and plants were forced a few days after the shade was removed, average fruit weight was half that of the unshaded control. No reduction in fruit weight occurred if the shade was removed and plants were grown in full sun for about 6 weeks before forcing. In a Tanzanian study, NI increased with shade and the effect increased with increasing levels of shade with 37% light transmission producing the greatest percentage of flowering (Chipungahelo *et al.*, 2007). Clearly shading has a significant effect on NI but the mechanisms that account for the responses are not understood.

Leaf trimming is used on farms in Costa Rica to help control NI when the risk is high (J. Vásquez Jiménez, Costa Rica, 2017, personal communication). Approximately the top 30 cm of plant leaves are pruned. Workers are required to disinfect their pruning tools before moving to an unpruned block. Vásquez Jiménez stated that the effect will reduce NI for only about 15 days, so areas selected for pruning must be at high risk for NI. The loss of some of the most photosynthetically active leaf tissue on the plant, both in terms of leaf age as well as exposure, will sharply reduce plant net assimilation and so likely will reduce the average leaf C:N ratio. Removal of leaves will most likely also reduce the plants ability to sense daylength and can be expected to have other hormonal effects. Before leaf pruning is done plant apex samples should be taken to confirm that plants are vegetative because leaf pruning done after NI has occurred will likely reduce average fruit weight. In a study in Hawaii, removal of 2–12 D and younger leaves from 'Smooth Cayenne' plants in October reduced NI from 90% to 0% with 20% NI when six leaves were removed. Similar results were obtained when leaves were removed in November, however, removal of 10 leaves only reduced NI to 30% (Sanford, 1964).

Plant growth regulators

A number of plant growth regulators were evaluated over many years. Kist *et al.* (2011) reported that 23 mg l^{-1} diquat with urea delayed NI in the Brazilian cultivar 'Pérola' without compromising fruit size. No further studies of this mixture on NI were found.

By the early 1990s it was clear that NI resulted from an increase in ethylene biosynthesis or increased plant sensitivity to it. Miller-Watt (1981) showed that silver thiosulphate, which blocks ethylene action, reduced NI of pineapple. Aminoethoxyvinylglycine hydrochloride (aviglycine), which inhibits ethylene biosynthesis by inhibiting the enzyme ACC synthase, prevented spontaneous flowering of an ornamental bromeliad (Mekers *et al.*, 1983). Aviglycine became available in commercial quantities in 1997 and Bartholomew (2001) found that aviglycine significantly reduced NI of 'Smooth Cayenne' pineapple. Kuan *et al.* (2005) reported that aviglycine reduced NI of 'Tainung 17', and later it was reported that 7.5 ml of a solution containing 150 to 250 mg l^{-1} aviglycine placed in the leaf whorl (Tsai and Kuan, 2013) or 200 mg l^{-1} at 14-day intervals (Tsai *et al.*, 2015) controlled NI while untreated plants exhibited >80% NI. Maia *et al.* (2017) reported that

1-methylcyclopropene, which blocks ethylene action by binding to ethylene receptors, reduced NI of 'Pérola' and 'Vitória' pineapples from 37.4% to 14.6%.

Good control of NI of an 'MD-2' clone and 'CO-2' was accomplished by weekly broadcast sprays of 100 mg l^{-1} a.i of aviglycine in 2340 l ha^{-1} of water (Bartholomew *et al.*, 2011) and similar results were obtained with 'Queen' in South Africa (Rabie *et al.*, 2011, 2013). Aviglycine must be applied before NI has occurred to avoid severe fruit deformities. The cost of aviglycine is high, but it is a standard practice in Hawaii with 93 ha of 'CO-2' treated in 2016 because it prevented NI, synchronized fruiting of the mother plant crop and sucker development for a first ratoon. Plant crop fruit are harvested on schedule and ratoon fruit boxes ha^{-1} are estimated to be double what would have been harvested without aviglycine.

Breeding and genetic engineering for NI control

Dole Food Company has produced a pineapple cultivar, Dole 14, by conventional breeding that is reported to have reduced sensitivity to NI (Young and Gonzales, 2010). Pineapple plants transformed to produce less ACC synthase (AcAcs2) have also been developed in Australia and field trials have demonstrated a flowering delay (Trusov and Botella, 2006a,b). Del Monte Fresh Produce Company patented the genetically transformed 'MD-2' pineapple, 'Rosé' (Firoozabady and Young, 2013), which has reputedly reduced sensitivity to NI. The US Food and Drug Administration (FDA, 2016) concluded that there were no unresolved safety or regulatory questions.

Molecular biology of flowering

Advances in molecular biology have substantially improved our understanding of plant processes in recent years. In particular, the application of high throughput Illumina sequencing with a transcriptomic approach has provided new insight into molecular pathways such as those associated with flowering.

Work by Li *et al.* (2016a) revealed 56 genes are differentially expressed (DEGs) between the flowering and non-flowering bromeliad, *Aechmea fasciata*. Similarly, Hanqing *et al.* (2013) found 60 differentially expressed genes in *A. fasciata* stimulated to flower with ethylene. Liu and Fan (2016) found 13 DEGs associated with floral induction by ethylene in pineapple. These genes are spread across several pathways giving us the first glimpse of a complex network of flowering controls likely in pineapple.

Ethylene biosynthesis

Endogenous ethylene is known to be a key stimulant of flowering in pineapple (Trusov and Botella, 2006a,b). The ethylene response in untreated plants starts with ethylene biosynthesis, a simple two-step pathway involving the enzymes ACC synthase and ACC oxidase. However there are multiple copies of these genes with specificity in stimuli and seven copies were found in mature 'MD-2' fruit (Redwan *et al.*, 2016). At least two ACC synthase genes have been characterized in pineapple. AcAcs1 is induced by wounding, AcAcs2 is induced by low-temperature and auxin. AcAcs2 is likely the gene responding in most cases of NI (Cazzonelli *et al.*, 1998; Botella *et al.*, 2000; Trusov and Botella 2006a,b) in cool environments.

There are many abiotic stresses and environmental cues that will promote ethylene biosynthesis in plants including pathogen attack, temperature, daylength, illumination, wounding, salinity and drought, which are all associated with different but interrelated pathways.

Ethylene signalling

Ethylene signalling is more complex than ethylene biosynthesis with several proteins involved through either positive or negative regulation. This ethylene signalling pathway is deduced from studies in several species, mostly *Arabidopsis*, with some supporting work in pineapple. Ethylene synthesized in the plant or absorbed from external applications is initially detected by ethylene receptors. There are four ERS (ethylene response sensor) and ETR genes (ethylene receptor) genes in 'Pérola' which are associated with flowering (Li *et al.*, 2016b). Research with *Arabidopsis* (Chang, 2016) showed that ethylene receptors are negative regulators that suppress ethylene responses and this suppression is removed when ethylene binds to them. These receptors suppress the negative regulator (CTR1 – constitutive response 1 kinase). In the absence of ethylene, the CTR gene prevents EIN2 (ethylene insensitive gene) function. Ethylene therefore causes a rise in EIN2, which in turn activates EIN3. Accumulation of EIN3 results in a cascade of other transcriptional genes leading to an increase in the AP2 transcription factor and ultimately a rise in the FT gene (Flowering locus-T) and flowering. A reduction in ethylene receptor activity through ethylene binding can therefore increase a plants sensitization to further ethylene.

Diurnal rhythms, daylength and circadian clock responses

A study in the bromeliad *Tillandsia usneoides* demonstrated that ACC production is light-dependent (Befler *et al.*, 1998) thus causing diurnal rises in endogenous ethylene.

The photoperiod pathway regulates the plant response to daylength and is particularly important in pineapple because of its potential implications in NI. The principle genes demonstrated in *Arabidopsis* and other plants are GIGANTEA (GI), CONSTANS (CO), early flowering 3 (ELF3), and circadian time-keeper (CTK). GI is expressed soon after ethylene flower induction in the bromeliad *A. fasciata* (Li *et al.*, 2016a). CO and CTK are also upregulated in pineapple in response to ethylene treatment (Liu and Fan, 2016).

GI promotes flowering in *Arabidopsis* under long days by regulating CO expression. CO is a key protein which integrates environmental and endogenous signals to eventually trigger flowering. Stability and cycling of CO is controlled by several proteins including blue light photoreceptor cryptochromes to ensure CO only accumulates during specific daylengths (Shrestha *et al.*, 2014). The photoreceptor cytochrome CRY2, which improves CO stability, is upregulated in adult bromeliad versus juveniles. GI and CRY2 are therefore seen as key photoperiodic genes which through their action on CO expression and subsequently FT, promote flowering in the bromeliad, *A. fasciata*.

So several genes regulate CO only in adult bromeliad plants. CO can be stimulated by photoperiod and by ethylene (Li *et al.*, 2016a).

FT-like (Flowering locus T) genes are considered key regulators of flowering in all angiosperms. FT expression is high in pineapple after induction by ethylene (Lv *et al.*, 2012). The FT gene sits at the end of the pathways stimulated by environmental cues and as such is believed to integrate flowering signals from various pathways including ethylene. FT interacts with the transcription factor, FD in the plant apex, to increase expression of the transcription factor AP1 (activator protein) and initiate flowering. The sugar trehalose-6-phosphate (Tre6P), a proposed proxy for plant carbohydrate status, appears necessary for FT expression (Li *et al.*, 2016a), thus demonstrating a link between carbohydrates and flowering. FT was stimulated by high temperature in *Narcissus tazetta* (Noy-Porat *et al.*, 2013).

Low temperature

Pineapple is known to flower in response to extended exposure to lower temperatures around 20°C (vernalization) and by cold shock (Chang *et al.*, 2011). In other plant species, the vernalization pathway genes, Vernalization (VRN1), Frigida (FRI) and LTI respond to extended periods of low temperature. Another gene, FLC (flowering locus C) is the main repressor of flowering in *Arabidopsis*. FRI increases the expression of FLC, thus both act as floral inhibitors. In monocots, the vernalization gene VRN, accumulates with exposure to low temperature and increases expression of FT and hence flowering through its effect on the vernalization pathway (Liu and Fan, 2016). VRN expression was shown to increase and FRI decrease following ethylene treatment of pineapple. It is thought that ethylene, through its action on these genes, might mimic the vernalization process (Liu and Fan, 2016). It is likely that the vernalization pathway is stimulated in pineapple by exposure to ≈20°C temperatures.

Plants sense sudden low temperature differently than gradual decreasing temperature. For example, a sudden low temperature increases the rigidity of the cell membrane and an associated influx of the signalling ion calcium into the cytosol. The calcium binding activator calmodulin responds to a rapid decrease in temperature in *Arabidopsis* but not a gradual decline as occurs in the vernalization pathway genes. Evidence shows that different plants use different pathways to respond to a decline in temperature depending on the severity of the decline (Kidokoro *et al.*, 2017). It is not clear how pineapple responds at a molecular level to sudden low temperature and how this relates to flowering. However, one or two cold treatments of the shoot apex of 'Tainung 17' pineapple using ice crystals doubled apex ethylene production but did not induce flowering (Lin *et al.*, 2009) while four applications did force the plants (Maruthasalam *et al.*, 2010). Soler *et al.* (2006) reported cold water treatment was more effective on plants most sensitive to NI, a result confirmed by Chang *et al.* (2011) who found that plants were not forced by ice-cold water when night temperatures were ≥25°C.

Calcium signalling

Calcium plays important roles in plants in maintaining cell wall integrity and in signalling of environmental stresses such as temperature, drought, salinity or pathogen attack. This signalling is achieved through regulation of the intracellular calcium concentration which is usually low (Ranty *et al.*, 2016; Zeng *et al.*, 2015). The movement of calcium across the cell membrane is associated with stress, and tonoplast and plasma membrane ATPase actively remove calcium from the cytosol. In the cytosol calcium binds to the highly conserved protein calmodulin (CaM) which is the calcium messenger protein. CaM is therefore an integral part of stress signalling in plants. Work with the bromeliad, *Guzmania* sp., demonstrated that the addition of a calcium accelerator with the ethylene inductant advanced the flowering response and CaM expression levels (Li *et al.*, 2012). CaM also interacts with a cytochrome P450 protein which is involved, among other processes, in daylength responses. Cytochrome P450 is one of the many genes upregulated in the bromeliad *A. fasciata* following floral induction with ethylene (Li *et al.*, 2016a). So it appears calmodulin is involved in the flowering response through its involvement in the calcium signalling pathway.

Age dependency

Pineapple and related Bromelia appear to exhibit age dependency for flowering (Hanqing *et al.*, 2013; Lei *et al.*, 2016; Li *et al.*, 2016a). Both juvenile and adult bromeliad plants produce ACC synthase after ethylene treatment but only adult plants appear sensitive to the ethylene (Li *et al.*, 2016a).

The transcription factor AfAP2-1 (AP2), which suppresses flowering in the bromeliad *Aechmea fasciata*, was shown to be age-dependent. Transcriptome analysis of juvenile and adult plants of this bromeliad has revealed transcripts of the AP2 gene are expressed more in juvenile than in adult plants. Overexpression of this gene in *Arabidopsis* delays flowering by suppressing the FT gene. Ap2 is thought to be a key gene in the transition of bromeliads from flowering insensitive to susceptible. This gene prevents flowering but its expression declines with age (Lei *et al.*, 2016) and treatment with ethylene in *A. fasciata* (Lei *et al.*, 2016). In other plant species, AP2s have also been implicated in defense and stress responses such as drought, cold and salinity (Phukan *et al.*, 2017).

A circadian clock transcription regulator gene (LHY) is down-regulated in a juvenile bromeliad (Li *et al.*, 2016a) but is upregulated in adult plants in response to ethylene (Avila Espinosa *et al.*, 2017). LHY is known to stimulate GI and hence CO and ethylene production.

It therefore appears that expression of at least two proteins associated with flowering are likely age-dependent.

Genotype differences

While there are no reported genotypic-based differences in any of these pathways in pineapple, there are many differences reported in other species. Small changes in levels of any of the contributing proteins or pathways could potentially result in a different response to ethylene or the environment.

The involvement and complexity of many pathways which respond to different stimuli and their subsequent effect on the flowering response of pineapple could explain why the flowering response is variable. There is also likely cross-talk between some of these pathways. As we unravel the contributions of each pathway and their interactions we hope to be better able to predict and control flowering.

Inflorescence and Fruit Development and Yield

Fruit yield for processing is measured in megagrams ha^{-1} (mg ha^{-1}) while fresh fruit yield is commonly boxes ha^{-1}. Whether mg or boxes, fruit yield is determined by plants ha^{-1}, number of fruit harvested from those plants, and average fruit weight. At a given planting density, fruit available for harvest is determined by the number of plants that produce an inflorescence that matures. The percentage of fruit actually harvested typically is higher when fruit are processed than when they are packed for the fresh market and harvest costs for both can determine the percentage of fruit harvested. The greatest losses are often associated with fields where plant size or forcing, or both, are not uniform. Regardless of field or farm or farm size it can be uneconomical to harvest every fruit where fruit maturation is spread. As a result, losses of about 13% are typical, with a range of 5–30%, depending on the spread of maturity (Wassman, 1982). Uniformity of plant size at forcing is critical and losses are least when fruit size and fruit maturation are uniform. The practice of degreening of fruit at a fruit SSC of 12–14% is done to reduce harvesting costs, not to please consumers. In the very competitive pineapple market, cost is the main driver of some practices and degreening fruit sharpens the harvest peak and increases recovery of fruit from the field, but reduces average fruit quality.

Within a given environment, fruit weight at harvest is determined in large part by plant weight at forcing (Gaillard, 1969; Py *et al.*, 1987; Tan and Wee, 1973; Malézieux 1988; Malézieux and Sébillotte, 1990a). However, other critical factors that affect average fruit weight include the success of inflorescence induction, the quality of induction in terms of the number of fruitlets initiated per unit of plant weight, and the fruit filling process. At maturity, average fruit weight is determined by the number of fruitlets formed at induction and the average fruitlet weight.

Resource allocation among fruit, slips and suckers

Few studies of carbohydrate allocation or reallocation in pineapple during reproductive development were found. During vegetative growth of 'Smooth Cayenne', and presumably for most pineapple cultivars, the dominant carbohydrate sinks for photosynthate are the young leaves, the stem, which also is a storage organ, and the roots, and there is an allometry between the leaves and the stem (Malézieux, 1988). Once reproductive development begins, there is little increase in stem length or diameter but stem dry matter (Py *et al.*, 1987) and starch accumulation increase rapidly after flower initiation (Fig. 6.16) (Pineapple Research Institute of Hawaii, 1965, unpublished data; Malézieux, 1993; Malézieux, unpublished data). In Hawaii, starch in the stem, dry weight basis, was about 15%, 9 months after planting to >60% midway through the fruit development period 10 months later. Leaf starch over the same time period increased from <2.0% to about 8.0%. While the increase in leaf starch is relatively small, when multiplied by the large leaf mass, it represents a large storage reserve. Starch accumulation decreases as the average temperature increases so there is little starch storage where the average night temperature is 25°C or greater.

In subtropical environments, rapid starch accumulation during early reproductive development likely results from little or no change in photosynthesis in conjunction with a decline in photosynthate demand because no new leaves are initiated and the fruit is not yet a significant sink for carbohydrates.

Mobilization of stored starch, which is indicated by a decrease in stem dry matter, begins about 100 days after forcing in Hawaii and sooner in Côte d'Ivoire, where temperatures are warmer and fruit growth is more rapid (Fig. 6.16; Pineapple Research Institute of Hawaii, unpublished data). The mobilization of starch from the stem helps meet high resource demand from the fruit and, where photosynthate supply and reserves are sufficient, developing suckers. In Hawaii, leaf dry matter also decreases at this time indicating that leaf starch also is mobilized to meet the increasing demand from the rapidly developing fruit. However, even when starch reserves are present, reallocation of assimilates from the stem and leaves to the fruit may depend on climate (drought, irradiance, temperature) during fruit filling and on carbon balance between source (assimilates produced) and sinks. Reallocation was greater for large than for small plants in Côte d'Ivoire (Malézieux, 1993), likely because with a higher dry matter content, and probably higher starch content, there was more substrate to reallocate.

Data from Australia, Côte d'Ivoire, and Hawaii shows that the developing fruit is a stronger sink than are suckers and slips. When reproductive development begins, apical dominance over axillary buds is broken and destructive sampling of 'Smooth Cayenne' plants in Hawaii showed that at least one small sucker was present 2 weeks after forcing. The extent to which the fruit and suckers develop in parallel is determined by the resources available and 'CO-2' and 'MD-2' appear to respond similarly. In Hawaii, early slip removal from 'CO-2' has little impact on fruit development or quality while crown removal significantly increased fruit weight (Table 11.2). Other significant effects were the reduction in fruit colour and the greater fruit translucency rating, possibly due to reduced shading and the resulting higher fruit temperatures. Crown and slip removal had a variable effect on TA and no effect on SSC.

In Cameroon, 'Smooth Cayenne' plants with visible suckers 6 weeks after harvest increased from 20% near sea level to 70% at 1000 m elevation (Py *et al.*, 1987). In the warm tropics net photosynthate is only allocated to developing suckers after fruit harvest. Since few or no slips are formed in such environments (Py *et al.*, 1987) the primary source of planting material is suckers so relatively large areas must be devoted to sucker production (see Chapter 7).

In subtropical areas, seasonal variation in irradiance at the date of forcing has some effect on the size of the fruit. Forced induction in the winter months when irradiance can be 50% of the summer maximum is a time of reduced photosynthesis so resources would limit flower number relative to number produced during periods of high irradiance. These smaller fruit develop and mature on plants with a leaf area that is large relative to the size of the developing fruit during a time of high irradiance and warm temperatures. The high photosynthetic capacity of the leaves exceeds carbohydrate demand by the developing fruit so surplus carbohydrate is available to support the growth of slips and suckers and any excess accumulates in the stem. In southern Queensland, Australia, suckers on plants from which fruit is harvested in summer can become large enough to shade the fruit. In Hawaii, delayed harvesting of slips on 'CO-2' plants after fruit harvest will cause them to shade and suppress the growth of developing suckers, the source of a ratoon crop. Similar results were found with 'Smooth Cayenne' (Sanford, 1964).

The converse of this situation occurs when forcing occurs in January-February in Australia and in the midsummer months in Hawaii. Fruit initiated during the summer are large relative to plant size, presumably because these photosynthate-rich plants provide the resources required to initiate large numbers of flowers. These relatively large fruit reach maximum size and accumulate sugars, a period of maximum resource demand, during cool weather when irradiance is approaching minimum levels. Sucker development on such plants is delayed, presumably because there is insufficient photosynthetic capacity to simultaneously support the high demand of a maturing fruit and developing suckers. In a time-of-planting trial in Australia, the days after induction when suckers weighing 5–40 g were first seen depended on the month when forcing occurred. Forcing in

Table 11.2 Effect of crown and slip removal on 'CO-2' fruit characteristics. The sampled area was forced on 15 August 2016, slips and crowns were removed on 19 January 2017 and fruit were harvested 28 February 2017. From R.E. Paull, personal communication. Values in the same column with a different letter in superscript are significantly different at the $P<0.05$ level.

Treatment	Fruit weight (kg)	Fruit colour rating	Translucency rating	Acidity (%)	SSC (%)
Control	1.59^{B}	96.4^{A}	9.80^{B}	0.56^{BC}	13.2^{A}
Slips removed	1.66^{AB}	96.7^{A}	11.80^{B}	0.62^{A}	12.9^{A}
Crown removed	1.78^{A}	90.6^{B}	32.58^{A}	0.60^{AB}	12.9^{A}
Crown and slips removed	1.60^{B}	92.5^{B}	36.88^{A}	0.54^{C}	12.3^{A}

January (summer) resulted in the longest interval before suckers appeared (295 days) while plants forced in September and October produced suckers in about 100 days (Sinclair, 1992a).

Defoliation experiments show the importance of leaf area and weight to the developing fruit, slips, and suckers. The weight of fruit, slips, and suckers all increased linearly the later defoliation occurred (Fig. 11.5; Hepton, unpublished data). It can be concluded that anything that reduces plant net assimilation reduces fruit weight and the size and number of shoots available for propagation.

Temperature

As noted earlier, variations in temperature associated with latitude, elevation and season (Aubert *et al.*, 1973; Fleisch, 1988; Glennie, 1981; Wassman, 1986; C. Olende, Thika, Kenya, 2017, personal communication) can result in large differences in the length of the crop cycle and in the days from forcing to fruit harvest. Rates of vegetative growth are easily monitored and as long as scheduling of the fruit supply does not override the target plant mass for forcing, days from planting to forcing will be determined by days required to reach some average plant mass. Once that target is reached, plants will be forced and the period of fruit development begins.

The rate of fruit growth over time and the duration of the development phases described previously are mainly determined by fruit temperature (Zhang *et al.*, 1997; Dorey *et al.*, 2015). However, cold temperatures that don't prevent induction but do prevent development can impact days from forcing to harvest. In cool southern Queensland, Australia, fruit development of plants forced with ethephon over 13 weeks in winter was delayed by cold temperatures so all fruit matured over a 3-week span the following summer. Thus fruit development periods ranged from 274 to 204 days (Sinclair, 1992a). Assuming temperature is high enough to permit growth, days from forcing to the plant open heart stage (about 2.5 cm) are well correlated with cumulative thermal time (Fleisch and Bartholomew, 1987; Fleisch, 1988; Lechaudel *et al.*, 2010). Thermal time is calculated using a base temperature (T_b) that theoretically represents the temperature at which no development occurs. For 'Smooth Cayenne' the T_b ranged from 13.6°C for early phenophases to 11.7°C for later ones (Fleisch and Bartholomew, 1987) and in a more comprehensive study ranged from 4° to 12.5°C, depending on the phenophase

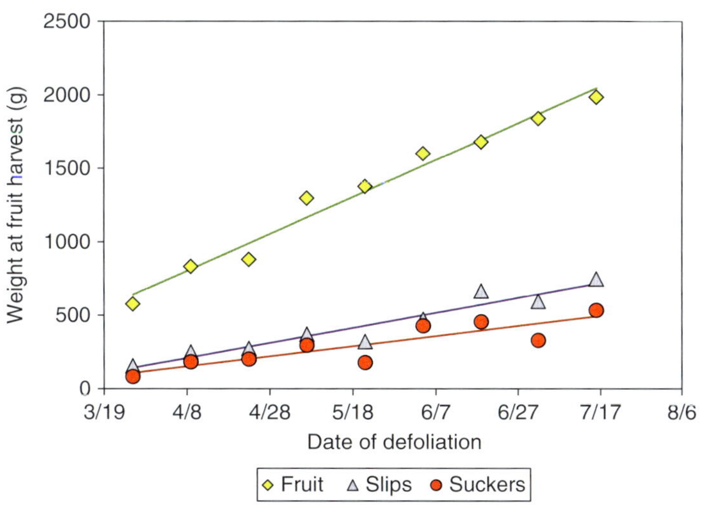

Fig. 11.5. Effect of date of leaf defoliation on mean fruit, slip, and sucker mass at the date of fruit harvest. For fruit, y = 636.4 + 12.57x, R_2 = 0.98; for slips, y = 141.2 + 5.11x, R_2 = 0.92; for suckers, y = 106.5 + 3.42x, R_2 = 0.76; n = 9 for all. From A. Hepton, personal communication.

(Flesich, 1988). A T_b of 9.24°C was obtained for the phenophase 'half-flowering' (50% of plants with at least one open corolla) to 50% of fruit mature for 'Queen' pineapple in Reunion (Lechaudel *et al.*, 2010).

In Hawaii, development after inflorescence emergence was better related to fruit thermal time, which was partially dependent on irradiance (Malézieux *et al.*, 1994). Exposure to direct sunlight elevates fruit temperatures, and such temperatures in the exposed section of the fruit can exceed 50°C with an ambient air temperature of about 30°C (Nightingale, 1942; Van Lelyveld, 1957; Soler, 1992). The difference in surface temperature between the sunny and shaded faces of the fruit can reach 13°C (Teisson, 1973). These high temperatures and temperature gradients can hasten the rate of fruit development (Zhang, 1992), though in hot environments, very high fruit temperatures may retard fruit development.

In Ghana, days from forcing to maturity were 130 for 'Smooth Cayenne' and 135 for 'MD-2' throughout the year (D. Bartholomew, 2006, personal communication of local grower) while at 1490 m in Kenya 'Smooth Cayenne' fruit mature in 213–232 days while 'MD-2' fruit mature in 234–258 days (C. Olende, 15 September 2016, Thika, Kenya). In Côte d'Ivoire (Malézieux and Lacoeuilhe, 1991) and Cameroon (Gaillard, 1970) days from forcing to maturity for 'Smooth Cayenne' range from 140 days for plants forced in February to a maximum of 160 days for plants forced in June. In Hawaii, days from forcing to harvest for 'Smooth Cayenne' grown from near sea level to 200 m elevation ranged from about 175 to 220 while at 800 m the range was 220 to 270 (Fleisch, 1988). In southern Queensland (27° 30′ to 26° S latitude), days from forcing to harvest can range from about 180 (September induction–April harvest) to 280 (April induction–January harvest) (Wassman, 1990; Sinclair, 1992a) and similar variation occurs in South Africa (Dalldorf *et al.*, 1975).

A fruit developmental disorder of 'Smooth Cayenne' and 'CO-2' referred to as 'prickly eye' (at maturity fruitlets are conical in shape; Swete Kelly and Bartholomew, 1993) is due to failure of fruitlets to fill out normally because plants flowered during periods of cool weather. Such fruit have substantially reduced weight. Frost can also cause injury, the extent of which determines whether or not a fruit can be processed or marketed fresh (Swete Kelly and Bartholomew, 1993).

Irradiance, slope, and aspect

Though the data is sparse, it indicates that fruit weight is positively correlated with increasing solar irradiance (Bartholomew and Malézieux, 1994). Shading reduced fruit temperature and slowed fruit development in Hawaii and in Taiwan (Sideris *et al.*, 1936; Lin *et al.*, 2015). Shading fruit from flowering to harvest markedly reduced fruit weight (Dull, 1966). Covering the fruit and crown reduced starch accumulation in the crown stem relative to stems of uncovered crowns (Sideris, 1933), indicating that crowns become autotrophic and sun exposure is required for stem starch accumulation, an important factor in plant establishment. Shading due to interplant competition reduces the fruit weight of plants in the centre row of 3-row plantings below that of adjacent rows (Bartholomew and Malézieux, 1994). Increased mutual shading at high planting densities reduces average fruit weight and a plant-removal experiment in Côte d'Ivoire illustrated the extent of the effect. Plantings with densities of 50,000 and 100,000 plants ha^{-1} were established and every other plant was removed in half the plots just before forcing (Malézieux and Sébillotte, 1990c). Plant removal significantly increased fruit weight 50% at the high density and by 14% at the lower one. Plant weight was the same in all plots at forcing so improved exposure of plants in plots where plants were removed increased fruit weight and the increase resulted from increased fruitlet weight.

In Côte d'Ivoire where temperature is relatively stable and irradiance fluctuates significantly with seasonal changes in cloud cover and rainfall, potential plant weight at forcing and fruit weight at harvest were correlated with irradiance intercepted during the vegetative growth period and for the entire cycle (Fig. 11.6) (Malézieux, 1988). In these unirrigated conditions, yield was depressed when drought was prolonged. Seasonal changes in irradiance increase with distance north and south of the equator and the effect of such changes on pineapple growth

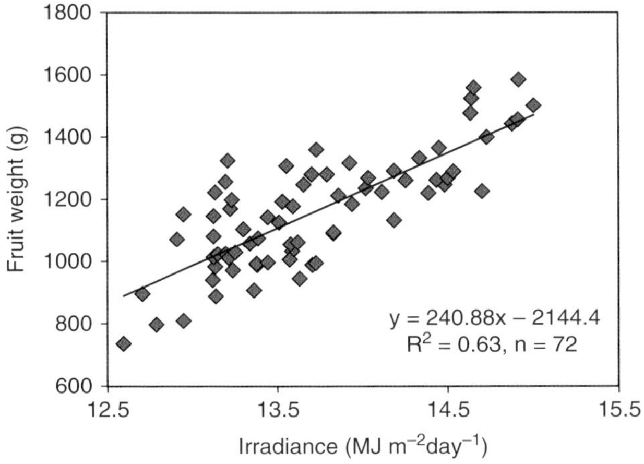

Fig. 11.6. Effect of mean irradiance from planting to harvest on average fruit weight of 'Smooth Cayenne' pineapple. Data are means of 60 fruit harvested from 72 monthly plantings in Côte d'Ivoire during a 6-year period. All plants were forced 8 months after planting. Redrawn from Bartholomew and Malézieux, 1994.

and yield are confounded with simultaneous changes in average temperature. Southern Queensland, Australia and South Africa are further from the equator than most pineapple growing regions and large seasonal effects on crop and fruit characteristics are observed in both regions. In winter when the sun angle and total irradiance are low, plant layout in the field can speed fruit maturity. Christensen (Queensland, Australia, 1999, personal communication) observed that where two row beds were planted in an east–west orientation, plants in the north-facing row make greater growth and shade plants in the south-facing row of two-row beds. Differences in the rate of fruit development in fields forced on the same day that have both north- and south-facing slopes helps growers manage peak harvests because areas with north-facing slopes mature about 2 weeks earlier than those with south-facing slopes (D. Christensen, Queensland, Australia, 1999, personal communication). While there is only limited direct evidence of an effect of irradiance on yield, a wide range of experience indicates growth, yield, and fruit quality are all poorer when irradiance is limiting.

Fruit and crown injury

Crown and fruit sunburn occur during periods of high irradiance and hot days, often with low or no wind. Exposed mature fruit skin temperatures of ~50°C for two or more hours have been recorded in Hawaii (Paull, unpublished data, 2016). Fruit are most susceptible to crown injury in the several weeks after emergence in the leaf whorl and when fruit are nearly mature (Newett and Rigden, 2015). Fruit and crown sunburn are more prevalent for 'CO-2', which has a long peduncle, than for 'MD-2' (D.P. Bartholomew, Hawaii, 2016, unpublished data). High irradiance during summer can reduce the size of 'CO-2' fruit and cause russeting of the skin, a relatively mild symptom (Sanewski, 2007; Newett and Rigden, 2015). Protection of emerging inflorescences or fruit with various covers or paper bags increased fruit size (larger fruitlets), improved fruit shape and reduced russeting of fruit. In Australia, irrigation has been used to reduce russeting and sunburn injury.

Increased planting density reduced the incidence of multiple crowns (Linford and Mehrlich, 1934; Norman, 1977; Scott, 1992), presumably due to greater mutual shading in the dense plantings. Sunburn is more common on field borders than inside the field and the highest incidence is on the west side of affected fields (Newett and Rigden, 2015). The west side of a field is more affected because the surfaces of already warm fruit are exposed to maximum air temperatures and nearly direct-beam radiation that can elevate exposed fruit surface temperatures to levels that cause injury. Overhead irrigation of developing fruit reduced the incidence of multiple crowns (Py *et al.*, 1987).

Sunburn during fruit development can asymmetrically reduce fruitlet size by 1.0 cm or more, which can produce fruit that are severely misshapen, a type of injury that is more prevalent on calm days (Swete Kelly and Bartholomew, 1993).

Crown heat injury symptoms vary from no crown due to injury soon after inflorescence emergence in the leaf whorl, multiple shoots that emerge after the crown apical tissue has been killed to stubby crowns due to growth suppression. Crown injuries affect fresh fruit marketability because fruit with no crown or large multiple shoots do not meet minimum marketing standards. When mature fruit are severely sunburned, the injured area becomes translucent and may become desiccated. Lodged fruit, more common in ratooned crops, are more likely to be injured than plant-crop fruit (Py *et al.*, 1987; Swete Kelly and Bartholomew, 1993). A more general 'cooking' of the fruit is associated with very high air temperatures, usually about 40°C. Fruit cooking, which has only been reported in Queensland, Australia, results in fruit that are pale white or translucent around the core and the flesh is spongy. Sunburned and cooked fruit have no market.

Practices to reduce the incidence of sunburn include tying leaves together above fruit, and covering fruit with paper or plastic covers. Spraying sun protectants (Lopes *et al.*, 2014; Rabie and Mbatha, 2016), and covering entire blocks with 30% black (a commercial practice in Honduras) to 40% white (Lopes *et al.*, 2014) saran shade cloth also reduced the percentage of sunburned fruit.

Drought and water excess

For a general discussion of the effects of water availability on pineapple growth and development, see Chapters 6 and 7, this volume. Few quantitative studies of the effects of drought on fruit development were found. Malézieux (E. Malézieux, Côte d'Ivoire, 1992, unpublished data) found that expected fruitlet number was reduced by drought. If stress occurs during fruit development, the peduncle may shrivel causing the fruit to lodge or break off (Swete Kelly and Bartholomew, 1993). Lopes *et al.* (2014) reported that fruit from plants irrigated at 77% of Class A pan evaporation, 6.4 mm day^{-1}, with 40% white shade cloth had higher SSC than unshaded fruit with no effect on fruit size or quality.

In Côte d'Ivoire, the fruit weight of plants grown in the field was related to the percentage of water requirement provided by irrigation between flowering and harvest and fruit weight was decreased 17% when only 20% of the plant water requirement was provided (Combres, 1980). A dry period in the month after forcing can reduce the number of fruitlets and a drought that occurs at the end of fruit development reduces fruitlet weight (E. Malézieux, 1992, unpublished data). In Côte d'Ivoire, yields were 12–15 mg ha^{-1} greater with irrigation in a normal year (Combres, 1979; Malézieux and Sébillotte, 1990b) and 25–30 mg ha^{-1} greater when drought was severe (Combres, 1979). While Huang and Lee (1969) reported that irrigation had no effect on fruit quality, Linford and Magistad (1933) found translucence was greater and acidity was decreased in fruit from irrigated plots. Beyond the effects on vegetative growth and fruit development, drought can also reduce sucker production (Medcalf, 1950) and reductions of up to 50% were observed at the end of the dry season (Combres, 1979). The greatest impact of irrigation on pineapple appears to be on vegetative growth (Carr, 2012) with only minor effects, if any, on fruit quality (Lopes *et al.*, 2014). Irrigation might however, in conjunction with good calcium levels, minimize the incidence of translucency because of the dependency of calcium uptake on water uptake.

Plants irrigated at 140% of the water requirement for the period between flowering and harvest reduced fruit yield by 15% (about 10 mg ha^{-1} of exportable fruit) in Côte d'Ivoire (Combres, 1983). Lopes *et al.* (2014) reported that SSC was significantly reduced by irrigation at 150% of Class A pan evaporation (6.4 mm day^{-1}).

Wind

Wind can transport large amounts of heat away from large fruit, which would help to reduce their internal temperature and may be very important in reducing the incidence of sunburn during periods of high irradiance and temperature. Wind could also reduce the incidence of malformed or multiple crowns. No reports on the detrimental effects of wind on pineapple leaves or fruit were found and no injury occurs in persistent trade wind conditions in Hawaii where velocities can exceed of 15 m s^{-1} (about

56 km h^{-1}). However, hurricane (typhoon) force winds may cause some injury, including lodging of exposed fruit, and such winds can carry debris, including sand, that can severely damage plants and fruit.

Plant population density and plant size

Effect of plant population density on average fruit size and yield

The effects of planting density, and thus of competition for light and nutrients, on average fruit weight and yield have been demonstrated many times (see Chapter 7) and plant nutrient requirements increase as green tissue mass increases (see Chapter 8). Fruit diameter (Treto *et al.*, 1974; Zhang, 1992) and fruit length (Norman, 1978) decreased as the planting density increased. Decreases in fruit weight were due mainly to a decrease in average fruitlet weight rather than in fruitlet number (Sanford, 1962; Pinon, 1981; E. Malézieux, 1992, unpublished data). All pineapple cultivars examined respond similarly to increasing plant population density, although the slopes of the lines fitted to the data were different for 'Queen', 'Spanish' and 'Smooth Cayenne' (see Chapter 7, Figs 7.7 and 7.8). The differences in slope are likely to be due in part to differences in the harvest index while within cultivar differences such as those for 'Smooth Cayenne' grown at different locations are probably due largely to differences in environment or cultural practices.

Effect of plant population density on crop duration and spread in maturity

The period from forcing to harvest is prolonged as planting density increases (Py *et al.*, 1987; Scott, 1992; Zhang, 1992). In Hawaii when 'Smooth Cayenne' plants were forced in September, no delay in maturity was found at plant population densities below 75,000 plant ha^{-1} (Zhang, 1992). Christensen (1995) found that there was about a one-day delay in maturity for every one-thousand plant ha^{-1} increase in density, though presumably there is a lower threshold where this effect is not observed. Christensen (1995) did not measure peak harvest at 93,700 plant ha^{-1} because plants segregated into large, unshaded fruit that matured early and small, shaded fruit that matured much later. This segregation was likely due to increased plant-to-plant variability that can occur when small variations in weight at planting are magnified as density increases, causing large plants to overtop smaller ones. Development of small fruit buried in the canopy is delayed because such fruit are shaded, which likely would retard development by lowering the average fruit temperature (Malézieux *et al.*, 1994) or by reducing the supply of assimilates allocated to the developing fruit.

Many of the issues related to planting density have been well resolved since the first edition of this book was published. Presently even small-scale commercial growers have adopted planting densities in the range of about 63,000 to 70,000 plant ha^{-1} for 'MD-2' (Leon and Kellon, 2012) and 88,000 to 120,000 for 'Queen' (Dorey *et al.*, 2015; E. Rabie, 2005, Kawzulu Natal, South Africa, personal communication).

Plant weight–fruit weight relationships

Plant weight, D-leaf weight and stem diameter at the time of forcing and fruit weight at harvest consistently are highly correlated for a given variety of pineapple within a given environment (Malézieux, 1988, 1993; Zhang and Bartholomew, 1997; Vilela *et al.*, 2015). However, the relationship between plant weight at forcing and fruit weight at harvest is complex and not always predictable. The strength of the relationship between plant or D-leaf weight and fruit weight depends on the growing conditions prevailing during a specific crop and hence it cannot be extrapolated. For example a study in Hawaii showed that stem weight per flower was 1.65 g for plants from one field and 10.45 g from a different field suggesting that the factors that determine flower numbers may not be proportional to the plants' ability to carry its fruitlets to maturity (Linford, 1933).

In regions near the equator where the environment is relatively uniform throughout the year, the correlation between plant or D-leaf weight at forcing and fruit weight at harvest might be expected to be high during most months of the year. However, in Côte d'Ivoire, variability in fruit weight within a particular field might or might not be well correlated with plant weight at forcing (Malézieux, 1988). The primary effect of plant weight is to determine fruitlet number rather than fruitlet weight at harvest (Malézieux, 1988). In an experiment where plants of different ages, and consequently plant weights, were forced at the same date (Malézieux and Sébillotte, 1990a), the increase in plant weight at forcing was associated with an increase in fruitlets per fruit (Fig. 11.7). LAI in this experiment ranged from 2.0 to 10 and the number of fruitlets was linearly correlated with plant growth during the month following forcing (Fig. 11.8).

A positive and significant relationship between plant weight at forcing and fruitlet number was obtained where plots planted monthly where forced 8 months later (Malézieux and Sébillotte, 1990b). Any reduction in expected fruitlet number could be related to drought and low radiation in the month following forcing (Fig. 11.9). There was no relationship between plant weight at forcing and fruitlet weight at harvest in these experiments.

Statistical relationships between plant weight at forcing and fruit weight at harvest (harvest index, HI) also may be established. In Côte d'Ivoire, the mean HI for the data obtained from the monthly-planting trial previously referred to (Malézieux, 1993) was 0.61 with a CV of 12.1%, a typical value for equatorial climates. A linear relationship was found between plant weight at forcing and fruit weight at harvest (Fig. 11.10) and. part of the residual variation may be explained by climatic conditions after forcing that influenced fruitlet number and

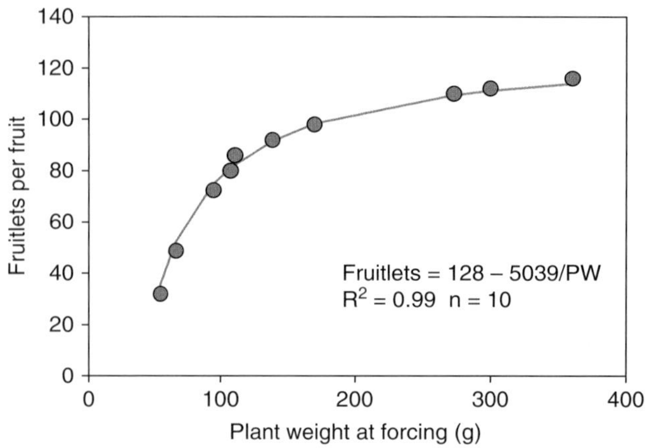

Fig. 11.7. Effect of plant weight at forcing, dry mass basis, on fruitlets per 'Smooth Cayenne' fruit grown in Côte d'Ivoire. From Malézieux and Sébillotte, 1990a.

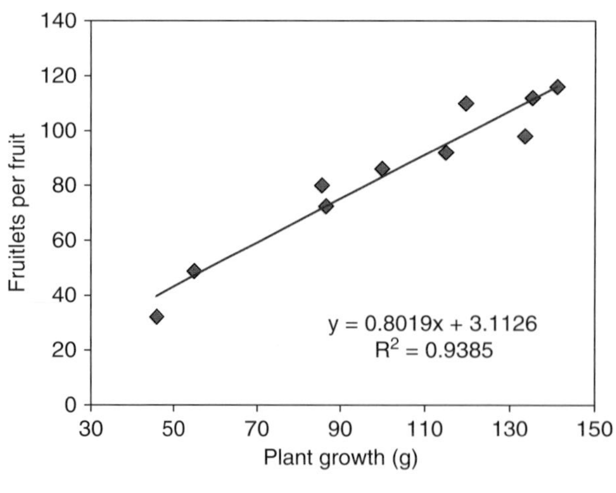

Fig. 11.8. Relationship between plant growth in the month following forcing and fruitlets initiated per fruit in Côte d'Ivoire. From Malézieux and Sébillotte, 1990a.

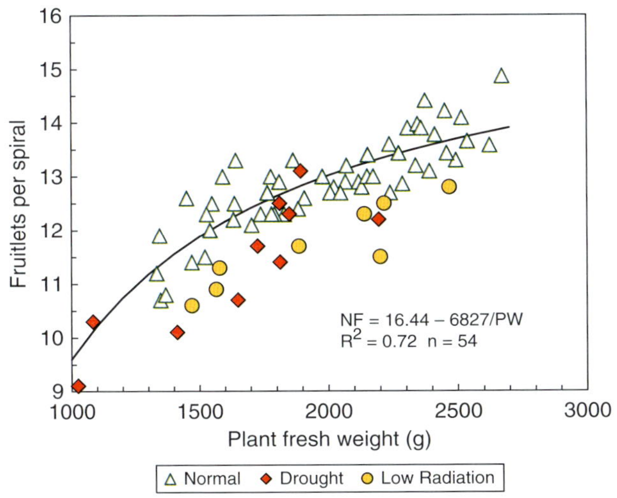

Fig. 11.9. Relationship between plant fresh weight and fruitlets per long spiral for 'Smooth Cayenne' pineapple grown in Côte d'Ivoire. From Malézieux and Sébillotte,1990b. In the equation, NF is number of fruitlets per long spiral (normal data only) and PW is plant fresh weight.

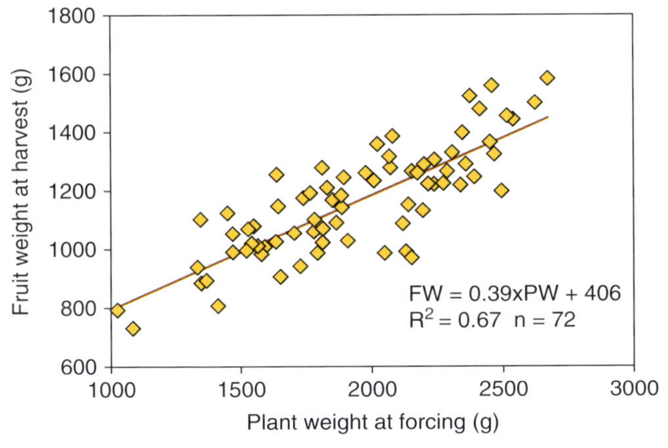

Fig. 11.10. Relationship between plant fresh weight at forcing and fruit weight at harvest (harvest index) for 'Smooth Cayenne' pineapple in Côte d'Ivoire. In the equation, FW is fruit weight at harvest and PW is plant weight at the time of forcing. Reproduced with permission of the International Society of Horticultural Science from Malézieux, E. (1993) Dry matter accumulation and yield elaboration of pineapple in Côte D 'Ivoire. *Acta Horticulturae* 334, 149–157. DOI:10.17660/ActaHortic.1993.334.15. https://doi.org/10.17660/ActaHortic.1993.334.15

fruitlet filling. Data points that fall below the regression line are due to inadequate fruit filling for plantings made in May, June and July because of drought and low irradiance during fruit development while data points located above the line are due to good growing conditions during fruit development for plantings made in January and February. This relationship is not universal, but depends on a variety of factors including the climatic conditions prevailing after forcing, the mineral status of the plant, and the quality of pest and disease control.

In a time-of-planting trial conducted in Queensland, Australia (Sinclair, 1992b) that included multiple dates of forcing, the relationship between plant weight at forcing and fruit weight at harvest was not significant if no account were taken of season of induction. Seasonal influence was a major determinant of fruit weight in this experiment. Fruit weight was mainly determined by climatic factors, primarily temperature that occurred during the period from flower induction to fruit development rather than plant weight at forcing (Sinclair, 1992b). In this trial, plants weighing 3.0 to nearly 4.5 kg that were forced in autumn produced small (1.0 kg) fruit (HI ranged from 0.22 to 0.33) at harvest because fruit development occurred during the winter. Plants weighing only about 2.5 kg that were forced in spring produced fruit with a fresh weight of about 2.0 kg at harvest (HI = 0.80).

Genetic effects

Some genetic variation in the length of the fruit development period exists between cultivars (Chan and Lee, 2000), though it apparently was not great enough to significantly shorten time from forcing to fruit maturity. In Queensland, a comparison of four 'Smooth Cayenne' clones indicated 'C10' matured approximately 7 days earlier than the 'Smooth Cayenne' clone 'F180'. The Pineapple Research Institute of Hawaii cultivar 53-116 matured approximately 14 days later than 'C10' (G. Sanewski, 1996, unpublished data). In the same environment, cultivars of Queen typically mature 4–6 weeks earlier than 'Smooth Cayenne'.

Crown Development

Crown size

Little work has been done on the growth of the crown and the factors that influence it, especially where pineapple fruit mature in all months of the year. Crown length, weight and leaf number reach a maximum at fruit maturity, presumably because translocation of water and nutrients through the peduncle ceases. Crown weights vary with the environment and are relatively small on fruit that develop in the tropics (Py *et al.*, 1987), and this effect was attributed to the relatively short interval between forcing and fruit harvest. Crowns are reduced in size when plants are deficient in nitrogen (Py *et al.*, 1987). As noted previously, crowns are smaller on fruit that mature in the winter and spring in subtropical environments and are largest when fruit mature in summer and autumn. In subtropical environments the largest crowns are produced in summer when high irradiance and warm temperatures sustain high rates of photosynthesis and growth. Conversely, much lower irradiance and the cool temperatures during the winter would reduce the photosynthate supply. When carbohydrate resources are limiting, the fruit, which is the strongest sink, would develop at the expense of the crown.

Multiple crowns

Multiple crowns is a common disorder or injury that can be of genetic or environmental origin. Collins (1960) states that fasciation, an abnormal growth resulting in two to numerous crowns, is relatively uncommon in 'Smooth Cayenne' and 'Queen,' but common in 'Singapore Spanish' and 'Pernambuco' (*syn.* 'Perola'). Collins (1960) also reported the existence of mutant clones of Smooth Cayenne that produced 50% or more multiple crowns. Fasciation can increase the size of the fruit core and result in flattening of the upper portion of the fruit, which reduces the value of the fruit for the fresh market and for canning.

While some cultivars can be more prone to multiple crowns, some cultivars such as CO-2, do not develop multiple crowns regardless of disorders or injury but will instead develop crownlets (crown slips).

Multiple crowns are associated with high fertility and rapid growth following a period of prolonged drought if such conditions occur about the time of inflorescence initiation (Collins, 1960; Py *et al.*, 1987). Multiple crowns can also be associated with calcium deficiency (Swete Kelly and Bartholomew, 1993). The increased incidence of multiple crowns during periods of high irradiance and high temperature was discussed above.

Fruit Enlargement

Synthetic auxins such as NAA, β-napthoxyacetic acid, and 2-(3-chlorophenoxy) propionic acid (3-CPA) applied after inflorescence emergence, and most often after flowering, increase fruit weight at harvest (Py *et al.*, 1987) and fruit yields can be increased by up to 40% (Williams, 1987). Apparently only 3-CPA was used on a commercial scale and a stated benefit for fresh fruit production was that it reduced crown size. However, no evidence was found that it is still used on pineapple.

Yamane and Ito (1970b) reported that untreated 'Smooth Cayenne' fruit averaged 1860 g while fruit sprayed with 150 mg l^{-1} of gibberellic acid (GA_3) weighed 1905 g and if sprayed with 300 mg l^{-1} fruit weighed 2123 g and maturation was delayed 18 days. Spraying 'Queen' fruit with 50 mg l^{-1} of GA_3 at 15 days after flowering increased fruit weights by about 20%. Trials in Hawaii which applied GA_3 to 'CO-2' during mid- to late-flowering did not show any consistent increase in fruit weight or quality. There is a consistent, but not significant, decrease in fruit soluble solids content (SSC) of 6% when fruit was sprayed with GA_3 (Li *et al.*, 2011). The results of six trials with GA_3 (RyzUp® 40SG) in Costa Rica showed that spraying 'MD-2' plants at 14 and 16 weeks after forcing with 2400 l ha^{-1} of a solution containing 400 mg l^{-1} GA_3 delayed fruit maturity by 7–10 days (Villalobos *et al.*, 2013). Crownless fruit weight gains ranged from 86 to 339 g, mostly due to an increase in fruit circumference. The greatest response was obtained with ratoon crop fruit and both height and circumference increased. Due to the delay in maturation, treated fruit had significantly lower translucency in five of six trials while titratable acids and SSC differed among the trials and the variation was attributed to physiological differences in fruit at the time of harvest rather than a consistent trend (Villalobos *et al.*, 2013).

Predicting Fruit Development and Yield

Growth of 'Smooth Cayenne' and 'Queen' fruit has been studied intensively and development is highly predictable from forcing to emergence in the leaf whorl by relatively simple heat-unit models (Fleisch and Bartholomew, 1987; Dorey *et al.*, 2015). However, a much more complex model was required to accurately predict fruit development in diverse environments of Hawaii (multiple planting dates at three elevations), Ivory Coast (multiple plantings), Queensland, Australia and Thailand (Malézieux *et al.*, 1994). The process-oriented ALOHA Pineapple V. 2.1 (Zhang and Bartholomew, 1993; Zhang *et al.*, 1997), which incorporated the improved fruit maturity simulation described by Malézieux *et al.* (1994) accounted well for the thermal variability across seasons, elevations and locations. The model was integrated into early versions of Decision Support System for Agrotechnology Transfer (DSSAT, 2017). In 2017, Dole Philippines contracted with the University of Florida to update and modularize ALOHA Pineapple and merge it with the current Cropping System Model (CSM) format for all crops that are a part of DSSAT 4.7 (G. Hoogenboom, Gainesville, FL, 2017; R. Umali, Polomolok, Philippines, 2017, personal communications). DSSAT 4.7 is open source and can be downloaded from the website.

The pineapple crop model SIMPÑA and submodels for the prediction of SSC and TA were specifically developed for 'Victoria' ('Queen') in Réunion Island (Dorey *et al.*, 2015, 2016a,b). Fruit development after emergence in the leaf whorl was dealt with in the SIMPIÑA model (Dorey *et al.*, 2015) by using a different base temperature for each fruit development stage while ALOHA Pineapple used heat units based on estimated fruit temperature (Malézieux *et al.*, 1994), which is determined by air temperature, irradiance and wind velocity, to improve prediction accuracy in a wider array of environments.

Factors Influencing Fruit Quality at Harvest

Fruit quality in pineapple is subjective because consumer preferences vary. However, among the generally accepted factors influencing fruit quality include fibrosity, porosity, opacity, translucence, acidity, SSC, esters, aroma, taste, and colour (Sideris and Krauss, 1933b). Other measures of quality include blackheart incidence, chilling injury and fruit nitrate levels. Black heart can occur in 'Queen' and 'Smooth Cayenne' fruit that mature during the winter months in subtropical Queensland, Australia and in South Africa (Newett and Rigden, 2015; Louw, 2017)

However, because there are only limited numbers of pineapple cultivars sold fresh or processed, consumers have little opportunity to discriminate pineapple fruit on the basis of quality.

In general, superior quality fruit have high SSC and relatively low TA, usually less than 0.7% as citric acid. However, very low acidity can also lower quality (Sideris and Krauss, 1933a) as the fruit will be perceived as being too bland, especially for consumers accustomed to fruit with higher acidity. Fruit of 'MD-2' are not susceptible to endogenous browning and 'CO-2' is reasonably resistant. High fruit nitrate levels result from excessive nitrogen fertilization and are only a concern where fruit are to be processed.

Changes in irradiance apparently can affect fruit SSC and TA, but there have been few definitive studies, particularly since growth regulator-induced flowering became a common practice. Sideris *et al.* (1936) reported that fruit specific gravity increased from 1.005 at 50% light to 1.034 at ambient levels. Total SSC in juice was unaffected by light level. Titratable acidity, expressed as citric acid, increased from 0.74 to 1.4% as light was decreased from ambient to 50% of ambient levels. Hamner and Nightingale (1946) reported that SSC was unaffected by 50% shade but juice acidity as citrate increased from 0.79 for control fruit to 1.12% for fruit that ripened in 50% light. The increased acidity was attributed to the effect of shade on fruit temperature rather than available light (Hamner and Nightingale, 1946). Mutual shading associated with higher planting densities decreased the proportion of translucent fruit and fruit specific gravity (E. Malézieux, 1992, unpublished data; Pinon, 1981). While shading materials or reflectant coatings can protect against fruit sunburn (Py *et al.*, 1987; Swete Kelly and Bartholomew, 1993), complete shading can reduce SSC (Py *et al.*, 1987), thus reducing fruit quality.

Fruit TA increased with increasing plant density (Py *et al.*, 1973; Chadha *et al.*, 1974), presumably due to increased mutual shading at the higher planting densities. Souza *et al.* (2013) reported that 'Smooth Cayenne' fruit SSC increased linearly with increasing irrigation and fruit acidity decreased over the range 50–125% of crop evapotranspiration. Titratable acidity can also vary within a fruit in relation to exposure. Flesh TA under the skin was less in sunburned areas than in unaffected parts of the fruit (Teisson, 1979a). When only part of the fruit was painted black to reduce its albedo, TA was significantly less in the painted parts than in the remainder of the fruit (Teisson, 1979b). Fruit TA varied relatively more than SSC and levels were highly negatively correlated with irradiance (Fig. 11.11) (Malézieux and Lacoeuilhe, 1991). 'Smooth Cayenne' fruit grown in tropical Thailand where average temperatures are high have such low TA that citric acid must be added to the fruit to permit low-temperature processing.

The seasonal variation in TA limited fresh 'Smooth Cayenne' fruit acceptability in winter so it was rapidly replaced by the low-acid 'MD-2' when it was marketed in quantity in the mid-1990s. No data were found on variation In SSC and TA in 'low acid' 'MD-2' fruit across environments where it is grown. Fruit of 'MD-2' dominate fresh fruit markets except in Brazil where 'Pérola' is the main cultivar and in China where 'Queen' dominate the markets (Sun, 2011). As with 'Smooth Cayenne', TA of 'MD-2' fruit is

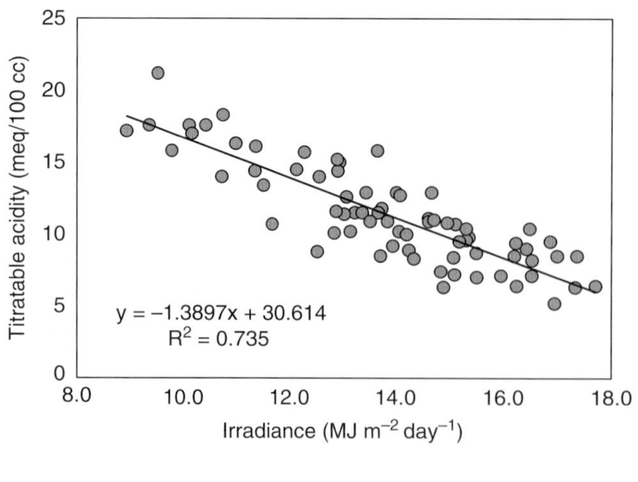

Fig. 11.11. Effect of mean irradiance for the 2 weeks prior to harvest on the titratable acidity of 'Smooth Cayenne' fruit juice. Data are means of 30 fruit harvested from 120 plantings made monthly in Côte d'Ivoire during a 10-year period. Malézieux and Lacoeuilhe, 1991.

lower in summer than in winter in Hawaii and much lower in Ghana than in Hawaii (D. Bartholomew, personal observation). While the maximum SSC for 'MD-2' is about 17, it is standard practice to degreen fruit with an acidified ethephon spray when SSC is between 12%, the minimum required to meet international grading standards, and 14% because riper fruit are more susceptible to bruising and have a reduced shelf life. The SSC of a representative sample of fruit is evaluated in the field to assure that they conform to an international or a higher company standard.

Dorey *et al.* (2016a,b) modelled SSC and TA of 'Queen' pineapples grown in Reunion. Fruit SSC ranged from 17 to 19.4% and was lowest when plants were grown under a nitrogen deficit and highest when fruit matured in unirrigated fields without a nitrogen deficit. Fruit TA ranged from 11 to 21 (units not given) and, similar to Fig. 11.11, was lowest when fruit were harvested when days were sunny.

A physiological disorder of 'Smooth Cayenne' fruit referred to as green-ripe fruit is characterized by translucent flesh and a green skin (Py *et al.*, 1987). The green-ripe disorder generally begins in the part of the fruit exposed to the sun and the incidence decreases with shading (Teisson, 1979a). The percentage of green-ripe fruit decreased from 10% with no shade to 7% with partial shade and 4% when fruit were completely shaded (Teisson, 1979a). The incidence of the disorder is not predictable and its physiological basis is not fully understood. Fruit become more translucent as the fruit ripen and air cavities in the fruit flesh become filled with juice. Fruit translucence increases from the base of the fruit to the top and translucent fruit were considered superior to opaque fruit for canning; however, in the fresh market translucency is considered a pre-harvest disorder because fruit are more fragile and prone to bruising than are opaque fruit. Some data show that translucency is reduced by calcium supply (Silva *et al.*, 2006) and post-flowering sprays of GA3 (Villalobos *et al.*, 2013). See Chapter 13 for more information on green-ripe.

References

Aubert, B. (1977) Etude du cycle de floraison naturalle de l'ananas 'Victoria' a la Réunion. *Fruits* 32, 25–41.

Aubert, B., Gaillard, J.P., Lossois, P., Py, C. and Marchal, J. (1973) Influence de l'altutude sur le comportement de l'ananas 'Cayenne Lisse'. Essais realises au pied du mont Cameroun. *Fruits* 28, 203–214.

Avila Espinosa, M.E., Moreira, R.O., Lima, A.A., Sagio, S.A., Barreto, H.G. *et al.* (2017) Early histological, hormonal, and molecular changes during pineapple (*Ananas comosus* (L.) Merrill) artificial flowering induction. *Journal of Plant Physiology* 209, 11–19.

Bartholomew, D.P. (1977) Inflorescence development of pineapple (*Ananas comosus* (L.) Merr.) induced to flower with ethephon. *Botanical Gazette* 138, 312–320.

Bartholomew, D.P. (2001) Control of natural flowering. *Pineapple News* 8, 23. Available at: http://www.ishs-horticulture.org/workinggroups/pineapple/PineNews08.pdf (accessed 14 March 2018).

Bartholomew, D.P. (2009) 'MD-2' Pineapple transforms the World's pineapple fresh fruit export industry. *Pineapple News* 16, 2–5. Available at: http://www.ishs-horticulture.org/workinggroups/pineapple/PineNews16.pdf (accessed 14 March 2018).

Bartholomew, D.P. (2014) History and perspectives on the role of ethylene in pineapple flowering. *Acta Horticulturae* 1042, 269–283.

Bartholomew, D.P. and Malézieux, E. (1994) Pineapple. In: Schaffer, B. and Anderson, P. (eds) *Handbook of Environmental Physiology of Fruit Crops*, Vol. II. Boca Raton, FL: CRC Press, Inc., pp. 243–291.

Bartholomew, D.P., Uruu, G., Lopez, J.A. and Leep, D. (2011) Refining aviglycine treatments to improve control of natural induction of 'MD-2' pineapple. *Acta Horticulturae* 902, 211–219.

Befler, B., Schmitgen S., Kuhnemann F., Gabler, R. and Urban, W. (1998) Light-dependent production of ethylene in *Tillandsia usneoides* L. *Planta* 205(1), 140–144.

Botella, J.R., Cavallaro, A.S. and Cazzonelli, C.I. (2000) Towards the production of transgenic pineapple to control flowering and ripening. *Acta Horticulturae* 529, 115–122.

Bourke, R.M. (1976) Seasonal influences on fruiting of rough leaf pineapples. *Papua New Guinea Agriculture Journal* 27, 103–106.

Burg, S.P. and Burg, E.A. (1966) Auxin-induced ethylene formation: its relations to flowering in the pineapple. *Science* 152, 1269.

Carr, M.K.V. (2012) Pineapple. *Advances in Irrigation Agronomy: Fruit Crops.* Cambridge: Cambridge University Press, pp. 265–278.

Cazzonelli, C.I., Cavallaro, A.S. and Botella, J.R. (1998) Cloning and characterisation of ripening-induced ethylene biosynthetic genes from non-climacteric pineapple [*Ananas comosus*] fruits. *Australian Journal of Plant Physiology* 25, 513–518.

Chadha, K.L, Melanta, K.R and Shikhamany, S.D. (1974) High density planting increases pineapple yields. *Indian Horticulture* 18, 3–5.

Chan, Y.K. and Lee, H.K. (2000) Breeding for early fruiting in pineapple. *Acta Horticulturae* 529, 139–146.

Chang, C. (2016) Q&A: how do plants respond to ethylene and what is its importance? *BMC Biology* 14, 7.

Chang, J.C., Maruthasalam, S., Liu, Y.L., Sun, C.M., Shih, W.S. *et al.* (2011) Forcing of 'Tainon17' pineapple with calcium carbide (CaC2) and/or ice-cold stress under field conditions. *Acta Horticulturae* 902, 327–335.

Chipungahelo, G.S., Ngereza, A., Kawamala, P. and Kwileka, T. (2007) Effects of light regimes on different crops, sweet potato (*Ipomoea batatas* L. Lam), cowpea (*Vigna unguiculata* L Walp) and pineapple (*Ananas comosus* L. Merr.). *8th African Crop Science Society Conference*, El-Minia, Egypt, 27–31 October 2007, pp. 467–471.

Christensen, D. (1995) Quality, efficiency and yield. In: *Pineapple Industry Farm Committee Pineapple Field Day Book.* Brisbane: Growcom, pp. 15–19.

Collins, J.L. (1960) *The Pineapple: Botany, Cultivation and Utilization.* New York: Interscience Publishers Inc.

Combres, J.C. (1979) *Economie d'eau ou irrigation intensive.* Réunion Annuelle IRFA, Doc. Interne No. 57.

Combres, J.C. (1980) *Essai ETM. AN-CI-EXT 528.* Doc. interne IRFA.

Combres, J.C. (1983) *Bilan énergétique et hydrique de l'ananas. Utilisation optimale des potentialités climatiques.* Doc. Interne MENRS/IRFA.

Conway, M.J. (1977) The effects of age, temperature and duration of exposure to temperature on susceptibility of pineapple to floral induction with ethephon. MS thesis, University of Hawaii at Manoa, Honolulu, USA.

Cox, K.J. (1979) Decomposition rates of ethrel at elevated pHs and temperatures. Queensland Department of Primary Industries Refnote R63/Oct 79.

Dalldorf, D.B. (1979) Flower induction of pineapples. *Farming in South Africa.* Department of Agricultural Technical Services, Pretoria, South Africa.

Dalldorf, D.B., Keetch, D.P. and Webster, G. (1975) Pineapples: forecasting of harvesting time in the Eastern Cape. Citrus and Subtropical Fruit Research Institute. *Information Bulletin* 39, 4–5.

Dass, H.C., Randhawa, G.S., Singh, H.P. and Ganapathy, K.M. (1976) Effect of pH and urea on the efficacy of ethephon for induction of flowering in pineapple. *Scientia Horticulturae* 5, 265–268.

Dorey, E., Fournier, P., Lechaudel, M. and Tixier, P. (2015) Validity of the pineapple crop model SIMPIÑA across the climatic gradient in Réunion Island. *European Journal of Agronomy* 62, 1–12.

Dorey, E., Fournier, P., Léchaudel, M. and Tixier, P. (2016a) Modeling sugar content of pineapple under agro-climatic conditions on Reunion Island. *European Journal of Agronomy* 73, 64–72.

Dorey, E., Fournier, P., Léchaudel, M. and Tixier, P. (2016b) A statistical model to predict titratable acidity of pineapple during fruit developing period responding to climatic variables. *Scientia Horticulturae* 210, 19–24.

DSSAT (2017) Decision Support System for Agrotechnology Transfer. Available at: http://www.dssat.net (accessed 12 July 2017).

Dull, G.G. (1966) Seasonal variation of internal fruit temperatures. *PRI News* 14, 92–95.

Dunsmore, J.R. (1957) The pineapple in Malaya. *Malaysian Agricultural Journal* 40, 159–187.

Evans, H.R. (1959) The influence of growth promoting substances on pineapples. *Tropical Agriculture, Trinidad* 36, 108–117.

FDA (2016) FDA concludes consultation on pink flesh pineapple. US Food and Drug Administration. Available at: https://www.fda.gov/food/newsevents/constituentupdates/ucm533075.htm (accessed 10 May 2017).

Firoozabady, E. and Young, T.R. (2013) Pineapple plant named Rosé (EF2-114). Available at: https://www.google.com/patents/US20130326768 (accessed 17 July 2017).

Fleisch, H. (1988) Modeling pineapple plant growth and inflorescence development. PhD dissertation, University of Hawaii at Manoa, Honolulu, USA.

Fleisch, H. and Bartholomew, D.P. (1987) Development of a heat unit model of pineapple ('Smooth Cayenne') fruit growth from field data. *Fruits* 42, 709–715.

Friend, D.J.C. and Lydon, J. (1979) Effects of daylength on flowering, growth, and CAM (Crassulacean acid metabolism) of pineapple (*Ananas comosus* (L.) Merrill). *Botanical Gazette* 140, 280–283.

Gaillard, J.P. (1969) Influence of planting date and weight of slips on the growth of pineapples in Cameroon. *Fruits* 24, 75–87.

Gaillard, J.P. (1970) Recherche d'un equilibre K/N dans la production de l'ananas frais au Cameroun. *Fruits* 25, 11–24.

Giacomelli, E.J., Py, C. and Lossois, P. (1984) Estudo sobre o ciclo natural do abacaxizeiro 'Cayenne' no planalto paulista. *Bragantia Campinas* 43, 629–642.

Glennie, J.D. (1979) The effect of temperature on flower induction of pineapple with Ethrel. In: *Pineapple Industry Farm Committee Pineapple Field Day Book*. Brisbane: Growcom, pp. 8–13.

Glennie, J.D. (1981) Induction with Ethrel. In: *Pineapple Industry Farm Committee Pineapple Field Day Book*. Brisbane: Growcom, pp. 34–36.

Gortner, W.A., Dull, G.G. and Krauss, B.H. (1967) Fruit development, maturation, ripening and senescence: a biochemical basis for horticultural terminology. *HortScience* 2, 141–144.

Gowing, D.P. (1961) Experiments on the photoperiodic response in pineapple. *American Journal of Botany* 48, 16–21.

Gowing, D.P. and Leeper, R.W. (1955) Induction of flowering in pineapple by beta-hydroxyethylhydrazine. *Science* 122, 1267.

Gowing, D.P. and Leeper, R.W. (1959) Further data comparing BOH and SNA as forcing agents. *Pineapple Research Institute of Hawaii Research Report* 63.

Guyot, A. and Py, C. (1970) Controlled the flowering of pineapples with Ethrel, a new growth regulator (conclusion). *Fruits* 25, 427–445.

Hamner, K.C. and Nightingale, G.T. (1946) Ascorbic acid contents of pineapples as correlated with environmental factors and plant composition. *Food Research* 11, 535–541.

Hanqing C, Zhiying. L. and Li X. (2013) Characterizing developmental and inducible differentiation between juvenile and adult plants of *Aechmea fasciata* treated with ethylene by transcriptomic analysis. *Plant Growth Regulation* 69, 247–257.

Huang, C.C. and Lee, L.C. (1969) Effect of irrigation on pineapples. *Taiwan Agriculture Quarterly* 5, 50–58.

Iglesias, R. (1979) Influence of a mixture of ethrel with urea and sodium carbonate on flowering, fruit quality and slip production in the pineapple cultivar Espanola Roja. *Cultivos Tropicales* 1, 117–130.

Kerns, K.R. (1936) Field results of acetylene on pineapple plants. *Pineapple Quarterly* 6, 96–114.

Kerns, K.R., Collins, J.L. and Kim, H. (1936) Developmental studies of the pineapple *Ananas comosus* (L.) Merr. I. Origin and growth of leaves and inflorescence. *New Phytologist* 35, 305–317.

Kidokoro, S., Koshi, Y., Hironori, T., Takahashi, F., Shinozaki, K. and Yamaguchi-Shinozaki, K. (2017) Different cold-signaling pathways function in the responses to rapid and gradual decreases in temperature. *Plant Cell* 29(4), 760–774.

Kist, H.G.K., Ramos J.D., Pio, R. and dos Santos, V.A. (2011) Diquat and urea in the natural flowering management of 'Pérola' pineapple. *Revista Brasileira de Fruticultura* 33, 1048–1054.

Kuan, C., Yu, C., Lin, M., Hsu, H., Bartholomew, D.P. and Lin, C.-H. (2005) Foliar application of aviglycine reduces natural flowering in pineapple. *HortScience* 40, 123–126.

Leal, F. and Coppens d'Eeckenbrugge, G. (1996) Pineapple. In: Janick, J. and Moore, J.N. (eds) *Fruit Breeding. I. Tree and Tropical Fruits*, Vol. I. New York: John Wiley & Sons, Inc., pp. 515–557.

Lechaudel, M., Damour, G., Fournier, P., Joas, J. and Jahiel, M. (2010) How to predict the harvest date of tropical fruit: from simple methods to complex models. *Acta Horticulturae* 880, 175–182.

Lei, M., Li, Z.Y., Wang, J.B., Fu, Y.L., Ao, M.F. and Xu, L. (2016) AfAP2-1, An age-dependent gene of *Aechmea fasciata*, responds to exogenous ethylene treatment. *International Journal of Molecular Sciences* 17, 303–303.

Leon, R.G. and Kellon, D. (2012) Characterization of 'MD-2' pineapple planting density and fertilization using a grower survey. *HortTechnology* 22, 644–650.

Li, Y., Wu, Y., Wu, B., Zou, M., Zhang, Z. and Sun, G. (2011) Exogenous gibberellic acid increases the fruit weight of 'Comte de Paris' pineapple by enlarging flesh cells without negative effects on fruit quality. *Acta Physiologiae Plantarum* 33, 1715–1722.

Li, Z., Yi, Z., Cong, H. and Xu, L. (2012) Regulation of Ca2+ regulators on the CaM gene expression during ethylene-induced flowering in *Guzmania* 'Amaranth'. *Genomics and Applied Biology* 31, 35–39.

Li, Z., Wang, J., Zhang, X., Lei, M., Fu, Y. *et al.* (2016a) Transcriptome sequencing determined flowering pathway genes in *Aechmea fasciata* treated with ethylene. *Journal of Plant Growth Regulation* 35, 316–329.

Li, Y.H., Wu, Q.S., Huang, X., Liu, S.H., Zhang, H.N. *et al.* (2016b) Molecular cloning and characterization of four genes encoding ethylene receptors associated with pineapple (*Ananas comosus* L.) flowering. *Frontiers in Plant Science* 7, 710–710.

Lin, C.H., Maruthasalam, S., Shiu, L.Y., Lien, W.C., Loganathan, M., Yu, C.W., Hung, S.H., Ko, Y. and Chen, Y.Y. (2009) Physical and chemical manipulation of flowering in pineapple. *Acta Horticulturae* 822, 117–123.

Lin, M., Chen, M., Lin, T., Kuan, C., Lee, C. and Yang, W. (2015) Prevention of natural flowering in pineapple (*Ananas comosus*) by shading and urea application. *Horticulture, Environment and Biotechnology* 56, 9–16.

Linford, M.B. (1933) Fruit quality studies. II. Eye number and eye weight. *Pineapple Quarterly* 3, 185–188.

Linford, M.B. and Magistad, O.C. (1933) Fruit quality studies. III. Soil moisture as a factor limiting the development and quality of pineapple fruits. *Pineapple Quarterly* 3, 196–206.

Linford, M.B. and Mehrlich, F.P. (1934) Fruit quality studies. VIII. Some influences of plant spacing in Kauai Pineapple Company experiment 17. *Pineapple Quarterly* 4, 23–37.

Liu, C.H. and Fan, C. (2016) De novo transcriptome assembly of floral buds of pineapple and identification of differentially expressed genes in response to ethephon induction. *Frontiers in Plant Science* 7, 203–203.

Loeillet, D., Dawson, C. and Paqui, T. (2011) Fresh pineapple market: from the banal to the vulgar. *Acta Horticulturae* 902, 587–594.

Lopes, O.P., Maia, V.M., Santos, S.R.D., Mizobutsi, G.P. and Pegoraro, R.F. (2014) Protection against sunburn of pineapple fruits submitted to different irrigation levels. *Revista Brasileira de Fruticultura* 36, 748–754.

Louw, J. (2017) Hluhluwe queen pineapples. Available at: http://www.pineapples.co.za/content%20page/production.aspx (accessed 18 July 2017).

Lv, L., Duan, J., Xie, J., Wei, C., Liu, Y., Liu, S. and Sun, G. (2012) Isolation and characterization of a FLOWERING LOCUS T homolog from pineapple (*Ananas comosus* (L.) Merr). *Gene* 505, 368–373.

Maia, V.M., Lopes, O.P., Santos, D.O., Prudencio dos Santos, J.R., Pegaaro, R.F. and Aspiazu, I. (2017) Inhibition of natural flowering of pineapple with the use of 1-MCP. *Pineapple News* 24, 15–16. Available at: http://www.ishs-horticulture.org/workinggroups/pineapple/PineNews24.pdf (accessed 14 March 2018).

Malézieux, E. (1988) Croissance et elaboration du rendement de l'ananas (*Ananas comosus* L. Merr.) [Growth and yield development of pineapple (*Ananas comosus* L. Merr.)]. Doctoral thesis, INA-PG, Paris.

Malézieux, E. (1993) Dry matter accumulation and yield elaboration of pineapple in Côte D'Ivoire. *Acta Horticulturae* 334, 149–157.

Malézieux, E. and Lacoeuilhe, J.J. (1991) Analyse de la variabilite des rendements chez l'ananas (*Ananas comosus* (L.) Merr). I. Caractérisation dans les conditions de Côte d'Ivoire, des fluctuations saisonnières du rendement et de certains facteurs liés à la qualité du fruit. *Fruits* 46, 227–239.

Malézieux, E. and Sébillotte, M. (1990a) Relations entre les processus d'accumulation de la matiére séche et le rendement chez l'ananas (*Ananas comosus* L. Merr.). II. L'effet de la biomasse présente l'induction florale. Note de réunion annuelle IRFA, doc. int. Montpellier.

Malézieux, E. and Sébillotte, M. (1990b) Relations entre les processus d'accumulation de la matire sche et le rendement chez l'ananas (*Ananas comosus* L. Merr.). IV. L'effet du climat. Note de réunion annuelle IRFA, doc. int. Montpellier.

Malézieux, E. and Sébillotte, M. (1990c) Relations entre les processus d'accumulation de la matire sche et le rendement chez l'ananas (*Ananas comosus* L. Merr.). III. L'effet de la competition intraspecifique. Note de réunion annuelle IRFA, doc. int. Montpellier.

Malézieux, E., Zhang, J., Sinclair, E. and Bartholomew, D.P. (1994) Predicting pineapple harvest date in different environments, using a computer simulation model. *Agronomy Journal* 86, 609–617.

Maruthasalam, S., Shiu, L.Y., Loganathan, M., Lien, W.C., Liu, L.L. *et al.* (2010) Forced flowering of pineapple (*Ananas comosus* cv. Tainon 17) in response to cold stress, ethephon and calcium carbide with or without activated charcoal. *Plant Growth Regulation* 60, 83–90.

Medcalf, J.C. (1949) I. Some use of climatologic data in plantation operations. Pineapple Research Institute of Hawaii *PRI Seminar Summary*, 4 March, pp. 3–22. Library, University of Hawaii Manoa, Honolulu, USA.

Medcalf, J.C. (1950) Hawaiian Pineapple Co.'s irrigation studies on Lanai. Pineapple Research Institute of Hawaii *Abstract of Monthly Seminar Summary*, pp. 16–28. Library, University of Hawaii Manoa, Honolulu, USA.

Mekers, O., De Proft, M. and Jacobs, L. (1983) Prevention of unwanted flowering of ornamental Bromeliaceae by growth regulating chemicals. *Acta Horticulturae* 137, 217–224.

Millar-Watt, D. (1981) Control of natural flowering in the Smooth Cayenne pineapple, *Ananas comosus* (L.) Merr. *Citrus and Subtropical Fruit Research Institute Information Bulletin* 110, 17–19.

Min, X.-J. (1995) Physiological effects of environmental factors and growth regulators on floral initiation and development of pineapple, *Ananas comosus* (L.) Merr. PhD dissertation, University of Hawaii at Manoa, Honolulu, USA.

Min, X.-J. and Bartholomew, D.P. (1997) Temperature affects ethylene metabolism and fruit initiation and size of pineapple. *Acta Horticulturae* 425, 329–338.

Neild, R.E. and Boshell, F. (1976) An agroclimatic procedure and survey of the pineapple production potential of Colombia. *Agricultural Meteorology* 17, 81–92.

Newett, S. and Rigden, P. (2015) *The Pineapple Problem Solver Field Guide*. Brisbane: Queensland Government.

Nightingale, G.T. (1942) Nitrate and carbohydrate reserves in relation to nitrogen nutrition of pineapple. *Botanical Gazette* 103, 409–456.

Norman, J.C. (1977) The effect of plant density on the growth, development and yield of pineapple (*Ananas comosus* (L.) Merr. cv. Smooth Cayenne) in Ghana. *Acta Horticulturae* 53, 349–354.

Norman, J.C. (1978) Responses of 'Sugarloaf' pineapple, *Ananas comosus* (L.) Merr to plant population densities. *Gartenbauwissenschaft* 43, 237–240.

Noy-Porat, T., Cohen, D., Mathew, D., Eshel, A., Kamenetsky, R. and Flaishman, M.A. (2013) Turned on by heat: differential expression of FT and LFY-like genes in *Narcissus tazetta* during floral transition. *Journal of Experimental Botany* 64, 3273–3284.

Okimoto, M.C. (1948) Anatomy and histology of the pineapple inflorescence and fruit. *Botanical Gazette* 110, 217–231.

Palmer, R.L., Lewis, L.N., Hield, H.Z. and Kumamoto, J. (1967) Abscission induced by β-hydroxyethylhydrazine: Conversion of β-hydroxyethylhydrazine to ethylene. *Nature* 216, 1216–1217.

Phukan, U.J., Jeena, G.S., Tripathi, V. and Shukla, R.K. (2017) Regulation of Apetala2/Ethylene Response Factors in plants. *Frontiers in Plant Science* 8, 150–150.

Pinon, A. (1981) Compte rendu des essais 355 et 356. Unpublished document, IRFA.

Py, C. (1964) Apercu sur le cycle de l'ananas en Martinique. *Fruits* 19, 133–139.

Py, C., Giacomelli, E.J., and Lossois, P. (1973) Essai densitè. Rèunion Annuelle IRFA, doc. interne No. 45.

Py, C., Lacoeuilhe, J.J. and Teisson, C. (1987) *The Pineapple. Cultivation and Uses*. Paris: Editions G.-P. Maisonneuve.

Rabie, E.C. and Mbatha, B.W. (2016) Evaluation of the efficacy of Eclipse® in reducing sunburn in 'Queen' pineapple of South Africa. *Acta Horticulturae* 1111, 241–248.

Rabie, E.C., Mbatha, B.W. and Tustin, H.A. (2011) The effect of aviglycine application rate and frequency on the inhibition of natural flowering of 'Queen' pineapple in South Africa. *Acta Horticulturae* 902, 281–290.

Rabie, E.C., Mbatha, B.W. and Tustin, H.A. (2013) The effect of plant size and spray volume on the efficacy of Aviglycine in the inhibition of natural flowering of 'Queen' pineapple (*Ananas comosus*) in South Africa. *Acta Horticulturae* 1007, 837–848.

Ranty, B., Aldon, D., Cotelle, V., Galau, J.P., Thubau, P. and Mazars, C. (2016) Calcium sensors as key hubs in plant responses to biotic and abiotic stresses. *Frontiers in Plant Science* 7(327). DOI: 10.3389/fpls.2016.00327.

Redwan, R.M., Saidin, A. and Kumar, S.V. (2016) The draft genome of MD-2 pineapple using hybrid error correction of long reads. *DNA Research* 23(5), 427–439.

Rodriquez, A.R. (1932) Influence of smoke and ethylene on the fruiting of pineapple (*Ananas sativus*) Schult. *Journal of Agriculture, University of Puerto Rico* 26, 5–18.

Rohrbach, K.G. and Taniguchi, G. (1984) Effects of temperature, moisture, and stage of inflorescence development on infection of pineapple *Ananas comosus* by *Penicillium funiculosum* and *Fusarium miniliforme* var. *subglutinans*. *Phytopathology* 74, 995–1000.

Samuels, G. (1969) Pineapple cultivars – 1970. *Proceedings of the Tropical Region American Society for Horticultural Science* 17, 13–24.

Sanewski, G.M. (2007) Skin russeting in the Pineapple Variety 73-50. *Pineapple News* 14, 9–12. Available at: http://www.ishs-horticulture.org/workinggroups/pineapple/PineNews14.pdf (accessed 21 March 2018).

Sanewski, G.M., Sinclair, E., Jobin-Décor, M. and Dahler, G. (1998) Preliminary studies into the effects of temperature on flower initiation of Smooth Cayenne in south-east Queensland. Abstract, *Third International Pineapple Symposium*. Bangkok, Thailand: Pattaya, Horticultural Research Institute.

Sanford, W.G. (1962) Plant crop results from recent planting density trials with Smooth Cayenne. Pineapple Research Institute of Hawaii *Research Report* 89. Library, University of Hawaii at Manoa, Honolulu, USA.

Sanford, W.G. (1964) Influence of the removal of one or more parts of the pineapple plant on the growth and yield of the remaining parts. *Pineapple Research Institute of Hawaii Research Report* 110.

Scott, C. (1992) The effect of plant density on spring plant crop/autumn first ratoon crop cycles; test number 718. *Pineapple Industry Farm Committee Pineapple Field Day Book*. Brisbane: Growcom, pp. 37–45.

Shrestha, R., Gomez-Ariza, J., Brambilla, V. and Fornara, F. (2014) Molecular control of seasonal flowering in rice, Arabidopsis and temperate cereals. *Annals of Botany* 114(7), 1445–1458.

Sideris, C.P. (1933) Physiology: plant defoliation studies. *Pineapple News* 7, 117–119.

Sideris, C.P. and Krauss, B.H. (1933a) Physiochemical variations in pineapple fruits and their importance in the quality of fresh and canned product. *Proceedings Hawaii Academy of Science, 8th Annual Meeting*, Honolulu, pp. 24–25.

Sideris, C.P. and Krauss, B.H. (1933b) Physiological studies on the factors influencing the quality of pineapple fruits. I. Physiochemical variations in the tissues of ripe pineapple fruits. *Pineapple Quarterly* 3, 82–98.

Sideris, C.P. and Krauss, B.H. (1938) Growth phenomena of pineapple fruits. *Growth* 2, 181–196.

Sideris, C.P., Krauss, B.H. and Young, H.Y. (1936) Effects of light intensity on the time of fruit harvest, yields and quality. *Pineapple Quarterly* 6, 56–66.

Silva, J.A., Hamasaki, R., Paull, R., Ogoshi, R., Bartholomew, D.P., Fukuda, S., Hue, N.V., Uehara, G. and Tsuji, G.Y. (2006) Lime, gypsum, and basaltic dust effects on the calcium nutrition and fruit quality of pineapple. *Acta Horticulturae* 702, 123–131.

Sinclair, E.R. (1991) Progress in ethrel flower induction studies by QDPI/CSIRO. *Pineapple Industry Farm Committee Pineapple Field Day Book*. Brisbane: Growcom, pp. 39–46.

Sinclair, E.R. (1992a) Seasonal growth rates in pineapple. *Pineapple Industry Farm Committee Pineapple Field Day Book*. Brisbane: Growcom, pp. 9–21.

Sinclair, E.R. (1992b) Yield of pineapple in relation to plant size and season. *Pineapple Industry Farm Committee Pineapple Field Day Book*. Brisbane: Growcom, pp. 23–29.

Sinclair, E.R. (1997) Weather summary for 1996/97. *Pineapple Industry Farm Committee Pineapple Field Day Book*. Brisbane: Growcom, pp. 29.

Sinclair, E.R. (1998) Weather summary for 1997/98. *Pineapple Industry Farm Committee Pineapple Field Day Book*. Brisbane: Growcom.

Sinclair, E.R. (1999) Weather summary for 1998/99. *Pineapple Industry Farm Committee Pineapple Field Day Book*. Brisbane: Growcom.

Soler, A. (1992) Metabolism de l'ethephon dans l'epiderme de l'ananas (*Ananas comosus* (L.) Merr.). *Fruits* 47, 471–477.

Soler, A., Teisson, C. Dole, B. and Marie-Alphonsine, P.-A. (2006) Forcing in pineapples: what is new? *Pineapple News* 13, 27–31. Available at: http://www.ishs-horticulture.org/workinggroups/pineapple/PineNews13.pdf (accessed 14 March 2018).

Souza, O.P.D., Zanini, J.R., Torres, J.L.R., Barreto, A.C. and Souza, E.L.C. (2013) Yield of pineapple juice and chemical quality in irrigation frequency and levels. *Bioscience Journal* 29, 1971–1980.

Sun, G.M. (2011) Pineapple production and research in China. *Acta Horticulturae* 902, 79–85.

Swete Kelly, D. and Bartholomew, D.P. (1993) Other disorders. In: Broadley, R.H. Wassman, R.C.I. and Sinclair, E. (eds) *Pineapple Pests and Disorders*. Brisbane: Queensland Department of Primary Industries, pp. 43–52.

Tan, K.M. and Wee, Y.C. (1973) Influence of size of pineapple slips on plant growth, fruit weight and quality in graded and mixed plantings. *Tropical Agriculture, Trinidad* 50, 139–142.

Tay, T.H. (1974) Effect of water on growth and nutrient uptake of pineapple. *MARDI Research Bulletin* 2, 31–49.

Teisson, C. (1972) Etude sur la floraison naturelle de l'ananas en Côte d'Ivoire. *Fruits* 27, 699–704.
Teisson, C. (1973) Essai ombrage du fruit. Etat nutritif et qualité du fruit. Réunion Annuelle IRFA, doc. interne No. 47 & 141.
Teisson, C. (1979a) Le developpment du fruit. Réunion Annuelle IRFA, doc. interne No. 47.
Teisson, C. (1979b) Essais 519 et 549: effets d'un ombrage excessif sur la qualité du fruit. Unpublished document.
Traub, H.P., Cooper, W.C. and Reece, P.C. (1940) Inducing flowering in the pineapple, *Ananas sativus*. *Proceedings of the American Society for Horticultural Science* 37, 521–525.
Treto, E., Gonzales, A. and Gomez, J.M. (1974) Etude de differentes densites de plantation chez la variete d'ananas Espanola Roja. *Fruits* 29, 279–284.
Trusov, Y. and Botella, J.R. (2006a) Delayed flowering in pineapples (*Ananas comosus* (L.) Merr.) caused by co-suppression of the ACACS2 gene. *Acta Horticulturae* 702, 29–36.
Trusov, Y. and Botella, J.R. (2006b) Silencing of the ACC synthase gene ACACS2 causes delayed flowering in pineapple [*Ananas comosus* (L.) Merr.] *Journal of Experimental Botany* 57(14), 3953–3960.
Tsai, H. and Kuan, C. (2013) Effect of aminoethoxyvinylglycine application on delaying natural flowering and fruit quality in 'Tainung No. 17' pineapple. *Journal of the Taiwan Society for Horticultural Science* 59, 183–190.
Tsai, H., Tang, C. and Kuan, C. (2015) Effect of planting time and flower forcing on fruit development and quality in pineapple 'Tainung No. 17'. *Crop, Environment & Bioinformatics* 12, 155–165.
Turnbull, C.G.N., Sinclair, E.R., Anderson, K.L., Nissen, R.J., Shorter, A.J. and Lanham, T.E. (1999) Routes of ethephon uptake in pineapple (*Ananas comosus*) and reasons for failure of flower induction. *Journal of Plant Growth Regulation* 18, 145–152.
Van de Poel, B., Ceusters, J. and De Proft, M.P. (2009) Determination of pineapple (*Ananas comosus*, MD-2 hybrid cultivar) plant maturity, the efficiency of flowering induction agents and the use of activated carbon. *Scientia Horticulturae* 120, 58–63.
Van Lelyveld, L.J. (1957) Pineapples. Sun scald-most dreaded enemy. *Farming in South Africa* 32, 28–32.
Van Overbeek, J. (1946) Control of flower formation and fruit size in the pineapple. *Botanical Gazette* 108, 64–73.
Van Overbeek, J. and Cruzado, H.J. (1948a) Note on flower formation in the pineapple induced by low night temperatures. *Plant Physiology* 23, 282–285.
Van Overbeek, J. and Cruzado, H.J. (1948b) Flower formation in the pineapple plant by geotropic stimulation. *American Journal of Botany* 35, 410–412.
Vilela, G.B., Pegoraro, R.F. and Maia, V.M. (2015) Predicting the production of 'Vitória' pineapple from phytotechnical and nutritional characteristics. *Revista Ciencia Agronomica* 46, 724–732.
Villalobos, M., Alfaro, K. Carvajal, C., Castillo, R., Kaiser, R., López, A., López, J. and Tolentino, P. (2013) RyzUp® 40SG delays fruit maturity and increases fruit weight in pineapple cv. MD-2 under Costa Rican growing conditions. *Pineapple News* 20, 34–37. Available at: http://www.ishs-horticulture.org/workinggroups/pineapple/PineNews20.pdf (accessed 14 March 2018).
Wassman, R.C. (1982) The importance of selected clones in pineapple production. In: *Pineapple Industry Farm Committee Pineapple Field Day Book*. Brisbane: Growcom.
Wassman, R.C. (1986) Sequential planting studies. Golden Circle Cannery Test 715. Brisbane: Golden Circle Ltd.
Wassman, R.C. (1990) Effects of seasonal temperature variations on pineapple scheduling for canning in Queensland. *Acta Horticulturae* 275, 131–138.
Wassman, R.C. (1991) The high cost of inadequate fruit induction. Summer forcing for spring harvest. *Pineapple Industry Farm Committee Pineapple Field Day Book*. Brisbane: Growcom, pp. 37–38.
Wee, Y.C. (1978) Natural flowering and fruiting of the Singapore Spanish pineapple in Johore, West Malaysia. *Journal of the Singapore National Academy of Science* 7, 9–14.
Wee, Y.C. and Ng, J.C. (1968) Some observations on the effect of month of planting on the Singapore Spanish variety of pineapple. *Malaysian Agricultural Journal* 46, 469–475.
Wee, Y.C. and Ng, J.C. (1970) Flower induction in pineapple cultivation. *Journal of the Singapore National Academy of Science* 2, 69–73.
Wee, Y.C. and Rao, A.N. (1977) Flowering response of Singapore Spanish pineapple to monthly applications of acetylene and NAA. *Malaysian Agricultural Journal* 51, 154–166.
Williams, D.D.F. (1987) History and development of fruit differentiation, growth and ripening control in pineapple. *Proceedings Plant Growth Regulation Society of America 14th Annual Meeting*, 413–422.

Yamane, G.M. and Ito, O. (1969) ACP 66–329, a potential forcing agent. Pineapple Research Institute of Hawaii *PRI News*, 17, 1–27. Library, University of Hawaii at Manoa, Honolulu, USA.

Yamane, G.M. and Ito, O. (1970a) Recent developments in Ethrel forcing. Pineapple Research Institute of Hawaii *PRI News* 18, 13–23. Library, University of Hawaii at Manoa, Honolulu, USA.

Yamane, G.M. and Ito, O. (1970b) Gibberellic acid-Ethrel effect on fruit quality. Pineapple Research Institute of Hawaii *PRI News* 18, 69–72.

Young, R. and Gonzales, J. (2010) Pineapple plant named 'DOLE-14'. United States patent number US PP20,885 P3.

Yow, Y.L. (1959) The time of maturity for the summer crop of pineapples in relation to climatic and cultural conditions. *Journal of the Agricultural Association of China* 27, 26–46.

Zeng, H., Xu, L., Singh, A., Wang, H., Du, L. and Poovaiah, B.W. (2015) Involvement of calmodulin and calmodulin-like proteins in plant responses to abiotic stresses. *Frontiers in Plant Science* 6, PMC4532166.

Zhang, J. (1992) Computer simulation of pineapple growth, development, and yield. PhD dissertation, University of Hawaii at Manoa, Honolulu, USA.

Zhang, J. and Bartholomew, D.P. (1993) Simulation of pineapple growth, development, and yield. *Acta Horticulturae* 334, 205–219.

Zhang, J. and Bartholomew, D.P. (1997) Effect of plant population density on growth and dry matter partitioning of pineapple. *Acta Horticulturae* 425, 363–376.

Zhang, J., Malézieux, E. and Bartholomew, D.P. (1997) ALOHA-Pineapple v.2.1: a computer model to predict the growth, development and yield of pineapple. *Acta Horticulturae* 425, 287–296.

Zhu, J. (1996) Physiological responses of pineapple (*Ananas comosus* (L.) Merr.) to CO_2 enrichment, temperatures and water deficit. PhD dissertation, University of Hawaii at Manoa, Honolulu, USA.

Zhu, J., Goldstein, G. and Bartholomew, D. (1999) Gas exchange and carbon isotope composition of *Ananas comosus* in response to elevated CO_2 and temperature. *Plant Cell & Environment* 22, 999–1007.

12 Pests, Diseases and Weeds*

Brent Sipes[1] and Aristoteles Pires de Matos[2]

[1]University of Hawaii at Manoa, Honolulu, USA; [2]Embrapa Mandioca e Fruticultura (Embrapa Cassava and Fruits), Cruz das Almas, Brazil

Introduction

Pineapple, like most cultivated plants, is susceptible to a variety of pests and pathogens. However, pineapple has relatively fewer pest and disease problems than many of the major crops like rice, wheat, maize or soybean (Rohrbach and Apt, 1993) (Table 12.1). The limited number of serious pests and pathogens could be related to pineapple's xeric nature. The pineapple's tough leaves and ability to move water on the leaves to the base are preformed defences that deter many pests and pathogens.

The major pests and disease of pineapple are commonly encountered wherever pineapple is cultivated commercially. Many of the pests and diseases of pineapple are widely distributed across all areas of cultivation. Pineapple mealybug wilt-associated virus and mealybugs are easily transported on planting material and their distributions reflect the exchange of shoot material for planting. Soil-borne diseases like *Phytophthora* rots and plant-parasitic nematodes are ubiquitous in agricultural fields and have wide host ranges that include pineapple. Arthropod pests such as symphylids, ants, scales and fruit-piercing moths occur worldwide in pineapple but the species depends upon the locality. Some pests and diseases have limited distributions but are locally devastating such as fusariosis and bacterial heart rot. Soil-borne pests and diseases show more location-specific distributions because crowns and slips used as planting material are less likely to be contaminated with soil-borne organisms. Several pests, such as bud moths, and diseases, such as lethal yellows, that are encountered in pineapple are opportunistic as pineapple is not a preferred host. The pests and pathogens feed on or infect pineapple because of a disturbance in the preferred host that is grown in proximity to the pineapple.

The ubiquitous nature of so many pests and diseases of pineapple is probably reflective of our vegetative propagation of the crop and our establishment of new plantings. Moving large volumes of planting material from one commercial operation to newly established plantings carries the risk of also unintentionally bringing pests and pathogens that contaminate the planting material. Equipment moved from old fields to new plantings may carry soil that can introduce soil-borne pests and pathogens. Rigorous inspections and strong quarantines can limit the introduction of pests and pathogens into new plantings. As an example, bacterial heart rot was not introduced in to Hawaii for nearly 100 years because of inspections, quarantine and phytosanitary measures (Sueno *et al.*, 2014).

Many approaches have been taken in presenting the pests and pathogens of pineapple. Some authors follow a taxonomic approach (Collins, 1960; Py *et al.*, 1987). In the first edition of this book, Rohrbach and Johnson

* A revision of the chapter by Kenneth G. Rohrbach and Marshall W. Johnson. University of Hawaii, USA.

Table 12.1. Pineapple pests, diseases and weeds, with their distribution in commercial pineapple production areas.

Pest, disease or weed	Distribution	Notes
Viral diseases		
Mealybug Wilt associated Virus	Worldwide	
Tomato Spotted Wilt Virus	Worldwide	
Bacterial diseases		
Bacterial heart rot	Limited	Central America, Philippines, Hawaii, Malaysia
Marbling disease	Worldwide	
Pink disease	Worldwide	
Stramenopile diseases		
Phytophthora root and heart rot	Worldwide	
Pythium root rot	Worldwide	
Fungal diseases		
Butt and black rot	Worldwide	
Fruitlet core rot	Worldwide	
Fusariosis	Limited	Most severe in Brazil
White leaf spot	Worldwide	
Yeasty fermentation	Worldwide	
Nematodes		
Lesion nematode	Common	Most severe in Africa
Reniform nematode	Common	Most severe in Hawaii, Asia and the Caribbean
Root-knot nematode	Worldwide	
Arthropods		
Symphylids	Worldwide	
Red mites	Worldwide	
Insects		
Ants	Worldwide	Species vary by location
Bud moths	Common	
Fruit flies	Limited	Reported in Paraguay and Peru
Fruit-piercing moths	Common	Species vary by location
Grubs	Worldwide	Species vary by location
Mealybugs	Worldwide	
Scales	Worldwide	Species vary by location
Souring beetles	Worldwide	
Weeds		
Broadleaf	Worldwide	Species vary by location
Grass	Worldwide	Species vary by location

presented the pests and pathogens of pineapple as encountered during the crop cycle. Each approach has its merits and shortcomings. Rohrbach and Schmitt (2003) presented the diseases of pineapple based upon the plant tissue infected. This chapter will present the pests and pathogens based upon the portion of the pineapple plant attacked – roots, shoots and fruit. Weeds, a serious constraint to yield, will be treated as a soil-borne problem, since their control depends on pre-plant control like many soil-borne pests and pathogens. The chapter will conclude with management of the pests and pathogens common to pineapple production.

Soil-borne Pests and Pathogens of the Roots

Bacterial heart rot

Bacterial heart rot of pineapple, caused by *Dickeya* sp. (Marrero *et al.*, 2013; Sueno *et al.*, 2014), has a limited distribution and is found only in Brazil, Costa Rica, Hawaii, Malaysia and the Philippines. Infected pineapple has water-soaked areas on the young leaves in the centre whorl that form brown streaks and blister-like lesions (Johnston, 1957; Kaneshiro *et al.*, 2008) (Fig. 12.1). Within a few days of infection by the bacteria, the pineapple meristem is dead. Juice of infected collapsed fruit constitutes an important sources of inoculum. Ants, wind and wind-blown rain introduce the bacteria into the stomata of nearby pineapple plants (Lim, 1985). Injury from mite feeding or chemical burns can also be an entry point for the bacteria (Rohrbach and Johnson, 2003). The disease has spread from Malaysia to Central and South America, the Philippines, and Hawaii through the exchange of infected planting material (Melo *et al.*, 1974; Chinchilla *et al.*, 1979; Rohrbach and Johnson, 2003; Kaneshiro *et al.*, 2008). Losses to bacterial heart rot vary by pineapple cultivar with 'Josapine' and 'MD-2' showing high susceptibility (Ramachandran *et al.*, 2015), 'Smooth Cayenne' being more resistant than the Spanish type cultivars (Rohrbach and Johnson, 2003), and 'CO-2' apparently highly resistant (A. Alvarez, personal communication).

Fig. 12.1. Bacterial heart rot of pineapple caused by *Dickeya* showing late plant symptoms. Photos R. Lucas.

Phytophthora root and heart rot

Root and heart rots are caused by several Stramenopiles (Oomycetes) including *Phytophthora nicotianae*, *P. cinnamomi*, and *P. palmivora* (Rohrbach and Schmitt, 2003). *Phytophthora cinnamomi* and *P. nicotianae* cause root rots. *Pythium* spp. are also causal agents of root rot in pineapple. This disease is usually severe in poorly drained soils, at low temperatures, during prolonged wet seasons, as well as in soils with a pH higher than about 5.5 (Py *et al.*, 1987). Under such conditions, Phytophthora root rot could be a very severe disease and may lead to a complete loss of crops. Additionally, root rot incidence varies according to the cultivar. The cultivar MD-2 is highly susceptible to *Phytophthora cinnamomi* (Chan *et al.*, 2003; Sanewski *et al.*, 2016). Infection occurs from zoospores, a swimming spore produced by sporangia. The zoospores are attracted by root exudates, reach the root surface, germinate and penetrate the root tissue inciting rot development. Root rots destroy the pineapple roots thus resulting in symptoms typical of drought and nutrient deficiency including a reddening of the pineapple leaves with yellowing and necrosis at the leaf margins. Plant anchorage becomes poor following the loss of roots. Heart rots caused by these stramenopiles (Fig. 12.2) are associated with wet environmental conditions and losses may reach 100% under severe disease pressure (Rohrbach and Schmitt, 2003). Heart rot was first reported in Hawaii but currently occurs in most of the pineapple growing areas around the world. *Phytophthora* spp. propagules such as chlamydospores, oospores and zoospores, among others, may be disseminated by water runoff, as well as

Fig. 12.2. *Phytophthora cinnamomi* heart rot of 'MD-2' pineapple. Photo G. Sanewski.

Fig. 12.3. Fungal butt rot of pineapple caused by *Chalara paradoxa* (previously *Ceratocystis paradoxa* and *Thielaviopsis paradosa*); Photo from Cooke, T., Persley, D and House, S. (2009). *Diseases of Fruit Crops in Australia*. CSIRO Publishing.

in soil attached to machinery, tools and worker footwear. Infection occurs from zoospores that may be splashed up into pineapple heart where the zoospores cause rot in the basal white tissues. In infected plants, the new leaves in the centre of the whorl fail to elongate and the infection quickly reaches the point where the leaves are easily pulled from the plant. The growing point of the pineapple simply rots away. If no control measures are implemented, Phytophthora heart rot may significantly reduce pineapple yield and may lead to a complete crop failure (Green and Nelson, 2015).

Fungal butt and black rot

The causal fungus for butt and black rot is *Ceratocystis paradoxa*. The fungus is widespread in the tropics and infects many other tropical crop in addition to pineapple (Rohrbach and Johnson, 2003; Hubert *et al.*, 2014). Butt rot is associated with a soft rot and blackening of the basal portion of the pineapple stem tissue (Fig. 12.3b). The bottom portion or the butt of pineapple plants become soft and black with infection by *C. paradoxa*. Butt rot is most serious on pineapple planting material (Rohrbach, 1983). Under moist conditions, such as that found in piles of planting material, an infection may lead to rot of the entire planting piece or the entire planting material pile. The growth of moderately infected planting material is slow and plants will be stunted due to loss of stem tissue that contains carbohydrate reserves and the initial roots (Rohrbach and Johnson, 2003). Since *C. paradoxa* resides in the soil, when uncured planting material is planted in fields with high inoculum levels, butt rot incidence may reach 100%.

Nematodes

Over 100 species of plant-parasitic nematodes have been associated with pineapple. The most damaging species of nematodes are *Rotylenchulus reniformis*, *Meloidogyne* spp., *Pratylenchus brachyurus*, and the spiral nematodes (*Helicotylenchus*, *Scutellonema*, and *Rotylenchus* spp.). Nematode infection of pineapple is often cryptic and should be suspected if plant growth is poor in localized regions or symptoms of stress are observed on the leaves in spite of satisfactory climatic and agronomic conditions. In some cases, simple observation of the roots may permit diagnosis of nematode infection, but soil and root sampling followed by laboratory assay is usually required to diagnose nematode infections in pineapple.

The reniform nematode, *R. reniformis*, is the most economically damaging nematode on pineapple in Hawaii (Caswell and Apt, 1989; Sipes and Schmitt, 2000), Asia and the Caribbean (Zem *et al.*, 1980) (Figs 12.4 and 12.5). The related *R. parvus* has been observed on pineapple in southern Africa but has not been associated with economic loss. *R. reniformis* is able to survive in an anhydrobiotic stage for at least 2 years in dry soil making management challenging (Radewald and Takeshita, 1964; Stoyanov, 1967; Tsai, 1978). *R. reniformis* can

Fig. 12.4. Reduced growth of pineapple in the field due to infection by *Rotylenchulus reniformis*. Photo Duane P. Bartholomew.

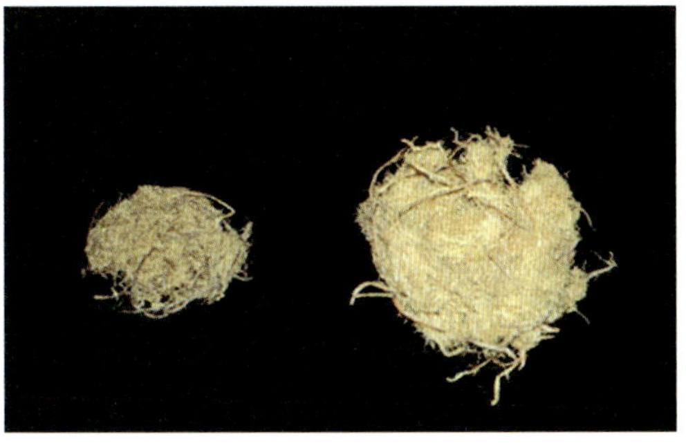

Fig. 12.5. Reduced root growth of pineapple caused by *Rotylenchulus reniformis* (left, root mass from infected plant; right, from healthy). Photo Brent Sipes.

reduce marketable pineapple yields by up to 38% and 60% during the first and second harvest, respectively (Sipes and Schmitt, 1996).

Root-knot nematodes, *Meloidogyne* spp., are distributed worldwide and have been reported as damaging to pineapple in Australia, Puerto Rico, Mexico, South Africa, Zimbabwe, the Ivory Coast, Hawaii, Thailand and some areas of the Philippines (Rohrbach and Apt, 1986; Sipes *et al.*, 2005). *Meloidogyne javanica* and *M. incognita* are the main species associated with pineapple. Infection by *Meloidogyne* slows root growth and usually results in terminal club-shaped galling (Godfrey and Oliveira, 1932) (Fig. 12.6). Brooming is not common but may occur if multiple non-terminal galls form (Godfrey, 1936). As with many soil-borne pests, root-knot nematodes stunt the pineapple root system resulting in poor anchorage and plants are more susceptible to moisture and nutrient stress. Damage thresholds are low (1 nematode/cm^3 soil) (Keetch, 1982; Sipes and Schmitt, 2000; Stirling and Kopittke, 2000). These low damage thresholds not only make control difficult but imperative if root-knot nematodes are found in a field.

Lesion nematode, *Pratylenchus brachyurus*, is most common on pineapple grown in low latitude tropical regions (Rohrbach and Schmitt, 2003; Sipes *et al.*, 2005). Lesion nematode is a migratory endoparasitic nematode that destroys the root cortical tissue and secondary roots. Pineapple growth is stunted and shoots become chlorotic due to the loss of functional roots associated with infection by lesion nematodes.

Spiral nematodes (*Helicotylenchus*, *Scutellonema* and *Rotylenchus*) are also commonly associated with pineapple (Sipes *et al.*, 2005). *Helicotylenchus dihystera* has only been associated with damage to pineapple in greenhouse

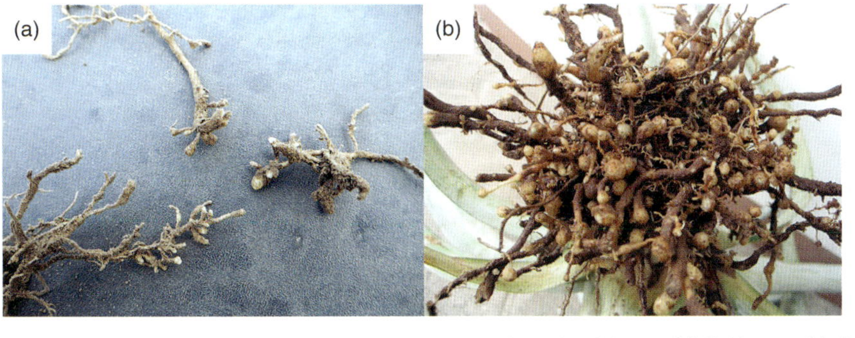

Fig. 12.6. Pineapple root 'clubbing' caused by *Meloidogyne javanica*. Photos: (a) S. Newett; (b) Col Scott.

experiments but not in the field (Ko and Schmitt, 1993). In South Africa, *Helicotylenchus*, *Scutellonema* and *Rotylenchus* have been reported to cause damage to pineapple (Keetch and Purdon, 1979). In general, spiral nematodes do not cause substantial injury to pineapple unless the population densities are in the thousands per 250 cm^3 soil.

Symphylids

Symphylids are soil arthropods distantly related to insects (Shear and Edgecombe, 2010). Adult symphylids measure 5–8 mm long, with a white to yellowish colour and a translucent integument (Fig. 12.7b). Symphylids spend their entire life in the soil feeding on microorganisms, plants, animals, and organic matter. Several species of symphylids are associated with pineapple including *Hanseniella unguiculata*, *H. ivorensis*, *Scutigerella sakimurai* and *Symphylella tenella* (Py *et al.*, 1987; Waite, 1993). Symphylids feed on the young roots and root hairs of pineapple and cause serious damage (Py *et al.*, 1987) often resulting in a 'witch's broom' or clubbing symptom of the roots (Rohrbach and Johnson, 2003). Symphylid feeding causes plants to be stunted, poorly anchored and lodge in the field (Rohrbach and Johnson, 2003). Symphylid feeding limits the plants uptake of nutrients and water, as well as causing wounds that allow entry of pathogens (Sakimura, 1966) (Fig. 12.7a). The fruit of affected plants are small and do not ripen synchronously thus leading to yield losses, especially when soil moisture is limited (Kehe, 1979). With the widespread use of soil fumigants, deep tillage, and long-lived halogenated hydrocarbon pesticides in the 1960s and 1970s, symphylids were of little concern. Commercial practices have now changed with a greater emphasis on less tillage and more consideration for the environment. The more effective, persistent pesticides and inexpensive fumigants are no longer available and symphylid damage has become more of a concern (G. Taniguchi, personal communication).

Mealybugs

Mealybugs are serious pests of pineapple because they are associated with mealybug wilt. *Dysmicoccus brevipes*, the pink mealybug, commonly feeds on the roots, leaves, stems, fruit, and crowns of pineapple (Jahn and Beardsley, 2000). The root feeding by *D. brevipes* is generally not of concern rather its role in mealybug wilt of Pineapple. Mealybugs are discussed further under mealybug wilt later in this chapter.

Grubs and other insects

The grubs or larvae of many scarab beetles can feed on pineapple roots (Fig. 12.8). Grub feeding can lead to stunting, wilting and yellowing of the plant (Petty, 1978; Smith *et al.*, 1995). Grub species vary by location. In Australia the Christmas beetle, *Anoplognathus porosus*, as well as many species of cane grub (*Antitrogus mussoni*, *Rhopaea magnicornis*, *Lepidiota squamulata*, *Lepidiota noxia* and *Lepidiota gibbifrons*) have been reported (Waite, 1993). Grubs found on pineapple roots in South Africa include *Asthenopholis*

Fig. 12.7. Pineapple (cultivar CO-2) showing field symptoms of damage from symphylid feeding (a) and a close-up of a symphylid (b). Photos S. Newett.

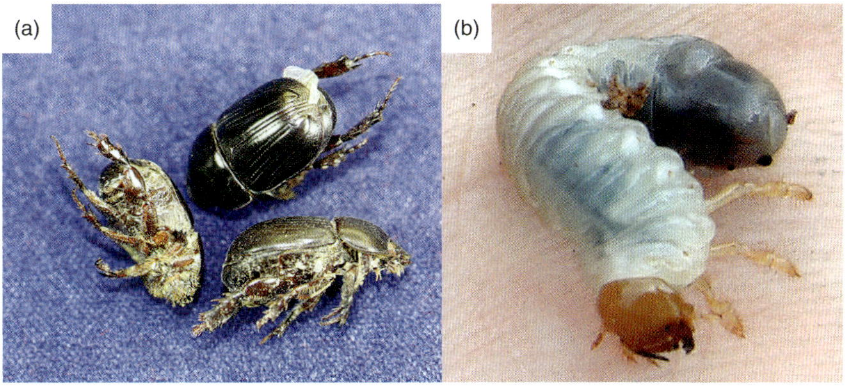

Fig. 12.8. Species of cane beetle and larvae found in pineapple fields in Australia. Photos: (a) G. Sanewski; (b) S. Newett.

subfaciata, *Adoretus ictericus*, *Trochalus politus*, and *Macrophylla ciliata* (Smith *et al.*, 1995). The Chinese rose beetle, *Adoretus sinicus*, and the anomala beetle, *Anomala orientalis* feed on pineapple roots in Hawaii (Carter, 1967). In India, the black spot beetle, *Metamasius dimidiatipennis*, and the fig beetle, *Cotinis mutabilis*, also damage pineapple roots (Camino-Lavín *et al.*, 1996). While grub feeding can damage pineapple, grubs are generally not a major pest but are only problematic when pineapple is planted into pastures, near sugar cane fields, or abandoned fields (Carrillo, 2011).

Weeds

Weeds are serious problems in the establishment and early growth of pineapple (Rohrbach and Johnson, 2003; Eshetu *et al.*, 2007). Weeds are problematic because they compete with the newly planted pineapple for light, water, and nutrients. Pineapple grows slowly and can be shallowly rooted without proper setting and irrigation (Reinhardt *et al.*, 2002).These factors aggravate the competition from weeds. Depending on the severity of the weed problem, grasses, broadleaf species, or even volunteer pineapple plants may affect the crop yield. Once the pineapple closes canopy, weeds are less of a problem. Additionally, weeds growing in or near pineapple fields can serve as alternate hosts to insects and pathogens of pineapple, making weed control very important.

Weed species vary by location and are not as common across all production areas as are some diseases and pests. As an illustrative example, 100 species of weeds were identified in a pineapple field in Côte d'Ivoire (Mangara *et al.*, 2016). Weeds in the *Poaceae*, *Astereaceae*, *Euphorbiaceae* and *Cyperaceae* predominated. Similar weed situations occur in other growing areas with weed species that are endemic to the area. *Cyperus* sp., *Digitaria* sp. and *Bidens pilosa* are common weeds in pineapple grown commercially throughout the world. These plants demonstrate better nutrient extraction from the soil as compared to pineapple (Catunda *et al.*, 2005; Maia *et al.*, 2012) and consequently are very important to control.

Manual weed control is difficult and an expensive activity to be undertaken in large scale commercial pineapple production. Workers must be protected from any spines on the leaves. Furthermore, manual weeding can also damage the plant and provide wounds that allow white tip to develop.

Pests and Pathogens of the Shoot

Yellow spot

Yellow spot of pineapple occurs in all production areas. Yellow spot is caused by the Tomato Spotted Wilt Virus (TSWV). TSWV is in the genus *Tospovirus* in the Bunyaviridae. TSWV is a negative sense RNA virus that is vectored by infected thrips. The thrips can become infected with the TSWV during their larval stage and transmit the virus for life. Pineapple is not a preferred host for thrips, but when thrips move into pineapple fields from adjacent weeds or fields transmission occurs inadvertently. Transmission of the virus to the pineapple plants occurs during this period as the thrips search for new host plants (Rohrbach and Johnson, 2003). Initial infection by TSWV results in a slightly raised yellowish spot with a darkened centre. Secondary spots arise and progress to a basil leaf and stem rot. Infection usually occurs on young pineapple plants but mature plants and their fruit and crown may also become infected (Linford, 1943; D.P. Bartholomew, 2017, Hawaii, personal observation). TSWV infection always results in the death of the pineapple plant (Fig. 12.9). Pineapple yellow spot is managed by controlling weeds in and surrounding pineapple fields. Eliminating weeds that serve as TSWV reservoirs and are hosts to thrips reduces inoculum and vector presence, thus protecting the pineapple crop (Rohrbach and Johnson, 2003).

Fusariosis

The fungus *Fusarium guttiforme* incites fusariosis, the most serious constraint of the Smooth Cayenne in Brazil (Kimati and Tokeshi, 1964). Fusariosis also occurs on 'Red Spanish' clones. *Fusarium guttiforme* is a member of the *Gibberella fujikuroi* species complex. The fungus infects approximately 40% of the planting material and kills about 20% of the pineapple plants prior to harvest. The causal agent has been reported in Bolivia (Matos *et al.*, 1992), Chile (Montealegre and Luchsinger, 1990), Cuba (Borras *et al.*, 2001), Hawaii (Rohrbach, 1994), and Martinique (Berasconi, 1994).

Symptoms incited by *F. guttiforme* are characterized by alterations such as a bent stem, usually to the side where the lesion is located; disrupted phyllotaxy; shortened stem and leaves; reduced overall plant development; death of the apical meristem and chlorosis, and exudation of a gum like substance (Matos, 1995). Infection of stem and base of leaves reduces water flow. The pineapple leaves turn red, and the plant ceases growth, wilts and dies (Fig. 12.10). Symptoms are similar in Brazil and

Fig. 12.9. Yellow spot of pineapple caused by the Tomato Spotted Wilt virus. Symptoms are shown for the plant (a), internal fruit of cv. '73-50' (b), crown of cv. '73-50' (c) and fruit external (d). Photos A and D are from Cooke, T., Persley, D and House, S. (2009). Diseases of Fruit Crops in Australia. CSIRO Publishing); B and C are by G. Sanewski.

Bolivia but are different in Cuba, Martinique and Hawaii. Fusariosis has been reported in pineapple fruits marketed in Chile (Montealegre and Luchsinger, 1990). In Malaysia, fusariosis is caused by *F. semitectum* and *F. chlamidosporum* (Ibrahim *et al.*, 2016).

Infection of slips, suckers and crowns is hypothesized to occur through natural growth cracks, insect injuries or wounds due to development of lateral buds and adventitious roots. Cultivars differ in susceptibility and evidence suggest it may be related to the scutiform

Fig. 12.10. Symptoms incited by *Fusarium guttiforme* in pineapple plants: (a) open heart; (b) chlorosis; (c) disrupted phyllotaxy and shortened leaves; (d) reddened leaves and ceased growth; (e) wilt; (f) lesion on stem and leaf base. Photos A.P. de Matos.

trichomes on pineapple leaves (Aquije *et al.*, 2014). 'Pérola' and 'Smooth Cayenne' are susceptible and have more trichomes than 'BRS Vitoria' that is resistant and has fewer trichomes on its leaves. Incidence and severity of fusariosis decreases as the time interval from wounding to inoculation increases, and no infection occurs 8 days after wounding. Decreased susceptibility in 'Smooth Cayenne' is associated with development of a wound periderm that contains lignin, suberin, callose and phenolic compounds (Matos and Mourichon, 1993). *F. guttiforme* survives on suckers, slips and crowns that are infected while still attached to the mother plant. Infected planting material can easily be brought into a field, thus constituting the initial inoculum for the new planting area. The survival of *F. guttiforme* in buried crop debris decreases gradually but does not reach the lowest level until 8 months after the crop is removed. Once introduced into an area, fusariosis is disseminated to new plants by rain splash, wind (Matos *et al.*, 1981) and insects (Aguilar and Sanches, 1982; Chalfoun and Cunha, 1984; Costa and Lordelo, 1988). Infected and susceptible healthy plants can be found in a same area at same time thus favouring epidemics.

White leaf spot

The causal fungus for white leaf spot is *Ceratocystis paradoxa*, which also causes butt rot. The fungal anamorph is *Chalara paradoxa* (previously named *Thielaviopsis paradoxa*). White leaf spot occurs on the tips of the pineapple shoots and is largely cosmetic (Fig. 12.11). Injury from wind or even movement in the row damages the leaves and predisposes the pineapple leaf to infection by *C. paradoxa* if accompanied by showery weather.

Fig. 12.11. White leaf spot in 'Smooth Cayenne'. Photo G. Sanewski.

Other fungal pathogens

Several other fungi cause minor or localized problems in pineapple. Anthracnose on pineapple is found worldwide. The causal agents associated with anthracnose on pineapple are *Colletorichum ananas* and *C. gloeosprioides* (Luo *et al.*, 2010). *Cochliobolus eragrostidis* and *Exserohilum rostratum* have also been reported to cause leaf spots (Luo *et al.*, 2012). These fungi are cosmopolitan and can be found worldwide. Fortunately, while the pathogen may be present in a pineapple production area, the environmental conditions needed to incite disease development may not be present or favourable, thus substantial disease does not occur except in certain areas.

Mealybug wilt of pineapple

Mealybug wilt is among the most serious and devastating diseases of pineapple because it is found in all pineapple production areas (Rohrbach and Johnson, 2003). Mealybug wilt symptoms include severe tip dieback, reddening of leaves, and leaf wilting that can lead to total collapse of mature plants (Carter, 1933) (Fig. 12.12). It is possible for plants to recover from wilt symptoms if the mealybugs are controlled.

Mealybug wilt is a complex and multicomponent disease involving mealybugs, ants, and a virus. Initially, ants were thought responsible but Illingworth (1931) associated mealybugs with wilting of pineapple plants (Jahn *et al.*, 2003). The mealybugs on pineapple are tended by the ants. The ants feed on the mealybug honeydew, protect the mealybugs from natural enemies, and transport the mealybugs to new areas on the same plant or to nearby plants. A closterovirus from infected pineapple plants was identified over 50 years after mealybugs were associated with wilt (Gunasinghe and German,

Fig. 12.12. Plants with the reddened and wilted leaves symptom of mealybug wilt of pineapple. Photos: (a) J. Hu; (b) Cooke, T., Persley, D and House, S. (2009). Diseases of Fruit Crops in Australia. CSIRO Publishing.

1989) and named pineapple mealybug wilt-associated virus (PMWaV). An additional decade passed before mealybugs were confirmed to transmit the virus (Sether *et al.*, 1998). In addition to the virus and the mealybugs, ants increase the virus transmission rate (Sether and Hu, 1998). Mealybug feeding alone, regardless of the numbers of mealybugs, will not cause mealybug wilt. Infection with PMWaV-2, one of several related viruses found in pineapple, is required for mealybug wilt to develop (Sether and Hu, 2002). Yield loss from PMWaV infected plants can be up to 13.4% under drought or reduced irrigation conditions even when no wilt symptoms were present (Sether and Hu, 2001). Infection by PMWaV-1, one of the associated viruses, increased the number of early, asynchronously maturing fruit by 30% in the plant crop and reduced average fruit weight in the ratoon crop by 10% (Sipes *et al.*, 2002b).

Mealybug wilt-associated viruses

The causal agent of mealybug wilt of pineapple is the pineapple mealybug wilt-associated virus (PMWaV) (Sether *et al.*, 2005). Several related closteroviruses (PMWaV-1, PMWaV-2, PMWaV-3, etc.) have been identified from symptomatic pineapple plants (Sether and Hu, 2002; Sether *et al.*, 2005; Gambley *et al.*, 2008; Dey *et al.*, 2015). These viruses are in the genus *Ampelovirus*. All PMWaVs are transmitted semi-persistently by mealybugs. Presence of the virus alone does not result in mealybug wilt. Mealybug feeding is necessary for the wilt to develop and removal of the mealybug allows the pineapple to recover (Sether *et al.*, 2001).

Mealybugs

The most common insects of pineapple are mealybugs. The most important species of mealybugs on pineapple are *Dysmicoccus brevipes* and *D. grassii*, the pink and grey mealybugs, respectively. Both mealybug species vector PMWaV (Sether and Hu, 1998). Mealybug feeding, in addition to transferring of the virus, is required for the development of mealybug wilt of pineapple (Sether and Hu, 2002). The mealybugs are believed to inject an agent while feeding that suppresses the pineapple plant's tolerance to the virus (Sether and Hu, 2002). Thus, control of the mealybugs can result in the recovery of plants suffering from MWP (Hu, unpublished).

Pink pineapple mealybugs (Fig. 12.13) commonly feed on the roots, leaves, stems, fruit, and crowns of pineapple, whereas grey pineapple mealybugs infest only the aerial roots, stems, fruit and crowns of pineapple (Jahn and Beardsley, 2000). Where they enter and feed in blossum cups, a brown spot will be seen in the fruit flesh beneath. The mealybugs have relatively wide host ranges with the pink pineapple mealybugs feeding on banana, sugarcane, rhodes grass, natal grass, guinea grass, common sand

Fig. 12.13. Pink mealybug on pineapple fruit. Photo G. Sanewski.

burr and vasey grass (Beardsley, 1959). Grey pineapple mealybugs have a slightly narrower host range. Other mealybug species reported on pineapple include *Ferrisia virgata*, *Planococcus citri*, *Planococcus minor*, *Pseudococcus viburni* and *Praelongorthezia praelonga*. The wide host range of mealybugs makes their management challenging and weed control imperative in commercial pineapple plantings.

Ants

Ants are important pests in pineapple not because they damage the pineapple directly but because the ants are associated with mealybugs and pineapple mealybug wilt. At least 28 different species of ants tend mealybugs on pineapples (Jahn *et al.*, 2003; Rohrbach and Johnson, 2003). *Pheidole* and *Solenopsis* are the genera most commonly associated with pineapple mealybugs throughout the world (Jahn and Beardsley, 2000). The ants protect mealybugs from their natural enemies (Rohrbach and Mau, 2005). Ants remove honeydew from mealybugs, thus preventing fungi from attacking mealybugs (Rohrbach and Johnson, 2003). In addition, the ants transfer mealybugs on a plant and from plant to plant, that may result in PMWaV transfer from one plant to another plant (Illingworth, 1931; Rohrbach and Schmitt, 2003).

Bud moths

The immature stages of several moths and butterflies (Lepidoptera) attack pineapple. The larvae, or caterpillars, feed on the stems, buds and fruit by boring or chewing into the pineapple (Rohrbach and Johnson, 2003; Vorsino *et al.*, 2005). At high population densities, *Strymon megarus* may attack crowns and slips (Sanches, 1999; Lacerda *et al.*, 2007). *Porthesia taiwana*, the Taiwan tussock moth, was recorded on pineapple in Taiwan (Huang and Lin, 2014). *Opogona sacchari*, generally a sugarcane pest, has attacked low-acid pineapple cultivar planting material and fruit in Hawaii (Vorsino *et al.*, 2005). The larvae of *O. sacchari* have even destroyed newly planted pineapple slips (Wright, 2012). *Opogona*, *Elaphria* and *Porthesia*, appear to be pests of low-acid hybrid pineapple. The fruit deformation and gummosis resulting from Lepidopteran larval feeding can lead to fruit rejection in the packing house and loss of marketable fruit.

Pests and Pathogens of the Fruit

Pink disease

Pink disease is caused by the bacteria *Tatumella citrea*. Pink disease is only a problem in canned pineapple (Fig. 12.14) (Cha *et al.*, 1997). Pineapples sold fresh market are unaffected by the disease because the bacteria are undetectable until the fruit is subjected to the high temperatures of canning. Pink disease is most likely vectored to open pineapple flowers by insects or

Fig. 12.14. Pink disease of processed pineapple. Healthy (left) and affected (right). Photo from Cooke, T., Persley, D and House, S. (2009). *Diseases of Fruit Crops in Australia*. CSIRO Publishing.

mites (Kado, 2003; Rohrbach and Johnson, 2003), therefore, management of insects and mites reduces the frequency of pink disease in canned fruit (Rohrbach and Pfeiffer, 1976b).

Marbling disease

The acetic acid bacteria *Acetobacter peroxydans* and *Pantoea ananatis* cause marbling disease in pineapple. Marbling disease is sporadic in the hot, humid lowland tropics. Symptoms include a brown, granular appearance in the fruit (Fig. 12.15). The bacteria are ubiquitously present on the plant and infection by bacteria occurs through the open flowers. Once in the flower, the bacteria remain latent until approximately 1 month before fruit maturity (Rohrbach, 1989). Management is incumbent upon limiting bacterial introduction into the flower by insects, wind or splash.

Fig. 12.15. Marbling disease of pineapple. Photo G. Sanewski.

Fusariosis

On fruits, *F. guttiforme* incites a soft rot of the flesh. Gum accumulates in the locules of the ovary and gum exudation from the infected fruitlet appears. Open pineapple flowers are the main infection sites (Bolkan *et al*., 1979). Fusariosis is more severe during the early stages of flowering (Matos and Souto, 1985). As the disease progresses on the fruit, infected fruitlets change to a reddish brown colour and become sunken (Fig. 12.16). Under favourable environmental conditions, a pinkish fungal sporulation is evident around infected fruitlets (Matos, 1995). Incidence of fusariosis in fruits varies according to the time of harvest. As the number of hours below 23°C during flowering increases with high rainfall from forcing to harvest, the higher the disease incidence on fruits. On the other hand, increasing the number of hours of temperature over 28°C decreases fusariosis incidence on pineapple fruits (Matos *et al*., 2000). The seasonal effect makes possible to establish a forcing programme that enables pineapple fruit development and harvest under unfavourable conditions for disease development. In Malaysia, *F. semitectum* and *F. chlamidosporum* also incite fusariosis on fruit (Ibrahim *et al*., 2016). In South Africa and in China, a pineapple fruit rot has been associated with *Fusarium ananatum* (Jacobs *et al*., 2010; Gu *et al*., 2015).

Black rot, soft rot, water blister and water rot

A black rot due to infection by *Ceratocystis paradoxa* may form in harvested fruit maintained at

Fig. 12.16. Pineapple fruit infected by *Fusarium guttiforme* showing external and internal symptoms: (a) gum exudation; (b) sunken fruitlets; (c) rot of the pulp. Photos A.P. de Matos.

room temperature for more than a few days. Infection takes place through stem cutting and wounds on the fruit surface generated during harvest, picking and packing. Usually, the disease will develop 3–4 days after harvest. Infected fruit initially shows a dark yellow soft rot of the flesh that turns to black with disease progression. When the disease is fully developed, the pineapple flesh becomes watery and juice exudes. Black rot damages may be severe when harvest occurs under high rainfall and in presence of high inoculum potential in the field. *Ceratocystis paradoxa* incites rots during transport and storage of pineapple fruits whenever refrigeration is not available. *Ceratocystis paradoxa* is discussed earlier as the causal agent of a disease of planting material.

Fruitlet core rot, leathery pocket and inter-fruitlet corking

Pineapple fruitlet core rot, leathery pocket, and inter-fruitlet corking are caused primarily by the fungus *Talaromyces stollii* (anamorph *Penicillium funiculosum*) and to a lesser degree by species in the *Gibberella fujikuroi* species complex such as *F. subglutinans*, *F. ananatum* and *F. verticillioides* (Rohrbach and Pfeiffer, 1976a; Gu *et al.*, 2015; Barral *et al.*, 2017). Open flowers are the principal infection sites for the fungus, however injury from the pineapple tarsonemid mite, *Steneotarsonemus ananas*, also allows entry of the fungus into the fruit (Petty, 1989). Fungal infection leads to a browning or black discoloration of a fruitlet (Fig. 12.17). Low-acid pineapple cultivars are more susceptible to fruitlet core rot, leathery pocket and inter-fruitlet corking than the high acid, canning cultivar Smooth Cayenne (Rohrbach and Johnson, 2003). The incidence of fruitlet core rot caused by *P. funiculosum* can be high at 16–21°C during the 6 weeks following forcing (Rohrbach and Johnson, 2003). On the other hand, infection by *Fusarium* sp. is favoured by warm temperatures. The interaction between the pathogen and tarsonemid mite makes management of the disease very difficult. Management is dependent therefore incumbent upon mite control and cultivar selection.

Fig. 12.17. Fruitlet core rot. Photo G. Sanewski.

Yeasty fermentation and glassy spoilage

Yeasty fermentation and glassy spoilage are postharvest diseases of pineapple caused by the yeast fungi *Hanseniaspora valbyensis* and *Candida intermedia* var. *alcoholophila* (Rohrbach and Johnson, 2003). Pineapple fruit harbour many yeasts and bacteria and in damaged or overripe fruit, these yeasts begin to grow and ferment the pineapple juice resulting in foaming or bubbling from affected tissues (Fig. 12.18). Losses can be minimized by reducing sunburn, avoiding mechanical injury, and harvesting fruit before it becomes overripe. Glassy spoilage is caused by *Candida guilliermondii* and has been associated with fruitlet core rots (Rohrbach and Johnson, 2003). As with yeasty fermentation, losses can be minimized by harvesting fruit before it is overripe.

Fig. 12.18. External symptoms of yeasty rot in 'Smooth Cayenne'. Photo from Cooke, T., Persley, D and House, S. (2009). *Diseases of Fruit Crops in Australia*. CSIRO Publishing.

Mites

Several species of mites occur as pests on pineapple. The pineapple red or flat mite, *Dolichotetranychus floridanus*, and the pineapple leathery pocket or fruit mite, *Steneotarsonemus ananas*, are most common (Petty *et al.*, 2002). Mites generally do not cause significant problems in pineapple but heavy infestations can kill a plant (Rohrbach and Johnson, 2003). The pineapple red mite is most damaging to newly planted pineapple where feeding causes lesions on the leaf bases which in turn cause the leaves to become dehydrated. Infestations are generally more common at warmer temperatures than cooler temperatures (Petty *et al.*, 2002). Dry conditions also favour their multiplication. The pineapple leathery pocket mite is associated with the postharvest diseases of inter-fruitlet corking, leathery pocket, and fruitlet core rot. The mites feed on and injure the trichomes, bracts and sepals of the pineapple flower. These injuries predispose the plant to infection by *Talaromyces stollii* (anamorph *Penicillium funiculosum*), the causal agent of inter-fruitlet corking, leathery pocket, and fruitlet core rot (Rohrbach and Apt, 1986).

Scales

Scale insects (Coccoidea) can affect shoots but are of most concern on the fruit. The primary scale insect of pineapple is *Diapis bromelia* (Waite, 1993) with *Aspidiotus nerii*, *Diaspis boisduvalii*, *Melanaspis smilacis*, *Pinnaspis strachani*, *Pseudaonidia trilobitiformis*, *Unaspis citri*, *Coccus viridis*, *Phenacoccus madeirensis*, *Pseudococcus jackbeardsleyi* and *Pseudococcus longispinus* also feeding on pineapple (Rohrbach and Johnson, 2003). The first instars are mobile and walk or are carried by the wind to new sites (Beardsley and Gonzalez, 1975; Jahn and Beardsley, 2000). Adults are usually sedentary as adults and covered in a protective wax, making them a challenge to control. Scale tends to be most severe on ratoon crops (Rohrbach and Johnson, 2003). Scale insects are usually found on the protected underside of pineapple leaves and at low population densities, may cause yellow spotting where they feed. Scale insects often do not reduce yield but affect fruit quality (Waite, 1993), however at high population densities, scales will kill the pineapple plants (Carter, 1967; Petty, 1987; Py *et al.*, 1987). At high populations on the fruit, scales cause unwanted and undesirable cracks between the fruitlets (Py *et al.*, 1987). Additionally, scale-infested fruit may be subjected to quarantine restrictions or simply be undesirable in the fresh fruit market (Rohrbach and Johnson, 2003).

Pineapple fruit-piercing moths and fruit borer butterfly larvae

Nearly 100 species of fruit-piercing moths (Noctuidae) feed on pineapple. These moths have wide host ranges and only the adults of the fruit-piercing moths are pests of pineapple (Rohrbach and Johnson, 2003). The moths have strong mouthparts allowing them to penetrate the pineapple skin and feed on the pineapple juice. Fruit damage from feeding may be exacerbated by secondary fungal and bacterial infections. The fruit deformation and gummosis can lead to fruit rejection in the packing-house and loss of marketable fruit.

Several butterfly species in the genus *Strymon* feed on pineapple (Schmid *et al.*, 2010). *Strymon megarus*, *S. ziba* and *S. oreala* larvae commonly attack pineapple in Central and South America moving from pineapple flowers and boring into young fruit (Sanches, 1999). Adults lay eggs in the flowers before anthesis (Harris, 1927; Py *et al.*, 1987; Sanches, 1999; Noronha *et al.*, 2009, 2016). The larvae enter the developing fruit creating holes that exude a gum which hardens and becomes resin-like (Fig. 12.19). Adult borers can act as a vector or facilitate *Fusarium guttiforme* infection and development of fusariosis (Rorhbach and Johnson, 2003). Third instar larvae are the most voracious stage causing severe damage to fruits (Velasco *et al.*, 1968). *Elaphria nucicolora*, the sugarcane midget, is a pest of pineapples in Mexico, Central and South America. The larvae of *E. nucicolora* feed on the skin of the pineapple fruit, scraping the surface resulting in malformation of the fruit while deeper lesions cause gummosis on the fruit (Gonzalez and Vásquez, 2007). Like *Opogona* and *Porthesia*, *Elaphria* has become a pest of low-acid pineapple cultivars.

Fig. 12.19. Fruit borer damage to pineapple: (a) adult fruit borer butterfly; (b) an egg laid on the developing pineapple fruit; (c) early damage caused by the boring of the larvae into the fruit; (d) entry point of the larvae into the pineapple fruit; (e) internal damage caused by the fruit borer larvae. Photos: (a, b) N.F. Sanches; (c–e) A.P. de Matos.

Pineapple fruit flies

Pineapple fruit flies, *Melanoloma canopilosum* and *M. viatrix*, are found mainly in Paraguay and Peru (Martínez *et al.*, 2000). The female lays her eggs in the pineapple fruit. The eggs hatch and the feeding larvae form coalescing cavities in the fruit. The fruit will ferment and not be marketable (Bello *et al.*, 1997).

Fig. 12.20. Souring beetles on decaying pineapple fruit. Photo G. Sanewski.

Souring beetles (dried fruit beetle)

Souring beetles feed mainly on decaying pineapple and overripe fruit (Fig. 12.20). These beetles are attracted to decomposing pineapple trash and overripe fruit that remain in the field after harvest. *Carpophilus humeralis*, *C. hemipterous* and *Haptoncus ocularis* are the most common souring beetle species found in pineapple (Carter, 1967; Py *et al.*, 1987). The souring beetle larvae feed on rotting fruit and on the exposed butts and stalks of planting material. Souring beetle feeding does not result in economically significant injury to pineapple but can cause marketing problems. The souring beetles are problematic with agrotourism activities near pineapple fields or in suburban developments near pineapple fields.

Pest and Pathogen Management

Pests, diseases including plant-parasitic nematodes, and weeds may impact pineapple production and sustainability. Special characteristics of the pineapple plant and crop, such as its clonal propagation, leaf shape and leaf distribution around the stem, year-round fruit, and fields with plants in several developmental stages of both the plant and the ratoon crops, facilitate the incidence and spread of these pests, diseases and weeds. Consequently, pest and disease management in pineapple generally involves an integrated approach. Because many pests of pineapple have limited distributions, exclusion through quarantines or inspections can be effective. Exclusion has been effective in limiting the spread of bacterial heart rot and pineapple fruit flies. However, no pineapple pathogen or pest has been eradicated from an area once detected. Detection may not occur until the pest or pathogen is well established, therefore growers must rely upon other management tactics. Pineapple growers can employ a range of cultural, biological and chemical methods to limit losses to pests and pathogens.

Cultural

Cultural controls for pest and pathogen management span the entire pineapple cropping cycle. Cultural practices can provide substantial pest and disease control when employed and reduce the need for pesticides.

Cultural management of diseases and pests begins with the initial decision to plant pineapple. Field history can be an important guide in the selection of practices to be used in the field. When a field is selected for pineapple, distance from other fields, the forest or weedy fields is important for the management of the pineapple fruit borer, ants, mealybugs and yellow spot (Sanches, 1999; Noronha *et al.*, 2009; Noronha *et al.*, 2016). As a field is prepared for planting, adjusting the soil pH to 4.5–5 helps to manage heart rot caused by *Phytophthora*. Improving soil drainage through raised beds or building ditches to enhance drainage reduces disease incidence (Pegg, 1969). Covering the planting beds with black plastic mulch helps to control weeds. Ants may be managed through row placement, control of weeds along field borders, tillage, and elimination of volunteer plants during the intercycle period (Jahn *et al.*, 2003). The avoidance of weeds that can serve as host to mealybugs is important in reducing the establishment of mealybugs and the subsequent development of mealybug wilt in the pineapple crop (Pandey and Johnson, 2006).

Cultivar and planting material selection can help to control many pests and diseases. For bacterial heart rot, fusariosis, mites, scale and mealybugs, planting material should not be taken from an area having high incidence of these pests. The selection of planting material that has not contacted soil can reduce lesion nematode in planting material as well as the incidence of heart rot. Properly cured planting material reduces the incidence of butt rot. Curing the planting material on the mother plants provides good air circulation and minimizes exposure to inoculum in the soil. Planting material should also be kept dry and not left piled in the field for long periods to reduce fungal and bacterial rots. Some cultivars are less susceptible and in some cases resistant to diseases and pests such as fusariosis, some Lepidoptera larvae or *Phytophthora* so growers may choose to avoid planting susceptible cultivars (Chan *et al.*, 2003; Rohrbach and Schmitt, 2003; Vorsino *et al.*, 2005).

Modification of planting and forcing dates can avoid conditions that historically have favoured development of diseases such as heart rot, inter-fruitlet corking, leathery pocket, or fruitlet core rot in a field (Petty *et al.*, 2002). During the growth cycle, roguing of symptomatic plants helps to control fusariosis and bacterial heart rot. When roguing, it is important that infected plants be destroyed or removed from the field. Hand weeding is common until the pineapple canopy closes. However, mechanical leaf damage caused when entering the field for weeding or crop logging, should be minimized to control bacterial heart rot spread and white tip (Rohrbach and Johnson, 2003). Black rot of fruit can be managed by minimizing handling injury.

Good agricultural practices are as important at the end of a cropping cycle as at the beginning of the cycle. At the end of the cropping cycle, incorporation of pineapple trash, decreases disease inoculum. To reduce the nuisance factor associated with souring beetles, fruit should not be left to rot in the field. Good agricultural practices are the most important factor in the management of souring beetles. Deep tillage helps to control root rots, nematodes, symphylids and grubs. Long, dry intercycle fallow periods reduce symphylid and nematode populations and the inoculum of many fungi (Py *et al.*, 1987; Sipes *et al.*, 2005). Tillage dries out the soil and exposes soil borne pests to UV light. Tillage, however, may increase soil erosion which is undesirable. Consequently, intercycle cover crops, such as sunn hemp, can be planted and grown to reduce populations of insect pests, nematodes, and weeds (Wang *et al.*, 2001, 2002, 2003). As the cropping cycle returns to planting, soil solarization with a thin film (25-μm-thick, UV-stabilized) transparent mulch can reduce weed, fungi, nematode, symphylid and insect populations before planting (Wang *et al.*, 2010; Gamliel and Katan, 2012).

Chemical

Chemical pest and disease controls have been widely used in pineapple cultivation. Pesticides have provided effective and predictable control of many pests and diseases of pineapple. Chemicals can be used at all stages of production.

Pre-plant pesticide applications are made for insect, disease and weed control in pineapple. Since the 1940s, pineapple producers in Hawaii have applied fumigant compounds to control nematodes (Caswell and Apt, 1989; Schneider *et al.*, 1995; Sipes, 2006). Pre-plant soil fumigation can also be used against white grubs (Christensen, 2004). Pre-emergence herbicides can be used to manage weeds in the pineapple field (Rohrbach and Johnson, 2003; Sipes and Wang, 2016).

Planting material is also frequently treated with pesticides to manage a variety of pest and disease problems. Applications of fosetyl aluminum or phosphorus acid formulations, as a pre-plant dip or spray application to planting material provides control for *Phytophthora* and *Pythium* (Rohrbach and Schenck, 1985; Sipes *et al.*, 2011). Butt rot can be controlled with pre-plant fungicidal dips (Rohrbach, 1994). Inducers of systemic acquired resistance applied to pineapple have reduced crown loss to heart rot (Sipes *et al.*, 2011) and to nematodes (Chinnasri *et al.*, 2006). Application of insecticides to planting material, usually through dips, are effective in protecting pineapple from Lepidopteran larvae, mealybugs, scales and mites (Rohrbach and Apt, 1986; Jahn and Beardsley, 2000; Rohrbach and Johnson, 2003; Vorsino *et al.*, 2005).

Pesticides are also commonly applied after pineapple planting. Post-plant nematicide application protects pineapple roots that emerge later in the cropping cycle (Apt and Caswell, 1988; Caswell and Apt, 1989). Post-plant insecticides are applied to control mealybugs or heavy infestations of scale. Miticides and fungicides are effective in protecting against inter-fruitlet corking, leathery pocket, and fruitlet core rot (Rohrbach and Apt, 1986; Petty *et al.*, 2002). Ant control is often achieved with chemical treatments applied to the entire field or along the perimeter of a field (Rohrbach *et al.*, 1988; Taniguchi, 2011). Bait traps can also be used but cost more (Taniguchi *et al.*, 2005). Hydramethylnon is very effective for ant control allowing natural biological control agents to attack the mealybugs (Reimer and Beardsley, 1990; Rohrbach and Mau, 2005).

Post-emergence herbicide applications of diuron and bromacil can be made as directed sprays in the interrow after pineapple has been planted (Rohrbach and Johnson, 2003). Directed sprays of glyphosate or specific monocot herbicides are also used to control weeds that might have escaped earlier herbicide treatments.

The timing of post-plant pesticide applications is extremely important. For inter-fruitlet corking, leathery pocket, and fruitlet core rot management, the pesticides must be applied at the appropriate stage of anthesis. Application too soon or too late provide no control. The avoidance of overhead irrigation during anthesis might also help with control. Post-plant nematicide applications also depend upon appropriate timing which is dependent upon the behaviour of the nematicide (nematicidal vs nematistatic), plant growth, and nematode populations. Insecticides provide greater flexibility because of greater product numbers and modes of actions available to the grower. With post-plant applications, growers must be aware of harvest intervals to comply with fruit pesticide maximum residue limits. In general, most pesticide applications cease 6 months before the anticipated harvest date to avoid pesticide residues on the fruit.

Postharvest pesticide applications can also be used on fruit. A fungicidal dip followed by refrigeration at 9°C provides control for black rot (Rohrbach and Johnson, 2003). Acetic acid dips and hot water treatments can also be deployed to control black rot of fruit (Wijeratnam *et al.*, 2005; Wijesinghe *et al.*, 2011). Edible waxes may also be applied to fruit to provide additional control of black rot.

Biological

Biological control has proven to be effective against many insect pests of pineapple but has been less successful against diseases. A substantial amount of research has been directed to providing biological control against plant-parasitic nematodes and fruit diseases. Biological alternatives for the control of plant-parasitic nematodes, other soil-borne diseases, fungal diseases of the fruit, and bacterial diseases of pineapple will likely become available in the future.

Insect pests have been successfully controlled in pineapple through introduction of parasites, predators and parasitoids as well as the use of bacterial toxins. The BT toxin from *Bacillus thurigiensis* is effective against Lepidopteran pests and can be applied to control these larvae (Rorhbach and Johnson, 2003). The fungus *Beauveria bassiana* has provided control for the white grub in Central America (Carrillo, 2011). When ants are controlled, the parasitoid, *Anagyrus ananatis*, and other biological control agents maintain populations of mealybug below economic thresholds protecting the pineapple from mealybug wilt (Rohrbach and Mau, 2005). Parasites, predators, and fungi generally provide sufficient control of bud moths and fruit-piercing moths in the field limiting the need for applications of insecticides (Heu *et al.*, 1985). Natural enemies including parasitic wasps and lady beetles (ladybirds) successfully manage scale insects in many pineapple growing areas (Waite, 1993; González-Hernández *et al.*, 1999; Rohrbach and Johnson, 2003). The biological control provided by these wasps and lady beetles can be quite successful if the scale population is low at planting (Waite, 1993; Rohrbach and Johnson, 2003).

A successful example of biological control of a fungal disease is the use of the fungus *Trichoderma* to control black rot of fruit (Wijeratnam *et al.*, 2006; Wijesinghe *et al.*, 2011). Other fungi have shown potential to control fusariosis when applied to the fruit at the appropriate time (Korres *et al.*, 2011). These fungal biological control agents compete with the pathogen and prevent infection.

Genetic

Genetic control for fungal, stramenophyle, and nematode diseases has been attempted in pineapple. Both traditional breeding through crossing and genetic engineering have been undertaken in pineapple. Breeding goals of the Pineapple Research Institute (PRI) in Hawaii included resistance to *Phytophthora*, mealybug wilt, nematodes and pink disease as goals (Chan *et al.*, 2003). The PRI breeders identified moderate to high resistance to *P. cinnamomi* and/or *P. nicotianae* in several cultivars including 10388 *syn.* Spanish Jewel, 59-656, 63-475, 52-323, and 61-2223 (Smith, 1965; Williams and Fleisch, 1993; Rohrbach and Johnson, 2003; PRI Plantation Tests 63-475). The popular low-acid cultivars MD-2 and CO-2 are more susceptible to heart rot than even 'Smooth Cayenne' (Anderson *et al.*, 2012; Sanewski *et al.*, 2016). Sanewski *et al.* (2017) identified a single broad loci on chromosome five associated with resistance to *P. cinnamomi*. Brazilian scientists identified resistance to fusariosis and using traditional breeding techniques developed pineapple resistant cultivars such as BRS Imperial (Cabral *et al.*, 2009) and BRS Vitória (Ventura *et al.*, 2009). Many commercial cultivars of pineapple have tolerance to infection by root-knot and reniform nematodes although genetic markers have yet to be identified (Sipes and Schmitt, 1994). For the commercial cultivars with tolerance to nematodes, indirect selection of clones for planting material by the plantations likely increased tolerance to nematode infection (Sipes and Schmitt, 1994). No pineapple cultivar has been identified that is entirely resistant to nematode infection. Breeding for resistance to pink disease has shown promise (Kado, 2003). The seed parent to 'MD-2', 'PRI 58-1184' is reportedly resistant to pink disease (Williams and Fleisch, 1993).

Genetic transformation has been undertaken in pineapple to confer resistance to pathogens and herbicides (Soneji and Nageswara Rao, 2009). The genes for *chitinase* and *ap24* were transformed into pineapple to confer resistance to *P. nicotianae* (Yabor *et al.*, 2006). General fungal resistance was attempted via transformation with a magainin gene analogue (Mhatre *et al.*, 2009). The efficacy of these transformations against the pathogens has not been reported to date. A modified rice cystatin (*ubi9-d86*) protease inhibitor was ingressed into pineapple to convey resistance to reniform and root-knot nematodes (Sipes *et al.*, 2002a). None of the transformed pineapple lines expressed more nematode resistance than the untransformed controls (Sipes, 2011). The coat protein of the PMWaV-2 was engineered into pineapple and several lines were reported to be resistant to the virus (Perez *et al.*, 2006). An RNA-mediated resistance was also engineered into pineapple and found to prevent wilt symptom development (Hu *et al.*, 2005). Tolerance to the herbicide glufosinate-ammonium has been transformed into pineapple thus allowing application of the herbicide to control weeds without damaging the pineapple. Pineapple transformed with the *bar* gene were resistant to the herbicide (Sripaoraya *et al.*, 2006; Yabor *et al.*, 2010). The *surB*, conveying resistance to chlorsulfuron, has also been successfully transformed into pineapple (Firoozabady *et al.*, 2006).

References

Aguilar, J.A.E. and Sanches, N.F. (1982) Disseminação de *Fusarium moniliforme* var. *subglutinans* do abacaxizeiro pela *Trigona spinipes* (Fabr. 1973) (Hymenoptera: Apidae). Cruz das Almas, Brazil: Embrapa/CNMF. *Comunicado Técnico* 2.

Anderson, J.M., Pegg, K.G., Scott, C. and Drenth, A. (2012) Phosphonate applied as a pre-plant dip controls Phytophthora cinnamomi root and heart rot in susceptible pineapple hybrids. *Australasian Plant Pathology* 41, 59–68.

Apt, W.J. and Caswell, E.P. (1988) Application of nematicides via drip irrigation. *Annals of Applied Nematology (Journal of Nematology 20, Supplement)* 2, 1–13.

Aquije, G.M. de F.V., Korres, A.M.N., Ventura, D.S., Fernandes, P.M.B. and Fernandes, A.A.R. (2014) Effects of leaf scales of different pineapple cultivars on the epiphytic stage of *Fusarium guttiforme*. *Crop Protection* 30, 375–378.

Barral, B., Chillet, M., Schorr-Galindo, S. and Léchaudel, M. (2017) Diagnostic survey on the occurrence of pineapple fruitlet core rot in Réunion Island. *IVth International Symposium on Postharvest Pathology*. Skukuza, Kruger National Park, South Africa.

Beardsley, J.W. (1959) On the taxonomy of pineapple mealybugs in Hawaii, with the description of previously unnamed species (Homoptera: Pseudococcidae). *Proceedings of the Hawaiian Entomological Society* 17, 29–37.

Beardsley Jr, J.W. and Gonzalez, R.H. (1975) The biology and ecology of armored scales. *Annual Review of Entomology* 20, 47–73.

Bello, A., Otiniano, J.A. and León, V. H. (1997) Mancha de la fruta de piña tipo galerias asociada a *Melanoloma canopilosum* Hendel. *Acta Horticulturae* 425, 493–500.

Berasconi, B. (1994) Maladies et ravageurs du fruit de l'ananas. *Regional Workshop on Pineapple, Martinique*, pp. 74–78.

Bolkan, H.A., Dianese, J.C. and Cupertino, F.P. (1979) Pineapple flowers as principal infection sites of *Fusarium mojiliforme* var. *subglutinans. Plant Disease Reporter* 3, 655–657.

Borras, O., Santos, R., Matos, A.P. de, Cabral, J.R.S. and Arzola, M. (2001) A first attempt to use a *Fusarium subglutinans* culture filtrate for the selection of pineapple cultivars resistant to fusariose disease. *Plant Breeding* 120, 435–438.

Cabral, J.R.S. and Matos, A.P. de. (2009) 'Imperial', a new pineapple cultivar resistant to fusariosis. *Acta Horticulturae* 822, 47–50.

Camino-Lavín, M., Jiminez-Perez, A., Castrejon-Gomez, V., Castrejon-Ayala, F. and Figueroa Brito, R. (1996) Performance of a new trap for melolonthine scarabs, root pests. *Southwestern Entomologist* 21, 325–330.

Carrillo, E.V. (2011) *Guide for Integrated Pest Identification and Management in Pineapple.* Available at: http://cep.unep.org/repcar/proyectos-demostrativos/costa-rica-1/publicaciones-proagroin/Guia%20manejo%20integrado%20en%20pina%20ENG.pdf (accessed 20 March 2018).

Carter, W. (1933) The pineapple mealy bug, *Pseudococcus brevipes*, and wilt of pineapples. *Phytopathology* 23, 207–242.

Carter, W. (1967) *Insects and Related Pests of Pineapple in Hawaii.* Pineapple Research Institute of Hawaii, Hawaii.

Caswell, E.P. and Apt W.J. (1989) Pineapple nematode research in Hawaii: past, present, and future. *Journal of Nematology* 21, 147–157.

Catunda, M.G., Freitas, S.P., Oliveira, J.G. and Silva, C.M.M. (2005) Efeitos de herbicidas na atividade fotossintética e no crescimento de abacaxi (*Ananas comossus*). *Planta daninha* 23, 115–121.

Cha, J-S., Pujol, C., Ducusin, A.R., Macion, E.A., Hubbard, C.H. and Kado, C.I. (1997) Studies on *Panotea citrea*, the causal agent of pink disease of pineapple. *Journal of Phytopathology* 145, 313–319.

Chalfoun, S.M. and Cunha, G.A.P. da (1984) Relação entre a incidência da broca-do-fruto e fusariose do abacaxi. *Pesquisa Agropecuária Brasileira* 19, 423–426.

Chan, Y.K., Coppens d'Eeckenbrugge, G. and Sanewski, C.M. (2003) Breeding and variety improvement. In: Bartholomew, D.P., Paull, R.E. and Rohrbach, K.G. (eds) *The Pineapple: Botany, Production, and Uses.* Wallingford, Oxfordshire: CAB International, pp. 33–55.

Chinchilla, C.M., González, L.C. and Morales, F. (1979) Pudrición bacteriana del cogollo de la piña en Costa Rica. *Revista Agronomía Costarricense* 3(2), 183–185.

Chinnasri, B., Sipes, B.S. and Schmitt, D.P. (2006) Effects of inducers of systemic acquired resistance on reproduction of *Meloidogyne javanica* and *Rotylenchulus reniformis* in pineapple. *Journal of Nematology* 38, 319.

Christensen, D. (2004) White grubs and other incidental pests of pineapple. *Pineapple News* 11, 14.

Collins, J.L. (1960) *The Pineapple: Botany, Cultivation and Utilization*: New York: World Crops Book.

Costa, J.L da S. and Lordelo, S. (1988) Papel da entomofauna na disseminação da fusariose do abacaxizeiro. *Fitopatologie Brastleira* 13, 63–65.

Dey, K.K., Borth, W.B., Melzer, M.J., Wang, M.-L. and Hu, J.S. (2015) Analysis of pineapple mealybug wilt associated virus –1 and –2 for potential RNA silencing suppressors and pathogenicity factors. *Viruses* 7, 969–995.

Eshetu, T., Tefera, W. and Kebede, T. (2007) Effect of weed management on pineapple growth and yield. *Ethiopian Journal of Weed Management* 1, 29–40.

Firoozabady, E., Heckert, M. and Gutterson, N. (2006) Transformation and regeneration of pineapple. *Plant Cell, Tissue and Organ Culture* 84, 1–16.

Gambley, C., Steele, V., Geering, A. and Thomas, J. (2008) The genetic diversity of ampeloviruses in Australian pineapples and their association with mealybug wilt disease. *Australasian Plant Pathology* 37, 95–105.

Gamliel, A. and Katan, J. (2012) *Soil Solarization: Theory and Practice.* St Paul, MS: American Phytopathological Society.

Godfrey, G.H. (1936) The pineapple root system as affected by the root-knot nematode. *Phytopathology* 26, 408–428.

Godfrey, G.H. and Oliveira, J.M. (1932) The development of the root-knot nematode in relation to root tissues of pineapple and cowpea. *Phytopathology* 22, 325–348.

González, C.S. and Vásquez, D.G. (2007) *Elaphria nucicolora* Guenée (Noctuidae: *Amphipirinae*), nueva plaga de la piña en el caribe costarricense. *Proyecto de graduación (licenciatura), Universidad EARTH.* Available at: http://usi.earth.ac.cr/glas/sp/dpg/01-2007.pdf (accessed 15 June 2015).

González-Hernández, H., Reimer, N.J. and Johnson, M.W. (1999) Survey of the natural enemies of *Dysmicoccus* mealybugs on pineapple in Hawaii. *BioControl* 44, 47–58.

Green, J. and Nelson, S. (2015) *Heart and Root Rots of Pineapple.* College of Tropical Agriculture and Human Resources. Plant Disease, PD-106.

Gu, H., Zhan, R.L., Zhang, L.B., Gong, D.Q. and Jia, Z.W. (2015) First report *of Fusarium ananatum* causing pineapple fruitlet core rot in China. *Plant Disease* 99, 1653.

Gunasinghe, U.B. and German, T.L. (1989) Purification and partial characterization of a virus from pineapple. *Phytopathology* 79, 1337–1341.

Harris, W.V. (1927) On a lycaenid butterfly attacking pineapples in Trinidad, B.W.I. *Bulletin of Entomological Research* 18, 183–188.

Heu, R., Teramoto, K., Shiroma, E., Kumashiro, B. and Dinker, R. (1985) *Hawaii Pest Report* 5, 4–5.

Hu, J.S., Sether, D.M., Metzer, M.J., Pérez, E., Gonsalves, A., Karasev, A.V. and Nagai C. (2005) Pineapple mealybug wilt associated virus and mealybug wilt of pineapple. *Acta Horticulturae* 666, 209–212.

Huang, S.H. and Lin, C.Y. (2014) Distribution and control of pink pineapple mealybug and survey of insect pests on pineapple. *Journal of Taiwan Agricultural Research* 63, 68–76.

Hubert, J., Fourrier, C., Laplace, D. and Loos, R. (2014) First report of pineapple black rot caused by *Ceratocystis paradoxa* on *Ananas comosus* in French Guiana. *Plant Disease* 98, 1584.

Ibrahim, N. F., Mohd, M.H., Nor, N.M.I.M. and Zakaria, L. (2016) Pathogenicity of *Fusarium semitectum* and *Fusarium chlamydosporum* associated with pineapple. *Malaysian Journal of Microbiology* 12, 164–70.

Illingworth, J.F. (1931) Preliminary report on evidence that mealy bugs are an important factor in pineapple wilt. *Journal of Economic Entomology* 24, 877–889.

Jacobs, A., Van Wyk, P.S., Marasas, W.F., Wingfield, B.D., Wingfield, M.J. and Coutinho, T.A. (2010) *Fusarium ananatum sp. nov.* in the *Gibberella fujikuroi* species complex from pineapples with fruit rot in South Africa. *Fungal Biology* 114, 515–527.

Jahn, G.C. and Beardsley, J.W. (2000) Interactions of ants (Hymenoptera: Formicidae) and mealybugs (Homoptera: Pseudococcidae) on pineapple. *Proceedings of the Hawaiian Entomological Society* 34, 181–185.

Jahn, G.C., Beardsley, J.W. and González-Hernández, H. (2003) A review of the association of ants with mealybug wilt disease of pineapple. *Proceedings of the Hawaiian Entomological Society* 36, 9–28.

Johnston, A. (1957) Bacterial heart rot of the pineapple. *Malaysian Agricultural Journal* 40, 2–8.

Kado, C.I. (2003) *Pink Disease of Pineapple.* DOI: 10.1094/APSnetFeature-2003-0303.

Kaneshiro, W.S., Burger, M., Vine, B.G., de Silva, A.S. and Alvarez, A.M. (2008) Characterization of *Erwinia chrysanthemi* from a bacterial heart rot of pineapple outbreak in Hawaii. *Plant Disease* 92, 1444–1450.

Keetch, D.P. (1982) Nematode pests of pineapple. In: Keetch, D.P. and Heyns, J. (eds) *Nematology in South Africa.* Pretoria, South Africa: Department of Agriculture and Fisheries, pp. 9–29.

Keetch, D.P. and Purdon, K.T. (1979) Effect of systemic nematicides on pineapples. *Citrus Subtropical Fruit Research Institute Information Bulletin* 82, 3–4.

Kehe, M. (1979) Les symphyles en culture d'ananas en Côte-d'Ivoire. In: *Congrès sur Lutte contre les Insectes en Milieu Tropical.* Marseille: Chambre de Commerce et d'Industrie de Marseille, pp. 441–445.

Kimati, H. and Tokeshi, H. (1964) Nota sobe a ocorrência de *Fusarium* sp. causando resinose fúngica em abacaxi. *Revista de Agricultura* 39, 131–133.

Ko, M.P. and Schmitt, D.P. (1993) Pineapple inter-cycle cover crops to reduce plant-parasitic nematode populations. *Acta Horticulturae* 334, 373–382.

Korres, A.M.N., Buss, D.S., Ventura, J.A. and Fernandes, P.M.B. (2011) *Candida krusei* and *Kloeckera apis* inhibit the causal agent of pineapple fusariosis, *Fusarium guttiforme. Fungal Biology* 115, 1251–1258.

Lacerda, J.T., Carvalha, R.A. and Oliveira, E.F. (2007) Broca-do-fruto trymon megarus um problema para a abacaxicultura do Brasil. *Tecnologia and Ciência Agropecuária* 1, 25–30.

Lim, W.H. (1985) Diseases and disorders of pineapples in Peninsular Malaysia. *MARDI Issue* 97, 53.

Linford, M.B. (1943) Influence of plant populations upon incidence of pineapple yellow spot. *Phytopathology* 33, 408–410.

Luo, Z.W., Fan, H.Y., He, F. and Hua, M. (2010) Identification and biological characteristics of pineapple anthracnose in Hainan province. *Plant Diseases and Pests* 1, 21–25.

Luo, Z.W., He, F., Fan, H.Y., Wang, X.H., Hua, M., Hu, F.C., Li, X.H., Liu, Z.X. and Yu, N.T. (2012) First report of leaf spot disease caused by *Exserohilum rostratum* on pineapple in Hainan province, China. *Plant Disease* 96, 458.

Maia, L.C.B., Maia, V.M., Lima, M.H.M., Aspiazú, I. and Pegoraro, R.F. (2012) Growth, production and quality of pineapple in response to herbicide use. *Revista Brasileira de Fruticultura* 34. DOI: 10.1590/S0100-29452012000300020.

Mangara, A., Soro, K., Kouame, N. and Thérèse, D.M. (2016) Floristic diversity of weeds in pineapple (*Ananas comosus* (L.) Merr.) crop in the locality of N'douci in Côte d'Ivoire. *International Journal of Agriculture and Crop Sciences* 9, 1–6.

Marrero, G., Schneider, K.L., Jenkins, D.M. and Alvarez, A.M. (2013) Phylogeny and classification of *Dickeya* based on multilocus sequence analysis. *International Journal of Systemic Evolution and Microbiology* 63, 3524–3539.

Martínez, N.B. de, Rosales, C.J. and Godoy, F. (2000) La mosca del fruto la piña *Melanoloma viatrix* (Diptera: Richardiidae) nuevo insecto plaga en Venezuela. *Agronomia Tropical* 50, 135–140.

Matos, A.P. de. (1995) Pathological aspects of the pineapple crop with emphasis on the Fusariosis. *Revista de la Facultad de Agronomía* 21, 179–197.

Matos, A.P. de, and Mourichon, X. (1993) Development of resistance to infection by *Fusarium moniliforme* var. *subglutinans* in wounds of pineapple plantlets. *Acta Horticulturae* 334, 423–428.

Matos, A.P. de and Souto, G.F. (1985) Reaction of pineapple, cultivars Pérola and Smooth Cayenne, to inoculation with *Fusarium moniliforme* Sheld. var. *subglutinans* WR & RG. *Fruits* 40, 641–645.

Matos, A.P. de, Aguilar, J.A.E. and Neiva, L.P.A. (1981) Método para determinar a disseminação de *Fusarium moniliforme* var. *subglutinans* no abacaxizeiro. *Pesquisa Agropecuária Brasileira* 16, 337–339.

Matos, A.P. de, Mourichon, X. and Pinon, A. (1992) Occurrence of *Fusarium moniliforme* var. *subglutinans* on pineapple in Bolivia. *Fruits* 47, 33.

Matos, A.P. de, Cabral, J.R.S., Sanches, N.F. and Caldas, R.C. (2000) Effect of temperature and rainfall on the incidence of *Fusarium subglutinans* on pineapple fruits. *Acta Horticulturae* 529, 265–272.

Melo, S.A.P., Araujo, E. and Suassuna, J. (1974) Occorrencia de *Erwinia* sp. em frutos de abacaxi (*Ananas comossus* (L.) Merr.) na Paraiba. *Fitopatologia* 62.

Mhatre, M., Nagi, L. and Ganapathi, T.R. (2009) *Agrobacterium*-mediated transformation of pineapple (*Ananas comosus* L.Merr.) leaf bases with MSI–99, a magainin analogue. *International Journal of Fruit Science* 9, 106–114.

Montealegre, J.R. and Luchsinger, L.E. (1990) Postharvest rot of pineapple fruits marketed in Chile. *Fitopatologia* 25, 51–53.

Noronha, A.C. da S., Silva, T. de A., Tavares, G. dos S., Kramer, V.R. and Matos, A.P. de (2009) A broca do fruto do abacaxi *Strymon megarus* no Estado do Pará. In: *Seminario Braxileiro de Producao Integrada de Frutas. Petrolina: Embrapa Semi-Árido, Fitossanidade* 1.

Noronha, A.C.S., Lemos, W.P., Faxolin, M., Sanaches, N.F. and Garcia, M.V.B. (2016) Abacaxi. In: Silva, N.M., Adaime, R. and Zucchi, R.A. (eds) *Pragas agrícolas e florestais na Amazônia*. Brasília, Brazil: Embrapa.

Pandey, R.R. and Johnson, M.W. (2006) Weeds adjacent to Hawaiian pineapple plantings harboring pink pineapple mealybugs. *Environmental Entomology* 35, 68–74.

Pegg, K.G. (1969) Pineapple top rot control. *Queensland Agriculture Journal* 95, 458–459.

Perez, E.A., Ma, H., Melzer, M.J., Sether, D.M., Borth, W.B. *et al.* (2006) Development of transgenic pineapple plant with resistance to PMWaV-2. *Pineapple News* 13, 41–42.

Petty, G. (1978) *H.12 Pineapple Pests: White Grubs in Pineapple. Pineapple Series H: Diseases and Pests*. Pretoria, South Africa: Government Printer.

Petty, G.J. (1987) Control of mealybugs and scale on pineapples by methyl bromide fumigation of planting material. *Phytophylactica* 19, 255–258.

Petty, G.J. (1989) Effect of endosulfan on leathery pocket disease in pineapples, and evidence for the association of the mite, *Steneotarsonemus ananas* (Acarina: Tarsonemidae) with leathery pocket. *Acta Horticulturae* 275, 663–670.

Petty, G.J., Stirling, G.R. and Bartholomew, D.P. (2002) Pests of Pineapple. In: *Tropical Fruit Pests and Pollinators: Biology, Economic Importance, Natural Enemies, and Control* 704(13), 157.

Py, C., Lacoeuilhe, J.J. and Teisson, C. (1987) *The Pineapple. Cultivation and Uses*. Paris: G.-P. Maisonneuve et Larose.

Radewald, J.D. and Takeshita, G. (1964) Desiccation studies on five species of plant-parasitic nematodes of Hawaii. *Phytopathology* 54, 903–904.

Ramachandran, K., Manaf, U.A. and Zakaria, L. (2015) Molecular characterization and pathogenicity of *Erwinia* spp. associated with pineapple [*Ananas comosus* (L.) Merr.] and papaya (*Carica papaya* L.). *Journal of Plant Protection Research* 55, 396–404.

Reimer, N.J. and Beardsley, J.W. (1990) Effectiveness of hydramethylnon and fenoxycarb for control of bigheaded ant (Hymenoptera: Formicidae), an ant associated with mealybug wilt of pineapple in Hawaii. *Journal of Economic Entomology* 83, 74–80.

Reinhardt, D.H., Cabral, J.R.S., Souza, L.F.S., Sanches, N.F. and Matos, A.P. (2002) Pérola and Smooth Cayenne pineapple cultivars in the state of Bahia, Brazil: growth, flowering, pests and diseases, yield and fruit quality aspects. *Fruits* 57, 43–53.

Rohrbach, K.G. (1983) Pineapple diseases and pests and their potential for spread. In: Singh, K.G. (ed.) *Exotic Plant Quarantine Pest and Procedures for Introduction of Plant Materials*. Selangor, Malaysia: ASEAN Plant Quarantine Center and Training Institute, pp. 145–171.

Rohrbach, K.G. (1989) Unusual tropical fruit diseases with extended latent periods. *Plant Disease* 73, 607–609.

Rohrbach, K.G. (1994) Fusariosis. In: Ploetz, R.C., Zentmyer, G.A., Nishijima, W.T., Rohrbach, K.G. and Ohr, H.D. (eds) *Compendium of Tropical Fruit Diseases.* St Paul, MN: American Phytopathological Society.

Rohrbach, K.G. and Apt, W.J. (1986) Nematode and disease problems of pineapple. *Plant Disease* 70, 81–87.

Rohrbach K.G. and Apt, W.J. (1993) Diseases of pineapple (*Ananas comosus* (L.) Merr.). Available at: https://www.apsnet.org/publications/commonnames/Pages/Pineapple.aspx (accessed 20 March 2018).

Rohrbach, K.G. and Johnson, M.W. (2003) Pests, diseases and weeds. In: Bartholomew, D.P., Paull, R.E. and Rohrbach, K.G. (eds) *The Pineapple: Botany, Production and Uses.* Wallingford, Oxfordshire: CAB International, pp. 203–251.

Rohrbach, K.G. and Mau, R. (2005) Pineapple integrated pest management in Hawaii. *Acta Horticulturae* 666, 205–208.

Rohrbach, K.G. and Pfeiffer, J.B. (1976a) Susceptibility of pineapple cultivars to fruit diseases incited by *Penicillium funiculosum* and *Fusarium moniliforme. Phytopathology* 66, 1386–1390.

Rohrbach, K.G. and Pfeiffer, J.B. (1976b) The interaction of four bacteria causing pink disease of pineapple with several pineapple cultivars. *Phytopathology* 66, 396–399.

Rohrbach, K.G. and Schenck, S. (1985) Control of pineapple heart rot, caused by *Phytophthora parasitica* and *P. cinnamomi*, with metalaxyl, fosetyl Al, and phosphorous acid. *Plant Disease* 69, 320–323.

Rohrbach, K.G. and Schmitt, D. (2003) Diseases of pineapple. In: Ploetz, R.C. (ed.) *Diseases of Tropical Fruit Crops.* Wallingford, Oxfordshire: CAB International, pp. 443–464.

Rohrbach, K.G., Beardsley, J.W., German, T.L., Reimer, N.J. and Sanford, W.G. (1988) Mealybug wilt, mealybugs, and ants of pineapple. *Plant Disease* 72, 558–565.

Sakimura, K. (1966) *A Brief Enumeration of Pineapple Insects in Hawaii.* XI Pacific Science Congress, pp. 1–7.

Sanches, N.F. (1999) Pragas e seu controle. In: Cunha, G.A.P., Cabral, J.R.S. and Souza, L.F.S. (eds) *O abacaxizeiro: cultivo, agroindústria e economia.* Brasília, Brazil: Embrapa Comunicação para Transferência de Tecnologia, pp. 307–341.

Sanewski, G., Ko, L., DeFaveri, J. and Kilian, A. (2016) Genetic resistance to the root rot pathogen *Phytophthora cinnamomi* in Ananas. *XXIX International Horticultural Congress on Horticulture: Sustaining Lives, Livelihoods and Landscapes (IHC2014): IV 1111*, pp. 281–286.

Sanewski, G., Ko, L., Innes, D., Kilian, A., Carling, J. and Song, J. (2017) DArTseq molecular markers for resistance to *Phytophthora cinnamomi* in pineapple (*Ananas comosus* L.). *Australasian Plant Pathology* 46, 499–509.

Schmid, S., Schmid, V.S., Kamke, R., Steiner, J. and Zillikens, A. (2010) Association of three species of *Strymon* Hübner (Lycaenidae: Theclinae: Eumaeini) with bromeliads in southern Brazil. *Journal of Research on the Lepidoptera* 42, 50–55.

Schneider, R.C., Sipes, B.S., Oda, C.H., Green, R.E. and Schmitt, D.P. (1995) Management of 1,3-dichloropropene in pineapple for efficacy and reduced volatile losses. *Crop Protection* 14, 611–618.

Sether, D.M. and Hu, J.S. (1998) Corollary analyses of the presence of pineapple mealybug wilt associated virus and the expression of mealybug wilt symptoms, growth reduction, and/or precocious flowering of pineapple. *Phytopathology* 88, S80.

Sether, D.M. and Hu, J.S. (2001) The impact of pineapple mealybug wilt-associated virus-1 and reduced irrigation on pineapple yield. *Australasian Plant Pathology* 30, 31–36.

Sether, D.M. and Hu, J.S. (2002) Closterovirus infection and mealybug exposure are necessary for the development of mealybug wilt of pineapple disease. *Phytopathology* 92, 928–935.

Sether, D.M., Ullman, D.E. and Hu, J.S. (1998) Transmission of pineapple mealybug wilt-associated virus by two species of mealybug (*Dysmicoccus* spp.). *Phytopathology* 88, 1224–1230.

Sether, D.M., Karasev, A.V., Okumura, C., Arakawa, C., Zee, F. *et al.* (2001) Differentiation, distribution, and elimination of two different pineapple mealybug wilt-associated viruses found in pineapple. *Plant Disease* 85, 856–864.

Sether, D.M., Melzer, M.J., Busto, J., Zee, F. and Hu, J.S. (2005) Diversity and mealybug transmissibility of ampeloviruses in pineapple. *Plant Disease* 89, 450–456.

Shear, W.A. and Edgecombe, G.D. (2010) The geological record and phylogeny of the Myriapoda. *Arthropod Structure Development* 39, 174–190.

Sipes, B.S. (2006) Nematode control for the early 21st century. *Acta Horticulturae* 702, 163–166.

Sipes, B.S. (2011) Evaluation of a bioassay for screening resistance of pineapple to root-knot and reniform nematode. *Acta Horticulturae* 902, 185–191.
Sipes, B.S. and Schmitt D.P. (1994) Evaluation of pineapple, *Ananas comosus*, for host-plant resistance and tolerance to *Rotylenchulus reniformis* and *Meloidogyne javanica*. *Nematoropica* 24, 113–121.
Sipes, B.S. and Schmitt, D.P. (1996) Control of *Rotylenchulus reniformis* on pineapple with emulsifiable 1,3-dichloropropene. *Plant Disease* 80, 571–574.
Sipes, B.S. and Schmitt, D.P. (2000) *Rotylenchulus reniformis* damage thresholds on pineapple. *Acta Horticulturae* 529, 239–246.
Sipes, B. and Wang. K.H. (2016) Pests, diseases and weeds. In: Lobo, M.G. and Paull, R.E. (eds) *Handbook of Pineapple Technology: Production, Postharvest Science, Processing and Nutrition*, pp. 62–88.
Sipes, B., Nagai, C., McPherson, M., Atkinson, H., Christopher, D. *et al.* (2002a) Pineapple genetically modified for resistance to plant-parasitic nematodes. *Phytopathology* 92, S76.
Sipes, B.S., Sether, D.M. and Hu, J.S. (2002b) Interactions between *Rotylenchulus reniformis* and pineapple mealybug wilt associated virus–1 in pineapple. *Plant Disease* 86, 933–938.
Sipes, B.S., Caswell-Chen, E.P., Sarah, J.L. and Apt, W.J. (2005) Nematode parasites of pineapple. In: Luc, M., Sikora, R.A. and Bridge, J. (eds), *Plant Parasitic Nematodes in Subtropical and Tropical Agriculture.* Wallingford, Oxfordshire: CAB International, pp. 709–731.
Sipes, B.S., Taniguchi, G. and Radovich, T. (2011) Vermicompost tea and BTH effect on pineapple heart rot. *Acta Horticulturae* 902, 373–376.
Smith, J.B (1965) Plantation variety tests 1965. *PRI News* 13, 255–262.
Smith, T.J., Petty, G.J. and Villet, M.H. (1995) Description and identification of white grubs (Coleoptera: Scarabaeidae) that attack pineapple crops in South Africa. *African Entomology* 3, 153–166.
Soneji, J.R. and Nageswara Rao, M. (2009) Genetic engineering of pineapple. *Transgenic Plant Journal* 3, 47–56.
Sripaoraya, S., Keawsompong, S., Insupa, P., Power, J.B., Davey, M.R. and Srinives, P. (2006) Genetically manipulated pineapple: transgene stability, gene expression and herbicide tolerance under field conditions. *Plant Breeding* 125, 411–413.
Stirling, G.R. and Kopittke, R. (2000) Sampling procedures and damage thresholds for root-knot nematode (*Meloidogyne javanica*) on pineapple. *Australian Journal of Experimental Agriculture* 40, 1003–1010.
Stoyanov, D. (1967) Additions to host records of *Meloidogyne* sp., *Helicotylenchus multicintus* and *Rotylenchus reniformis. Nematologica* 13, 173.
Sueno, W.S.K., Marrero, G., de Silva, A.S., Sether, D.M. and Alvarez, A.M. (2014) Diversity of *Dickeya* strains collected from pineapple plants and irrigation water in Hawaii. *Plant Disease* 98, 817–824.
Taniguchi, G. (2011) Clumped versus broadcast applications of bait to control big-headed ants, *Pheidole megacephala* (F.), (Hymenoptera: Formicidae) in pineapple fields. *Sociobiology* 57, 285–290.
Taniguchi, G., Thompson, T. and Sipes, B. (2005) Control of the big-headed ant, *Pheidole megacephala* (Hymenoptera: Formicidae) in pineapple cultivation using Amdro in bait stations. *Sociobiology* 45, 1–7.
Tsai, B.Y. (1978) Anhydobiosis of the reniform nematode: survival and coiling. MS thesis, University of Hawaii at Manoa, Honolulu, HI.
Velasco, P.H., Ibarra, D.G.E., Ceballos, R.E. and Sifuentes, A.J.A. (1968) Biologia y control del gusano barrenador *Thecla* sp. de la pina em el sureste de México. *Agricultura Técnica en México* 2, 426–429.
Ventura, J.A., Costa, H., Cabral, J.R.S. and de Matos, A.P. (2009) 'Vitória': new pineapple cultivar resistant to fusariosis. *Acta Horticulturae* 822, 51–56.
Vorsino, A., Taniguchi, G.Y. and Wright, M.G. (2005) *Opogona sacchari* (Lepidoptera: Tineidae), a new pest of pineapple in Hawai'i. *Proceedings of the Hawaiian Entomological Society* 37, 97–98.
Waite, G.R. (1993) Pest. In: Broadley, R.H., Wassman III, R.C. and Sinclair, E. (eds) *Pineapple Pests and Disorders.* Queensland, Australia: Department of Primary Industries.
Wang, K.H., Sipes, B.S. and Schmitt, D.P. (2001) Suppression of *Rotylenchulus reniformis* by *Crotalaria juncea*, *Brassica napus*, and *Tagetes erecta*. *Nematropica* 31, 235–249.
Wang, K.H., Sipes, B.S. and Schmitt, D.P. (2002) Management of *Rotylenchulus reniformis* in pineapple, *Ananas comosus*, by intercycle cover crops. *Journal of Nematology* 34, 106–114.
Wang, K.H., Sipes, B.S. and Schmitt, D.P. (2003) Enhancement of *Rotylenchulus reniformis* suppressiveness by *Crotalaria juncea* amendment in pineapple soils. *Agriculture, Ecosystems and Environment* 94, 197–203.
Wang, K.H., Sipes, B.S. and Hooks, C.R.R. (2010) Sunn hemp cover cropping and solarization as alternatives to soil fumigants for pineapple production. *Acta Horticulturae* 902, 221–232.

Wijeratnam, R.S.W., Hewajulige, I.G.N. and Abeyratne, N. (2005) Postharvest hot water treatment for the control of *Thielaviopsis* black rot of pineapple. *Postharvest Biology and Technology* 36, 323–327.
Wijeratnam, R.S.W., Hewajulige, I.G.N. and Perera, M.G.D.S. (2006) The control of black rot disease and the application of a novel wax formulation for extending storage life of pineapple. *Acta Horticulturae* 702, 185–189.
Wijesinghe, C.J., Wijeratnam, R.S.W., Samarasekarab, J.K.R.R. and Wijesundera, R.L.C. (2011) Development of a formulation of *Trichoderma asperellum* to control black rot disease on pineapple caused by *Thielaviopsis paradoxa. Crop Protection* 30, 300–306.
Williams, D.D.F. and Fleisch, H. (1993) Historical review of pineapple breeding in Hawaii. *Acta Horticulturae* 334, 67–76.
Wright, M.G. (2012) Pineapple insects: ecology and control. In: *Encyclopedia of Pest Management.* New York: Taylor & Francis, pp. 1–3.
Yabor, L., Arzola, M., Aragón, C., Hernández, M., Arencibia, A. and Lorenzo, J. C. (2006) Biochemical side effects of genetic transformation of pineapple. *Plant Cell, Tissue and Organ Culture* 86, 63–67.
Yabor, L., Valle, B., Carvajal, C., Aragón, C., Hernández, M., González, J. and Lorenzo, J.C. (2010) Characterization of a field-grown transgenic pineapple clone containing the genes chitinase, AP24, and bar. *In Vitro Cellular and Developmental Biology – Plant* 46, 1–7.
Zem, A.C., Sanches, N.F. and Reinhardt, D.H.R.C. (1980) Niveis de infestacao de nematoides na abacaxicultura do Estado da Bahia. *Revista de Agricultura, Piracicaba, Brazil* 55, 25–32.

13 Postharvest Physiology, Handling and Storage

Robert E. Paull[1] and Ching-Cheng Chen[2]
[1]University of Hawaii at Manoa, Honolulu, USA; [2]National Chung Hsing University, Taiwan, Republic of China

The shipment of fresh pineapples began soon after the fruit was discovered in the new world in the 16th century. However, reliable supplies of fresh fruit were not widely available until refrigeration of fresh produce became common early in the 20th century and even then fruit quality was inconsistent. Now, fresh fruit of good quality is regularly shipped to markets throughout the developed world. Demand for all pineapples grew from 15.1 million tons in 2000 to over 25.4 million tons in 2014 (FAO, 2017a). Much of the growth is likely attributable to increased fresh fruit production. Supplies of pineapple fruit year-round became possible when forced induction of fruiting was widely adopted and dramatic increases in fresh fruit shipments occurred after commercial quantities of 'MD-2' (Gold) fruit became available after 1996 (Bartholomew, 2009).

Though 'Smooth Cayenne' is the major cultivar worldwide, it has deficiencies as a fresh fruit. These deficiencies include high acidity, low ascorbic acid, sometimes poor flavour and a susceptibility to translucency. Cultivars better suited for fresh fruit are now planted (Paull, 1992; Taniguchi *et al.*, 2008). For the fresh market, 'MD-2' (PRI 73-114) is mainly grown in the warm tropics, 'CO-2' (PRI 73-50; MD-1) is grown in the cooler tropics (Taniguchi *et al.*, 2008), 'Pérola' is popular in Brazil and 'Queen' is grown in Africa, Australia and China.

Fruit Development

Natural induction (NI) of inflorescence development disrupts harvest scheduling, increases the number of passes necessary to harvest a field, and if small plants are induced, leads to smaller fruit. Techniques to prevent NI, including transgenic transformation, are reviewed by Wang and Paull (forthcoming) and in Chapter 11. Fruit development is also covered in Chapters 2 and 11.

Due to the fused nature of the tissue within a single fruitlet, the flesh of the fruit is not sterile but contains fungi, yeasts and bacteria (Rohrbach and Apt, 1986) though the population of microorganisms declines with fruit development (Chen and Paull, unpublished results). Fruit development studies (Sideris and Krauss, 1938) have shown that fruit weight and its components (core, fruitlets, the collective flesh, fruit shell) increase in a continuous sigmoid fashion (Fig. 13.1a) once the inflorescence has been initiated (Singleton, 1965; Gortner *et al.*, 1967; Teisson, 1973). Fruit mass increases about 20-fold from the time of flowering until maturation (Singleton, 1965; Teisson and Pineau, 1982). The number of fruitlets comprising a fruit mainly varies with the cultivar and fruit size. A typical 'Smooth Cayenne' fruit has about 150 fruitlets and produces a mature fruit weighing about 2.2 kg (Tay, 1977). The final fruit size/weight is affected by the number of fruitlets and individual fruitlet cell number and cell size; the cell number is the most significant factor (Li *et al.*, 2010).

Fruit dry matter content can vary with the conditions prevailing during fruit development, while the crown probably indirectly affects the growth of the fruit (Senanayake and Gunasena, 1975; Py *et al.*, 1987; Chen, 1999; Liu *et al.*, 2017). Crown growth commences 30–45 days after fruit growth has commenced and is significantly influenced by the growing environment and fertilization (Py *et al.*, 1987). Crown removal early in fruit development in some studies does increase fruit weight. Crown size can be reduced by the plant growth regulator chloroflurenol when applied at the flowering stage (Dalldorf, 1981; Py *et al.*, 1987). NAA and 3-CPA have also been used to reduce crown size (Bartholomew and Criley, 1983). Preliminary work also suggests that the crown may play a role in fruit translucency development (Paull and Reyes, 1996; Chen, 1999).

Crown size is an aesthetic character of economic concern for packing, and is generally part of the grading standard (Table 13.1). Crown size can vary greatly and the variation has not been carefully studied but apparently is not highly correlated with fruit size or stem starch. Crown photoassimilates seem to be derived from its own photosynthesis (Hepton, 2003).

Fruit Physiology

Most fruit physiology studies referred to below are with 'Smooth Cayenne'. The half-yellow stage is regarded as ripe and is near the maximum fruit weight if still on the plant (Wardlaw, 1937). Fruit development (Fig. 13.1a) and composition changes (Fig. 13.1b) during growth have been reviewed (Gortner *et al.*, 1967; Bartholomew and Paull, 1986; Py *et al.*, 1987; Paull, 1993). The most marked changes in flesh composition occur 3–7 weeks prior to the half-yellow shell colour stage (Dull *et al.*, 1967; Dull, 1971; Tay, 1977; Teisson and Pineau, 1982; Chen and Paull, 1995). Just prior to the half-yellow stage, fruit translucency can start to develop until maturation, with this development continuing after harvest in some fruit.

Senescence related loss of membrane integrity leads to water-soaked, translucent flesh that tends to be softer than non-translucent fruit. Gortner and Leeper (1969) showed that amides, nitriles, simple esters and salts of phenoxy acids, naphthalene compounds, phenyl acid and trichloro phenoxyacetic acids can delay fruit senescence. Naphthaleneacetic acid and phenoxyacetic acid are the most effective on delaying shell yellowing when applied as brief postharvest dips (Gortner, 1969). These chemicals induce little change in soluble solids content (SSC), acidity, carotenoid pigments or vitamin C content, but are phytotoxic to the crown. There are no marked changes in fruit texture during ripening. Water loss can lead to some reduction in fruit firmness.

Shell chlorophyll

There is little change in shell chlorophyll until the final 10–15 days before full shell yellowing,

Table 13.1. Subjective rating used for fresh pineapple fruit evaluation.

Criteria	Scale	Description
Shell colour	0 to 6	Immature green to overripe 0–12%, 13–37%, 38–62%, 63–87%, 88–100% yellow eyes
Shell appearance	1 to 3	Brown or black discoloration to normal, fresh glossy
Crown	1 to 3	Dry, limp, brown to turgid normal colour and fresh
Whole fruit texture	1 to 3	Soft to firm: feel of fruit
Flesh texture	1 to 3	Limp to firm, crisp or fibrous texture
Translucency (slice 1/3 from base)	1 to 5	Opaque to fully translucent
Disease and disorders		Number affected fruit and severity in each fruit
Flavour	1 to 3	Fermented, bitter aftertaste Very high acid, very high esters, very bland Normal

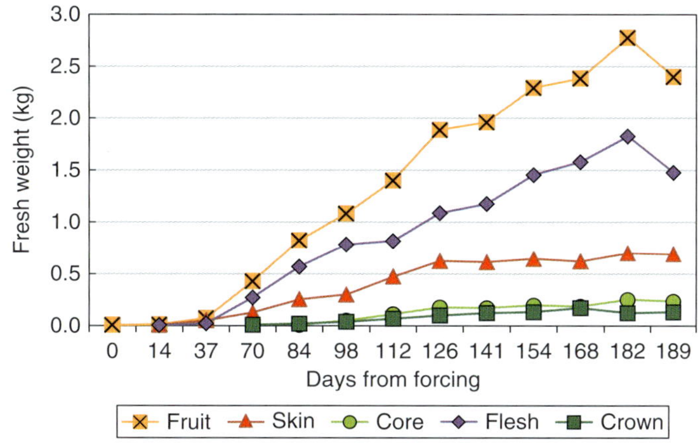

Fig. 13.1a. Changes in the fruit during development. The sequence in the development of different fruit components (flesh, skin, core and crown) from anthesis. Figure by R.E. Paull.

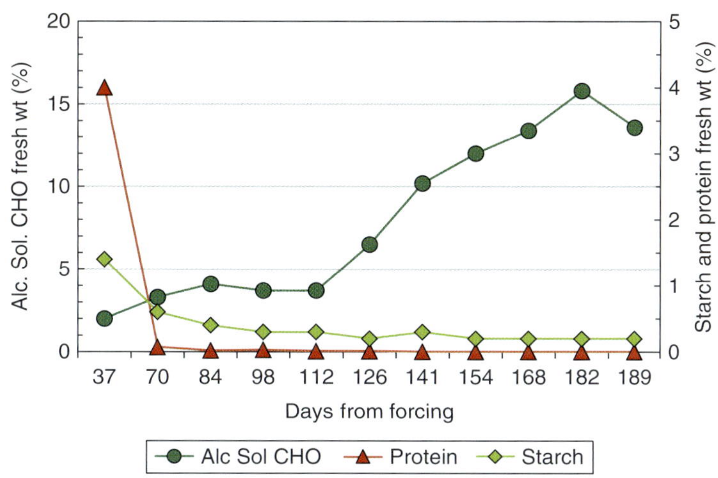

Fig. 13.1b. Changes in the flesh starch, protein and alcohol soluble carbohydrates as a percentage of the flesh fresh weight. The alcohol-soluble carbohydrates are mainly the sugars that contribute to fruit sweetness. Figure by R.E. Paull.

after which it declines (Gortner, 1965; Py *et al.*, 1987; Abdullah *et al.*, 2002). Shell carotenoid pigments also remain reasonably constant during this last phase, declining slightly before rising again as the fruit senesces. Flesh carotenoids increase during the final 10 days before the full ripe stage (Gortner, 1965; Teisson and Pineau, 1982; Py *et al.*, 1987). A similar decrease in shell chlorophyll and an increase in flesh carotenoids occur in harvested fruit (Dull *et al.*, 1967; Chen and Paull, 1995). Chlorophyll in the crown leaves and the fruit skin also decreases during postharvest storage (Abdullah *et al.*, 2002).

Respiration

The non-climacteric pineapple fruit produces around 22 ml kg^{-1} h^{-1} of CO_2 at 23°C with no dramatic respiratory change during ripening (Dull *et al.*, 1967). The rate of respiration of the fruit increased while that of the crown decreased during storage at ambient temperature (Abdullah *et al.*, 2002; Shiomi *et al.*, 2002). Ethylene production in the fruit increased during ripening (Shiomi *et al.*, 2002), but had no pronounced peak (Dull *et al.*, 1967; Shiomi *et al.*, 2002). Cazzonelli *et al.* (1998) reported that

ACC synthase and ACC oxidase mRNAs accumulated during pineapple fruit ripening, but the role of ethylene during pineapple fruit ripening remains unclear. Exogenous ethylene application stimulates only respiration rate when there is some chlorophyll remaining in the shell, and may also open the crown leaf stomata.

Sugar metabolism

Sugar content plays an important role in the flavour characteristics and commercial assessment of pineapple fruit quality (Py *et al.*, 1987). SSC (TSS), mainly sugars, is often used as an indicator of fruit maturity and quality (Paull, 1993). SSC can vary by 40 g l^{-1} from the more mature, sweeter basal tissue to the crown end of the fruit (Sideris and Krauss, 1933a,b; Miller and Hall, 1953), and decline only slightly after harvest (Paull and Rohrbach, 1982; Chen and Paull, 1995). Starch is not accumulated as the fruit ripens (Dull, 1971), though high during early fruit growth, which could explain the absence of dramatic postharvest changes in SSC.

The major sugars in mature fruit are sucrose, glucose and fructose (Gawler, 1962) and the peak sucrose concentration is attained at full-yellow stage then declines. Fruit sugars continued to increase through to senescence, unless the fruit is harvested (Kelly, 1911). Chen (1999) showed that total SSC was low during fruit growth and composed mainly of glucose and fructose. The glucose concentration was slightly higher than fructose during early fruit development. Sucrose accumulated rapidly 6 weeks before commercial harvest (Lodh *et al.*, 1972; Py *et al.*, 1987), and ultimately exceeded the glucose and fructose concentration (Chen, 1999). Fructose and glucose continue to increase postharvest (Singleton and Gortner, 1965; Tay, 1977). In addition, sucrose accumulated more in the fruitlet than in the inter-fruitlet tissue until the last 2 weeks of fruit development, when sucrose accumulation rate in the inter-fruitlet tissue was greater than in the fruitlet (Chen, 1999).

The enzymes sucrose synthase (SS), sucrose phosphate synthase (SPS), and invertase are thought to control sugar accumulation in fruit tissue. The activity of SS is high in younger pineapple fruit and declines to very low level 6 weeks before harvest, while the activity of SPS is relatively low and constant throughout fruit development when precautions are taken to avoid invertase activity during sampling (Chen, 1999). The activities of acid, neutral, and cell-wall invertases are high in the younger fruit and decline to low levels 6 weeks before harvest when sucrose starts to accumulate. The activity of cell-wall invertase (CWI) does increase 4 weeks before harvest, mainly in fruitlet tissue, while the activities of neutral invertase (NI) and acid invertase (AI) remain low, concomitant with the accumulation of sucrose, indicating that these enzymes may be a prerequisite for sucrose accumulation in pineapple fruit flesh. The apoplastic liquid volume and sugar concentrations increase along with fruit maturation (Chen, 2010). The high activity of CWI, favouring apoplastic phloem unloading, may play a role in sugar accumulation in pineapple fruit flesh at the latter stages of fruit development (Chen, 1999).

Organic acid metabolism

Juice pH declines from 3.9 to 3.7 as fruit approach the full-yellow stage (Teisson and Pineau, 1982; Saradhuldhat and Paull, 2007), and increases as the fruit senesces, with titratable acidity showing the opposite trend (Gortner and Singleton, 1965; Teisson and Pineau, 1982; Chen and Paull, 1995; Saradhuldhat and Paull, 2007). The flesh acidity increases distally from the central core (4 meq 100 ml^{-1}) outwards to 10 meq 100 ml^{-1} (Huet, 1958b), and 65–70% of the total nonvolatile acids, which are citric and malic (Chan *et al.*, 1973), occur as free organic acids (Chan *et al.*, 1973; Teisson and Combres, 1979). Malic acid can vary from 18% to 30% of total acids in pineapple and does not vary markedly between cool and warm season crops (Chan *et al.*, 1973), but can fluctuate threefold with the weather (Gortner, 1963).

Fruit acid accumulation is strongly affected by cultivar (Saradhuldhat and Paull, 2007) and temperature during early fruit growth, radiation later, and rainfall throughout (Dorey *et al.*, 2016). A canning 'Smooth Cayenne' clone with high acid content, gradually increased in fruit acidity to peak a week before harvest (Fig. 13.2). In contrast, a low acid cultivar increased in acidity 6–8 weeks after flowering, reaching a peak 15 weeks after flowering and then sharply declined in 2 weeks as the fruit ripened (Saradhuldhat and Paull, 2007). The increase in titratable fruit acidity paralleled changes in citric acid content of both clones. Citric acid (28–66% of total acids) is lower in 'Smooth Cayenne' fruit harvested in the warm season and tends to vary primarily with stage of fruit development (Gortner, 1963; Singleton and Gortner, 1965). Malic acid accumulates when sunlight and evapotranspiration are low (Singleton and Gortner, 1965), there are no marked changes in malic acid just before harvest and changes in fruit potassium were significantly correlated with fruit acidity. Citrate synthase (EC 4.1.3.7) activity increased the week before harvest and the increase coincided with the peak in citric acid in the high-acid clone. Just before harvest aconitase (EC 4.2.1.3) activity increased as the acidity declined in the low-acid cultivar (Saradhuldhat and Paull, 2007). Phosphoenolpyruvate carboxylase, malate dehydrogenase and malic enzyme may contribute to these changes (Saradhuldhat and Paull, 2007; Liu, 2009).

Fruit malic acid levels do not change after harvest or during and after storage (Chen and Paull, 1995). During storage at 7°C, citric acid increases about 25%, then declines slightly when fruit are held at 22.5°C. The citric acid content of non-stored fruit does not change but titratable acidity declines slightly after harvest.

Ascorbic acid content varies significantly with cultivars (Kerns *et al.*, 1936; Singleton, 1955; Teisson and Combres, 1979) and can vary from 200 mg l^{-1} in 'Smooth Cayenne' to 710 mg l^{-1} in 'Spiny Guatemala' (Singleton, 1955). Ferreira *et al.* (2016) found a significant positive correlation between ascorbic acid concentration and antioxidant activity among Brazilian cultivars. Ascorbic acid is also correlated with a cultivar's acidity (Hamner and Nightingale, 1946), though it does not contribute substantially to titratable acidity, and is 25% higher near the surface of the fruit than near the core. The ascorbic acid level increases with increasing solar radiation (Gortner and Singleton, 1965) and air temperature. The level of fruit ascorbic acid at harvest has been negatively related to the intensity of internal browning symptoms associated with postharvest chilling injury (CI) (Miller and Heilman, 1952; Teisson and Combres, 1979; Paull and Rohrbach, 1982). Internal browning is a minor problem if fruit ascorbic acid content is greater than 500 µM (Teisson and Combres, 1979).

Fruit cell walls

Overall the non-cellulosic walls of pineapple fruit are intermediate between the unlignified *Poaceae* and typical dicotyledon cell walls. Glucurono-arabinoxylans are the major component of the non-cellulose fraction. Xyloglucans

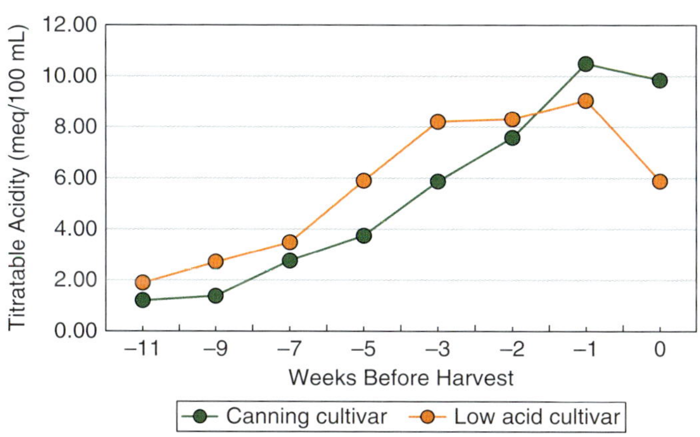

Fig. 13.2. Changes in titratable acidity of a low-acid hybrid and a higher-acid canning cultivar, Smooth Cayenne during fruit growth and development. Redrawn from Saradhuldhat and Paull, 2007.

along with smaller amounts of pectic polysaccharides and glucomannans are present. The cell walls of pineapple fruit parenchyma are unlignified, but do contain esterified ferulic acid (Smith and Harris, 1995). The ferulic acid is ester-linked to glucurono-arabinoxylans as is typical of the primary walls of *Poaceae* species (Smith and Harris, 2001). Glucuro-arabinoxylans have been isolated from lignified cell walls of pineapple leaf (Bhaduri *et al.*, 1983; Jarvis *et al.*, 1988).

Pineapple juice contains predominantly galactomannans (Chenchin and Yamamoto, 1978) and no acid sugars are detected, suggesting little pectin hydrolysis during ripening. This juice neutral polysaccharide forms a gum on processing equipment that is readily hydrolysed by commercial cellulase, hemicellulase and pectinase preparations (Chenchin *et al.*, 1984). The neutral polysaccharide extracted from pineapple fruit cell wall is composed of xylose, arabinose, glucose, galactose, and minor quantities of mannose (Bartolomé and Rupérez, 1995). Glucan ($1\rightarrow3,1\rightarrow4$) linked are absent in contrast to *Poaceae* cell walls.

Volatiles

Nearly 300 different volatile compounds have been identified in pineapple (Dupaigne, 1970; Flath and Forrey, 1970; Flath, 1986; Takeoka *et al.*, 1989; Wu *et al.*, 1991; Umano *et al.*, 1992; Elss *et al.*, 2005; Liu *et al.*, 2011; Zheng *et al.*, 2012; Žemlička *et al.*, 2013). Nevertheless, only some of them have been identified as key contributors of the pineapple aroma (Žemlička *et al.*, 2013). The aroma volatile compounds of pineapple vary with varieties (Gray, 1953; Liu *et al.*, 2008; Zhang *et al.*, 2009; Zheng *et al.*, 2012; Pino, 2013; Žemlička *et al.*, 2013), growing seasons (Haagen-Smit *et al.*, 1945; Silverstein, 1971; Liu *et al.*, 2011); different parts of the same fruit (Wei *et al.*, 2011a; Zheng *et al.*, 2012; Steingass *et al.*, 2015b), being higher in the pulp than the core, stage of fruit development (Umano *et al.*, 1992), ripening stages of the fruit (Wei *et al.*, 2011b; Steingass *et al.*, 2014, 2015a, 2015c, 2015d), and postharvest storage conditions (Wei *et al.*, 2011b; Liu and Liu, 2014; Steingass *et al.*, 2016).

Esters are the main volatile compounds detected in the fruit flesh (Umano *et al.*, 1992; Wei *et al.*, 2011a), which increase both on the plant and after harvest (Flath, 1986). Translucent fruit are higher in ester content, such as ethyl acetate, which is very low in opaque fruit (Gray, 1953).

Lipids and amino acids

Total lipids decline at maturity, along with squalene content, while phospholipids, total sterol and amino acid values increase (Selvaray *et al.*, 1975). Free amino acids in the juice are at a minimum midway in fruit growth (Gortner and Singleton, 1965). The exception is free methionine, found at low levels until the onset of ripening when it increases to 0.7 mM at senescence.

Phenols

Pineapple fruit are rich in phenolics and are a good source of antioxidants (Haripyaree *et al.*, 2010; Hossain and Rahman, 2011). The phenolic content is higher in the skin than in the stem and the flesh (Rasheed *et al.*, 2012: Steingass *et al.*, 2015b). Eight phenolic compounds (gallic acid, gentisic acid, syringic acid, vanillin, ferulic acid, sinapic acid, isoferulic acid and o-coumaric acid) are commonly found in pineapple fruit flesh (Yapo *et al.*, 2011). The major polyphenols in pineapple peels are gallic acid, catechin, epicatechin, and ferulic acid (Li *et al.*, 2014).

Enzymes

Peroxidase activity falls steadily during fruit development (Gortner and Singleton, 1965) reaching a minimum of one-third the initial value during ripening. Acid peroxidase (pH optimum 5.0) does not appear to be related to CI symptom development (Teisson, 1977), though it does decline during storage (Teisson and Martin-Prevel, 1979). The two chilling related enzymes, peroxidase and polyphenol oxidase (PPO) have been purified (Teisson, 1977) and the genes characterized (Stewart *et al.*, 2001;

Zhou *et al.*, 2003a,b; Raimbault *et al.*, 2011). When the PPO gene is down-regulated in pineapple chilling induced tissue browning is suppressed (Ko *et al.*, 2013).

Van Lelyveld and de Bruyn (1977) found no ascorbic acid oxidase; however, ammonium sulfate precipitated protein contained an ascorbic acid oxidase with a pH optimum >8 and no activity below pH 6.0 (Teisson, 1977). Different assay conditions may explain the discrepancy. The PPO optimum pH is ≈5.0 and optimum temperature near 45°C (Teisson, 1977). Das *et al.* (1997) found three PPO isoforms with the major isoform being a tetramer of identical subunits (ca. 25 kDa), and had optimum activity between pH 6 and 7. The difference in pH may be due to the different extraction methods used or different isoforms. PPO is stable to heat when extracted, but loses >50% of its activity following 20 minutes exposure to 60°C *in vivo* (Teisson, 1977) and thiols and bisulphite inhibit activity. Catalase and indole acetic acid oxidase have also been detected but changes in activity during ripening or storage have not been reported. Indole acetic acid oxidase has a pH optimum of 3.5 which is at variance with other fruit (Teisson, 1977).

Acetone powders of pineapple stem extracts contain the proteinase bromelain and a family of polypeptide inhibitors of this enzyme (Reddy *et al.*, 1975). Proteinase activity abruptly appears after flowering and remains high during fruit development (Gortner and Singleton, 1965) and declines during ripening (Gortner and Singleton, 1965; Lodh *et al.*, 1972). The major cysteine proteinase is fruit bromelain (EC 3.4.22.33), a nonglycosylated proteinase that is immunologically distinct from the glycosylated stem bromelain (EC 3.4.22.32) (Rowan *et al.*, 1990; de Lencastre Novaes *et al.*, 2016). Fruit bromelain has an estimated molecular mass between 23 to 31 kDa by different laboratories (Yamada *et al.*, 1976; Ota *et al.*, 1985; Rowan *et al.*, 1990). The fruit bromelain is an acidic protein and its isoelectric point (pI) is 4.6, different from the basic stem bromelain whose pI is 9.6 (Yamada *et al.*, 1976). The amino terminal end of the fruit bromelain has an additional alanine (Yamada *et al.*, 1976). The sequence around the reactive cysteine is the same. Fruit bromelain has a 20% cross-reactivity with anti-stem bromelain (Sasaki *et al.*, 1973).

Fruit Harvesting and Handling

Quality criteria

Pineapple fruit quality refers to the sum of characteristics that makes a fruit most palatable and desirable to consumers (Singleton, 1955; Paull, 1993). However, this definition is not a measurable standard because consumer tastes, ethnicity and maybe even price paid can affect consumer preferences. If multiple fruit quality criteria (Table 13.1) were measured, the difficulty is how to sum the criteria and what normal range of variation would be used in the application of this scale.

Quality criteria used have been appearance (size, condition, shape), colour (shell and flesh), taste (sugars, acids), aroma, flesh translucency, texture and fibre content (Table 13.1). Skin colour is the most common measure of physiological maturity and consumer expectations of quality (Dull, 1971; Soler, 1992a). For some consumers, food value and vitamin content are also considerations. The changes in these parameters during pineapple fruit growth and development have been summarized (Gortner *et al.*, 1967) and these data indicate the complexity of the changes that occur. For fruit quality purposes, sugars are estimated as SSC by refractometry as percentage (g/100 ml juice) mainly sugars or the less preferred °Brix. Problems soon arise in applying these criteria; SSC to acidity ratio is often used as a measure of flavour, though the ratio is more impacted by acid variation than SSC. A 20% SSC with 1% acidity will not taste the same as 10% SSC and 0.5% acidity. The problem of SSC and acidity is compounded by the effect of cultivars, climate (tropical vs subtropical), season, maturity stage at harvest, position in the fruit and fruit development conditions (Smith, 1993; Chen, 1999; Saradhuldhat and Paull, 2007). The ascorbic acid content in 'Smooth Cayenne' varies from 130 mg l^{-1} at the base of the fruit to 280 mg l^{-1} at the less mature top part of the fruit, a similar variation occurs for titratable acidity (Singleton, 1955), while SSC can be 19% SSC at the base and 15% at the top. Seasonal variation in sugar and acid levels can be significant, acid levels are higher during cool season and sugars (SSC) are lower.

Flesh SSC of 'Smooth Cayenne' was reported as having the highest correlation with taste panel eating quality ($R^2 = 0.70$, Prob < 0.01) and to be the most suitable year-round index (Smith, 1988a). The introduction of the low-acid 'MD-2' and 'CO-2' have greatly diminished concerns about fruit quality and a panel in Malaysia preferred 'MD-2' over other cultivars because of its aroma, sweetness and colour (Syahrin, 2011). 'Smooth Cayenne' pineapple is now mainly processed, which allows processors to adjust sugars and acid to meet consumer preferences. The minimum flesh standard for low acid cultivars is a SSC of 12% (FAO, 2017b) and the worldwide acceptance of 'MD-2' indicates that few other criteria are important. The FAO standards also cover 'Queen' fruit imported into EU countries.

Specific gravity (SG) could be a useful intact fruit grading index for eating quality of 'Smooth Cayenne' fruits (Smith, 1988b) but for 'MD-2' fruit is mainly a method of identifying translucent fruit that have poor shipping characteristics.

Fruit size and quality

Fruit size is set by plant size at forcing within a given environment, as plant size influences the number of florets per fruit developed. Plant size can be represented as plant weight, 'D' leaf weight or the number of leaves (Van Overbeek, 1946; Py *et al.*, 1987; Malézieux, 1993). The harvest index (fruit weight at harvest/plant weight at forcing for 'Smooth Cayenne' (Hepton, 2003) can vary from 0.5 in the humid tropics to 1.0 in the cool subtropics and 'MD-2' responds similarly (J. Vásquez Jiménez, 2017, personal communication). The subject is covered in detail in Chapters 7 and 11.

The subject of nutrition and fruit quality is covered in Chapter 8. In general, if nutrition is in the optimum range, there is no consistent effect of higher levels of most nutrients on fruit quality. The one exception appears to be potassium, which if applied during fruit development may increase fruit acdity and has also been reported to reduce fruit CI, especially when supplied as the chloride (Teisson *et al.*, 1979).

An increase in temperature from 25 to 27°C over 5 months before harvest is related to a decrease in acidity from 12 to 7 meq/100 ml of juice (Huet, 1958a). Sunlight can also cause a similar decline in acidity (Lacoeuilhe, 1978). An inverse relationship is found between evapotranspiration in the 1–2 weeks before harvest and malic acid content (Gortner, 1963), while citric acid levels are little affected and more related to fruit development. There is a lag of about 1 week between the increase in the evaporation (sunlight) and the decrease in malic acid content.

In addition to the decrease in fruit weight with increasing planting density, there is also a decline in translucency that may not be related to light competition and fruit photosynthate supply (Chen and Paull, 2000). Planting density has little effect on fruit SSC or acidity though they tended to be slightly higher (Sanford, 1963). Fruit esters do tend to decrease in fruit grown at higher planting densities. Though probably unrelated to density, ratoon crops tend to have smaller fruit with higher sugars, less acid and more flavour.

Fruit availability

On large plantations in favourable climates, planting, forcing and harvesting occur throughout the year. Adjustment in area planted per month can allow for lower fruit demand in some months of the year. When planting throughout the year, days from planting to forcing to first flower and first flower to harvest will vary at different times of the year. The variation is relatively small near the equator with near-constant average temperatures. As production moves away from the equator, adjustments in planting date and forcing date need to be made to assure harvestable fruit at a certain time of the year.

Accurate estimates of harvest date and flowering uniformity in a field improves labour scheduling and fresh fruit marketing. Days from forced induction to various stages of fruit development is highly temperature-dependent and predictable at least in some environments. More information can be found in Chapter 11.

Preparation for harvest

Ethephon is used by growers 48 hours or more before harvest to accelerate shell degreening and foster 'ripening' (Poignant, 1971; Crochon *et al.*, 1981; Smith, 1991; Soler, 1992a,b). The advantage of this treatment is that fruit yellow (shell degreens) more uniformly in the carton, than untreated fruit, hence presenting a better overall carton appearance. This accelerated shell degreening is due to the destruction of chlorophyll, giving the shell a more uniform yellow colour. The application of acidified ethephon should occur just as natural shell yellowing begins, to assure good fruit quality. In large commercial fields of 'MD-2' and 'CO-2', degreening is done when average fruit SSC in representative field samples reaches 12%, though some growers closer to markets may use higher levels. See Chapter 7 for more information.

Fruit maturity and quality at harvest

Fruit maturity is evaluated on the extent of fruit 'eye' flatness and shell yellowing in the 'Smooth Cayenne' clones (Paull, 1993). Consumers similarly judge fruit quality by shell colour and aroma. A minimum of 12% SSC is required for fresh fruit in Hawaii (Standards, 1968), while others have suggested 14% (Smith, 1988b; Soler, 1992a). The SSC (TSS) determined by refractive index needs to be adjusted for acidity to arrive at sugar percentage. The formula is: sugar (%) = SSC (% refractive index) – 0.192 (ml of 1 M NaOH to neutralize 100 ml juice). This formula applies generally, except for very translucent or overripe fruit. A rule of thumb value to obtain actual sugar content is to multiply the refractive index value by 0.85. This 0.85 factor may vary from 0.8 to 0.88 in opaque fruit and 0.86 to 0.96 in translucent fruit (Sideris and Krauss, 1933a,b). Fruit SSC could also be determined by near-infrared spectroscopy (Shiina *et al.*, 1993; Guthrie and Walsh, 1997) to within 1% SSC. A sugar to acid ratio of 20:40 is recommended in older literature while Soler (1992a) recommended 14:20, both are lower than recommended for canned product (30:50) (Sideris and Krauss, 1934). However, extremely high or low values can lead to an insipid flavour.

Full-yellow fruit of most cultivars are unsuitable for transporting to distant markets because they have a short shelf life. Therefore, less mature fruit are selected (Akamine, 1963; Cancel, 1974). Immature fruit are not shipped, since they do not develop good flavour, have low total soluble solids, and are more prone to CI (Rohrbach and Paull, 1982). Immature fruit, if held at 20°C, have better eating quality with time, approaching the quality of green mature fruit that do not change in eating quality after harvest (Smith, 1993). The same author found that one-quarter ripe fruit maintained eating quality for up to 15 days at 20°C.

Harvesting

Pineapple is hand harvested with pickers being directed as to stage or stages of shell colour required. Picking early in the morning and protection of harvested fruit from the sun can reduce heat load at cooling. In Hawaii, fruit are selected, broken from the peduncle by pickers, and placed on a conveyor belt running on a boom to transfer the fruit to truck field bins or to be field packed (Fig. 13.3a). If packed at a central packing shed, the fruit is placed in the bin crown down by hand to avoid injury. Field-packed fruit shows significantly less fruit bruising and crown injury. When the field bin with fruit arrives at the packing shed fruit may be unloaded by hand, by submerging the field bin in water (Fig. 13.3b), or allow the fruit to slide out of the field bin into water. Fruit with high translucency ('sinkers') are separated at this step. The sinkers are highly translucent fruit that are very fragile and have poor shelf life. High translucency is also associated with bacterial and yeasty fermentation and acetic souring during handling, shipping and marketing. Adjusting the specific gravity with salt in the dump water can be used to grade fruit as to translucency percentage. The water in the dump tank needs to be chlorinated and replaced frequently to prevent build-up of disease organisms.

Care is taken to avoid damage to the crown leaves. Gibberellic acid can help retain crown condition during storage (Liu, 1986), though it is not approved in the USA for postharvest use.

Fig. 13.3. (a) Pineapple harvest and postharvest handling steps. Selected hand-harvested fruit are placed on a boom that spans half the block, the belt on the boom brings the fruit to the lorry and the fruit is stacked in the lorry bin. (b) The loaded bin is submerged into chlorinated water and the fruit tipped out, translucent fruit that do not float are elevated out of the tank by the elevator on the left-hand side, the fruit that float are removed from the dump/wash tank by the right hand elevator. Both streams are rinsed with potable water. (c) The fruit are inspected and oriented on the conveyor belt. (d) Coating with fungicide is applied to the fruit body, not the crown. (e) Fruit after sorting by size are inspected and sorted by fruit colour then packed into cartons. (f) Palletized cartons are subjected to forced air cooling before shipping. Photos R.E. Paull.

Postharvest fungicides and waxing

Hawaiian pineapples are commercially treated with a fungicide in a dip or spray application (Fig. 13.3c) to control postharvest fruit rot, caused by the fungus *Chalara paradoxa (Ceratocystis paradoxa* (de Cynes) Moreau) (Chapter 12).

The shelf life of pineapple fruit can be prolonged by surface coating such as waxes (Ayernor *et al.*, 2010, 2011; Adetunji *et al.*, 2012), gibberellic acid (Mandal *et al.*, 2015) and salicylic acid treatment (Lu *et al.*, 2010; Mandal *et al.*, 2015). Fruit coatings frequently contain mixtures of polyethylene/paraffin or carnauba/paraffin and are applied with a fungicide with the coating imparting a more uniform fungicide covering. The major advantages of wax coatings are to reduce the internal browning symptoms of chilling injury (CI). Waxing also reduces postharvest water loss and improves fruit appearance (Paull and Rohrbach, 1985; Hu *et al.*, 2012; Benítez *et al.*, 2014). There is no worldwide uniformity in acceptance of wax components, so importing country restrictions need to be considered. If the wax injures the crown leaves, only the fruit body is waxed (Fig. 13.3d) by correctly orientating the fruit on the conveyor belt (Fig. 13.3c).

Controlled and modified atmosphere storage

As with other non-climacteric fruit, controlled atmospheres via decreased oxygen levels or increased carbon dioxide have shown only minimal effectiveness at extending pineapple shelf life (Akamine and Goo, 1971; Paull and Chen, 2018a). The fruit waxes currently used in Hawaii generate higher internal concentrations of carbon dioxide (5%), and reduced oxygen tensions (Paull and Rohrbach, 1982). This suggests that some beneficial effect might be gained from controlled atmosphere (4% oxygen) treatments in reducing CI symptom development. Low oxygen (2–12%) has, however, been shown to enhance water loss from the crown leaves. There is limited published work on lower oxygen levels with or without various controlled carbon dioxide concentrations (Paull and Chen, 2018a).

Cellophane and polyethylene bags have been tested on numerous occasions to extend postharvest life (Paull and Rohrbach, 1982, 1985). Both types of bags delay shell degreening when compared to non-bagged fruit. Polyethylene bagging is objectionable because of the condensation problem that leads to mould development, while cellophane bags allow moisture vapour exchange and the crown especially tends to dry out. Additional problems with all bags, are the development of off-flavours and the difficulty of avoiding puncturing the bag with the crown leaves. An over-wrap on the carton overcomes the puncturing problem but leads to problems in market inspection and the wrap would need to allow some gas exchange to prevent excessive CO_2 build-up that apparently leads to off-flavours.

Postharvest degreening of shell

Treating mature green pineapple fruit with ethylene postharvest leads to shell degreening occurring in *c.*40% of the time required if fruit are left on the plant (Singleton, 1957a). Postharvest use of ethephon has been tested in the Ivory Coast (Poignant, 1971; Crochon *et al.*, 1981), Hawaii (R.E. Paull, 1985, unpublished) and Australia (Smith, 1991). Treated fruit show uniform skin degreening; however, shelf life is shortened (Paull, unpublished; Smith 1991). Ethephon can be successfully incorporated into the fruit wax (Paull, unpublished). The ability to apply ethephon with the wax is useful, as waxing delays fruit degreening. In the USA, the absence of registration for ethephon's use postharvest precludes its use. The need to degreen is related to the consumer's perception that a ripe pineapple must have a yellow skin. Exposing waxed fruit to red light during storage at 8°C also accelerates the rate of shell degreening upon return to ambient temperature (Paull, unpublished). The degree of flesh translucency development postharvest is also less than that if the fruit is left on the plant to develop a fully yellow skin. Ethylene treatment has no significant effect on flesh SSC or acidity.

Fruit grading

The fruit to be packed needs to be mature, firm, well formed, free of defects, with flat eyes and a minimum SSC of 12% in Hawaii and EU countries (Standards, 1968; FAO, 2017b). Fruit are graded based upon recognized appearance characteristics: degree of skin coloration; size (weight); absence of defect and diseases; and other market needs before packing (Table 13.2). Crown size is a crucial grade component, with a minimum size, and ratio of crown to fruit length (0.33–1.5) for the higher grades. Crown developed during the summer in Hawaii tend to be

Table 13.2. USA fresh pineapple fruit standards. All grades have 10% limits on defects, 5% on serious damage, and on 1% decay at the shipping point. Fruit defects include bruising, sunburn, gummosis and internal breakdown.

Standards	Varietal characteristics	Maturity	Free from	Slips knobs	Crown to fruit ratio length
US Fancy	Similar	Mature, firm, dry, well formed, well-developed eyes	Decay, sunburn, injury, fresh cracks, bruising, well-cured butt	Single crown, no slips	Ratio less than 1.5, not less than 12 cm in size
US #1	Similar	Mature, firm, dry, well formed, well-developed eyes	Decay, sunburn, injury, fresh cracks, overripe, bruising, fairly well-cured butt	<5 slips	Ratio less than 2.0 and greater than 10 cm in size
US #2	Similar	Mature, firm, dry, fairly well formed	Serious damage, fresh cracks, overripe, sunburn, injury, bruising	Slips	Any ratio 2 or fewer crowns

larger and may require 'gouging' at harvest to meet the standard. This 'gouging' leaves a wound for possible pathogen entry and degrades overall appearance. 'Gouging' 2 months before harvest avoiding visible scarring is also practised to limit crown growth (Soler, 1992a).

Packaging

Fruit are normally packed into cartons of two different sizes. A large carton (20 kg) containing 10–16 fruit for surface or air shipment and a smaller carton (10 kg) for air shipment with five or six fruit (Fig. 13.3e). Tourist packs of two to four fruit are also prepared. Absorbent pads are used at the bottom of the carton and between layers, if fruit are alternated horizontally within the carton. In other packs, fruit are placed vertically.

Postharvest storage

Temperatures in the range of 7.5–12°C are recommended for storage, with relative humidity of 70–95%; the higher humidity significantly reduces water loss (Paull, 1997). The more recent recommendation is for 90–95% relative humidity. In large packing sheds, forced air cooling is utilized to reduce the temperature to about 10°C (Fig. 13.3f). At a temperature of 0–4°C, the fruit may be stored for weeks, but upon removal, the fruit fails to continue ripening and, depending on the cultivar, may show severe CI.

Quarter-yellow fruit at harvest, gain about one additional week's storage for every 6°C decrease in storage temperature (Dull, 1971). Half-ripe 'Smooth Cayenne' fruit can be held for about 10 days at 7.5–12.5°C and still have about a week of shelf life, with no chilling-induced internal browning. The maximum storage life at 7°C is about 4 weeks (Paull and Rohrbach, 1985); however, when removed to ambient conditions CI-induced internal browning develops within 2–3 days. CI symptom development is more rapid at 20–25°C. Storage at 7.5–12.5°C does not generally lead to dramatic changes in acidity, though both sugars and ascorbic acid may decline (Singleton, 1957b; Paull and Rohrbach, 1982).

Postharvest chilling injury (CI)

CI symptoms in 'Smooth Cayenne' may be caused by preharvest or postharvest exposure to temperatures less than 10–12°C (Akamine *et al.*, 1975; Keetch and Dalldorf, 1979). No information on CI was found for 'MD-2' or 'CO-2', though both appear to be resistant. The symptoms develop when the fruit is returned from chilling conditions to physiological temperatures (18–30°C) and the crown is more susceptible than the fruit (Paull and Rohrbach, 1985).

Postharvest changes in fruit quality

Both postharvest storage and waxing of pineapple at low temperature (8°C) can alter external and internal fruit characteristics (Chen and Paull, 1995). Shell colour increases slightly during storage at 8°C, while there is little change in overall shell appearance. The rate of skin yellowing upon removal of fruit from 8°C to 22°C and that of fruit continuously held at 22°C are similar. Fruit stored at 8°C and removed to 22°C quickly lose shell appearance, while fruit held at 22°C begin to lose appearance after 8 days at 22°C. Juice pH declines from 3.5 to 3.1 during 14 days storage at 8°C, while titratable acidity increases. Juice pH of non-stored and fruit removed from 14 days storage at 8°C to 22°C, declines in parallel to 3.3 from 3.5, in non-stored fruit and 2.8 in stored fruit. Citric acid content increases slightly, with little change in malic acid during storage at 8°C. There was no change in malic or citric acids in fruit held at 22°C. The titratable acidity of non-stored fruit does not change for 8 days at 22°C, then increases *c.*20%, before declining. Fruit previously stored then removed to 22°C show a continuous decline in titratable acidity from 135 to 85 meq l^{-1}. Waxing fruit prevents the storage decline in juice pH at 8°C and reduces the increase in titratable acidity. Ascorbic acid initially increases then decreases during storage at 8°C (Teisson and Martin-Prevel, 1979; Chen and Paull, 1995) and increases when held at 22°C. Storage for 2 weeks at 10°C has little effect of total soluble solids or the content of individual sugars (Paull, 1993; Chen and Paull, 1995). Changes in phenols, especially postharvest and during storage, has been associated with CI (Teisson, 1977; Paull and Rohrbach, 1985; Stewart *et al.*, 2001; Zhou *et al.*, 2003a,b; Raimbault *et al.*, 2011).

Postharvest insect disinfestation

Pineapple fruit ('Smooth Cayenne') irradiated with 300–500 Grays show delayed shell degreening with no effect on flavour, SSC or crown appearance (Upadhya and Brewbaker, 1966), while 'Queen' pineapple fruit showed browning of the skin and softening of flesh at doses greater than 250 Grays (Damayanti *et al.*, 1992). Rooting of crowns treated with greater than 150 Grays is inhibited.

Postharvest Disorders

Fruit related preharvest and postharvest disorders are covered in greater details by Paull and Chen (2018b). Many of the fruit disorders seen postharvest are significantly influenced by preharvest factors. The more commonly seen disorders are discussed below.

Fruit bruising

Fruit bruising is a major problem during harvesting and packing (Paull *et al.*, 2016). The bruised area leads to leakage of cell contents and provides openings for saprophytes and disease organisms (Fig. 13.4a). Bruising has been reported in 14% of inspected shipments arriving in New York (Cappellini *et al.*, 1988). Translucent fruit are highly susceptible to bruising, and leaking of cellular contents is common in bruised translucent fruit. The bruised flesh appears slightly straw-coloured, and becomes lead-grey with time (Keetch, 1978). A bruising test based on absorption spectral changes of carotenoid extracted in 95% alcohol: petroleum ether (1:1, 60–110°C) was developed, though not used commercially. The changes in carotenoids are associated with isomerization of the carotenoids at 466 and 425 nm (Singleton *et al.*, 1961), caused by the release of acids from bruised cells (Gortner and Singleton, 1961).

A 30 cm drop can cause some impact damage (Singleton, 1958; Singleton *et al.*, 1961), normally on the impact side of the fruit. Timm and Brown (1991) estimated the impacts experienced during unloading and grading of pineapple and found that high impacts most often occur after the fruit are moved from the

Fig. 13.4. Observed disorders on fruit in the market. (a) Mechanical impact injury of fruit, affected area turns greyish-brown. (b) Chilling injury starts normally at the base of the individual florets and expands in severe cases to affect all the flesh and skin. (c) Translucent flesh has a water-soaked appearance. (d) Non-pathogenic mould growth on the end of the broken peduncle frequently associated with more translucent fruit. (e) Brown discoloration due to mild sunburn in the middle of the fruit; in severe cases the area is black and frequently collapses. (f) Small black/dark brown seeds are sometimes found under the skin. Photos R.E. Paull.

water-filled dump tank on to unpadded inspection and singulator rollers made of steel or hard plastic. The drop into the sizer cups or padded collection conveyers were half the force of drops onto the rollers (115 versus 55G).

Field packing can significantly reduce harvest and central packing shed-related bruising to 25% of shed pack operations though washing to remove debris and dust is not generally possible.

Chilling injury (CI)

CI is common in 'Queen' and 'Smooth Cayenne' but has not been reported for other cultivars. The injury can occur with fruit exposure to low temperature in the field and in postharvest storage (Fig. 13.4b). The symptoms include: (i) wilting, drying and discoloration of crown leaves; (ii) failure of green-shelled fruit to yellow; (iii) browning and dulling of yellow fruit; and (iv) internal flesh browning (Lim, 1985; Paull and Rohrbach, 1985). The internal flesh browning has been called endogenous brown spot, physiological breakdown, blackheart and internal browning, with the latter being more common. CI symptom development can be divided into two events: the first relates to the exposure to temperature less than 10–12°C, while the second is tissue darkening, which can range from brown to black (Paull and Rohrbach, 1985). After exposure of fruit to 10–12°C, tissue darkening (Fig. 13.4b) is more pronounced between 17.5°C and 27.5°C (Paull and Rohrbach, 1985). The development of tissue darkening is also related to preharvest shading and low temperatures (Linford, 1932; Akamine *et al.*, 1975; Keetch and Dalldorf, 1979; Swete Kelly and Bragshaw, 1993), ascorbic acid content (Miller, 1951; Teisson and Martin-Prevel, 1979; Paull and Rohrbach, 1985; Abdullah *et al.*, 1986), fruit calcium content (Hewajulige *et al.*, 2003), lower sugars and opaque fruit (Abdullah and Rohaya, 1983; Paull and Rohrbach, 1985), storage atmosphere (Paull and Rohrbach, 1982; Hassan *et al.*, 1985), and cultivar (Wilson Wijeratram *et al.*, 1993; Sanewski and Giles, 1997; Stewart *et al.*, 2002; Hewajulige *et al.*, 2003; Weerahewa and Adikaram, 2005b; Raimbault *et al.*, 2011; Nukuntornprakit *et al.*, 2015). The symptoms are seen occasionally in Hawaii's cool season fruit (Akamine *et al.*, 1975) and can be more severe on the northern side of a bed than on the southern side.

The occurrence of CI following exposure to temperatures between 12 and 21°C (Smith, 1983) is difficult to reconcile with published findings and may be related to preharvest stress (Swarts, 1991; Swete Kelly and Bragshaw, 1993) or a delay in symptom development at these temperatures. Partial to complete control of CI symptom development has been achieved by pre-harvest applications of potassium (Nanayakkara *et al.*, 2005; Soares *et al.*, 2005), calcium (Selvarajah *et al.*, 1998a; Herath *et al.*, 2003; Nanayakkara *et al.*, 2005; Wilson Wijeratnam *et al.*, 2006; Youryon *et al.*, 2013; Pusittigul *et al.*, 2014), salicylic acid (Lu *et al.*, 2011), postharvest application of fruit skin coating (Rohrbach and Paull, 1982; Hu *et al.*, 2011, 2012), temperature preconditioning (Abdullah *et al.*, 2008), heat treatment (Akamine *et al.*, 1975; Akamine, 1976; Selvarajah *et al.*, 1998b; Acedo *et al.*, 2004; Weerahewa and Adikaram, 2005a; Su, 2008), controlled and modified atmosphere storage (Abdullah *et al.*, 1985; Paull and Rohrbach, 1985; Selvarajah *et al.*, 1997), short-term anoxia treatment (Phonyiam *et al.*, 2016; Techavuthiporn *et al.*, 2017), 1-methylcyclopropene (1-MCP) (Selvarajah *et al.*, 2001), ascorbic acid (Sun, 1971; Nanayakkara *et al.*, 2005), salicylic acid (Lu *et al.*, 2010, 2011), methyl jasmonate (Nilpraprruck *et al.*, 2008), and abscisic acid (ABA) (Zhang *et al.*, 2015).

The internal browning of pineapple fruit originates at the phloem of vascular bundles (Luengwilai *et al.*, 2016). Similar cases were also reported in other fruits, such as avocados (Kaiser, 1993) and apples (Argenta *et al.*, 2001). The initial symptom of internal browning is the appearance of water-soaked spots at the base of the fruitlets near the core that turns brown and spreads throughout the fruit flesh as the symptoms develop (Weerahewa and Adikaram, 2005b). In severe cases, the entire flesh and core of the pineapple turn completely black.

Internal browning cannot be recognized until the fruit are cut. Near Infrared transmittance spectroscopy is a nondestructive method that has potential to be used to predict this disorder in intact pineapples (Sukwanit and Teerachaichayut, 2013).

Ascorbic acid content and polyphenol oxidase (PPO) enzyme activity are the two major biochemical factors related to the occurrence of internal browning disorder. However, the initial ascorbic acid content and PPO activity before low temperature induction do not indicate fruit susceptibility to internal browning (Hassan *et al.*, 2010; Tseng, 2013). Cultivars containing higher ascorbic acid content do not necessarily have higher resistance to internal browning (Sanewski and Giles, 1997), though 'MD-2' and 'CO-2' fruits have high ascorbic acid contents and are highly and moderately resistant respectively to CI. There are no significant differences in ascorbic acid contents (Youryon *et al.*, 2008; Tseng, 2013) and PPO activity (Tseng, 2013) between the core and pulp tissues of pineapple. Total phenolics and tannin content may also contribute to the antioxidant capacity in pineapple fruit (Chiet *et al.*, 2014). The antioxidant activities, estimated by FRAP, in different pineapple cultivars before cold storage are not related to the resistance to CI (Tseng, 2013). Instead, the change in antioxidant activity before and after cold storage is highly correlated to the intensity of chilling injury in fruit flesh after cold storage, with a decrease in a highly susceptible cultivar and an increase in resistant one.

PPO is synthesized *de novo* in response to chilling, implicating a role for the enzyme in the development of internal browning (Stewart *et al.*, 2001, 2002). Maturity level significantly affects susceptibility of fruit to CI which is highly correlated to the response of PPO activity to chilling (Zhou *et al.*, 2003a). It is believed that an increase in PPO activity is associated to the symptom development of CI, both temporally and spatially (Zhou *et al.*, 2003a) but not the cause of it (Zhou *et al.*, 2003b). Up to six PPO isoforms have been identified in susceptible cultivars (Raimbault *et al.*, 2011). Transgenic 'Smooth Cayenne' with a down-regulated pineapple PPO gene showed resistance to internal browning (Ko *et al.*, 2013).

PPO catalyses the conversion of monophenols and diphenols to quinones in higher plants which can then undergo polymerization and bind covalently to nucleophilic amino acids to form black or brown pigments (Mayer and Harel, 1979). The preferred substrates are diphenols for PPO extracted from pineapple fruit (Das *et al.*, 1997). PPO is a copper-containing enzyme (Mayer, 2006) and a significantly positive correlation was found between fruit flesh copper content and the intensity of internal browning after cold storage for one cultivar and among different cultivars harvested from different production areas (Tseng, 2013). PPO is located in the plastids and the phenolic substrates for PPO are stored in the vacuole. When subcellular separation is lost due to CI of membranes, the enzymatic browning reaction occurs, leading to the internal browning symptom in fruit flesh.

Browning symptoms are observed in susceptible fruit flesh chilled at 10°C and are progressively less severe at lower temperatures (6°C and 0°C) (Stewart *et al.*, 2002). Nevertheless, the core of both susceptible and resistant (hybrid 53-116) fruit showed increased symptoms at lower temperatures and the symptoms were restricted to the core at 0°C (Stewart *et al.*, 2002). In canning, these symptoms are less important since the cores are discarded. A 6°C storage temperature maintained lower PPO activity and reduced internal browning symptoms and provided a longer shelf life than was obtained at 10°C (Hong *et al.*, 2013).

During low temperature storage, membrane permeability of pineapple fruit flesh increased (Zhou *et al.*, 2014a,b). The development of CI is correlated with an increase in membrane permeability and with lower total antioxidant capacity, while no correlation is found between CI severity and membrane fatty acid composition or fatty acid oxidation and the concentrations of hydrogen peroxide or ascorbic acid (Om-arun and Siriphanich, 2005; Nukuntornprakit *et al.*, 2015). However, low temperature does alter plasma membrane of lipid composition and decreased V-ATPase activity in pineapple fruit with chilling injury (Zhou *et al.*, 2014a,b).

CI in plants is associated with reactive oxygen species (ROS) that promotes the oxidation of lipids and proteins (Prasad, 1996; Anjum *et al.*, 2015). Increase of H_2O_2 was reported during blackheart development while ascorbic acid contents ot the core and pulp tissues were not different (Nukuntornprakit and Siriphanich 2005; Youryon *et al.*, 2008). Cell protection capability, such as ascorbate peroxidase (Zhou *et al.*, 2003a), calcium mobilization (Youryon *et al.*, 2013; Hewajulige *et al.*, 2003, 2006; Wilson Wijeratnam *et al.*, 2006) and balance of plant hormone abscisic acid (ABA) and gibberellins (Zhou and Tan, 1997; Pusittigul *et al.*, 2012) is also suggested to be involved. Endogenous ABA may be associated with the resistance response to chilling injury in pineapples (Pusittigul *et al.*, 2012) and exogenous application of ABA (Zhang *et al.*, 2015) could reduce the intensity of chilling injury.

Chilling injury could be induced by exogenous application of gibberellins (Zhou and Tan, 1997; Pusittigul *et al.*, 2012), suggesting that endogenous GA concentrations may play a role in the development of chilling injury in pineapple fruit (Pusittigul *et al.*, 2012). The involvement of gibberellin 2-oxidase and gibberellins catabolism gene (*AcGA2ox*) in regulation of GA levels and chilling injury has been reported (Zhang *et al.*, 2016). Two PPO genes having homologies to gibberellin response complexes and the sequences of both promoters have been isolated from pineapple (Zhou *et al.*, 2003b).

Proteases are responsible for protein degradation that is regulated by specific endogenous protein inhibitors (Kidrič *et al.*, 2014). In cherimoya fruit, a decrease in protease activity to barely detectable levels during the early phase of cold storage is found, the proteolytic activity then increased, and the endogenous proteins were broken down to free amino acid components (Montero *et al.*, 1995). Similar results have also been found in pineapple. The involvement of an aspartic acid protease (AcAP1) in postharvest chilling stress resistance in pineapple fruit has been reported (Raimbault *et al.*, 2013b). The control of bromelain (a protease) and AcCYS1 (cystatin, a protease inhibitor) expression levels directly correlates to the resistance to internal browning in 'MD-2' pineapple fruit (Raimbault *et al.*, 2013a). However, the regulating mechanism of proteases and protease inhibitors in chilling injury development in pineapple fruit is still unknown.

Flesh translucency

Translucency is a physiological disorder where the pineapple fruit flesh shows water soaking symptoms (Fig. 13.4c). Translucency cannot be detected by external symptoms though the broken peduncle does tend to stay wet after harvest and the fruit feels heavier and does not float in a water tank if severe. X-rays can be used to assess translucency severity (Haff *et al.*, 2006). Translucency, as opposed to opaqueness, lacks air in the intercellular spaces of the fruit flesh and has a higher specific gravity (Sideris and Krauss, 1933a). The specific gravity can be employed as a nondestructive method to detect translucent fruit and separate them into different translucency classes (Fig. 13.3b). Translucent fruit are extremely fragile, making them prone to mechanical injury during harvest and postharvest handling (Py *et al.*, 1987). In addition, translucent fruit are more susceptible to diseases (Gortner *et al.*, 1963) and preharvest sunburn (Keetch, 1977). Highly translucent fruit has flat and overripe off-flavours, and significantly lower edible quality (Bowden, 1967).

The incidence of peduncle leakage is correlated with translucence severity, and may lead to unsightly *Penicillium* and *Ceratocystis* mould growth (Fig. 13.4d) (Paull and Reyes, 1996). Translucency occurs before harvest (Bowden, 1969; Rohrbach and Paull, 1982; Paull and Reyes, 1996) and generally the basal flesh is the first to show symptoms; in severe cases the whole fruit is affected. Translucent fruit are more common in the cooler season. Waxing can reduce the rate of translucency development after harvest in low shell colour fruit. Soler (1993, 1994a) regards translucency as being due to early flesh senescence, however, it is apparently also related to the weather three or more months before harvest (Paull and Reyes, 1996). Fruit with larger crowns have a lower incidence and severity of translucency.

Fruit with increasing translucency have increased pH, TSS/acid ratio, and fruit weight, and decreased total esters and acids. The decline in organic acids during ripening is more pronounced in translucent (Sideris and Krauss, 1933b) than in opaque fruit, while sugars show little change. Translucent fruit usually have a higher SSC:TA ratio than opaque fruit due to lower acidity (Sideris and Krauss, 1934; Bowden, 1969).

The cause or causes of translucency are unknown. Translucency in Hawaii has been associated with the cultivar grown, high nitrogen, large vigorous plants, spring ripened fruit, treatment with fruit enlarging agents, irrigation rate, planting density, larger crowns, and environmental factors (Paull and Reyes, 1996). Paull and Reyes (1996) found that both crown weight and fruit translucency at harvest are correlated to the monthly average air temperature 2–3 months before harvest, and the correlation between crown weight and translucency severity was significantly negative. However, removing the crown either at an early or late stage of pineapple fruit development did not have any significant effect on fruit weight or translucency (Chen, 1999). The significantly negative correlation between the air temperature 2–3 months before harvest and translucency (Paull and Reyes, 1996) is possibly related to an increase in heat-tolerance of fruit flesh (Chen, 1999).

Pineapple fruit translucency has been suggested to be related to an increase in cell wall hydrolases (Soler, 1993) and membrane permeability (Soler, 1994a). High calcium concentration may decrease the secretion or activities of cell wall hydrolases (Huber, 1983) and membrane permeability, however, calcium concentration and the divalent cation binding capacity of cell walls in pineapple fruit flesh decline with development (Chen, 1999). Pineapple fruit translucency increases with increasing fruit weight (Bowden, 1969), possibly due to a decrease in calcium concentration since larger fruit may need more calcium to stabilize cell membranes. When fruit cannot acquire sufficient calcium, cell membranes may lose integrity and lead to leakage and translucency. Partial control of translucency occurrence has been associated with calcium application (Chen, 1999; Silva *et al.*, 2006).

Basal fruitlets show translucency first. These tissues have higher sugar content than the inter-fruitlet tissue and the flesh at the top of the fruit, respectively, suggesting that translucency is related to maturity. Chen (1999) reported that plant defoliation conducted 3–4 weeks before harvest decreased the pineapple fruit flesh soluble solids content (SSC) and translucency. There was a linear cause–effect relationship between percentage of defoliation and SSC and translucency. Defoliation did not significantly affect the membrane permeability of pineapple fruit flesh cells, suggesting that translucency caused by sugar unloading is different from translucency caused by heat stress, and the former was much more significant at the latter stages of fruit development.

Intercellular spaces filled with liquid in translucent pineapple fruit flesh suggests that a higher osmotic pressure exists in the apoplast. The apoplastic sugar content increases with fruit maturation (Chen, 2010) and could attract more water into the apoplast. Cell-wall invertase (CWI) activity in pineapple fruit flesh increases rapidly 4 weeks before harvest and is followed almost immediately by the first symptom of translucency. A positive correlation also exists between the CWI activity and the severity of translucency (Chen, 1999) which suggests that CWI in the pineapple fruit flesh could be a positive factor in the occurrence of translucency.

Translucency can occur when the fruit shell is still green (Py *et al.*, 1987) and green-shell translucency is found more frequently during high temperature periods (Green, 1963). High temperature may produce lesions in the cell membrane that then cause translucency. Soler (1993, 1994b) showed that green-shell translucency, at a biochemical level, is characterized by an increase in catalase, α- and β-galactosidase activities and ascorbic acid synthesis, that might be related to translucency by the modification of membrane galactolipids and permeability changes. Pineapple fruit flesh became susceptible to high fruit temperature-induced leakage and translucency at the latter stages of fruit development (Chen, 1999). Sideris *et al.* (1935) suggested that flesh translucency and shell degreening were independent phenomena associated with fruit senescence, with the flesh of green ripe fruit senescing before shell degreening started. Green-shell translucent fruits degreen rapidly when treated with ethephon.

The susceptibility of pineapple fruit to translucency and its severity is strongly under genetic control (Chen, 2010). In susceptible cultivars, the maturity-related changes including membrane leakage, sugar content, calcium concentration, apoplastic liquid volume and CWI activity all are correlated with translucency occurrence since the incidence and severity of translucency increases with fruit maturation. However, those maturity-related changes also occur in resistant varieties which do not develop translucency. There are no significant correlations between those maturity-related factors and translucency occurrence among cultivars with different degrees of susceptibility to translucency (Chen, 2010). Therefore, those significant correlations found in susceptible cultivars do not suggest a cause–effect relationship with translucency. Translucency is related to apoplastic liquid volume, intercellular free space, and the ratio of the above two, in the same cultivar harvested from different production areas and in different cultivars harvested from the same production area and from different production areas (C.-C. Chen and M.-L. Chen, unpublished results). A highly significant negative correlation was also found between translucency severity and locular cavity volume among different cultivars harvested from the same production area. It seems that the larger the intercellular free space and locular cavity provide more space for apoplastic liquid storage and evaporation of water inside the fruit, respectively, which reduce the chance of translucency occurrence. The role of intercellular space and locular cavity in translucency occurrence needs further investigation.

Crown dehydration

The crown is a continuation of the vegetative stem and the spirally arranged leaflets have similar morphology. The stomata in mature leaves open at night and are closed during the day (Sideris *et al.*, 1948). However, the stomata of young leaves of the crown may not show this same diurnal function during storage (Paull, 1985, unpublished) because water loss rate appears to be constant during storage. At 10°C stomatal opening and closing is slowed and CO_2 uptake is reduced compared to 17°C (Kent, 1967). Ethylene increases leaf respiration, causes stomatal opening and enhances dehydration (Dull and Staruszkiewicz, 1966). Opening of stomata requires illumination on the proceeding day but ethylene negates this requirement (Kent, 1967), which results in crown desiccation.

The crowns of fresh fruit soon lose their green healthy look and the leaves become dehydrated and necrotic especially if the crown leaves are broken or folded during postharvest handling. Careful handling is required to avoid mechanical injury to the crown leaves to retain crown appearance. The oldest and largest crown leaves are generally most affected, and this may involve stress ethylene production. Postharvest ethylene treatment leads to greater water loss than if crowns are untreated. Rate of dehydration is also related to the relative humidity of storage (Dull and Staruszkiewicz, 1966). A well-ventilated crown can lose up to 10% of its weight in 14 days while a crown treated with 200 μl l^{-1} ethylene can lose up to 64% of its fresh weight. Waxing of the crown frequently leads to phytotoxicity and excessive water loss, and some paraffin wax formulations can slightly reduce weight loss (Schappelle, 1941).

Sunburn

Sunburn is common during hotter periods (>35°C) (Keetch and Dalldorf, 1979) when the fruit is not shaded by leaves and especially in ratoon crops. Fruit flesh temperatures on an exposed side of the fruit in the field can be 15°C higher than the air temperature of about 27°C, and the crown leaves provide little protection. The condition is more prevalent in the outer rows of a field and when fruits are lodged. Drought in warm season can lead to weak peduncles and lodging of fruit, while fruit that develop in the cool season also tends to have longer peduncles that increase fruit lodging. Forcing immature plants to flower also leads to longer peduncles that have a greater tendency to lodge (Dalldorf, 1979). Sun scorched fruit first show a bleached yellow-white skin that turns pale grey and brown (Fig. 13.4e), with damage to the flesh underneath that softens and may extend to the core in severe cases (Lindford and Spiegelberg, 1933). Cracking of the shell and flesh can occur. These damaged areas are more susceptible to disease organisms (Lim, 1985) and severely sunburned fruit have no market. Sunburn can be managed by shading of individual fruits, entire rows, or the whole field.

Malformations

Knobs on the base of fruit is a mutation (Collins, 1960). Culling of these fruit crowns reduces the subsequent field incidence. These fruit are not consigned to the fresh market as trimming allows rots to develop where the knobs are removed. The other genetic off-type is multiple crowns (fasciation), with the fruit taking on a flattened appearance. High temperatures injury after inflorescence emergence and calcium deficiency also can increase the incidence of multiple crowns (Soler, 1992b). These fruit may not be marketable and it is not recommended to plant the crowns even though the condition is often not heritable (Linford and Spiegelberg, 1933). Fruit with pronounced 'eyes' or fruitlets are normally not accepted in Fancy Grades of fruit and the thicker skin and more porous flesh means lower flesh recovery. This condition is common in fruit that flower during cool weather.

Broken and hollow core

A transverse break in fruit flesh that occurs early in development is marked by a slight depression on the outside of the fruit (Lim, 1985). The break leaves a gap of 2–10 mm in the flesh and the tissue becomes corked and brown, sometimes accompanied by rot. The upper portion of the fruit usually ripens before the lower half. This condition is not seen in 'Smooth Cayenne', but is common in 'Singapore Spanish' and 'Tainung No. 13'. Boron deficiency might be involved in this disorder. Hollow core refers to a vertical crack in the core that becomes dry and leathery and it may extend to the peduncle. The fruit are usually opaque and the condition is thought to be due to water deficit.

Seediness

Seediness can sometimes be a problem, as the seeds are small brown and hard (Fig. 13.4f). This occurs with where different cultivars are planted in close proximity and flower together.

Shell surface pitting

Surface pitting is the shallow brown pitting of sepals and bracts that does not extend more than 6 mm into the fruit, and the affected tissue shrinks. This condition is most often found in opaque fruit with yellow shells in fields where a heavy dew or light shower frequently moistens the fruit (Lindford and Spiegelberg, 1933). Rough handling of fruit will also result in skin bruising, which presents as pitting in the more raised surfaces.

References

Abdullah, H. and Rohaya, M.A. (1983) The development of black heart disease in Mauritius pineapple (*Ananas comosus* cv. Mauritius) during storage at lower temperatures. *MARDI Research Bulletin* 11, 309–319.

Abdullah, H., Rohaya, M.A. and Zaipun, M.Z. (1985) Effect of modified atmosphere on black heart development and ascorbic acid contents in 'Mauritius' pineapple (*Ananas comosus* cv. 'Mauritius') during storage at low temperature. *ASEAN Food Journal* 1, 15–18.

Abdullah, H., Rohaya, M.A. and Zaipun, M.Z. (1986) Storage studies of pineapple (*Ananas comosus* W. Sarawaki) with special emphasis on black heart disorder. *MARDI Research Bulletin* 14, 132–138.

Abdullah, H., Rohaya, M.A., Latifah, M.N., Mohammed Selamat, M. and Underhill, S. (2002) Respiration rate, ethylene production and chlorophyll content of the fruit and crown of pineapple stored at low temperatures. *Journal of Tropical Agriculture and Food Science* 30, 99–107.

Abdullah, H., Rohaya, M.A. and Engku Hasmah, E.A. (2008) Increasing pineapple fruit resistance to chilling injury during storage by temperature preconditioning. *Acta Horticulturae* 768, 217–224.

Acedo Jr, A.L., Akinaga, T. and Tanabe, T. (2004) Inhibition of chilling injury and quality changes in pineapple fruit with prestorage heat treatment. *Journal of Food, Agriculture and Environment* 2, 81–86.

Adetunji, C.O., Fawole, O.B., Arowora, K.A., Nwaubani, S.I., Ajayi, E.S. *et al.* (2012) Effects of edible coatings from aloe vera gel on quality and postharvest physiology of *Ananas comosus* (L.) fruit during ambient storage. *Global Journal of Science Frontier Research Bio-Tech and Genetics* 12, 38–43.

Akamine, E.K. (1963) Fresh pineapple storage. *Hawaii Farm Science* 12, 1–4.

Akamine, E.K. (1976) Postharvest control of endogenous brown spot in fresh Australian pineapples with heat. *HortScience* 11, 586–588.

Akamine, E.K. and Goo, T. (1971) Controlled atmosphere storage of fresh pineapples (*Ananas comosus* (L.) Merr. 'Smooth Cayenne'). *Hawaii Agricultural Experiment Station Research Bulletin* 152.

Akamine, E.K., Goo, T., Steepy, T., Greidanus, T. and Iwaoka, N. (1975) Control of endogenous brown spot of pineapple in postharvest handling. *Journal of the American Society for Horticultural Science* 100, 60–65.

Anjum, N.A., Sofo, A., Scopa, A., Roychoudhury, A., Gill, S.S. *et al.* (2015) Lipids and proteins – major targets of oxidative modifications in abiotic stressed plants. *Environmental Science and Pollution Research* 22, 4099–4121.

Argenta, L., Fan, X. and Mattheis, J. (2001) Development of internal browning in Fuji apples during storage. *Washington Tree Fruit Postharvest Conference.* Available at: http://postharvest.tfrec.wsu.edu/pc2001c.pdf (accessed 21 March 2018).

Ayernor, G.S., Ohene Afoakwa, E., Bartels, P.K. and Budu, A.S. (2010) Effects of polymeric coating and anti-fungal treatment on the shrinkage characteristics of pineapple fruits during storage. *International Journal of Fruit Science* 10, 309–322.

Ayernor, G.S., Ohene Afoakwa, E., Bartels, P.K. and Budu, A.S. (2011) Effect of polymeric coating on the post-harvest quality characteristics of pineapple cv. 'Smooth Cayenne' fruits. *Fresh Produce* 5, 34–38.

Bartholomew, D.P. (2009) 'MD-2' pineapple transforms the world's pineapple fresh fruit export industry. *Pineapple News* 16(2), 2–5. Available at: http://www.ishs-horticulture.org/workinggroups/pineapple/PineNews16.pdf (accessed 16 March 2018).

Bartholomew, D.P and Criley R.A. (1983) Tropical fruit and beverage crops. In: Nickell, L.G. (ed.) *Plant Growth Regulating Chemical*, Vol. 2. Boca Raton, FL: CRC Press, pp. 1–11.

Bartholomew, D.P. and Paull, R.E. (1986) Pineapple. In: Monselise, S.P. (ed.) *CRC Handbook of Fruit Set and Development.* Boca Raton, FL: CRC Press, pp. 371–388.

Bartolomé, A.P. and Rupérez, P. (1995) Polysaccharides from the cell walls of pineapple fruit. *Journal of Agricultural and Food Chemistry* 43, 608–612.

Benítez, S., Soro, L., Achaerandio, I., Sepulcre, F. and Pujolá, M. (2014) Combined effect of a low permeable film and edible coatings or calcium dips on the quality of fresh-cut pineapple. *Journal of Food Process Engineering* 37(2), 91–99.

Bhaduri, S.K., Sen, S.K. and Dasgupta, P.C. (1983) Structural studies of an acidic polysaccharide isolated from the leaf fiber of pineapple (*Ananas comosus* Merr.). *Carbohydrates Research* 121, 211–220.

Bowden, R.P. (1967) Translucency as an index of ripeness in pineapple. *Food Technology Australia* 19, 424–427.

Bowden, R.P. (1969) Further studies on ripeness in pineapple. *Food Technology Australia* 21, 160–163.

Cancel, H.L. (1974) Harvesting and storage conditions for pineapples of the Red Spanish variety. *Journal of Agricultural Research, Puerto Rico* 58, 162–169.

Cappellini, R.A., Ceponis, M.J. and Lightner, G.W. (1988) Disorders in avocado, mango, and pineapple shipments in the New York market. *Plant Disease* 72, 270–273.

Cazzonelli, C.I., Cavallaro, A.S. and Botella, J.R. (1998) Cloning and characterisation of ripening-induced ethylene biosynthetic genes from non-climacteric pineapple (*Ananas comosus*) fruits. *Australian Journal of Plant Physiology* 25, 513–518.

Chan, H.T., Chenchin, E. and Vonnahme, P. (1973) Nonvolatile acids in pineapple juice. *Journal of Agriculture and Food Chemistry* 21, 208–210.

Chen, C.-C. (1999) Effects of fruit temperature, calcium, crown and sugar metabolizing enzymes on the occurrence of pineapple fruit translucency. PhD dissertation, University of Hawaii at Manoa, Honolulu, HI.

Chen, M.-L. (2010) Studies on physical-chemical characteristics, mineral elements and translucency disorder in pineapple (*Ananas comosus*) fruit during development. MSc thesis, National Chung Hsing University, Taichung, Taiwan.

Chen, N.J. and Paull, R.E. (1995) Effect of waxing and storage on pineapple fruit quality. *Proceedings of the International Symposium on Postharvest Science and Technology of Horticultural Crops*, Beijing, China, 27 June–1 July.

Chen, C.-C. and Paull, R.E. (2000) Sugar metabolism and pineapple fruit flesh translucency. *Journal of the American Society for Horticultural Science* 125, 558–562.

Chenchin, K.L. and Yamamoto, H.Y. (1978) Isolation characterization and enzymic hydrolysis of pineapple gum. *Journal of Food Science* 43, 1261–1263.

Chenchin, K.L., Yugawa, A. and Yamamoto, H.Y. (1984) Enzymic degumming of pineapple and pineapple mill juices. *Journal of Food Science* 49, 1327–1329.

Chiet, C.H., Zulkifli, R.M., Hidayat, T. and Yaakob, H. (2014) Bioactive compounds and antioxidant activity analysis of Malaysian pineapple cultivars. *AIP Conference Proceedings* 1589, 398. DOI: http://dx.doi.org/10.1063/1.4868827.

Collins, J.L. (1960) *The Pineapple*. London: Leonard Hill.

Crochon, M., Tisseau, R., Teisson, C. and Huet, R. (1981) Effet d'une application d'Ethrel avant la recolte sur la qualité gustative de ananas de Côte d'Ivoire. *Fruits* 36, 409–415.

Dalldorf, D.B. (1979) Flower induction of pineapple. Farming in South Africa Pineapple G.3/1979. Department of Agricultural Technical Services, Pretoria, South Africa.

Dalldorf, E.R. (1981) The effect of chloroflurenol methylester on the reduction of top size of Smooth Cayenne pineapples. *Subtropica* 2, 9–11.

Damayanti, M., Sharma, G.J. and Kundu, S.C. (1992) Gamma radiation influences postharvest disease incidence of pineapple fruits. *HortScience* 27, 807–808.

Das, J.R., Bhot, S.G. and Eowda, L.R. (1997) Purification and characterization of a polyphenol oxidase from the Kew cultivar of Indian pineapple fruit. *Journal of Agricultural and Food Chemistry* 45, 2031–2035.

de Lencastre Novaes, L.C., Jozala, A.F., Lopes, A.M., de Carvalho Santos-Ebinuma, V., Mazzola, P.G. and Pessoa Junior, A. (2016) Stability, purification, and applications of bromelain: a review. *Biotechnology Progress* 32(1), 5–13.

Dorey, E., Fournier, P., Léchaudel, M. and Tixier, P. (2016) A statistical model to predict titratable acidity of pineapple during fruit developing period responding to climatic variables. *Scientia Horticulturae* 210, 19–24.

Dull, G.G. (1971) The pineapple: general. In: Hulme, A.C. (ed.) *The Biochemistry of Fruits and Their Products*, Vol. 2. London: Academic Press, pp. 303–331.

Dull, G.G. and Staruszkiewicz, W. (1966) An observation on crown dehydration on stored fresh fruit bagged in cellophane. *PRI News* 14, 98–100 (private document).

Dull, G.G., Young, R.E. and Biale, J.B. (1967) Respiratory patterns in fruit of pineapple, *Ananas comosus* detached at different stages of development. *Physiology Plantarum* 20, 1059–1065.

Dupaigne, P. (1970) L'arome de l'ananas. *Fruits* 25, 793–805.

Elss, S., Preston, C., Hertzig, C., Heckel, F., Richling, E. and Schreier, P. (2005) Aroma profiles of pineapple fruit (*Ananas comosus* [L.] Merr.) and pineapple products. *LWT-Food Science and Technology* 38, 263–274.

FAO (2017a) Crops. Available at: http://www.fao.org/faostat/en/#data/QC (accessed 9 November 2017).

FAO (2017b) Codex standard for pineapples. Available at: http://www.fao.org/input/download/standards/313/CXS_182e.pdf (accessed 9 November 2017).

Ferreira, E.A., Siqueira, H.E., Boas, E.V.V., Hermes, V.S. and Rios, A.O. (2016) Bioactive compounds and antioxidant activity of pineapple fruit of different cultivars. *Revista Brasileira de Fruticultura* 38(3), e-146. DOI: 10.1590/0100-29452016146.

Flath, R.A. (1986) Pineapple. In: Nagy, S. and Shaw, P.E. (eds) *Tropical and Subtropical Fruits: Composition, Properties and Uses*. Westport, CT: AVI Publishing, Inc., pp. 157–183.

Flath, R.A. and Forrey, R.R. (1970) Volatile components of Smooth Cayenne pineapple. *Journal of Agricultural and Food Chemistry* 18, 306–309.

Gawler, J.H. (1962) Constituents of canned Malayan pineapple juices. I. Amino acids, non-volatile acids, sugars, volatile carbonyl compounds and volatile acids. *Journal of the Science of Food and Agriculture* 13, 57–61.

Gortner, W.A. (1963) A short term effect of weather on malic acid in pineapple fruit. *Journal of Food Science* 28, 191–192.

Gortner, W.A. (1965) Chemical and physical development of the pineapple fruit. IV. Plant pigment constituents. *Journal of Food Science* 30, 30–32.

Gortner, W.A. (1969) Relation of chemical structure to plant growth-regulator activity in the pineapple plant: Reloading senescence of pineapple fruit with applications of 2, 4, 5- Trichlorophynoxy acetic acid and 1-naphthalene acetic acid. *Journal of Food Science* 34, 577–580.

Gortner, W.A. and Leeper, R.W. (1969) Studies on the relation of chemical structure to plant growth regulatory activity in the pineapple plant. V. Postharvest delay of senescence of pineapple fruit. *Botanical Gazette* 130, 87–97.

Gortner, W.A. and Singleton, V.L. (1961) Carotenoid pigments of pineapple fruit. II. Influence of fruit ripeness, handling and processing on pigment isomerization. *Journal of Food Science* 26, 53–55.

Gortner, W.A. and Singleton, V.L. (1965) Chemical and physical development of the pineapple fruit. III. Nitrogenous and enzyme constituents. *Journal of Food Science* 30, 24–29.

Gortner, W.A., Spiegelberg, C.H., Dull, G.G. and Krauss, B.H. (1963) Field-fresh pineapple for export. *Pineapple Research Institute* 99, 24–30.

Gortner, W.A., Dull, G.G. and Krauss, B.H. (1967) Fruit development, maturation, ripening and senescence: a biochemical basis for horticultural terminology. *HortScience* 2, 141–144.

Gray, R.A. (1953) The volatile flavor constituents of pineapple. *Pineapple Research Institute, Research Report* #33 (private document).

Green, G.C. (1963) The pineapple plant. In: *The Effect of Weather and Climate upon the Keeping of Fruit.* Technical Note 53. Geneva: World Meteorological Organization, pp. 136–180.

Guthrie, J. and Walsh, K. (1997) Non-invasive assessment of pineapple and mango fruit quality using near infrared spectroscopy. *Australian Journal Experimental Agriculture* 37, 253–263.

Haagen-Smit, A.J., Kirchner, J.G., Prater, A.N. and Deasy, C.L. (1945) Chemical studies of pineapple (*Ananas sativa* Lindl.) I. The volatile flavor and odor constituents of pineapple. *Journal of the American Chemical Society* 67, 1646–1652.

Haff, R.P., Slaughter, D.C., Sarig, Y. and Kader, A. (2006) X-ray assessment of translucency in pineapple. *Journal of Food Processing and Preservation* 30, 527–533.

Hamner, K.C. and Nightingale, G.T. (1946) Ascorbic acid content of pineapples as correlated with environmental factors and plant composition. *Food Research* 11, 535–541.

Haripyaree, A., Guneshwor, K. and Damayanti, M. (2010) Evaluation of antioxidant properties of phenolics extracted from *Ananas comosus* L. *Notulae Scientia Biologicae* 2, 68–71. DOI: 10.15835/nsb224615.

Hassan, A., Atan, R.M. and Zain, Z.M. (1985) Effect of modified atmosphere on black heart development and ascorbic acid contents in 'Mauritius' pineapples (*Ananas comosus* cv. 'Mauritius') during storage at low temperature. *ASEAN Food Journal* 1, 15–18.

Hassan, A., Wills, R., Atan, R.M., Othman, Z., Fatt, L.P. and Smith, M. (2010) Blackheart disorder in fresh pineapple. *New Trends in Postharvest Management of Fresh Produce II. Fresh Produce* 4, 29–35.

Hepton, A. (2003) Cultural system. In: Bartholomew, D.P., Paull, R.E. and Rohrbach, K.G. (eds) *The Pineapple: Botany, Production and Uses.* Wallingford, Oxfordshire: CAB International, pp. 109–142.

Herath, H.M.I., Bandara, D.C. and Abeysinghe Banda, D.M.G. (2003) Effect of pre-harvest calcium fertilizer application on the control of internal browning development during the cold storage of pineapple 'Mauritius' (*Ananas comosus* (L.) Merr.). *Journal of Horticultural Science and Biotechnology* 78, 762–767.

Hewajulige, I.G.N., Wilson Wijeratnam, R.S., Wijesundera, R.L.C. and Abeysekere, M. (2003) Fruit calcium concentration and chilling injury during low temperature storage of pineapple. *Journal of the Science of Food and Agriculture* 83, 1451–1454.

Hewajulige, I.G.N., Wilson Wijeratnam, S. and Wijesundera, R.L.C. (2006) Pre-harvest application of calcium to control black heart disorder in Mauritius pineapples during low-temperature storage. *Journal of the Science of Food and Agriculture* 86, 420–424.

Hong, K., Xu, H., Wang, J., Zhang, L., Hu, H. *et al.* (2013) Quality changes and internal browning developments of summer pineapple fruit during storage at different temperatures. *Scientia Horticulturae* 151, 68–74.

Hossain, M.A. and Rahman, S.M. (2011) Total phenolics, flavonoids and antioxidant activity of tropical fruit pineapple. *Food Research International* 44, 672–676.

Hu, H., Li, X., Dong C. and Chen, W. (2011) Effects of wax treatment on quality and postharvest physiology of pineapple fruit in cold storage. *African Journal of Biotechnology* 10, 7592–7603.

Hu, H., Li, X., Dong, C. and Chen, W. (2012) Effects of wax treatment on the physiology and cellular structure of harvested pineapple during cold storage. *Journal of Agricultural and Food Chemistry* 60, 6613–6619.

Huber, D.J. (1983) The role of cell wall hydrolases in fruit softening. *Horticulture Review* 5, 169–219.

Huet, R. (1958a) La composition chimique de l'ananas. *Fruits* 13, 183–197.

Huet, R. (1958b) Distribution of vitamin C, acids and sugars in the Smooth Cayenne variety of pineapple. *Quality of Plant and Vegetable* 3/4, 244–255.

Jarvis, M.C., Forsyth, W. and Duncan, H.J. (1988) A survey of the pectic content of non-lignified monocot cell walls. *Plant Physiology* 88, 309–314.

Kaiser, C. (1993) Vascular and associated tissue of 'Hass' avocado (*Persea americana* Mill.) fruit. *South African Avocado Growers' Association Yearbook* 16, 22–27.

Keetch, D.P. (1977) Sunburn in pineapple. *Farming in South Africa: Pineapple Series* No. H.8/1977.

Keetch, D.P. (1978) Bruising of pineapples. *Farming in South Africa Bulletin* H11/1978.

Keetch, D.P. and Dalldorf, D.B. (1979) The incidence of certain pineapple fruit blemishes in the Eastern Cape and Border. *Citrus & Subtropical Fruit Journal* 551, 12–15.

Kelly, W.P. (1911) A study of the composition of Hawaiian pineapples. *Journal of Industry Engineering Chemistry* 3, 403–405.

Kent, M.J. (1967) Stomatal behavior in pineapple leaves. *Pineapple Research News* 15, 36–45 (private document).

Kerns, K.R., Collins, J.L. and Kim, H. (1936) Vitamin C in pineapple varieties and hybrids. *Pineapple Quarterly* 6, 38–47 (private document).

Kidrič, M., Kos, J. and Sabotič, J. (2014) Proteases and their endogenous inhibitors in the plant response to abiotic stress. *Botanica Serbica* 38, 139–158.

Ko, L., Eccleston, K., O'Hare, T., Wong, T., Giles, J. and Smith, M. (2013) Field evaluation of transgenic pineapple (*Ananas comosus* (L.) Merr.) cv. 'Smooth Cayenne' for resistance to blackheart under subtropical conditions. *Scientia Horticulturae* 159, 103–108.

Lacoeuilhe J.J. (1978) La fumire N-K de l'ananas en Cote d''Ivoire. *Fruits* 33, 341–348.

Li, Y.-H., Zhang, Z. and Sun, G.M. (2010) Changes in cell number and cell size during pineapple (*Ananas comosus* L.) fruit development and their relationship with fruit size. *Australian Journal of Botany* 58, 673–678.

Li, T., Shen, P., Liu, W., Liu, C., Liang, R., Yan, N. and Chen, J. (2014) Major polyphenolics in pineapple peels and their antioxidant interactions. *International Journal of Food Properties* 17, 1805–1817.

Lim, W.H. (1985) Diseases and disorders of pineapples in Peninsular Malaysia. *MARDI Report* #97. Kuala Lumpur, Malaysia: Malaysian Agricultural Research and Development Institute.

Linford, M.B. (1932) Endogenous or non-parasitic brown spot. *Pineapple Quarterly* 2, 46–58 (private document).

Linford, M.B. and Spiegelberg, C.H. (1933) Illustrated list of pineapple fruit diseases, blemishes and malformities. *Pineapple Quarterly* 3(4), 135–178 (private document).

Liu, Y.K. (1986) Method for treating pineapples. United States Patent No. 4,614,659.

Liu, M.-C. (2009) Organic acid metabolism in pineapple fruit. MSc thesis, National Chung Hsing University, Taichung, Taiwan.

Liu, C. and Liu, Y. (2014) Effects of elevated temperature postharvest on color aspect, physiochemical characteristics, and aroma components of pineapple fruits. *Journal of Food Science* 79, 2409–2414.

Liu, S.H., Wei, C.B., Sun, G.M. and Zang, X.P. (2008) Analysis of aroma components in 3 pineapple cultivars. *Food Science* 29, 614–617.

Liu, C., Liu, Y., Yi, G., Li, W. and Zhang, G. (2011) A comparison of aroma components of pineapple fruits ripened in different seasons. *African Journal of Agricultural Research* 6, 1771–1778.

Liu, J., He, C., Shen, F., Zhang, K. and Zhu, S. (2017) The crown plays an important role in maintaining quality of harvested pineapple. *Postharvest Biology and Technology* 124, 18–24.

Lodh, S.B., Selvaraj, Y., Chadha, K.L. and Melanta, K.R. (1972) Biochemical changes associated with growth and development of pineapple fruit variety Kew II. Changes in carbohydrate and mineral constituents. *Indian Journal of Horticulture* 29, 287–291.

Lu, X.H., Sun. D.Q., Xi, Y.W.M. and Sun, G.M. (2010) Effect of postharvest salicylic acid treatment on fruit quality and anti-oxidant metabolism in pineapple during cold storage. *Journal of Horticultural Science and Biotechnology* 85, 454–458.

Lu, X., Sun, D., Li, Y., Shi, W. and Sun, G. (2011) Pre- and post-harvest salicylic acid treatments alleviate internal browning and maintain quality of winter pineapple fruit. *Scientia Horticulturae* 130, 97–101.

Luengwilai, K., Beckles, D.M. and Siriphanich, J. (2016) Postharvest internal browning of pineapple fruit originates at the phloem. *Journal of Plant Physiology* 202, 121–133.

Malézieux, E. (1993) Dry matter accumulation and yield elaboration of pineapple in Côte d'Ivoire. *Acta Horticulturae* 334, 149–157.

Mandal, D., Lalremruata, Hazarika, T.K. and Nautiyal, B.P. (2015) Effect of post-harvest treatments on quality and shelf life of pineapple (*Ananas comosus* [L.] Merr. 'Giant Kew') fruits at ambient storage condition. *International Journal of Bio-resource and Stress Management* 6, 490–496.

Mayer, A.M. (2006) Polyphenol oxidases in plants and fungi: going places? A review. *Phytochemistry* 67, 2318–2331.

Mayer, A.M. and Harel, E. (1979) Polyphenol oxidases in plants. *Phytochemistry* 18, 193–214.

Miller, E.V. (1951) Physiological studies of the pineapple, *Ananas comosus* L. Merr. with special reference to physiological breakdown. *Plant Physiology* 26, 66–75.

Miller, E.V. and Hall, G.D. (1953) Distribution of total soluble solids, ascorbic acid, total acid, and bromelin activity in the fruit of the natal pineapple (*Ananas comosus* L. Merr.). *Plant Physiology* 28, 532–534.

Miller, E.V. and Heilman, A.S. (1952) Ascorbic acid and physiological breakdown in the fruits of the pineapple (*Ananas comosus* L. Merr). *Science* 116, 505–506.

Montero, L.M., Escribano, M.I., De La Plaza, J.L. and Merodio, C. (1995) Chilling temperature storage induces changes in protein patterns and protease activity in cherimoya fruit. *Postharvest Biology and Technology* 5, 251–260.

Nanayakkara, K.P.G.A., Herath, H.M.W. and Senanayake, Y.D.A. (2005) Effects of pre-harvest treatments of potassium, post-harvest treatments of calcium, potassium, abscisic acid and light on reducing internal browning in pineapple (*Ananas comosus* (L.) Merr. cv. Mauritius) under cold-storage. *Acta Horticulturae* 666, 321–327.

Nilprapruck, P., Pradisthakarn, N., Authanithee, F. and Keebjan, P. (2008) Effect of exogenous methyl jasmonate on chilling injury and quality of pineapple (*Ananas comosus* L.) cv. Pattavia. *Silpakorn University Science and Technology Journal* 2, 33–42. DOI: 10.14456/sustj.2008.9.

Nukuntornprakit, O. and Siriphanich, J. (2005) Hydrogen peroxide and ascorbic acid contents, superoxide dismutase and catalase activities in Smooth Cayenne and Queen pineapples during cold storage. *Acta Horticulturae* 682, 611–615.

Nukuntornprakit, O., Chanjirakula, K., van Doorn, W.G. and Siriphanicha, J. (2015) Chilling injury in pineapple fruit: fatty acid composition and antioxidant metabolism. *Postharvest Biology and Technology* 99, 20–26.

Om-arun, N. and Siriphanich, J. (2005) Hydrogen peroxide and ascorbic acid contents, superoxide dismutase and catalase activities in Smooth Cayenne and Queen pineapples during cold storage. *Acta Horticulturae* 682, 611–615.

Ota, S., Muta, E., Katahira, Y. and Okamoto, Y. (1985) Reinvestigation of fractionation and some properties of the proteolytically active components of stem and fruit bromelains. *Journal of Biochemistry* 98, 219–228.

Paull, R.E. (1992) Postharvest handling of Smooth Cayenne pineapple in Hawaii for the fresh fruit market. *Acta Horticulturae* 334, 273–285.

Paull, R.E. (1993) Pineapple and papaya. In: Seymour, G., Taylor, J. and Tucker, G. (eds) *Biochemistry of Fruit Ripening*. London: Chapman & Hall, pp. 291–323.

Paull, R.E. (1997) Pineapple. In: Mitra, S.K. (ed.) *Postharvest Physiology and Storage of Tropical and Subtropical Fruits*. Wallingford, Oxfordshire: CAB International, pp. 123–143.

Paull, R.E. and Chen, N.J. (2018a) Pineapple postharvest. In: Gil, M.I. and Beaudry, R.M. (eds) *Controlled and Modified Atmosphere Use for Fresh and Fresh-cut Produce*. New York: Elsevier.

Paull, R.E. and Chen, N.J. (2018b) Postharvest physiological disorders in pineapples. In: de Freitas, T. (ed.) *Postharvest Physiological Disorders in Fruits and Vegetables*. Boca Raton, FL: CRC Press, Taylor & Francis Group.

Paull, R.E. and Reyes, M.E. (1996) Preharvest weather conditions and pineapple fruit translucency. *Scientia Horticulturae* 66, 59–67.

Paull, R.E. and Rohrbach, K.G. (1982) Juice characteristics and internal atmosphere of waxed 'Smooth Cayenne' pineapple fruit. *Journal of the American Society for Horticultural Science* 107, 448–452.

Paull, R.E. and Rohrbach, K.G. (1985) Symptom development of chilling injury in pineapple fruit. *Journal of the American Society of Horticultural Science* 110, 100–105.

Paull, R.E., Chen, N.J. and Saradhuldhat, P. (2016) Pineapple harvesting and postharvest handling. In: Lobo, M.G. and Paull, R.E. (eds) *Handbook of Pineapple Technology: Production, Postharvest Science, Processing and Nutrition*. Wiley-Blackwell, New Jersey, 89–107.

Phonyiam, O. Kongsuwan, A. and Setha, S. (2016) Effect of short-term anoxic treatment on internal browning and antioxidant ability in pineapple cv. Phulae. *International Food Research Journal* 23, 521–527.

Pino, J.A. (2013) Odour-active compounds in pineapple (*Ananas comosus* [L.] Merril cv. Red Spanish). *International Journal of Food Science and Technology* 48, 564–570.

Poignant, A. (1971) La maturation controlee de l'ananas II – L'ethrel et son action au cours des phases ascendante et descendarte de la maturite. *Fruits* 26, 23–35.

Prasad, T.K. (1996) Mechanisms of chilling-induced oxidative stress injury and tolerance in developing maize seedlings: Changes in antioxidant system, oxidation of proteins and lipids, and protease activities. *Plant Journal* 10, 1017–1026.

Pusittigul, I., Kondo, S. and Siriphanich, J. (2012) Internal browning of pineapple (*Ananas comosus* L.) fruit and endogenous concentrations of abscisic acid and gibberellins during low temperature storage. *Scientia Horticulturae* 146, 45–51.
Pusittigul, I., Siriphanich, J. and Juntee, C. (2014) Role of calcium on internal browning of pineapples. *Acta Horticulturae* 1024, 329–338.
Py, C., Lacoeville, J.J. and Teisson, C. (1987) *The Pineapple, Cultivation, and Uses*. Paris: G.P. Maisonneuve et Larose.
Raimbault, A.-K., Marie-Alphonsine, P.-A., Horry, J.-P., Francois-Haugrin, M., Romuald, K. and Soler, A. (2011) Polyphenol oxidase and peroxidase expression in four pineapple varieties (*Ananas comosus* L.) after a chilling injury. *Journal of Agricultural and Food Chemistry* 59, 342–348.
Raimbault, A.-K., Zuily-Fodil, Y., Soler, A., Mora, P. and Cruz de Carvalho, M.H. (2013a) The expression patterns of bromelain and AcCYS1 correlate with blackheart resistance in pineapple fruits submitted to postharvest chilling stress. *Journal of Plant Physiology* 170, 1442–1446.
Raimbault, A.-K., Zuily-Fodil, Y., Soler, A. and Cruz de Carvalho, M.H. (2013b) A novel aspartic acid protease gene from pineapple fruit (*Ananas comosus*): cloning, characterization and relation to postharvest chilling stress resistance. *Journal of Plant Physiology* 170, 1536–1540.
Rasheed, A.A., Cobham, E.I., Zeighami, M. and Ong, S.P. (2012) Extraction of phenolic compounds from pineapple fruit. *The 2nd International Symposium on Processing & Drying of Foods, Vegetables and Fruits*, University of Nottingham, Malaysia Campus.
Reddy, M.N., Keim, P.S., Heinrickson, R.L. and Kezdy, F.J. (1975) Primary structural analysis of sulfhydryl protease inhibitors from pineapple stem. *Journal of Biological Chemistry* 250, 1741–1750.
Rohrbach, K.G. and Apt, W.J. (1986) Nematode and disease problems of pineapples. *Plant Diseases* 70, 81–87.
Rohrbach, K.G. and Paull, R.E. (1982) Incidence and severity of chilling induced browning of waxed 'Smooth Cayenne' pineapple. *Journal of American Society for Horticultural Science* 107, 453–457.
Rowan, A.D., Buttle, D.J. and Barrett, A.J. (1990) The cysteine proteinases of the pineapple plant. *Biochemistry Journal* 266, 869–875.
Sanewski, G.M. and Giles, J. (1997) Blackheart resistance in three clones of pineapple (*Ananas comosus* (L.) Merr.) in subtropical Queensland. *Australia Journal for Experimental Agriculture* 37, 459–461.
Sanford, W.G. (1963) Fruit characteristics and product recovery as affected by some agronomic practices. (II) Date of harvest and planting density. *Pineapple Research Institute Research Report* No. 97 (private document).
Saradhuldhat, P. and Paull, R.E. (2007) Pineapple organic acid metabolism and accumulation during fruit development. *Scientia Horticulturae* 112, 297–303.
Sasaki, M., Kato, T. and Iida, S. (1973) Antigenic determinant common to four kinds of thiol-proteases of plant origin. *Journal of Biochemistry* 74, 635–637.
Schappelle, N.A. (1941) A physiological study on the effects of waxing pineapples of different stages of maturity. *Puerto Rico Agricultural Experiment Station Research Bulletin* No. 3.
Selvarajah, S., Herath, H.M.W. and Liyanage, K.K. (1997) Effect of modified atmospheric storage on internal browning development in pineapple. *Tropical Agricultural Research* 9, 388–391.
Selvarajah, S., Herath, H.M.W., Bandara, D.C. and Abeysinghe Banda, D.M.G. (1998a) Effect of pre-harvest calcium treatment on post-harvest quality of pineapple. *Tropical Agricultural Research* 10, 214–224.
Selvarajah, S., Herath, H.M.W. and Bandara, D.C. (1998b) Physiological effects of pre-heat treatment on pineapple fruit stored at low temperatures. *Tropical Agricultural Research* 10, 417–419.
Selvarajah, S., Bauchot, A.D. and John, P. (2001) Internal browning in cold-stored pineapples is suppressed by a postharvest application of 1-methylcyclopropene. *Postharvest Biology and Technology* 23, 167–170.
Selvaray, Y., Dirakar, N.G., Subhas Chander, M., Chadha K.L. and Melanata, K.R. (1975) Biochemical changes associated with growth and development of pineapple fruit Kew IV. Changes in major lipid constituents. *Indian Journal of Horticulture* 32, 64–67.
Senanayake, Y.D.A. and Gunasena, H.P.M. (1975) A study on the influence of crown leaves on fruit growth of pineapple *Ananas comosus* (L.) Merr. cv. Kew. *Journal of Natural Agriculture Society, Ceylon* 12, 106–114.
Shiina, T., Ijiri, T., Matsuda, I., Sato, T., Kawano, S. and Ohoshiro, N. (1993) Determination of Brix value in pineapple fruits by near infrared spectrography. *Acta Horticulturae* 334, 261–272.

Shiomi, S., Chono, K., Nishikawa, M., Okabe, M. and Nakamura, R. (2002) Effect of 1-methylcyclopropene (1-MCP) on respiration, ethylene production and color change of pineapple fruit after harvest. *Food Preservation Science* 28, 235–241.

Sideris, C.P. and Krauss, B.H. (1933a) Physiological studies on the factors influencing the quality of pineapple fruits. I. Physico-chemical variations in the tissue of ripe pineapple fruits. *Pineapple Quarterly* 3, 82–98.

Sideris, C.P. and Krauss, B.H. (1933b) Physiological studies on the factors influencing the quality of pineapple. II. Physico-chemical variations in the tissue of ripe pineapple fruits. *Pineapple Quarterly* 3, 99–114.

Sideris, C.P. and Krauss, B.H. (1934) Physiological studies on the factors influencing the quality of pineapple fruits. III. Physico-chemical variations in the tissue of ripe pineapple fruits (continued). *Pineapple Quarterly* 4, 48–57.

Sideris, C.P. and Krauss, B.H. (1938) Growth phenomena of pineapple fruits. *Growth* 2, 181–196.

Sideris, C.P., Krauss, B.H. and Young, H.Y. (1935) The physiochemical basis of green-shell and yellow-shell flesh transluscent pineapple fruits. *Pineapple Quarterly* 5, 18–28.

Sideris, C.P., Young, H.Y. and Chun, H.H.Q. (1948) Diurnal changes and growth rate as associated with ascorbic acid, titratable acidity, carbohydrate and nitrogen fractions in the leaves of *Ananas comosus* (L.) Merr. *Plant Physiology* 23, 38–69.

Silva, J.A., Hamasaki, R., Paull, R., Ogoshi, R., Bartholomew, D.P. *et al.* (2006) Lime, gypsum, and basaltic dust effects on the calcium nutrition and fruit quality of pineapple. *Acta Horticulturae* 702, 123–132.

Silverstein, R.M. (1971) The pineapple: flavour. In: Hulme, A.C. (ed) *The Biochemistry of Fruits and Their Products.* London: Academic Press, pp. 325–331.

Singleton, V.L. (1955) Pineapple fruit composition, flavor and quality. *Pineapple Research Institute Report* No. 38. (private document).

Singleton, V.L. (1957a) Selection and handling of pineapple fruits for shipment fresh. *Pineapple Research Institute Research Reports* #49 (private document).

Singleton, V.L. (1957b) The nature of the separation produced by floatation of Cayenne pineapple fruits. *Pineapple Research Institute News* 5, 1–10 (private document).

Singleton, V.L. (1958) At test for the degree of bruising of pineapple flesh. *Pineapple Research Institute News* 6, 111–114 (private document).

Singleton, V.L. (1965) Chemical and physical development of the pineapple fruit. I. Weight per fruitlet and other physical attributes. *Journal of Food Science* 30, 98–104.

Singleton, V.L. and Gortner, W.A. (1965) Chemical and physical development of pineapple fruit III. Nitrogenous and enzyme constituents. *Journal of Food Science* 30, 24–29.

Singleton, V.L., Gortner, W.A. and Young, H.Y. (1961) Carotenoid pigments of pineapple fruits I. Acid catalyzed isomerization of the pigments. *Journal of Food Science* 26, 49–52.

Smith, L.G. (1983) Causes and development of black heart in pineapples. *Tropical Agriculture (Trinidad)* 60, 31–35.

Smith, L.G. (1988a) Indices of physiological maturity and eating quality in Smooth Cayenne pineapples. 1. Indices of physiological maturity. *Queensland Journal of Agriculture and Animal Science* 45, 213–218.

Smith, L.G. (1988b) Indices of physiological maturity and eating quality in Smooth Cayenne pineapples. 2. Indices of eating quality. *Queensland Journal of Agriculture and Animal Science* 45, 219–228.

Smith, L.G. (1991) Effects of ethephon on ripening and quality of fresh market pineapples. *Australian Journal of Experimental Agriculture* 31, 123–127.

Smith, L.G. (1993) Optimization of fresh-market pineapple eating quality in Queensland. *Acta Horticulturae* 334, 287–294.

Smith, B.G. and Harris, P.J. (1995) Polysaccharide composition of unlignified cell walls of pineapple (*Ananas comosus* (L.) Merr.) fruit. *Plant Physiology* 107, 1399–1409.

Smith, B.G. and Harris, P.J. (2001) Ferulic acid is esterified to glucuronoarabinoxylans in pineapple cell walls. *Phytochemistry* 56, 513–519.

Soares, A.G., Trugo, L.C., Botrel, N. and da Silva Souza, L.F. (2005) Reduction of internal browning of pineapple fruit (*Ananas comosus* L.) by preharvest soil application of potassium. *Postharvest Biology and Technology* 35, 201–207.

Soler, A. (1992a) *Pineapple.* Paris: CIRAD-IRFA.

Soler, A. (1992b) Metabolism de l'ethephon dans l'eprideme de l'ananas (*Ananas comosus* (L.) Merr). *Fruits* 47, 471–477.

Soler, A. (1993) Enzymatic characterization of stress induced translucence of pineapple flesh in the Ivory Coast. *Acta Horticulturae* 334, 295–304.

Soler, A. (1994a) Deviation de la maturation chez l'ananas le 'jaune' ou translucidite I. Caracteristiques physiques et chimiques du fruit translucide. *Fruits* 49, 5–15.

Soler, A. (1994b) Deviation de la maturation chez l'ananas le 'jaune' ou translucidite II. Caracterisation enzymatique de fruit translucide. *Fruits* 49, 83–91.

Standards (1968) *Wholesale Standards for Hawaiian-grown Pineapple.* Honolulu: Hawaii Department of Agriculture, Marketing Division.

Steingass, C.B., Grauwet, T. and Carle, R. (2014) Influence of harvest maturity and fruit logistics on pineapple (*Ananas comosus* [L.] Merr.) volatiles assessed by headspace solid phase microextraction and gas chromatography – mass spectrometry (HS-SPME-GC/MS). *Food Chemistry* 150, 382–391.

Steingass, C.B., Carle, R. and Schmarr, H.G. (2015a) Ripening-dependent metabolic changes in the volatiles of pineapple (*Ananas comosus* (L.) Merr.) fruit: I. Characterization of pineapple aroma compounds by comprehensive two-dimensional gas chromatography-mass spectrometry. *Analytical and Bioanalytical Chemistry* 407, 2591–2608.

Steingass, C.B., Glock, M.P., Schweiggert, R.M. and Carle, R. (2015b) Studies into the phenolic patterns of different tissues of pineapple (*Ananas comosus* [L.] Merr.) infructescence by HPLC-DAD-ESI-MSn and GC-MS analysis. *Analytical and Bioanalytical Chemistry* 407, 6463–6479.

Steingass, C.B., Jutzi, M, Müller, J., Carle, R. and Schmarr, H.G. (2015c) Ripening-dependent metabolic changes in the volatiles of pineapple (*Ananas comosus* (L.) Merr.) fruit: II. Multivariate statistical profiling of pineapple aroma compounds based on comprehensive two-dimensional gas chromatography-mass spectrometry. *Analytical and Bioanalytical Chemistry* 407, 2609–2624.

Steingass, C.B., Langen, J., Carle, R. and Schmarr, H.G. (2015d) Authentication of pineapple (*Ananas comosus* [L.] Merr.) fruit maturity stages by quantitative analysis of c- and d-lactones using headspace solid-phase microextraction and chirospecific gas chromatography – selected ion monitoring mass spectrometry (HS-SPME–GC–SIM-MS). *Food Chemistry* 168, 496–503.

Steingass, C.B., Dell, C., Lieb, V., Mayer-Ullmann, B., Czerny, M. and Carle, R. (2016) Assignment of distinctive volatiles, descriptive sensory analysis and consumer preference of differently ripened and post-harvest handled pineapple (*Ananas comosus* [L.] Merr.) fruits. *European Food Research and Technology* 242, 33–43.

Stewart, R.J., Sawyer, B.J.B., Bucheli, C.S. and Robinson, S.P. (2001) Polyphenol oxidase is induced by chilling and wounding in pineapple. *Australian Journal of Plant Physiology* 28, 181–191.

Stewart, R.J., Sawyer, B.J.B. and Robinson, S.P. (2002) Blackheart development following chilling in fruit of susceptible and resistant pineapple cultivars. *Australian Journal of Experimental Agriculture* 42, 195–199.

Su, C.-L. (2008) Effects of plant growth regulators and hot water dipping treatment on the occurrence of pineapple (*Ananas comosus* L. Merr.) blackheart disorder. MSc thesis, National Chung Hsing University, Taichung, Taiwan.

Sukwanit, S. and Teerachaichayut, S. (2013) Nondestructive prediction of internal browning in pineapple using transmittance short wavelength near infrared spectroscopy. *Acta Horticulturae* 989, 395–399.

Sun, S.-K. (1971) A study of black heart disease of the pineapple fruits. *Plant Protection Bulletin, Taiwan* 13, 39–48.

Swarts, D.H. (1991) Invloed van verskillen de verskeping, temperature op Queen pynappels. *Subtropica* 12, 12–16.

Swete Kelly, D.E. and Bragshaw, J. (1993) Effect of fruit handling and fruit coatings on black heart (internal brown spot) and other aspects of fresh pineapple quality. *Acta Horticulturae* 334, 305–316.

Syahrin, S. (2011) Consumer preferences towards pineapple cultivars in Malaysia. *Acta Horticulturae* 902, 595–599.

Takeoka, G., Battery, R.G., Flath, R.A., Teranishi, R., Wheeler, E.L. *et al.* (1989) Volatile constituents of pineapple (*Ananas comosus* [L.] Merr.). In: Teranishi, R., Buttery, R.G. and Shahidi, F. (eds) *Flavor Chemistry: Trends and Development.* Washington, DC: American Chemical Society, pp. 223–237.

Taniguchi, G., Sanewski, G.M., Bartholomew, D.P. and Paull, R.E. (2008) Characteristics of the pineapple research institute of Hawaii hybrids 73-50 and 73-114. *Pineapple News* 15(8), 27–33.

Tay, T.H. (1977) Fruit ripening studies on pineapple. *MARDI Research Bulletin* 4, 29–34.

Techavuthiporn, C., Boonyaritthongchai, P. and Suriyan Supabvanich, S. (2017) Physicochemical changes of 'Phulae' pineapple fruit treated with short-term anoxia during ambient storage. *Food Chemistry* 228, 388–393.

Teisson, C. (1973) Developpement et croissance de l'unflorescence d'*Ananas comosus* (cv. Cayenne lisse). *Fruits* 28, 433–439.
Teisson, C. (1977) Le brunissement interne de l'ananas [Internal browning of pineapple]. Doctoral thesis, University of Abidjan, Ivory Coast.
Teisson, C. and Combres, J.C. (1979) Le brunissement interne de l'ananas. III. Symptomatologie. *Fruits* 34, 315–339.
Teisson, C. and Martin-Prevel, P. (1979) Le brunissement interne de l'ananas. iv. Approache biochimiqe du phenomene. *Fruits* 34, 315–339.
Teisson, C. and Pineau, P. (1982) Quelques données sur les dernières phases du developpement de l'ananas. *Fruits* 37, 741–748.
Teisson, C., Lacoeuilhe, J.J. and Combres, J.C. (1979) Les brunissement interne de l'ananas. v. Recherches des moyen de lutte. *Fruits* 34, 399–415.
Timm, E.J. and Brown, G.K. (1991) Impacts recorded on avocado, papaya, and pineapple packing lines. *Applied Engineering in Agriculture* 7, 418–422.
Tseng, B.-S. (2013) Correlation between physical–chemical characteristics and blackheart disorder in pineapple (*Ananas comosus*) fruit during storage. MSc thesis, National Chung Hsing University, Taichung, Taiwan.
Umano, K., Hagi, Y., Nakahara, K., Shoji, A. and Shibamoto, T. (1992) Volatile constituents of green and ripened pineapple (*Ananas comosus* [L.] Merr.). *Journal of Agriculture and Food Chemistry* 40, 599–603.
Upadhya, M.D. and Brewbaker, J.L. (1966) Effects of gamma irradiation on the pineapple. *Hawaii Farm Science* 15(1), 8–9.
Van Lelyveld, L.J. and de Bruyn, J.A. (1977) Polyphenols, ascorbic acid and related enzyme activities associated with black heart in Cayenne pineapple fruit. *Agrochemophysica* 9,1–6.
Van Overbeek, J. (1946) Control of flower formation and fruit size in the pineapple. *Botanical Gazette* 108, 64–73.
Wang, M.L. and Paull, R.E. (forthcoming) Genetic transformation of pineapple. In: Ming, R. (ed.) *Genetics and Genomics of Pineapple*. New York: Springer.
Wardlaw, C.W. (1937) Tropical fruits and vegetables. An account of their storage and transport. *Tropical Agriculture* 24, 288–298.
Weerahewa, D. and Adikaram, N.K.B. (2005a) Heat-induced tolerance to internal browning of pineapple (*Ananas comosus* cv. 'Mauritius') under cold storage. *Journal of Horticultural Science and Biotechnology* 80, 503–509.
Weerahewa, D. and Adikaram, N.K.B. (2005b) Some biochemical factors underlying the differential susceptibility of two pineapple cultivars to internal browning disorder. *Ceylon Journal Science (Biological Science)* 34, 7–20.
Wei, C.B., Liu, S.H., Liu, Y.G., Lu, L.L., Yang, W.X. and Sun, G.M. (2011a) Characteristic aroma compounds from different pineapple parts. *Molecules* 16, 5104–5112.
Wei, C.B., Liu, S.H., Liu, Y.G., Zang, X.P., Lu, L.L. and Sun, G.M. (2011b) Changes and distribution of aroma volatile compounds from pineapple fruit during postharvest storage. *Acta Horticulturae* 902, 431–436.
Wilson Wijeratnam, R.S., Abeysekara, M. and Surjani, S. (1993) Studies on black heart disorder in pineapple varieties grown in Srilanka. *Acta Horticulturae* 334, 317–324.
Wilson Wijeratnam, R.S., Hewajulige, I.G.N., Wijesundera, R.L.C. and Abeysekere, M. (2006) Fruit calcium concentration and chilling injury during low temperature storage of pineapple. *Acta Horticulturae* 702, 203–208.
Wu, P., Kuo, M.C., Hartman, T.G., Rosen, R.T. and Ho, C.T. (1991) Free and glycosidically bond aroma compounds in pineapple (*Ananas comosus* L. Merr). *Journal of Agriculture and Food Chemistry* 39, 170–172.
Yamada, F., Takahashi, N. and Murachi, T. (1976) Purification and characterization of a proteinase from pineapple fruit, fruit bromelain FA2. *Journal of Biochemistry* 79, 1223–1234.
Yapo, E.S., Kouakou, H.T., Kouakou, L.K., Kouadio, J.Y., Kouamé, P. and Mérillon, J.-M. (2011) Phenolic profiles of pineapple fruits (*Ananas comosus* L. Merrill). Influence of the origin of suckers. *Australian Journal of Basic and Applied Sciences* 5, 1372–1378.
Youryon, P., Wongs-Aree, C., McGlasson, W.B., Glahan, S. and Kanlayanarat, S. (2008) Internal browning occurrences of 'Queen' pineapple under various low temperatures. *Acta Horticulturae* 804, 555–560.

Youryon, P., Wongs-Aree, C., McGlasson, W.B., Glahan, S. and Kanlayanarat, S. (2013) Alleviation of internal browning in pineapple fruit by peduncle infiltration with solutions of calcium chloride or strontium chloride under mild chilling storage. *International Food Research Journal* 20, 239–246.

Žemlička, L., Fodran, P., Kolek, E. and Pronayova, N. (2013) Analysis of natural aroma and flavor of MD2 pineapple variety (*Ananas comosus* [L.] Merr.). *Acta Chimica Slovaca* 6, 123–128.

Zhang, Q., Liu, Y., He, C. and Zhu, S. (2015) Postharvest exogenous application of abscisic acid reduces internal browning in pineapple. *Journal of Agricultural and Food Chemistry* 63, 5313–5320.

Zhang, Q., Rao, X., Zhang, L., He, C., Yang, F. and Zhu, S. (2016) Mechanism of internal browning of pineapple: the role of gibberellins catabolism gene (AcGA2ox) and GAs. *Scientific Reports* 6, 33344. DOI: 10.1038/srep33344.

Zhang, X.M., Du, L.Q., Sun, G.M., Liu, S.H., Wei, C.B. *et al.* (2009) Analysis of aromatic components in pineapple varieties. *Food Science* 30, 275–279.

Zheng, L.Y., Sun, G.M., Liu, Y.G., Lu, L.L., Yang, W.X. *et al.* (2012) Aroma volatile compounds from two fresh pineapple varieties in China. *International Journal of Molecular Sciences* 13, 7383–7392.

Zhou, Y.-C. and Tan, X.-J. (1997) Mechanism of blackheart development induced by low temperature and gibberellic acid in pineapple fruit. *Acta Horticulturae* 425, 587–593.

Zhou, Y., Dahler, J.M., Underhill, S.J.R. and Wills, R.B.H. (2003a) Enzymes associated with blackheart development in pineapple fruit. *Food Chemistry* 80, 565–572.

Zhou, Y., O'Hare, T.J., Jobin-Décor, M., Underhill, S.J., Wills, R.B. and Graham, M.W. (2003b) Transcriptional regulation of a pineapple polyphenol oxidase gene and its relationship to blackheart. *Plant Biotechnology Journal* 1, 463–478.

Zhou, Y., Pan, X., Qu, H. and Underhill, S.J. (2014a) Low temperature alters plasma membrane lipid composition and ATPase activity of pineapple fruit during blackheart development. *Journal of Bioenergetics and Biomembranes* 46, 59–69.

Zhou, Y., Pan, X., Qu, H. and Underhill, S.J. (2014b) Tonoplast lipid composition and proton pump of pineapple fruit during low-temperature storage and blackheart development. *Journal of Membrane Biology* 247, 429–39.

Index

Note: Page numbers in **bold** type refer to **figures**
Page numbers in *italic* type refer to *tables*